The Chemistry of Clay-Organic Reactions

The second edition of *The Chemistry of Clay-Organic Reactions* book provides a comprehensive and fully updated summary of the literature on the interactions of clay minerals with organic molecules, including reaction mechanisms and bonding modes together with their practical and industrial applications. The reader will gain an insight into the formation and properties of complexes between clay minerals and a variety of organic compounds and the use of such complexes as sorbents and carriers of organic pollutants, pesticides, dyes, and pharmaceuticals.

KEY FEATURES

- An authoritative resource providing a detailed synthesis of published data on clay-organic complexes and reactions.
- Authored by a globally recognized expert in the field.
- Describes developments in the interactions of organic compounds with fibrous and short-range order clay minerals.

This book is written for environmental and industrial chemists, organic geochemists, and soil scientists, and it will appeal to academics, researchers, industry professionals, and graduate students.

Benny K.G. Theng is an honorary research associate with Manaaki Whenua – Landcare Research in Palmerston North, New Zealand (NZ). Before joining the research staff of MW-LC in 1992, he was a senior scientist with the NZ Soil Bureau, Department of Scientific and Industrial Research (DSIR) in Lower Hutt (Wellington). The focus of his distinguished research career has been on the behaviour and reactivity of small and polymeric organic compounds at clay mineral surfaces. He was born in Indonesia but did all his undergraduate and postgraduate studies at the University of Adelaide, Australia, gaining a B.Ag.Sc. degree with first class honours, followed by a PhD degree in soil science. He was a research fellow, lecturer, mentor, and visiting scientist at several universities and research institutes in Australia, Belgium, Germany, Japan, France, Chile, China, South Korea, and Spain. He also did some consulting work for clay-based companies in America, Australia, and Europe. He is the recipient of a senior Humboldt fellowship (Germany), an Adam Hilger Prize (UK) for his first book, the ICI Prize (NZ) for chemical research, a Marsden Fund Award (NZ), a Japan Society for the Promotion of Science fellowship, and the Bailey Distinguished Member Award for outstanding contributions to clay science from The Clay Minerals Society (USA). He is a fellow of the NZ Institute of Chemistry, NZ Society of Soil Science, and the Royal Society of New Zealand. He likes reading historical books and thrillers, playing contract bridge, and listening to classical and chamber music.

The Chemistry of Clay-Organic Reactions

Second Edition

Benny K.G. Theng

CRC Press is an imprint of the
Taylor & Francis Group, an **informa** business

Cover credit: © Shutterstock

Second edition published 2024
by CRC Press
6000 Broken Sound Parkway NW, Suite 300, Boca Raton, FL 33487–2742

and by CRC Press
4 Park Square, Milton Park, Abingdon, Oxon, OX14 4RN

CRC Press is an imprint of Taylor & Francis Group, LLC

First edition published by Adam Hilger Limited, London (UK) 1974

Library of Congress Cataloging-in-Publication Data
Names: Theng, B. K. G., author.
Title: The chemistry of clay-organic reactions / Benny K.G. Theng.
Description: Boca Raton : CRC Press, [2024] | Includes bibliographical references and index.
Identifiers: LCCN 2023046362 (print) | LCCN 2023046363 (ebook) | ISBN 9780367530389 (hbk) | ISBN 9780367530419 (pbk) | ISBN 9781003080244 (ebk)
Classification: LCC QE389.625 .T47 2024 (print) | LCC QE389.625 (ebook) | DDC 549/.6015472—dc23/eng/20231026
LC record available at https://lccn.loc.gov/2023046362
LC ebook record available at https://lccn.loc.gov/2023046363

ISBN: 978-0-367-53038-9 (hbk)
ISBN: 978-0-367-53041-9 (pbk)
ISBN: 978-1-003-08024-4 (ebk)

DOI: 10.1201/9781003080244

Typeset in Times LT Std
by Apex CoVantage, LLC

To my daughters, Monika and
Tjun and my son, Andrew

Contents

Preface to the First Edition

Some years ago, G.F. Walker suggested that I should collaborate with him in writing a monograph on clay-organic systems which was to be a critical assessment and not just a neutral review of the literature. It was not until the spring of 1968, however, when I joined Walker's group at the Division of Applied Mineralogy, C.S.I.R.O., Melbourne, that I seriously began to sift through the published material dealing with clay-organic interactions. The untimely death of G.F. Walker in early 1970 not only deprived me of a co-author who had an intimate and incisive knowledge of the subject but also brought the project to a halt, at least temporarily. Its resumption was prompted by the need for a comprehensive treatment of clay-organic chemistry which would serve as a work of reference. Also, by writing about it I hoped to sort out my own ideas and, in doing so, to see the subject with greater clarity. In style and content, the finished product would, of course, have been different had Walker shared in its authorship. Further, by going it alone I am solely responsible for whatever merits and deficiencies the book may have.

The subject of clay-organic reactions has made appreciable advances over the past few years. In common with many other disciplines, both the volume and complexity of information in this area are growing at such a rate that we are facing 'the prospect of instant antiquity'*. Some selection of the literature was therefore necessary to avoid excessive length while retaining essential depth and detail. In collating and screening the published material and in preparing the text I have tried to be as objective as possible.

This book is directed to the interests of my fellow workers as well as teachers and students of soil science. Owing to the interdisciplinary nature of the subject, however, it may also be read to advantage by surface and colloid chemists, organic geochemists, and agronomists.

I am indebted to the many colleagues and authors who have sent me reprints of their papers and whose diagrams have served as a basis for the illustrations. In this connection, I acknowledge the various publishers and councils of scientific societies who gave permission to reproduce figures and diagrams. I wish especially to record my indebtedness to the late Dr. M. Fieldes, formerly Director of the Soil Bureau, for allowing me to work full time on the final stages of the book. Also, I thank those who typed the manuscripts and Mrs. M. Oliver, who drew the numerous diagrams. Finally, I would like to thank my wife for her patient support and encouragement.

Benny K.G. Theng
Lower Hutt
November 1972

* Lukasiewicz, J., 1970. The ignorance explosion. *Impact of Science on Society* 20: 251–253.

Preface to the Second Edition

The first edition of this book was published in August 1974 by Adam Hilger Limited of London, UK, and for this book I was awarded the Adam Hilger Prize. The idea of revising the book was suggested by many colleagues on the ground that much new information on the clay-organic interaction has since come to light. It was not until I began to search the relevant literature, however, that I realized the extent to which the discipline has developed and grown over the past four or so decades in terms of quantity, quality, and diversity.

The primary objective of writing a second edition, therefore, is to update the data and information contained in the first edition. In the process, I have modified some figures and added new diagrams and schemes. The general arrangement of the contents in the 1974 edition has been retained. However, the chapter dealing with clay-catalyzed organic reactions has been omitted since this topic has been comprehensively reviewed (Theng 2018). At the same time, I have included a chapter on the interactions of clay minerals with negatively charged organic species and one dealing with organic complexes of layer-ribbon and short-range order silicates. Following the first edition, each chapter here is presented in the form of a self-contained, critical review with its own complement of references to the published literature.

I am grateful to several colleagues for reviewing and supporting my proposal to CRC Press that a revision of the 1974 book is both warranted and timely. I am indebted to the library staff of Manaaki Whenua-Landcare Research for obtaining journal articles and reprints and especially to Cissy Pan for drawing up the numerous figures, diagrams, and schemes. I also wish to acknowledge the editorial/managing staff of scientific journals and publishers for granting permission to reproduce figures and illustrations. Last, but not least, I thank my wife, Judy Theng, for her encouragement, patience, and support.

Benny K.G. Theng
Palmerston North, New Zealand
July 2023

REFERENCE

Theng, B.K.G. 2018. *Clay Mineral Catalysis of Organic Reactions*. Boca Raton, FL: CRC Press.

1 Clays and Clay Minerals
Structures and Surface Properties

1.1 INTRODUCTION

Clay minerals are layer silicates (phyllosilicates) that dominantly make up the fine-grained fraction of soils and sediments. Although most phyllosilicates are now known to have long-range order, the 'clay-mineral concept', that is, the general acceptance of the crystalline nature of clays, dates back only to the 1930s (Mackenzie and Mitchell 1966; Grim 1968). In the soil science literature, the term 'clay' or 'clay fraction' refers to a class of minerals whose particles are smaller than 2 μm in equivalent spherical diameter (e.s.d.). In geology, sedimentology, and geo-engineering, however, the particle size limit is commonly set at 4 μm e.s.d., while in colloid science the value of < 1 μm e.s.d. is generally accepted. Furthermore, not all clay-size minerals in soils (and sediments) may be phyllosilicates, or even crystalline. Indeed, nano-size soil minerals ('nanoclays') of short-range order play an important role in the retention of plant nutrients, water, and organic matter (Theng and Yuan 2008; Calabi-Floody et al. 2009, 2011; Yuan and Theng 2012; Nanzyo and Kanno 2018).

Clay may be defined as a naturally occurring material of fine-grained texture that becomes plastic when wet and hardens when dried or fired, while the term *clay mineral* denotes a class of hydrated phyllosilicates and minerals that 'impart plasticity to clay and which harden upon drying or firing' (Guggenheim and Martin 1995). Unlike clay, clay mineral may therefore be synthetic. Nevertheless, the terms 'clay' and 'clay mineral' are often used interchangeably. Thus, the clay component in the clay-organic and clay-water interaction almost invariably denotes a specific clay mineral.

Because of their ultrafine grain size, the characterization and identification of clay minerals call for a combination of various techniques. In addition to the classic X-ray powder diffraction, chemical and thermal analyses, organic reaction/intercalation, electron microscopy, and various spectroscopic methods have been used for this purpose (Grim 1968; Gard 1971; Mackenzie and Caillère 1975; Sudo et al. 1981; Newman 1987; Dixon and Weed 1989; Madejová 2003; Bergaya et al. 2006; Gates et al. 2017). Even so, the occurrence of structural defects, layer stacking disorders, and interstratified structures make for practical difficulties. Progress in this area has been considerably aided by the application of modern, sophisticated instrumental methods, such as electron spin resonance (ESR), Mössbauer, nuclear magnetic resonance (NMR), neutron scattering, X-ray photoelectron (XPS) and X-ray absorption (XAS) spectroscopies, supplemented by computational studies, notably density functional theory and molecular dynamics simulation. Details of the underlying theories and the quality of information that these techniques can provide are given in the reviews and books written or edited by Stucki and Banwart (1980), Fripiat (1981), Thomas (1984), Wilson (1994), Bergaya and Lagaly (2013), and Yuan et al. (2016).

An added problem is the presence of 'associated minerals', such as carbonates, quartz, and the hydr(oxides) of iron and aluminium (Schwertmann 1979; Brown 1980; Guggenheim and Martin 1995; Dixon and Schulze 2002). These mineral 'impurities', in the form of either discrete entities or surface-adsorbed species, are difficult to detect because of their low concentration and nano-size dimension (Theng and Yuan 2008). Furthermore, some associated minerals are of short-range order in that their structural regularity does not extend beyond ~ 1 nm. Being X-ray amorphous, such minerals have also been referred to as non-crystalline or poorly crystalline (Brown et al. 1978; Parfitt 1990).

DOI: 10.1201/9781003080244-1

Only the basic features of the structures and surface properties of clay minerals will be described here, pointing out those peculiarities of architecture and organization that influence their behaviour towards organic compounds. Indeed, the clay-organic interaction can provide valuable insight – into the surface reactivity of the clay mineral component (Lagaly 1981; Lagaly et al. 2006; Johnston 2010; Theng 2012, 2018). For details of the mineralogy, structures, and properties of clay minerals, the reader is referred to the books and monographs written or edited by Grim (1968), Barrer (1978), Greenland and Hayes (1978), Bailey (1980), Brindley and Brown (1980), Dixon and Weed (1989), Moore and Reynolds (1997), Auerbach et al. (2004), Bergaya et al. (2006), Carretero and Pozo (2007), and Fiore et al. (2013).

A number of books and reviews dealing with some particular aspect of clay science are also available, notably that by Weaver and Pollard (1973) and Newman (1987) on the chemistry, by Farmer (1974) and Gates et al. (2017) on infrared and Raman spectroscopies, by Yariv and Cross (2002) and Theng (2008, 2012) on the clay-organic and clay-polymer interactions, by van Olphen (1977) on colloid chemistry, by Sudo et al. (1981) on electron microscopy, by Sposito (1984) on surface chemistry, by Adams (1987) and Balogh and Laszlo (1993) on synthetic organic chemistry, by Konta (1995) on the clay-human relationship, by Murray (2007) on the occurrence, processing, and industrial applications, and by Theng (2018) on organic catalysis.

1.2 STRUCTURAL AND SURFACE-CHEMICAL ASPECTS

The basic building block in the phyllosilicate structure is an alumino-silicate layer comprising a silica tetrahedral sheet and an alumina octahedral sheet, joined together in certain proportions. An individual tetrahedron consists of a central Si^{4+} cation coordinated to four oxygens (Figure 1.1a), while in an octahedron the central Al^{3+} (or Mg^{2+}) cation is coordinated to six hydroxyls (Figure 1.1c).

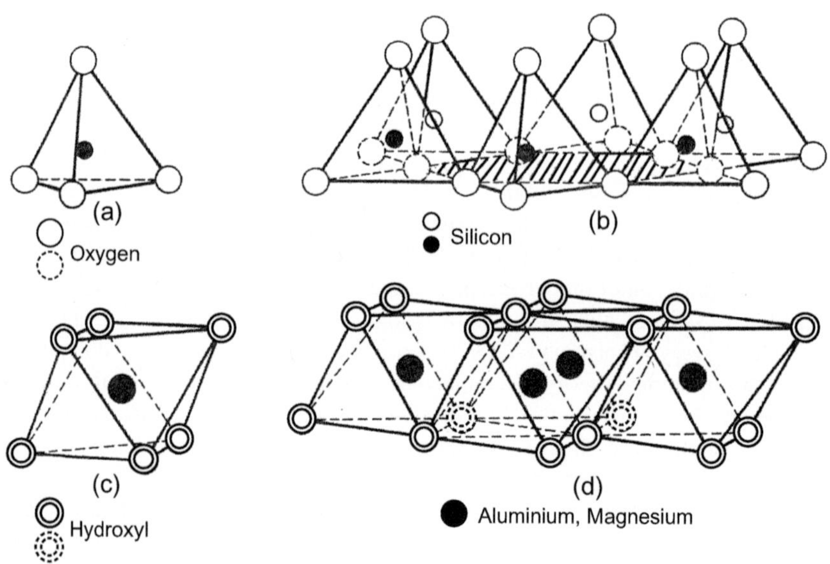

FIGURE 1.1 Diagram showing the basic building blocks of clay mineral structures: (a) a silica tetrahedron with the central silicon ion coordinated to four oxygens; (b) linkage between silica tetrahedra through corner-sharing yields a tetrahedral sheet with the basal oxygens forming a network of hexagonal cavities (hatched area); (c) an alumina octahedron with the central aluminium ion coordinated to six hydroxyls; and (d) an alumina octahedral sheet formed by linking individual octahedra through edge-sharing. (Reproduced with permission from *Formation and Properties of Clay-Polymer Complexes, 2nd edition* by B.K.G. Theng, Chapter 1, Copyright Elsevier 2012.)

A tetrahedral sheet is formed when individual tetrahedra are linked by sharing three corners (Figure 1.1b). As a result, the basal oxygens become approximately coplanar, forming an open hexagonal – in reality ditrigonal – network with the apical oxygens pointing in the same direction (Figure 1.3). On the other hand, individual octahedra in an octahedral sheet are linked by sharing edges (Figure 1.1d). When Al^{3+} is the central cation, only two out of every three octahedral sites are occupied to maintain electrical neutrality. The resultant structure, as found in the mineral gibbsite, is referred to as *dioctahedral.* When Mg^{2+} is the central cation, however, all three octahedral positions are filled. The brucite-like sheet structure that arises is accordingly designated as *trioctahedral.*

Combination of one tetrahedral (T) sheet with one octahedral (O) sheet gives rise to a *1:1 or (T-O) layer type* (Figure 1.2a) of which kaolinite represents the dioctahedral, and serpentine the trioctahedral, prototype. As a result, one basal plane consists of oxygens from the tetrahedral sheet ('siloxane surface'), while the other is made up of hydroxyls from the octahedral sheet ('aluminol/hydroxyl surface'). Likewise, combination of two tetrahedral sheets with one octahedral sheet where the octahedral sheet is sandwiched between two inward-pointing tetrahedral sheets, gives rise to a *2:1 or (T-O-T) layer type* (Figure 1.2b). The corresponding dioctahedral structure is represented by pyrophyllite (Figure 1.10), while the trioctahedral counterpart is exemplified by talc. In these structures both basal planes of individual silicate layers are of the siloxane type.

As already remarked on, the arrangement of the basal oxygens (of the tetrahedral sheet) is more ditrigonal than hexagonal. The departure from true hexagonal symmetry results from an internal adjustment between the larger tetrahedral sheet and its smaller octahedral counterpart. This mutual sheet adjustment largely occurs by the alternate rotation, in a clockwise and anti-clockwise fashion, of adjoining tetrahedra through angles of α about an axis normal to the basal oxygen plane (c-axis). As a result, the a- and b-dimensions of the sheet are shortened by approximately cos α, and the

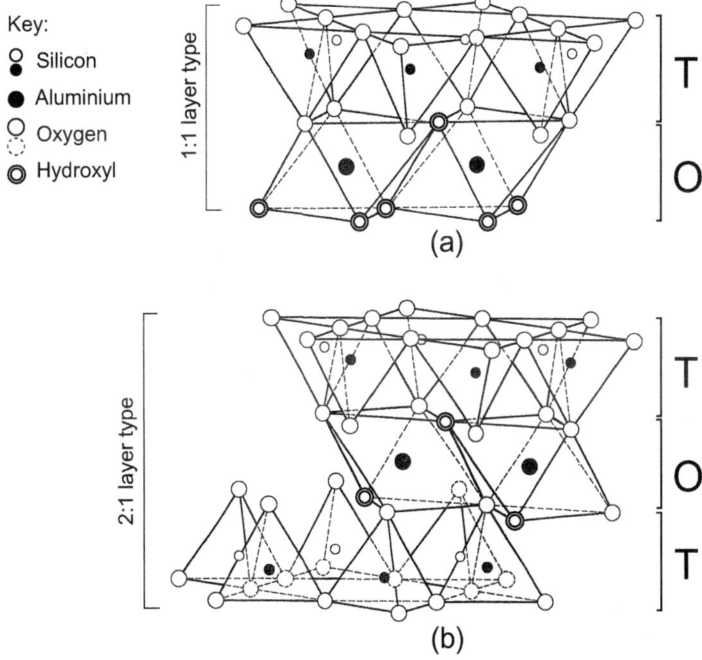

FIGURE 1.2 (a) 1:1 type layer structure formed by combining a silica tetrahedral sheet (T) with an alumina octahedral sheet (O); and (b) 2:1 type layer structure in which an alumina octahedral sheet (O) is sandwiched between two silica tetrahedral sheets (T).

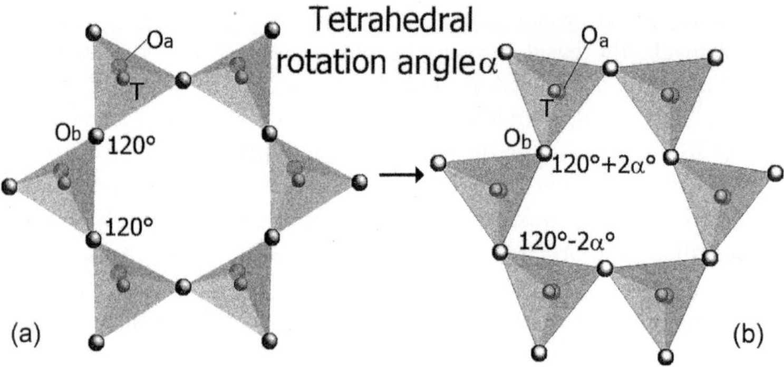

FIGURE 1.3 Diagram showing the principal adjustment adopted by the tetrahedral sheet to fit its larger lateral dimension to that of the smaller octahedral sheet through (a) rotation of adjoining tetrahedra as evaluated by the angle α (deviation from 120° of each angle in the ring) as shown in (b). As a result, the hexagonal cavity formed by the basal oxygens of the tetrahedral sheet becomes ditrigonal. (From Brigatti, M.F. et al., Structure and mineralogy of clay minerals, in *Handbook of Clay Science, 2nd edition*. Developments in Clay Science, Vol. 5A, (Eds.) F. Bergaya and G. Lagaly. Amsterdam, The Netherlands: Elsevier, pp. 21–81, 2013.)

(ideal) hexagonal geometry of the basal oxygen network becomes essentially ditrigonal, enclosing a cavity with a diameter of ~ 0.21 nm (Figure 1.3). Tetrahedral rotation, however, cannot fully compensate for the intrinsic misfit between the tetrahedral and octahedral sheets. Further sheet adjustment is made by means of tetrahedral flattening and tilting (Brigatti et al. 2013). As we will see later, the intercalation of organic compounds by clay minerals is often accompanied by the partial penetration ('keying') of the molecules into the ditrigonal cavities/depressions in the basal siloxane surface.

There is considerable scope in these structures for isomorphous substitution; that is, the partial replacement of Si^{4+} (in tetrahedral positions) and Al^{3+} (in octahedral sites), by cations of similar size and coordination but (usually) of lower valency. As a result, the layer acquires a negative charge which is commonly expressed as electron charge per formula unit (= half unit cell) and denoted by a positive number (x). Ranging in value from ~ 0 to ~ 2, x is therefore an important parameter in the classification of phyllosilicates. In the absence or near-absence of isomorphous substitution ($x \sim 0$), the interlayer space is essentially empty. For the majority of 2:1 type layer silicates, however, $x > 0$ and the resultant negative layer charge is balanced by sorption of extraneous cations the majority of which are located in the interlayer space.

These charge-balancing cations or *counterions* can generally be exchanged for other (inorganic or organic) cations in the external solution. Since the exchange process is stoichiometric, unless steric and chemical factors intervene, the terms *exchangeable cations* and counterions are interchangeable. Thus, most clay minerals have an inherent or permanent (pH-independent) cation exchange capacity (CEC). The SI unit of CEC, expressed as centimoles of charge ($cmol_c$) per kilogram of (calcined) clay, is numerically equal to the traditional value of milliequivalents (meq) per 100 g clay. The interlayer space may also be filled with a (positively charged) hydroxide sheet as in the chlorite group of phyllosilicates (Figure 1.14). These points are summarized in Table 1.1 showing a classification scheme for clay minerals, based on layer type, net layer charge, nature of the interlayer materials, and octahedral character (Martin et al. 1991; Guggenheim et al. 2006). We might add here that almost all the clay minerals, listed in Table 1.1, have been shown to occur on Mars (Du et al. 2023).

TABLE 1.1
Classification Scheme for Clay Minerals and Related Phyllosilicates Together with Some Structural-Chemical Properties.

A. Minerals with planar layer structures

Layer type	Group	Charge per formula unit (x)[1]	Octahedral character	Layer thickness (nm)	Interlayer materials: Cations	Interlayer materials: OH-sheet	Interlayer materials: Water	Representative species
1:1	Kaolin	Zero	dioctahedral	0.7	none	absent	absent	Kaolinite, dickite, nacrite
		Zero	dioctahedral	0.7–1.0	none	absent	up to $2H_2O$	Halloysite
	Serpentine	Zero	trioctahedral	0.7	none	absent	absent	Chrysotile
2:1	Pyrophyllite	Zero	dioctahedral	0.92	none	absent	absent	Pyrophyllite
	Talc	Zero	trioctahedral	0.9	none	absent	absent	Talc
	Smectite	0.2–0.6	dioctahedral	≥ 0.96	various	absent	variable	Montmorillonite, beidellite, nontronite
		0.2–0.6	trioctahedral	≥ 0.96	various	absent	variable	Saponite, hectorite, sauconite
	Vermiculite	0.6–0.9	dioctahedral	≥ 0.94	various	absent	variable	Dioctahedral vermiculite
		0.6–0.9	trioctahedral	≥ 0.94	various	absent	variable	Trioctahedral vermiculite
	Interlayer-deficient mica	0.6–0.85	dioctahedral	1.0	K	absent	variable	Illite[2]
	True mica	0.85–1.0	dioctahedral	1.0	K	absent	absent	Muscovite, celadonite
		0.85–1.0	trioctahedral	1.0	K	absent	absent	Biotite, phlogopite
	Brittle mica	1.8–2.0	dioctahedral	1.0	Ca	absent	absent	Margarite
		1.8–2.0	trioctahedral	1.0	Ca	absent	absent	Clintonite
	Chlorite	Variable	dioctahedral	1.4	None	present	absent	Donbassite
		Variable	di,trioctahedral	1.4	None	present	absent	Sudoite
		Variable	trioctahedral	1.4	None	present	absent	Chamosite, clinochlore

B. Minerals with Layer-Ribbon Structures

Layer type	Modulated Component	Charge per formula unit (x)[1]	Octahedral character	Layer thickness (nm)	Inter-ribbon materials: Cations	Inter-ribbon materials: OH-sheet	Inter-ribbon materials: Water	Representative species
2:1	Octahedral sheet	~ 0.1	di-trioctahedral	1.29 (c sinβ)	Ca (common)	absent	$2H_2O$	Palygorskite
	Octahedral sheet	~ 0.1	trioctahedral	1.34	Ca (common)	absent	$4H_2O$	Sepiolite

C. Minerals with Short-Range Order Structures[3]

Group	Layer thickness (nm)	Representative species
Allophane (variable Al/Si ratios)	0.7 – 1.0 (spherule wall)	Allophane
Imogolite (Al/Si = 2.0)	~ 1.5 (tubule wall)	Imogolite

Source: Theng, B.K.G., *Clay Mineral Catalysis of Organic Reactions.* CRC Press, Boca Raton, Florida, USA, pp. 5–6, 2018.

[1] net negative layer charge, expressed as a positive number

[2] now considered to be a 'series' name

[3] also denoted as 'non-crystalline' or 'poorly crystalline'

The particles of (non-interstratified) phyllosilicates are made up of a finite number of either 1:1 (T-O) or 2:1 (T-O-T) type layers that are continuous in the *a*- and *b*-directions and stacked in a more or less parallel manner in the *c*-direction perpendicular to the layer planes. A typical particle of kaolinite, for example, consists of 70–200 individual layers, stacked on top of each other (Wan and Tokunaga 2002; Johnston 2010). Such an arrangement offers scope for layer displacement, by rotation and/or translation, and hence for the occurrence of *polytypes*. Thus, dickite and nacrite have the same layer type and octahedral character as kaolinite but show different layer stacking sequences (Brigatti et al. 2013).

Layer stacking disorder in the kaolin group of minerals (Giese 1988; Bookin et al. 1989; Plançon et al. 1989) has been elegantly demonstrated by the combined application of high-resolution transmission electron microscopy (HRTEM), selected area electron diffraction (ED), and low-temperature Fourier Transform infrared spectroscopy (FTIR) (Kogure and Inoue 2005a, 2005b; Johnston 2010; Balan et al. 2014). As might be expected, individual clay particles ('crystals') may also contain different types of layers that are either regularly or randomly interstratified. Many pedogenic clay materials have this kind of mixed-layer structural arrangement. The formation, identification, and characterization of interstratified or mixed-layer clay minerals have been summarized by Sawhney (1989) and Fiore et al. (2013).

An important parameter in clay mineral identification by X-ray diffraction (XRD) is the repeat distance along the *c*-axis representing the first-order reflection from planes parallel to the layers. Referred to as the *basal* or *d*(001) *spacing*, this parameter corresponds to the thickness of an individual silicate layer plus the interlayer distance (Δ-value) separating successive layers within a particle (Figure 1.11). Thus, for non-expanding phyllosilicates having no interlayer materials (e.g., kaolinite), the basal spacing is equal to the layer thickness (Figure 1.4). On the other hand, in structures containing interlayer materials (cations, water), as in smectite and vermiculite, the basal

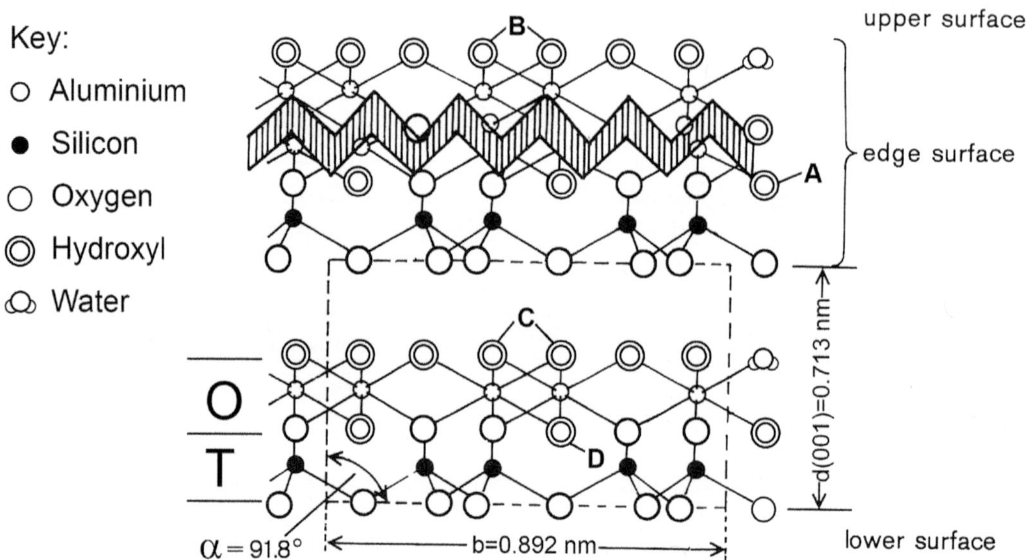

FIGURE 1.4 The structure of kaolinite, viewed along the *a*-axis, showing the superposition of adjacent layers within a particle. The upper particle surface consists of a hydroxyl plane, while the lower surface is made up of an oxygen plane. At the particle edge surface, each exposed aluminium ion normally carries a water molecule, while the edge silicon ion has an attached hydroxyl group. The outer hydroxyls are denoted by A and B, the inner-surface hydroxyls by C, and the inner hydroxyl by D. The unit cell dimension is indicated by the broken rectangle. (Reproduced with permission from *Formation and Properties of Clay-Polymer Complexes, 2nd edition* by B.K.G. Theng, Chapter 1, Copyright Elsevier 2012.)

spacing is invariably greater than the minimum van der Waals thickness of an individual silicate layer (~ 0.96 nm), even after complete dehydration because of the finite size of the interlayer counterions (Figure 1.11).

It is generally assumed that the unit structure composition of a given mineral is invariant, or at least homogeneous, throughout the particle. However, for interstratified structures (MacEwan and Ruiz-Amil 1975; Reynolds 1980; Sawhney 1989; Brigatti et al. 2013) where two or more layer types are present within a single particle, this assumption is clearly invalid. Rectorite is an example of a regularly interstratified phyllosilicate in which a mica layer (illite) alternates with a (dioctahedral) smectite layer in a 1:1 ratio (Sawhney 1989; Guggenheim et al. 2006; Brigatti et al. 2013). Even in the absence of interstratification, strict unit structure homogeneity is probably more the exception than the rule. Indeed, both the extent and location of isomorphous substitution in smectite particles tend to vary from layer to layer (Lagaly and Weiss 1970, 1976; Stul and Mortier 1974; Christidis and Dunham 1993; Janek et al. 1997; Christidis and Eberl 2003; Czímerová et al. 2006; Ertem et al. 2010). Another source of inhomogeneity arises from variations in layer stacking arrangement to which we have already referred (Brown et al. 1978; Brindley 1980; Brigatti et al. 2013).

The occupancy or otherwise of the interlayer space by extraneous materials provides another source of structural variation. In the smectite and vermiculite groups of phyllosilicates, this space contains various counterions and (commonly) water molecules. In mica the interlayer space is occupied by potassium ions, in brittle mica by calcium ions, and in chlorite by a hydroxide sheet (Table 1.1). The role of interlayer cations and water in the clay-water and clay-organic interactions will be described in Chapter 2.

Meunier and Velde (2004) have suggested that illite be regarded as a separate group of clay-size, dioctahedral mica-type minerals. Following Guggenheim et al. (2006), however, illite is listed here as a dioctahedral species in the interlayer-deficient mica group. We have also included palygorskite and sepiolite in Table 1.1 although these minerals have a layer-ribbon structure rather than a planar one. For the sake of completeness, allophane is included although the nano-size particles of this alumino-silicate are spheroidal and have short-range order. Nevertheless, allophane is regarded as a clay mineral because of past usage. The same argument applies to imogolite, a nano-size tubular alumino-silicate which shows long-range order along the tubule length.

Phyllosilicates with a (planar) layer structure comprise two broad groups having either a 1:1 (T-O) or 2:1 (T-O-T) layer type (Table 1.1). Among the 1:1 type layer silicates, *kaolinite* leads the field in terms of the clay-organic interaction, followed closely by *halloysite*, and very distantly by *chrysotile*. Within the 2:1 type layer silicates, the smectite group stands out in its capacity to 'host' to a wide range and variety of organic 'guest' molecules. Among the various species in this group (Table 1.1), *montmorillonite* has occupied the limelight because of its propensity and capacity for adsorbing and intercalating small and polymeric organic compounds (Mortland 1970; Yariv and Cross 2002; Lagaly et al. 2006; de Paiva et al. 2008; Theng 2008, 2012; Park et al. 2011; Yuan et al. 2013a; He et al. 2014; Zhu et al. 2016a; Li et al. 2018; Biswas et al. 2019). In this regard, *palygorskite*, *sepiolite*, and *vermiculite* are less reactive than montmorillonite or even kaolinite. The same comment applies to *allophane*, *chlorite*, *chrysotile*, *illite*, *imogolite*, and *micas*. For this reason, only a brief description of these clay mineral species will be given.

1.3 1:1 TYPE PHYLLOSILICATES WITH PLANAR LAYER STRUCTURES

1.3.1 Kaolinite

Figure 1.4 shows the kaolinite layer structure, viewed along the a-axis, together with the superposition of adjacent layers within a particle. The repeat distance along the c-axis of 0.713 nm represents the basal or $d(001)$ spacing. In the absence of interlayer materials, this value is equal to the thickness of an individual layer. The angle α of 91.8° arises from a displacement by $-\delta b$ of successive layers in the triclinic crystal. In kaolinite identical 1:1 layers are stacked with an interlayer shift of $2a/3$

whereas dickite and nacrite have a two-layer stacking sequence. The paper by Bailey (1963) has more details on the polymorphism in the kaolin group of minerals.

Also shown are the four types of hydroxyl groups, designated A through D. X-ray and neutron diffraction studies (Brindley and Nakahira 1958; Newnham 1961; Adams 1983; Suitch and Young 1983; Brigatti et al. 2013; Balan et al. 2014) indicate that the inner-surface hydroxyls (C) are nearly perpendicular to the layers, enabling the hydroxyl hydrogens of one layer to interact with the oxygens of an adjacent layer by hydrogen-bonding. The inner-surface hydroxyl groups are also responsible for the three absorption bands near 3697, 3670, and 3650 cm^{-1} in the IR spectrum of kaolinite (Ledoux and White 1966; Farmer 1974; Frost et al. 2000; Balan et al. 2014). The group responsible for the band at ca. 3650 cm^{-1} apparently lies close to the layer plane whereas the hydroxyls that give rise to the two higher frequency bands are nearly perpendicular to the silicate layer (Rouxhet et al. 1977). The fourth IR band at 3620 cm^{-1} is due to the stretching vibration of the inner-hydroxyl group (D). This hydroxyl lies centrally over a ditrigonal cavity, with its O-H bond pointing towards a vacant octahedral site and making a small angle to the (001) plane (Ledoux and White 1964; Giese and Datta 1973; Adams 1983; Frost et al. 2000).

Summation of the anionic and cationic charges in the ideal unit formula composition of $Al_2Si_2O_5(OH)_4$ shows that the layer structure is electrically neutral. In reality, however, most kaolinites carry a net negative charge, arising from either the presence of an alumino-silicate gel surface coating (Ferris and Jepson 1975) or the inclusion of mica and smectite layers within the particle (Lee et al. 1975; Lim et al. 1980; Jepson 1984; Ma and Eggleton 1999). In addition, chemical and spectroscopic (FTIR, Mössbauer, NMR, ESR, XANES) studies provide evidence for some isomorphous substitution within the kaolinite structure (Malden and Meads 1967; Angel and Hall 1973; Rengasamy et al. 1975; Bolland et al. 1976; McBride 1976; Komusiński et al. 1981; Fysh et al. 1983; Dixon 1989; Delineau et al. 1994; Newman et al. 1994; Gualtieri et al. 2000).

Thus, Fe^{3+} often substitutes for Al^{3+} in octahedral positions, while Al^{3+} may replace Si^{4+} in the tetrahedral sheet. For kaolinites from different sources, the resultant net negative charge, expressed in terms of CEC, ranges from 1.4 to 3.6 $cmol_c$/kg (Bolland et al. 1976, 1980). This range corresponds to a charge density of 0.073–0.145 C/m^2 which is comparable to that of smectite (montmorillonite) and not much lower than that shown by vermiculite (Table 1.2). The existence of a permanent negative layer charge (in kaolinite and halloysite) has also been inferred by Lee and Kim (2002) and Malek and Ramli (2015) from adsorption of hexadecyltrimethylammonium and cetylpyridinium bromide, respectively, and confirmed by Mbey et al. (2019) using potentiometric titration, electrokinetic measurements, and gas adsorption. Atomic force microscopy (Chang et al. 2021) also indicates that (at neutral pH) hydrated calcium ions are dominantly adsorbed to the basal siloxane surface through electrostatic interactions. In this context, we might add that the cation exchange sites in kaolinite are apparently confined to the tetrahedral basal surface rather than being uniformly distributed throughout the whole particle (Weiss and Russow 1963; McBride 1976; Tombácz and Szekeres 2006).

Of greater importance than the likely presence of a permanent negative layer charge is the variable charge that develops at the edge surface of kaolinite particles as the pH of the ambient solution is varied (Swartzen-Allen and Matijević 1974; van Olphen 1977). This pH-dependent charge arises from the protonation (under acid conditions) and deprotonation (under alkaline conditions) of incompletely coordinated or 'under-coordinated' aluminol groups exposed at particle edges (Schoonheydt and Johnston 2013). The pH at which the net edge charge is zero is referred to as the 'point of zero (net) charge' (PZC). The PZC for kaolinites ranges from 6 to pH 7.5 (Rand and Melton 1975; Herrington et al. 1992; Blockhaus et al. 1997), depending on sample history and origin, and mode of measurement. Acid-base titration of aqueous suspensions of Zettlitz kaolinite, for example, yielded a PZC of 6–6.5 for the edge surface (Tombácz and Szekeres 2006), while Liu et al. (2013) derived a pK_a value of 5.7 from first principle molecular dynamics (FPMD) simulations. Thus, the edge surface would be positively charged when the suspension pH falls below ~ 6.

TABLE 1.2
Cation Exchange Capacity (CEC), Layer Charge Density, and Swelling Characteristics of Some Common Clay Mineral Species as Affected by the Nature of the Interlayer Counterions.

		Mean charge density[2]					
Species	**CEC[1] ($cmol_c$/kg)**	**$\mu eq/m^2$**	**C/m^2**	**area/charge (nm^2)**	**Interlayer counterion**	**Basal spacing in water (nm)[3]**	**Swelling in water[4]**
Kaolinite	nil	nil	nil	n.a.	None	0.71	Absent
Pyrophyllite	nil	nil	nil	n.a.	None	0.92	Absent
Talc	nil	nil	nil	n.a.	None	0.93	Absent
Montmorillonite (Wyoming)	98	1.22	0.118	1.36	Li	2.20; ind.(> 4 nm)	restricted; extensive
					Na	1.90; ind.(> 4 nm)	restricted; extensive
					K	1.50	restricted
					Ca, Mg	1.90	restricted
Vermiculite (Kenya)	124	1.58	0.152	1.04	Li	1.50; ind (> 4 nm)	restricted; extensive
					Na	1.48	restricted
					K	1.10	restricted
					Ca	1.54	restricted
					Mg	1.48	restricted
Illite (Fithian)	26	2.80	0.271	0.60	K	1.0	Absent
Mica (muscovite)	< 5	3.54	0.342	0.46	K	1.0	Absent
Brittle mica (margarite)	< 5	7.26	0.702	0.23	Ca	1.0	Absent

[1] In reality, kaolinite has a small CEC due to isomorphous substitution in its structure. The CEC of illite, mica, and brittle mica is low because their interlayer counterions are practically non-exchangeable.
[2] The values for montmorillonite, vermiculite, and illite are derived on the basis of external and interlayer surface areas, while those for muscovite and margarite are calculated based on external particle areas; n.a. = not applicable.
[3] Basal spacing can be indeterminate (ind); i.e., larger than 4 nm.
[4] Extensive ('osmotic') swelling is associated with the development of diffuse electric double layers on interlayer surfaces; restricted swelling is often referred to as 'crystalline' swelling. (Reused with permission from *Clay Mineral Catalysis of Organic Reactions* by B.K.G. Theng, Table 1.2, p. 13, CRC Press, 2018.)

Thus, the edge surface of kaolinite particles is amphoteric, being positively charged when the solution pH is more acidic than the PZC, and negatively charged when the solution pH exceeds the PZC. In highly alkaline solutions (pH > 9), the edge silanol groups can even lose their protons, and hence contribute to the overall negative surface charge (Brady et al. 1996; Ma and Eggleton 1999; Schoonheydt and Johnston 2013). These points are illustrated in Figure 1.5, based on the early work by Schofield and Samson (1953). A similar scheme has been proposed by Alkan et al. (2005a) and Zhu et al. (2016b) based on electrokinetic measurements, potentiometric titration, ^{1}H NMR and FTIR spectroscopies, and field emission scanning microscopy.

When the suspension pH falls below the PZC, the positively charged edge surface of kaolinite particles can take up extraneous anions by electrostatic interactions or *anion exchange*. Being located outside the coordination layer, such anions (e.g., chloride) can exchange stoichiometrically for other species (e.g., nitrate) in solution. On the other hand, such anions as phosphate and sulfate can enter the coordination layer, and replace the hydroxyl groups attached to edge aluminium through *ligand exchange* or *specific adsorption* (Hingston et al. 1967, 1972) as shown in Equation 1.1.

$$[\text{edge}-\text{Al}(\text{H}_2\text{O})(\text{OH})]^0 + \text{H}_2\text{PO}_4^- \rightarrow [\text{edge}-\text{Al}(\text{H}_2\text{O})(\text{HPO}_4)]^{-1} + \text{H}_2\text{O} \qquad \text{(Equation 1.1)}$$

Unlike anion exchange, ligand exchange can occur even when the particle edge is negatively charged (Blockhaus et al. 1997). Besides releasing hydroxyl ions into solution, ligand exchange makes the surface charge more negative (McBride 1994). These points may be illustrated by the interaction of humic acid (HA) with allophane (Yuan et al. 2000). Here the carboxylate groups of HA exchange for hydroxyls of (HO)Al(OH_2) groups, exposed at wall perforations (Figure 1.20) as indicated by Equation 1.2 where ❃ represents an allophane spherule ('nanoball'). As sorption (by ligand exchange) progresses, the surface charge of the allophane-HA complex becomes more negative.

$$[❃-\text{Al}(\text{H}_2\text{O})(\text{OH})]^0 + \text{HA-COO}^- \rightarrow [❃-\text{Al}(\text{H}_2\text{O})(\text{COO-HA})]^{-1} + \text{OH}^- \qquad \text{(Equation 1.2)}$$

A similar bonding mechanism has been proposed for the interaction of salicylate with edge aluminium in illite (Kubicki et al. 1997), and citrate with edge =Al(OH) groups in montmorillonite (Ramos and Huertas 2014), using attenuated total reflectance Fourier transform infrared (ATR-FTIR) spectroscopy.

We might also add that edge aluminium can act as a Lewis acid or electron acceptor in various clay-organic reactions (Solomon 1968; Theng 1971, 2018). In an aqueous medium, however, this ability is largely suppressed because water, being a Lewis base, would attach to (under-coordinated) aluminium exposed at the particle edge surface (Figure 1.5) (Solomon and Hawthorne 1983). By the same token, the electron-accepting activity of edge aluminium is enhanced when water is excluded from the system through prior drying/dehydration of the clay mineral or by conducting the reaction in a non-polar organic solvent. Recent molecular dynamics and metadynamics simulation studies by

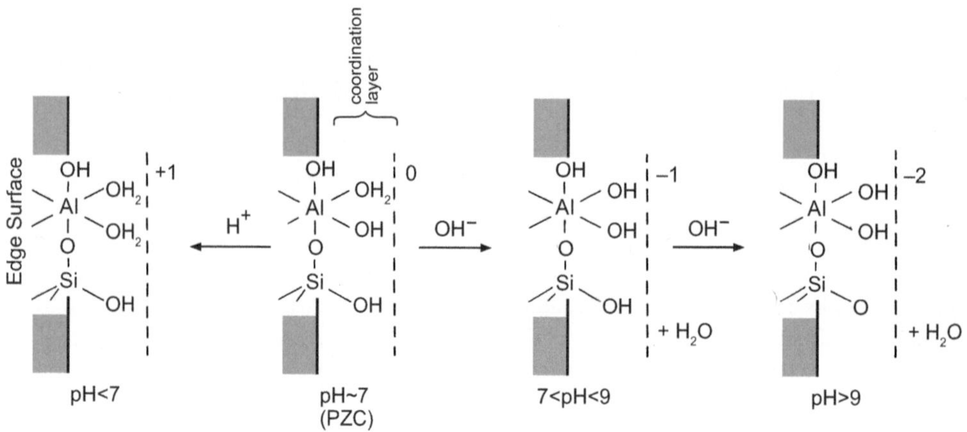

FIGURE 1.5 Diagram showing the development of positive and negative charges at the edge surface of kaolinite particles in aqueous suspension, in response to changes in medium pH. NB: the point of zero charge (PZC) is actually a little less than pH 7; the vertical broken line indicates the extent of the (imaginary) coordination layer. (Reproduced with permission from *Formation and Properties of Clay-Polymer Complexes, 2nd edition* by B.K.G. Theng, Chapter 1, Copyright Elsevier 2012.)

Zhang et al. (2023) also indicate that at low relative humidity water adsorption to the edge surface prevails and occurs earlier than to the basal surface.

Scanning electron micrographs of kaolinite show particles with a hexagonal outline, ranging in thickness from 0.05 to 2 μm, forming 'book-like' aggregates (Figure 1.6). As such, the edge area of kaolinite particles makes up a significant proportion (12–36%) of the total (external particle) surface area (Liétard et al. 1980; Lee et al. 1991; Yariv 1992; Brady et al. 1996; Wan and Tokunaga 2002) whereas the corresponding value for smectite seldom exceeds 5%. Accordingly, the kaolinite edge surface plays an important role in electron transfer and related colour reactions of adsorbed organic compounds as well as in silane grafting reactions (He et al. 2013). On the other hand, the bulk of organic reactions and transformations involving smectite take place on interlayer surfaces of the mineral.

The superposition of oxygen and hydroxyl planes in contiguous layers of a kaolinite particle (Figure 1.4) is conducive to O–H····O hydrogen-bond formation. The forces arising from interlayer H-bonding (and van der Waals interactions) must therefore be overcome if interlayer sorption of organic species is to occur. Not surprisingly, organic intercalation by kaolinite is difficult to achieve, and adsorption is generally confined to external particle surfaces. Nevertheless, kaolinite can form interlayer complexes with organic molecules that can form hydrogen bonds with the interlayer surface (e.g., amides, hydrazine, urea) or have high dipole moments (e.g., dimethylsulphoxide) in one

FIGURE 1.6 Scanning electron micrograph of kaolinite showing the parallel stacking of particles to form book-like aggregates. (Image reproduced from the 'Images of Clay Archive' of the Mineralogical Society of Great Britain & Ireland and The Clay Minerals Society [www.minersoc.org/images-of-clay.html].)

of three ways: (a) contacting the clay with the intercalant in the form of a liquid, a melt, or a concentrated aqueous solution; (b) first separating the silicate layers with an entraining agent; and (c) displacing a previously intercalated species (MacEwan and Wilson 1980; Lagaly et al. 2006; Johnston 2010; Detellier and Schoonheydt 2014). By comparison with halloysite and smectite, however, organic intercalation into kaolinite is slow and often incomplete (Olejnik et al. 1970; Theng et al. 1984; Lagaly et al. 2006). These points are described more fully in Chapter 6. We would also refer to the review by Kloprogge (2017) on the characterization of organic complexes with kaolinite (and halloysite) by Raman and IR spectroscopies.

1.3.2 Halloysite

Like kaolinite, halloysite is a dioctahedral 1:1 type phyllosilicate (Table 1.1), but here the individual layers are, or have been, separated by a monolayer of water molecules (Figure 1.7). Indeed, the presence of interlayer water in its natural state, or during its past history, is the single, most characteristic feature that distinguishes halloysite from kaolinite (Churchman and Carr 1975). Halloysites can occur with varying degrees of hydration. The fully hydrated form with a unit formula of $Al_2Si_2O_5(OH)_4.2H_2O$ and a basal spacing of 1.01 nm, is referred to as *halloysite-(10Å)*, while the completely dehydrated material with a unit formula of $Al_2Si_2O_5(OH)_4$ and a basal spacing of ~ 0.72 nm is designated as *halloysite-(7Å)*. The difference of 0.29 nm (in basal spacing) between the two end members corresponds to the thickness of the interlayer water sheet, being equal to the van der Waals diameter of a water molecule. Most of the interlayer water in halloysite-(10Å) is removed on standing in dry air, evacuation at room temperature, or mild heating (70°C) in air, reducing the basal spacing to 0.72 nm (Churchman and Carr 1972; Brindley 1980). Once dehydrated, halloysite does not rehydrate when exposed to water vapour, not even after prolonged contact with liquid water. Thus, halloysite-(10Å) should be stored under high-to-saturated humidity conditions (Joussein et al. 2007).

Like montmorillonite, halloysite-(10Å) can intercalate a wide variety of polar, uncharged organic compounds by displacement of its interlayer water. Halloysite, however, fails to intercalate more than a monolayer of water and organic molecules whereas montmorillonite can take up multiple layers (Lagaly et al. 2006). Like kaolinite, halloysite-(7Å) is capable of intercalating organic compounds that are highly polar or capable of forming hydrogen bonds with the interlayer surface (MacEwan 1948; Carr and Hwa Chih 1971; Churchman and Theng 1984; Joussein et al. 2005). Interlayer complex formation is accompanied by layer re-alignment as indicated by the appearance of new [*hkl*] reflections in the XRD pattern. This observation is consistent with the formation of a 2-layer stacking order rather than a kaolinite-like layer stacking sequence (Churchman et al. 2016).

Also noteworthy is that intercalation into halloysite takes place at a faster rate and to a more complete extent as compared with kaolinite (Theng et al. 1984). On this basis, a simple and rapid intercalation method using formamide has been developed for differentiating kaolinite from halloysite in mixtures, especially when the latter mineral is dehydrated (Churchman et al. 1984). The formamide test, however, can underestimate the concentration of halloysite in dehydrated, halloysite-rich soil fractions (Joussein et al. 2007). This test also fails when applied to some high-charged, potassium-selective soil halloysites because of surface contamination by 2:1 type layer silicates (Takahashi et al. 2001). We might also mention that addition of ethylene glycol to (slightly hydrated) halloysite causes the intensity of the (001) peak at ~ 0.72 nm to decrease and that of the (002) peak at ~ 0.358 nm to increase at the same time. The resultant narrowing of the (001)/(002) intensity ratio is diagnostic for halloysite. Furthermore, ethylene glycol solvation can detect as little as 20% of halloysite in mixtures with kaolinite (Hillier and Ryan 2002). The interactions of halloysite with organic compounds are described more fully in Chapter 6.

In the structure of halloysite-(10Å), proposed by Hendricks and Jefferson (1938), the interlayer water forms an almost rigid hexagonal network in which the molecules are hydrogen-bonded to each other and to opposing silicate layers (Figure 1.7). This concept of a rather uniform ice-like

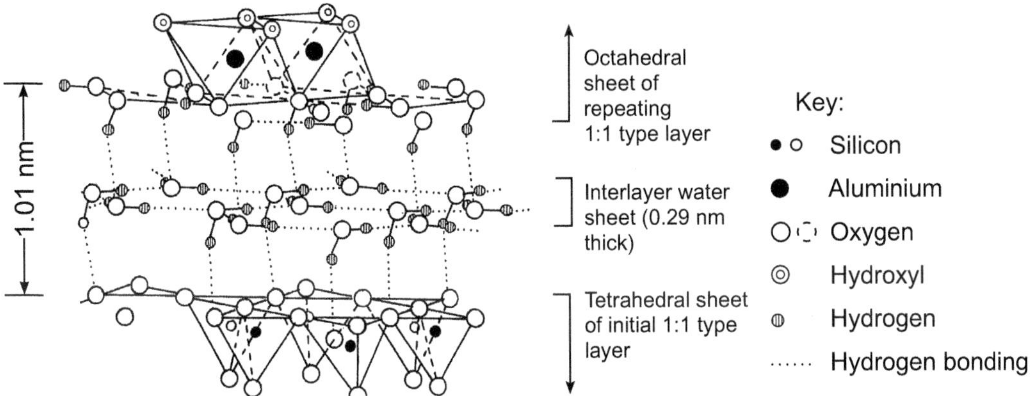

FIGURE 1.7 The structure of hydrated halloysite, commonly referred to as 'halloysite-(10Å)', according to Hendricks and Jefferson (1938), showing the presence in the interlayer space of a single sheet of water molecules that are hydrogen-bonded (dotted line) to each other and to the hydroxyls and oxygens of opposing silicate layers. (Reproduced with permission from *Formation and Properties of Clay-Polymer Complexes, 2nd edition* by B.K.G. Theng, Chapter 1, Copyright Elsevier 2012.)

structure for interlayer water, however, does not appear to accord with reality. Rather, the water molecules show a pronounced ordering and reduced mobility (Presti et al. 2016). Bailey (1990) has gone so far as to suggest that the interlayer water molecules in halloysite-(10Å) are coordinated to counterions that balance the negative layer charge arising from isomorphous substitution of Al^{3+} for Si^{4+} in the tetrahedral sheet. Subsequent solid-state ^{27}Al NMR spectroscopic analysis by Newman et al. (1994), however, indicates that the Al^{IV} content of many halloysites is like that of a standard kaolinite. These workers have further suggested that Fe^{3+} can substitute non-stoichiometrically for Al^{3+} in the octahedral sheet, giving rise to a positive charge deficiency which is balanced by intercalation of counterions with their complement of water molecules.

Subsequent XPS spectroscopic analysis of five halloysites from New Zealand has provided clear evidence for the substitution of Fe^{3+} for Al^{3+} in octahedral sites (Soma et al. 1992). In this process, approximately two Al^{3+} ions are lost for every Fe^{3+} ion gained, creating cation vacancies in the octahedral sheet. The positive correlation between the cation exchange capacity (5–15 $cmol_c$/kg) of the samples and their structural iron content is consistent with the formation of such vacancies (Soma and Theng 1998). For some tubular halloysites, Gray et al. (2016) have reported an average CEC value of 8.5 $cmol_c$/kg (at pH 7).

Like that of kaolinite (Figure 1.5), the edge surface of halloysite particles would carry a pH-dependent charge. Since halloysite can assume a variety of particle shapes, however, its capacity for sorbing anions (Equation 1.1) is also variable (Theng et al. 1982). Using rheological measurements, Theng and Wells (1995) derived a point of zero charge (PZC) of 6.0 and 7.1 for the edge surface of two New Zealand halloysites differing in particle shape.

Halloysite can occur in a variety of particle shapes, namely, tubular, platy, spheroidal, and prismatic (Kirkman 1977; Joussein et al. 2005; Churchman 2015). The hollow tubular form (Figure 1.8) has received special attention because of its potential usage in the synthesis of polymer-clay nanocomposites and its capacity to carry active agents for agricultural, environmental, and medical applications (Du et al. 2010; Rawtani and Agrawal 2012; Lvov and Abdullayev 2013; Tan et al. 2014; Abdullayev and Lvov 2015; Pasbakhsh and Churchman 2015; Yuan et al. 2015, 2016; Hanif et al. 2016; Liu et al. 2016; Pasbakhsh et al. 2016). The length of halloysite tubules ranges from 50 to 5000 nm, the outer diameter from 20 to 200 nm, the inner ('lumen') diameter from 5 to 70 nm, and the wall thickness from 5 to 100 nm (Joussein et al. 2005; Pasbakhsh et al. 2013; Keeling and

Pasbakhsh 2015; Yuan et al. 2015; Zhang et al. 2016). Interestingly, a spherical halloysite from New Zealand (Kirkman and Pullar 1978) shows a pronounced anti-inflammatory activity as indicated by its ability to inhibit oedema in mice (Cervini-Silva et al. 2016).

The isomorphous substitution of iron for aluminium in the halloysite structure is known to affect particle shape. Low structural iron contents tend to be associated with long tubular particles, intermediate contents with short and wide tubes ('laths'), and high contents with spheroidal and platy forms (Churchman and Theng 1984; Joussein et al. 2005; Churchman et al. 2016). Although halloysite particles from various geological settings are tubular in shape (Keeling 2015), a historical analysis of the literature indicates that tubular morphology *per se* cannot be taken as either a necessary or a sufficient characteristic of halloysite (Churchman and Carr 1975; Joussein et al. 2005).

Layer curling has been ascribed to the intrinsic misfit between the laterally larger silicon tetrahedral sheet and the smaller aluminium octahedral sheet (Bates et al. 1950; Bailey 1990). Although this mismatch is partly offset by the alternate rotation of tetrahedra in opposite directions, the silicate layers are under a certain amount of strain. When contiguous layers are separated by a water sheet, however, the resultant strain can be relieved by layer curling with the basal oxygen plane on the convex side (Radoslovich 1963) as shown in Figure 1.8. In kaolinite, this process does not take place because of H-bonding between adjacent layers within a particle (Hofmann 1968), but when this bonding is weakened by interlayer hydration, layer rolling can occur (Singh 1996).

In line with this hypothesis, Singh and Mackinnon (1996) were able to induce kaolinite layers to curl by repeated intercalation and removal ('deintercalation') of potassium acetate. The approach used by Gardolinski and Lagaly (2005) was to graft the interlayer surface of kaolinite with alkanols, diols, and glycol mono-ethers, followed by intercalation of *n*-alkylamines, and delamination in toluene. Similarly, Kuroda et al. (2011) and Yuan et al. (2013b) were able to transform platy kaolinite layers into halloysite-like 'nanoscrolls' by inserting quaternary ammonium salts into the interlayers of a methoxy-modified kaolinite (cf. Chapter 6). More recently, Wan et al. (2020) were able to prepare montmorillonite nanotubes by rolling up their nanolayer counterparts under hydrothermal conditions, using cetyltrimethylammonium bromide as a directing agent. The reviews by Guggenheim (2015) and Niu (2016) have more details on the 'rolling' of kaolinite layers to form tubular halloysite.

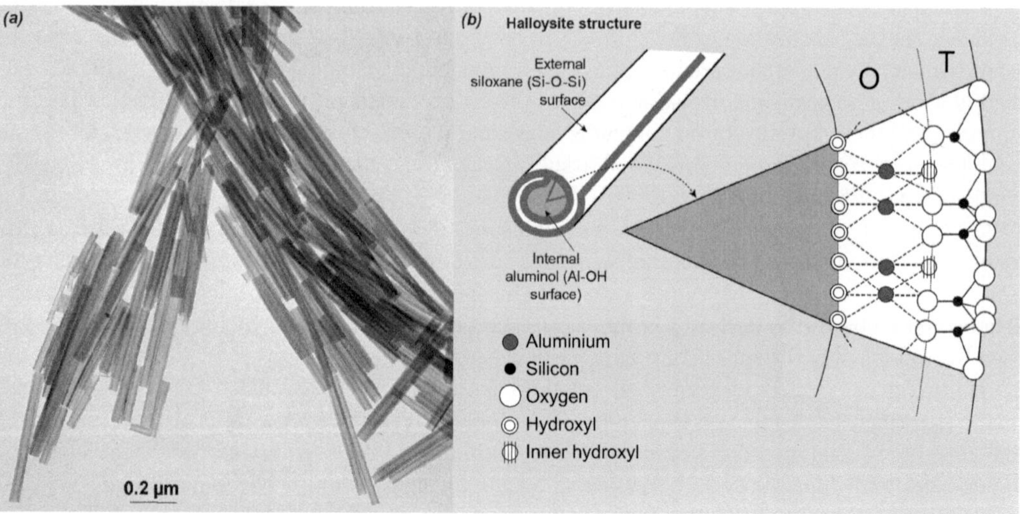

FIGURE 1.8 (a) transmission electron micrograph of tubular halloysite from Camel Lake, South Australia; and (b) diagram of a halloysite nanotube (HNT) made up of an external siloxane surface (tetrahedral sheet) and an internal aluminol surface (octahedral sheet). (From Keeling and Pasbakhsh 2015, Figure 1. Reproduced from *MESA Journal*.)

1.3.3 Chrysotile

Chrysotile is a trioctahedral mineral species in the serpentine group of 1:1 type layer silicates (Table 1.1). Its ideal layer structure composition of $Mg_3Si_2O_5(OH)_4$ indicates that all octahedral sites are occupied by magnesium. As a result, the lateral dimension of the octahedral sheet is appreciably larger than that of the tetrahedral sheet. Since there is neither tetrahedral rotation nor substitution of Al^{3+} for Si^{4+} in the tetrahedral sheet (to increase its lateral dimension), layer curling occurs to compensate for the misfit between its constituent sheets (Roy and Roy 1954; Bates 1959) which involves an out-of-plane tilting of tetrahedra so that the apical oxygen anion distance of the tetrahedral sheet can increase to form a common junction with the octahedral sheet (Guggenheim 2015). The result is the scroll or cylinder form. Unlike the situation in rolled forms of halloysite, it is the brucite-like octahedral sheet that forms the outer side, while the tridymite-like tetrahedral sheet makes up the inner side of the fibril (Figure 1.9). Like that of (ideal) kaolinite, the unit layer of chrysotile is uncharged ($x = 0$), and its interlayer space is devoid of counterions and water molecules (Table 1.1).

Whereas halloysite particles can adopt a variety of shapes, naturally occurring chrysotile almost invariably occurs as long (> 100 nm), hollow fibrils or tubules. The tubular morphology, however, was not recognized until the early 1950s when Bates et al. (1950) and Noll and Kircher (1951) published electron micrographs of chrysotile. Using high-resolution TEM, Yada (1967, 1971) subsequently confirmed that chrysotile tubes occurred as concentric cylinders (scrolls) of serpentine layers. An individual fibril consists of several superposed 1:1 (T–O) type layers, each of which is ca. 0.73 nm thick, curled into concentric cylinders or spirals about the fibril axis (Figure 1.9). Because of layer curling, long O–H····O hydrogen bonding between contiguous layers in a particle (as in kaolinite) is no longer possible. Instead, a rather complicated steric packing arrangement is adopted (Wicks and Whittaker 1975). For more details on the structure, chemistry, and

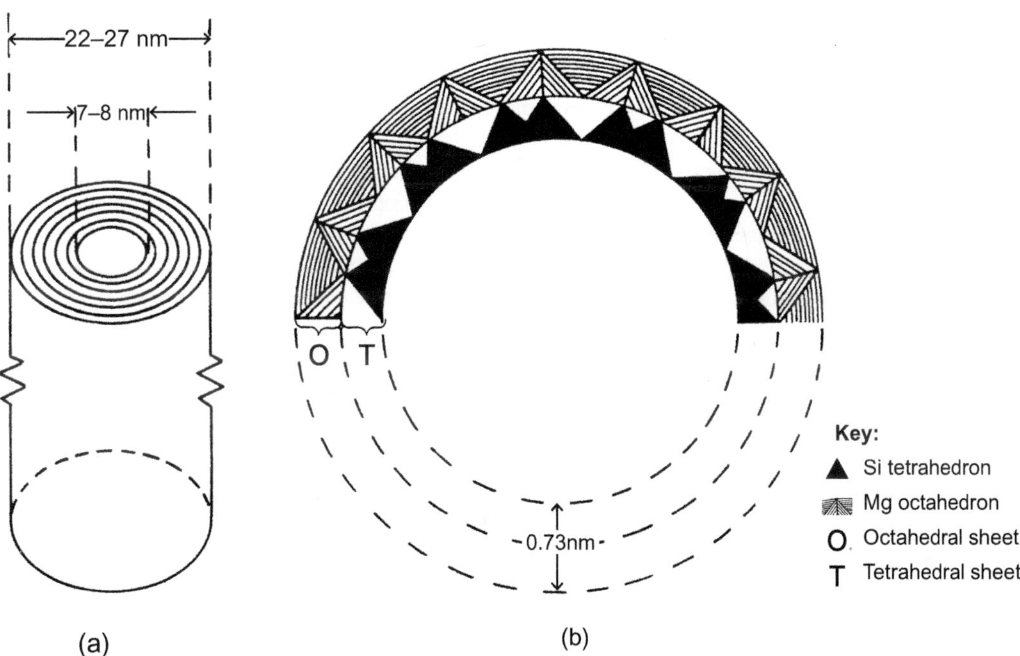

FIGURE 1.9 (a) hollow chrysotile fibril having an outer diameter of 22–27 nm, an inner diameter of 7–8 nm, and a length of > 100 nm, consisting of T–O type layers, curled into concentric cylinders; and (b) the layer structure of chrysotile viewed along the fibril axis. (Reused with permission from *Clay Mineral Catalysis of Organic Reactions*, by B.K.G. Theng, Figure 1.12, pp. 33, CRC Press, 2018.)

identification of chrysotile the reader is referred to the reviews by Guggenheim (2015) and Fiore and Huertas (2015), while the health risk of this mineral has been summarized by Bernstein et al. (2013).

Although the fibril thickness (cross-section) varies between samples, the most frequently occurring dimension for the outer and inner diameter, estimated from electron micrographs, is 22–27 and 7–8 nm, respectively (Yada 1971), as indicated in Figure 1.9. On the other hand, Fripiat and Della Faille (1967) measured average values of 13.6–21 and 2.4–4.9 nm for the outer and inner diameter, respectively, in seven samples of natural chrysotile. These values are comparable to those reported by Sprynskyy et al. (2011). Except for the sample from Coalinga (California), the intra-tubular pores contained (up to 50%) amorphous materials. For the Coalinga chrysotile, the surface area accessible to nitrogen and water was 53.4 and 94.4 m^2/g, respectively, indicating that the intra-tubular pore space ('lumen') was inaccessible to nitrogen. On this basis, approximately 19% and 22% of the void volume may be assigned to inter- and intra-tubular pores, respectively. The sample used by Sprynskyy et al. (2011), on the other hand, has a nitrogen area of 15.3 m^2/g which increases to 63.6 m^2/g following acid treatment, corresponding to an average pore diameter of 9.8 and 3.9 nm, respectively.

The presence of mineral and trace metal impurities, isomorphous substitution, and surface defects in natural samples has led many workers to synthesize 'stoichiometric' and metal-substituted chrysotiles under controlled hydrothermal conditions (Roy and Roy 1954; Noll et al. 1958; Swift 1977; Suzuki and Ono 1984; Falini et al. 2004, 2006; Foresti et al. 2005; Roveri et al. 2006; Piperno et al. 2007; Olson et al. 2008). Besides being relatively non-cytotoxic, synthetic chrysotile can potentially serve as an attractive alternative to inorganic and carbon nanotubes for some industrial and technological applications (Gazzano et al. 2005; Bloise et al. 2009; Foresti et al. 2009).

1.4 2:1 TYPE PHYLLOSILICATES WITH PLANAR LAYER STRUCTURES

1.4.1 Smectite

Within the planar 2:1 type phyllosilicate class (Table 1.1) only the minerals in the pyrophyllite and talc groups have no layer charge ($x \sim 0$). Thus, summation of the anionic and cationic charges in the unit formula of $Al_2Si_4O_{10}(OH)_2$ for dioctahedral pyrophyllite (Figure 1.10) shows that its layer structure is electrically neutral (uncharged). The same applies to trioctahedral talc with a layer composition of $Mg_3Si_4O_{10}(OH)_2$. Since the interlayer space does not contain counterions and/or water, both pyrophyllite and talc have no measurable CEC. More importantly, their external basal (siloxane) surface is essentially hydrophobic (Jaynes and Boyd 1991a; Atluri et al. 2019). By the same token, these minerals do not show interlayer swelling in water since the contiguous silicate layers within a particle are strongly held together by van der Waals and electrostatic attractive forces (Giese 1975). Nevertheless, pyrophyllite and talc can take up some aromatic hydrocarbons (e.g., benzene, toluene) from water to the same extent as, if not better than, tetramethylphosphonium- and tetramethylammonium-exchanged montmorillonite (Hashizume 2009, 2013). Pyrophyllite is similarly effective in taking up methylene blue and more so after acid treatment and ball milling (Sheng et al. 2009). Talc can adsorb chlorpheniramine on its external particle surface (Lv et al. 2012) which is also available for silylation ('grafting') with aminated trioxysilanes (da Fonseca and Airoldi 2001).

On the other hand, the *smectite* group of 2:1 type layer silicates has a negative layer charge (x) of 0.2–0.6 per formula unit (half unit cell) which is balanced by interlayer (hydrated) counterions. This group includes dioctahedral *montmorillonite, beidellite*, and *nontronite* together with *hectorite* and *saponite* which are trioctahedral (Table 1.1). Although all five smectite species have a high propensity for adsorbing and intercalating organic molecules, it is montmorillonite that has long been the focus of attention (Mortland 1970; Rausell-Colom and Serratosa 1987; Yariv and Cross 2002; Johnston et al. 2004; Lagaly et al. 2006; Theng 2012, 2018; Okada et al. 2014). Indeed, the next four chapters are largely concerned with the montmorillonite-organic interaction. Montmorillonite has

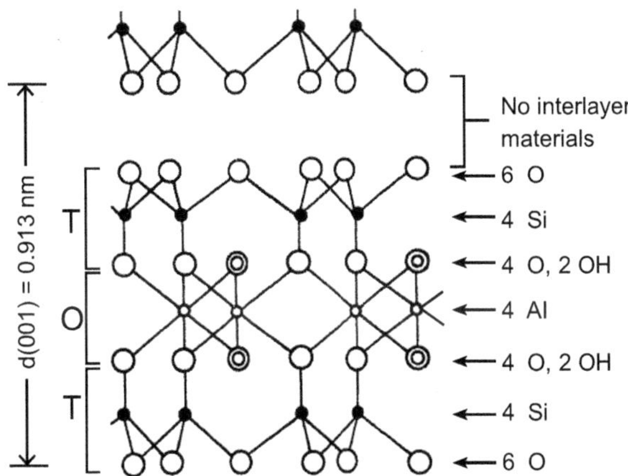

FIGURE 1.10 The layer structure of pyrophyllite showing the absence of isomorphous substitution and interlayer materials, giving a basal or *d*(001) spacing of 0.913 nm. (Reproduced with permission from *Formation and Properties of Clay-Polymer Complexes, 2nd edition* by B.K.G. Theng, Chapter 1, Copyright Elsevier 2012.)

also served as nature's own model for a planar, negatively charged colloid in terms of the clay-water interaction (van Olphen 1977).

The layer structure of montmorillonite (Figure 1.11) was deduced by Hofmann et al. (1933) on the basis of its similarity to that of pyrophyllite but with (isomorphous) substitution of Al^{3+} for Mg^{2+} in the octahedral sheet. Further, the stacking of successive layers within a particle is more or less random (Marshall 1935; Maegdefrau and Hofmann 1937; Hendricks 1942). The Hofmann-Endell-Wilm-Marshall-Maegdefrau-Hendricks structure, or 'Hofmann structure' for short, is now generally accepted. Montmorillonite has the ideal composition of $M^{+}_{x}.nH_2O$ $(Al_{2-x}Mg_x)(Si_4)O_{10}(OH)_2$, indicating that the (negative) layer charge is balanced by hydrated (monovalent) cations, $M^{+}_{x}.nH_2O$, occupying interlayer positions.

Under optimum conditions of dispersion in water, the layers of Li^+- and Na^+-montmorillonites may dissociate completely through extensive interlayer ('osmotic') swelling, exposing a surface area close to the calculated value of ca.760 m^2/g. Many clay-organic reactions, however, are carried out with montmorillonites that have previously been dehydrated. Under these conditions, the silicate layers within a particle collapse to a basal spacing of ca. 1 nm, and much of the exposed area is contained within interparticle pores. The extent of this area, commonly determined by nitrogen adsorption at 77 K and applying the Brunauer et al. (1938) (BET) equation, varies from 31 to 130 m^2/g, depending on particle size distribution, sample pretreatment, and counterion type. Unlike nitrogen, polar organic molecules (glycerol, ethylene glycol monomethyl ether, *p*-nitrophenol) can adsorb to both external particle and interlayer surfaces of dry and heat-collapsed smectites, yielding surface areas of 486–834 m^2/g (van Olphen and Fripiat 1979; Stul and Van Leemput 1982; Tiller and Smith 1990; Theng 1995; Kaufhold et al. 2010). Chapter 2 has more details on the measurement of specific surface areas by adsorption or retention of organic molecules.

Beidellite and nontronite (Figure 1.11) are the other two dioctahedral species in the smectite group of phyllosilicates (Table 1.1). The ideal composition of beidellite may be written as $M^{+}_{x}.nH_2O$ $(Al_2)(Si_{4-x}Al_x)O_{10}(OH)_2$, indicating that its layer charge is entirely due to substitution of Al^{3+} for Si^{4+} in the tetrahedral sheet. In nontronite with the ideal composition of $M^{+}_{x}.nH_2O$ $(Fe_2^{3+})(Si_{4-x}Al_x)$ $O_{10}(OH)_2$ there is total replacement of Al^{3+} by Fe^{3+} in the octahedral sheet as well as some tetrahedral substitution of Al^{3+} for Si^{4+}.

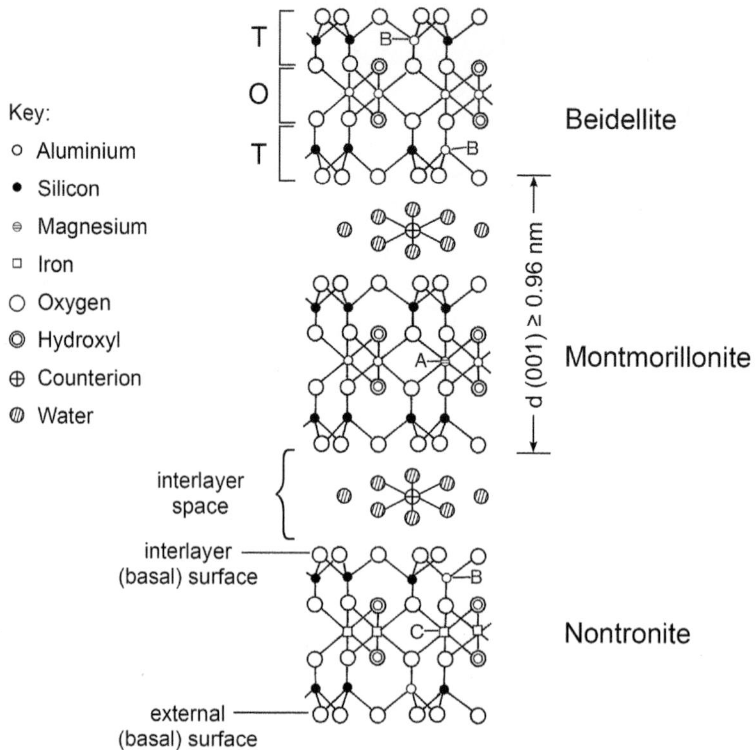

FIGURE 1.11 The layer structure of montmorillonite, beidellite, and nontronite viewed along the *a*-axis. The preponderant site of isomorphous substitution and the common substituting cation are indicated by the letters A, B, and C, respectively. The basal or *d*(001) spacing varies according to the nature of the interlayer counterion and the ambient relative humidity. Under conditions of complete dehydration and layer collapse, this spacing is close to 0.96 nm. (Reused with permission from *Clay Mineral Catalysis of Organic Reactions*, by B.K.G. Theng, Figure 1.13, p. 35, CRC Press, 2018.)

In reality, however, the composition of dioctahedral smectites is intermediate between that of the three end members, and the layer charge distribution is heterogeneous (Stul and Mortier 1974; Lagaly and Weiss 1976; Janek et al. 1997; Czímerová et al. 2006). Thus, montmorillonite nearly always contains some Al^{3+} and Fe^{3+} ions in tetrahedral and octahedral positions, respectively (Weaver and Pollard 1973).

Hectorite has an ideal composition of $M^{+}_{x}.nH_2O\ (Mg_{3-x}Li_x)(Si_4)O_{10}(OH)_2$ and may be regarded as the trioctahedral analogue of montmorillonite in that its layer charge is due to isomorphous substitution of Li^+ for Mg^{2+} in the octahedral sheet. Many naturally occurring hectorites have fluoride replacing hydroxyl in the structure. Zhang et al. (2019) have remarked on the rarity of hectorite deposits in nature and the ease of synthesizing this smectite species by a hydrothermal process under laboratory conditions (Strese and Hofmann 1941; Granquist and Pollack 1960; Barrer and Jones 1970; Ogawa et al. 2008; Vinci et al. 2020). Synthetic hectorite generally has a smaller particle size and a higher surface area than its natural counterpart (Carrado et al. 1997). We might also add that the presence of various organic species (alkylammonium ions, porphyrins, phtalocyanines, aromatic dyes) during hydrothermal synthesis promotes the formation of the respective interlayer hectorite-organic complexes (Carrado 2000). The review by Kloprogge et al. (1999) has more details about the synthesis of smectite clay minerals.

The first large-scale synthesis of a hectorite-like mineral was reported over 50 years ago (Neumann and Sansom 1970a, 1970b). Synthetic hectorite with the composition of $Na^{+}_{0.33}$

$(Mg_{2.67}Li_{0.33})(Si_4)O_{10}(OH)_2$ is marketed under the trade name Laponite® by Laporte Absorbents, UK and BYK-Chemie GmbH, Germany. A closely similar material is available from Kremer Pigmente, Germany. Synthetic *fluorohectorite* with fluoride replacing hydroxyl in the structure, has received much attention because of its layer charge homogeneity (Breu et al. 2001; Kalo et al. 2010), its capacity for capturing and storing CO_2 (Cavalcanti et al. 2018), and its use in preparing polymer-clay nanocomposites (Varghese and Karger-Kocsis 2004; Theng 2012). Earlier, Tamura and Nakazawa (1996) reported the intercalation of *n*-alkyltrimethylammonium ions into a synthetic lithium fluorohectorite with the ideal composition of $Li^+_{0.33}$ $(Mg_{2.67}Li_{0.33})(Si_4)O_{10}(F_2)$, available from Topy Industries Company, Japan.

Saponite is another trioctahedral species in the smectite group of phyllosilicates (Table 1.1) with an ideal composition of $M^+_x.nH_2O$ $(Mg_3)(Si_{4-x}Al_x)O_{10}(OH)_2$, indicating that the layer charge arises from isomorphous substitution of Al^{3+} for Si^{4+} in tetrahedral positions. In common with hectorite, there are no large deposits of saponite clay in nature. Furthermore, natural saponite contains surface impurities and its composition is highly variable (Utracki 2007; Zhou et al. 2019). For this reason, information about the surface properties of saponite, and its reactivity towards organic compounds, has mostly been obtained using synthetic saponite (Vogels et al. 2005; Tao et al. 2016; Zhang et al. 2017). A synthetic Na-saponite under the trade name 'Sumecton' is available from Kunimine Industries Co. Ltd., Japan.

As already remarked on, the layer charge distribution in montmorillonite generally varies among individual layers within a particle with 80–90% of the total charge being assigned to interlayer sites (Senkayi et al. 1985). Accordingly, the interlayers account for the bulk of exchange sites and, by extension, the total CEC of montmorillonite as well as its particle surface area. Besides being capable of taking up a wide range and variety of organic compounds, the interlayer space provides a peculiar micro-environment in terms of surface geometry, charge density, and retaining hydrated counterions (Ballantine et al. 1983; Li et al. 2020a). For all these reasons, smectites in general, and especially montmorillonite, are well suited for intercalating, stabilizing, and transforming organic molecules (Theng 2018).

By analogy with kaolinites, the under-coordinated aluminium ions exposed at the particle edge of (dioctahedral) smectites, can act as electron acceptors (Solomon 1968; Solomon and Hawthorne 1983). The Lewis acid character from this source, however, is not as pronounced as for kaolinites because the edge surface of smectite takes up only a small fraction (< 6%) of the total particle surface area (Wan and Tokunaga 2002; Macht et al. 2011). The small edge/basal surface ratio is related to layer aggregation in that smectites occur as *tactoids* or *quasi-crystals* in aqueous suspensions (Aylmore and Quirk 1971; Whalley and Mullins 1991; Laird 2006). A tactoid comprises a number of silicate layers in parallel ('face-to-face') arrangement, adjacent layers being separated by 0.9–1.0 nm thick water sheets. Thus, scanning electron micrographs of smectites commonly show curved, flexible, and flaky particles with warped edges (Mouzon et al. 2016) as shown in Figure 1.12. Under optimum conditions of dispersion, the layers of Li^+- and Na^+-montmorillonite tactoids may dissociate completely due to extensive interlayer ('osmotic') swelling (Arndt et al. 2017).

Like those in kaolinite (Figure 1.5), the aluminol groups at the edge surface of smectite particles can acquire or lose protons in response to fluctuations in suspension pH. The associated point of zero (net) charge varies from 6.5 to 8.1, depending on sample origin and pretreatment, and suspension ionic strength (Tombácz and Szekeres 2004; Tournassat et al. 2004; Rozalén et al. 2009). On the other hand, Kaufhold et al. (2011) measured an average pK of 4.73 for 36 samples of bentonite by titrating the clay suspensions (adjusted to pH 3) with alkali up to pH 12. Using first principle molecular dynamics (FPMD) simulations, Liu et al. (2013) deduced a pK_a of 7.0 and 8.3 for the edge $\equiv$Si–OH and =Al–OH_2OH group, respectively, of montmorillonite. All the same, the pH-dependent charge, associated with the edge surface of smectite particles, rarely exceeds 10% of the total charge or cation exchange capacity (Ras et al. 2007; García-García et al. 2009).

FIGURE 1.12 Scanning electron micrograph of montmorillonite from Miocene arkose, Madrid Basin, Spain, showing thin flakes with warped edges. (Image reproduced from the 'Images of Clay Archive' of the Mineralogical Society of Great Britain & Ireland and The Clay Minerals Society [www.minersoc.org/images-of-clay.html].)

1.4.2 Vermiculite

Vermiculites may occur as both micro- and macro-size particles. Fine-grained vermiculite or 'clay vermiculite', either dioctahedral or trioctahedral, are widely distributed in soils of different climatic regions (Douglas 1989). On the other hand, macroscopic vermiculite from which most of our information on structural and surface properties has derived, is invariably trioctahedral, and apparently does not occur in soil (Walker 1975).

Like smectite, vermiculite has a 2:1 (T-O-T) layer structure but with extensive replacement of Si^{4+} by Al^{3+} in tetrahedral positions to give a net negative charge (x) of 0.6–0.9 per formula unit (Table 1.1) and a cation exchange capacity of 150–250 $cmol_c/kg$ (Brown et al. 1978). In naturally occurring vermiculites the counterions balancing the negative layer charge are commonly magnesium. These points are illustrated by the composition of two well-known and much investigated vermiculites, namely, $Mg^{2+}_{0.32}\ 4.32H_2O\ (Mg_{2.36}Fe^{3+}_{0.48}Al_{0.16})(Si_{2.72}Al_{1.28})O_{10}(OH)_2$. for the specimen from Kenya, and $Mg^{2+}_{0.47}\ 4.72H_2O\ (Mg_{2.83}Fe^{3+}_{0.01}Al_{0.15})(Si_{2.86}Al_{1.14})O_{10}(OH)_2$ representing the specimen from Llano, Texas (Mathieson and Walker 1954; Shirozu and Bailey 1966; Douglas 1989). The basal spacing of this stable two-sheet hydrate of Mg^{2+}-vermiculite is 1.43–1.44 nm. A similar cation-water configuration has been proposed by Beyer and Graf von Reichenbach (2002) for the 1.485 nm hydrate of Na-vermiculite.

Figure 1.13 shows a structural diagram of hydrated Kenya vermiculite, deduced by Mathieson and Walker (1954) from single crystal X-ray diffraction analysis. The interlayer Mg^{2+} ions (occupying one-third of possible cation exchange sites) are located midway between every two opposing silicate layers. Under ambient conditions of humidity and temperature, the interlayer space also

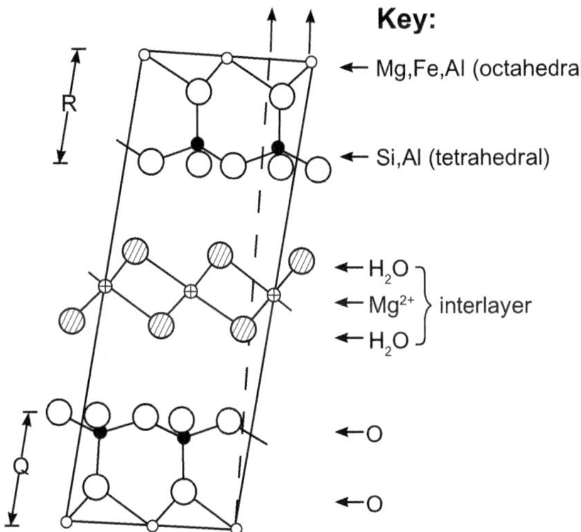

FIGURE 1.13 Projection normal to the *ac* plane for Mg^{2+}-vermiculite at the 1.436 nm phase of hydration-dehydration. The attendant stacking arrangement of successive silicate layers is indicated by the broken line and vertical arrows. NB: Q and R denote silicate half layers (From Walker 1975, Figure 4. Reproduced with permission from *Formation and Properties of Clay-Polymer Complexes, 2nd edition* by B.K.G. Theng, Chapter 1, Copyright Elsevier 2012.)

contains two superposed sheets of water molecules. These are arranged in a regular hexagonal network around each Mg^{2+} ion, and interact mutually, and with the surface oxygens, of opposing silicate layers by means of hydrogen bonding (Sposito 1984). As in smectite, the basal spacing of vermiculite is dependent on the nature of the (inorganic) counterion and the relative humidity of the atmosphere (Walker 1975; Huo et al. 2012). In common with smectite, the layer charge density in vermiculite is not homogeneous but varies among individual layers within a particle (Lagaly 1982).

Like montmorillonite, vermiculite can intercalate a variety of organic compounds (Steinfink et al. 1963; Susa et al. 1967; Walker 1967; Pérez-Rodríguez and Maqueda 2002). However, the extent to which vermiculite layers may be expanded by organic liquids is often less than that observed for the relatively low-charged montmorillonite. Thus, Mg-saturated vermiculite fails to expand beyond a basal spacing of ~ 1.45 nm when treated with glycerol, indicative of a single-layer complex whereas montmorillonite forms a double-layer complex with a basal spacing of ~ 1.78 nm (MacEwan 1961; Walker 1960). By the same token, the replacement of interlayer inorganic cations by alkylammonium ions in montmorillonite is complete after bringing the clay into contact with an aqueous solution of the appropriate alkylammonium chlorides for 2 hours at room temperature (Theng et al. 1967). Much longer periods of contact and elevated temperatures are usually required to achieve a similar interchange in vermiculite (Johns and Sen Gupta 1967; Walker 1967; Serratosa et al. 1970; Slade and Gates 2004). We might add here that some organic species, such as tylosine, can readily intercalate into montmorillonite (by cation exchange) but fails to do so with vermiculite (Call et al. 2019).

On the other hand, vermiculite has a distinct advantage over smectite in that macrocrystalline forms ('flakes') can be used for experimental purposes. Thus, such organic cations as monomethylammonium, dimethylammonium, tetramethylammonium, and tetramethyl-phosphonium (Vahedi-Faridi and Guggenheim 1997, 1999a, 1999b), anilinium (Slade et al. 1987), cetylpyridinium (Slade et al. 1978), hexadecyltrimethylammonium (Slade and Gates 2004), and trimethylsulfoxonium (Johns et al. 2013) are packed in a certain fashion when they enter the interlayer space. As a result, the layer stacking order in the corresponding (macrocrystalline) organo-vermiculites tends to increase, giving rise to sharp 'superlattice' reflections in the corresponding X-ray diffraction patterns (Garrett and Walker 1962, 1967; Slade et al. 1987; Vahedi-Faridi and Guggenheim 1997, 1999a, 1999b). The early literature (up to 1997) on the vermiculite-organic interaction has been summarized by Pérez-Rodríguez and Maqueda (2002), while the more recent review by Wang and Wang (2019) describes the exfoliation, organic intercalation, surface modification, and applications of vermiculite.

Among the inorganic forms of vermiculite, only the lithium-saturated variety can show extensive interlayer swelling in water (Table 1.2). As described in Chapter 4, vermiculites exchanged with *n*-propylammonium and *n*-butylammonium ions behave likewise when the complexes are immersed in water or dilute solutions of the corresponding *n*-alkylammonium chlorides (Walker 1960; Garrett and Walker 1962; Rausell-Colom 1964; Braganza et al. 1990; McCarney and Smalley 1995). Subsequently, Williams-Daryn and Thomas (2002) described the intercalation of various surfactants into (Eucatex) vermiculite, and the swelling in toluene of the resultant interlayer clay-organic complexes. Other examples of the vermiculite-organic interaction are given in Chapter 2.

1.4.3 Chlorite

Most chlorites are trioctahedral with a structure made up of repeating talc-like layers that are interleaved with a continuous magnesium hydroxide sheet. The thickness of the 2:1 type layer is ca.1.0 nm, while that of the brucite-like sheet is ca. 0.4 nm. The basal spacing of ~ 1.4 nm is, therefore, a characteristic feature of the chlorite group of minerals (Figure 1.14). This value is comparable to the *d*(001) spacing of a two-sheet hydrate of Mg^{2+}-vermiculite depicted in Figure 1.13. Unlike vermiculite, however, the basal spacing of chlorite remains unchanged on treatment with ethylene glycol or after heating at 500°C.

As in vermiculites, there is substitution of Al^{3+} for Si^{4+} in the tetrahedral sheet of the 2:1 layer of chlorite. The resultant negative charge is partly offset by substitution of Al^{3+} for Mg^{2+} in octahedral positions but most of it is balanced by the positively charged interlayer hydroxide sheet (Barnhisel and Bertsch 1989). With a composition of $[(Mg_{3-x}Al_x)(OH)_6]^{x+}$, the brucite-like sheet is strongly bound to opposing silicate layers by electrostatic attraction and hydrogen bonding. As a result, chlorites are highly stable to heating and fail to expand in water and organic liquids. The requirement

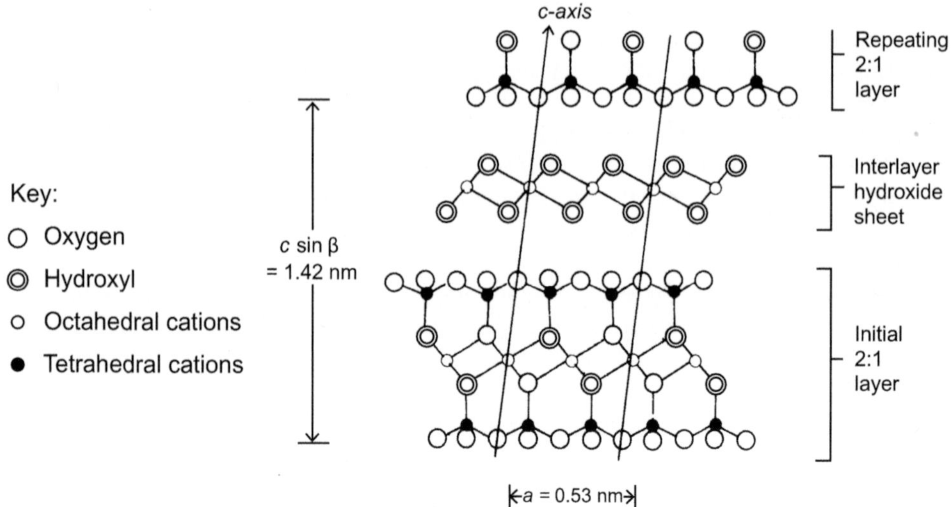

FIGURE 1.14 The structure of chlorite projected on the *ac* plane showing the interlayer hydroxide sheet sandwiched between two repeating (successive) 2:1 type layers to give a basal spacing of 1.42 nm. (From Bailey 1975, Figure 10a. Reproduced with permission from *Formation and Properties of Clay-Polymer Complexes, 2nd edition* by B.K.G. Theng, Chapter 1, Copyright Elsevier 2012.)

for (long) hydrogen bonding between the hydroxyl groups of the interlayer sheet and the oxygens of the tetrahedral bases of opposing silicate layers also restricts the number of possible layer stacking arrangements (Bailey and Brown 1962).

In addition to the main trioctahedral subgroup, represented by chamosite and clinochlore (cf. Table 1.1), there are chlorites in which both the 2:1 type layers and the interlayer hydroxide sheets are dioctahedral as in donbassite. The third chlorite subgroup is referred to as di, trioctahedral. Here the 2:1 type layers are dioctahedral, while the hydroxide sheets are trioctahedral as exemplified by the mineral sudoite (Table 1.1). In all these structures, Mg^{2+} (in the brucite-like sheet) may be replaced by Fe^{2+} and other suitable divalent cations, and Fe^{3+} may substitute for Al^{3+} in octahedral positions.

Ulian et al. (2020) have pointed out that the brucite-like sheet is hydrophobic, while the 2:1 type layer is hydrophilic. As such, chlorite would be able to take up organic compounds as well as catalyze their transformations. For example, DNA can adsorb to the positively charged brucite-like sheet but is repelled from the negatively charged talc-like layer (Valdrè 2007; Moro et al. 2016). On the whole, the chlorite-organic interaction has attracted little attention as the following summary would indicate.

Satterberg et al. (2003) reported that chlorite was second only to montmorillonite in its ability to adsorb organic matter from several phytoplankton species. The interactions of various amino acids and oligopeptides with chlorite have been investigated by Moro et al. (2015, 2019, 2020) using cross-correlated atomic force microscopy (AFM) and quantum mechanics/density functional theory simulations. Glycine was strongly adsorbed with its plane parallel to the (001) brucite-like sheet, forming monolayers with different patterns. Similarly, L-alanine formed stable (dot-like, agglomerate, and filamentous) structures on the brucite-like surface of clinochlore. In the case of di-L-alanine, the length of the filament-like structures could exceed 160 nm.

1.5 MICA, ILLITE, SYNTHETIC SWELLING MICA, FLUOROTETRASILICIC MICA, TAENOLITE

1.5.1 Muscovite, Biotite, Phlogopite

Dioctahedral *muscovite* together with trioctahedral *biotite* and *phlogopite* are the three well-known species within the true mica group (Table 1.1). The ideal (unit formula) composition of muscovite may be written as $K^+ Al_2(Si_3Al)O_{10}(OH)_2$, indicating that one out of every four Si^{4+} in the tetrahedral sheet is replaced by Al^{3+}. The resultant negative layer charge is balanced by interlayer potassium ions. Moreover, the K^+ ion can fit ('key') snugly into the space between two opposing ditrigonal cavities because its size is almost identical with the cavity diameter (Sposito 1984) (Figure 1.15, hatched). The characteristic basal spacing of 1.0 nm remains unchanged on treatment with ethylene glycol (or glycerol), or after heating at 500°C. The high energy of layer separation (Giese 1978), the failure of mica crystals to expand (swell) in water, and the low exchange capacity (as measured by conventional methods) are in keeping with strong interlayer bonding.

Nevertheless, the potassium counterions in many micas can (under certain conditions) be exchanged for long-chain alkylammonium ions (Weiss et al. 1956; Weiss 1963; Walker 1967; Mackintosh et al. 1971; Ferrow et al. 1999). The variability of interlayer exchange for octylammonium ions in phlogopites has been ascribed to chemical and structural inhomogeneities of the mica samples (Aldushin et al. 2007). A potassium sericite mica (CEC = 203 $cmol_c$/kg) could intercalate dodecylammonium ions, giving a basal spacing of 2.2 nm (Uno et al. 2009).

The adsorption and arrangement of long-chain alkylammonium ions on the muscovite (001) surface have been widely investigated using NEXAFS spectroscopy (Brovelli et al. 1999) and molecular dynamics and Monte Carlo simulations (Heinz et al. 2003; Klebow and Meleshyn 2011; Xu et al. 2014, 2015; Wang et al. 2016; Tsagkaropoulou et al. 2019). Neutron reflection indicates that didocecyldimethylammonium bromide is adsorbed on muscovite mica as a bilayer with a thickness of 2.4 nm (Browning et al. 2014). This double-chain cationic surfactant resists desorption by water but can be completely removed by adding sodium dodecylsulfate at its critical micelle concentration (Allen et al. 2018).

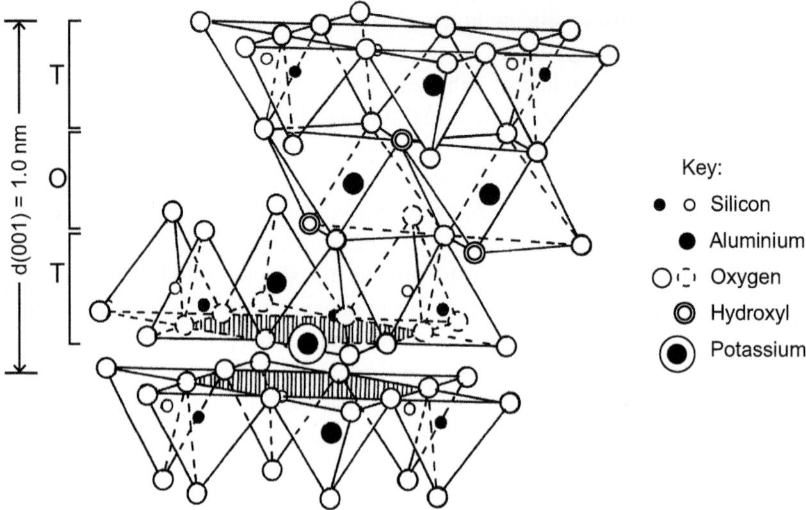

FIGURE 1.15 Diagram showing the idealized structure of muscovite mica with a *d*(001) or basal spacing of 1.0 nm. The charge-balancing interlayer K^+ ion fits snugly in the centre between two opposing ditrigonal cavities (hatched) becoming effectively coordinated to twelve oxygens. (Reused with permission from *Clay Mineral Catalysis of Organic Reactions*, by B.K.G. Theng, Figure 1.18, p. 42, CRC Press, 2018.)

Besides being coordinated to 12 oxygen atoms of two superposed ditrigonal rings, each interlayer K^+ ion is associated with two hydroxyl groups (belonging to the octahedral sheet), one lying directly above and one below the counterion. In muscovite, the O–H bond is deflected away from the K^+ ion toward the vacant octahedral site (Rothbauer 1971) whereas in phlogopite where there are no octahedral vacancies, the O–H bond is perpendicular to the silicate layer (Rayner 1974). The orientation of these hydroxyls (with respect to the layer plane) affects the strength with which the interlayer K^+ ions are held, and hence the rate of mica weathering (Norrish 1973; Fanning et al. 1989). By the same token, the substitution of the highly electronegative fluoride for hydroxyl in the mica structure has the effect of enhancing interlayer attraction by increasing the layer separation energy (Giese 1975). The inverse relationship between fluorine content and weathering rate, which applies to micas in general, may be explained in similar terms (Rausell-Colom et al. 1965; Newman 1969).

The ideal (unit formula) composition of biotite may be written as $K^+ (MgFe^{2+})_3 (Si_3Al)O_{10}(OH)_2$, and that of phlogopite by $K^+ (Mg)_3(Si_3Al)O_{10}(OH)_2$. Substitution of fluoride for hydroxyl in the phlogopite structure gives rise to fluorophlogopite (Brown et al. 1978). The XRD pattern of biotite, like that of muscovite, is characterized by an intense *d*(001) peak at ~ 1 nm which does not change after treatment with glycerol (Fanning et al. 1989). In this context, we should also mention *hydrobiotite* which consists of 1:1 regularly interstratified layers of biotite and vermiculite, giving a *d*(001) peak at 2.4 nm (Brindley et al. 1983; Kogure et al. 2012). Interestingly, cetyltrimethylammonium cations could intercalate into the vermiculite, but not the biotite, layer structure of hydrobiotite. Furthermore, intercalation apparently takes place in a layer-by-layer fashion rather than simultaneously and uniformly as in montmorillonite (Zhu et al. 2008).

1.5.2 Illite

Grim et al. (1937) introduced the term *illite* to denote a group of clay-size or fine-grained micaceous minerals with a basal spacing of ~ 1 nm that does not increase after treatment with ethylene glycol. A more precise description of illite, proposed by Środoń and Eberl (1984), is that of ‘a non-expanding, dioctahedral, aluminous, potassium mica-like mineral that occurs in the < 4 μm fraction' (of soils and sediments). Since illite particles often contain smectite layers that expand on glycolation (MacEwan et al. 1961), the term 'illite layer' is used for a 1.0 nm thick, non-expanding, potassium-rich layer which makes up a component of dioctahedral, mixed-layer structures while 'illitic material' becomes synonymous with the original 'illite' of Grim et al. (1937).

As the description suggests, illite is structurally similar to muscovite. Apart from its particle size range, however, illite differs from muscovite in having a smaller negative layer charge, less potassium, and more water. An 'average' composition for illite, derived from chemical analyses of a large number of samples (Weaver and Pollard 1973), may be written as $(Ca_{0.05}Na_{0.03}K_{0.63})_{int}(Al_{1.55}Fe^{3+}_{0.21}Fe^{2+}_{0.04}Mg_{0.28})_{oct}(Si_{3.41}Al_{0.59})_{tet}O_{10}(OH)_2$, indicating appreciable isomorphous substitution (of Al^{3+} for Si^{4+}) in the tetrahedral sheet, and some substitution (of Mg^{2+}, Fe^{3+}, and Fe^{2+} for Al^{3+}) in octahedral sites, with Ca^{2+}, Na^+, and K^+ occupying interlayer (int) positions. Any chemical formula for illite, however, may be misleading since many samples are essentially mixed-layer structures.

The negative layer charge (0.6–0.85 per formula unit) of illite falls within the range given by vermiculite (Table 1.1). Unlike vermiculite, however, the interlayer counterions (largely K^+) in illite are practically non-exchangeable. Accordingly, the CEC of illite is much smaller than their layer charge deficit would suggest (Table 1.2). In this respect, and in showing a basal spacing of ca. 1.0 nm that does not change on heating, illite is more akin to mica than to vermiculite. Since illite contains more water than mica, the terms 'hydromica' and 'hydrous mica' have sometimes been used as alternatives. The 'extra' water in illite apparently occurs in the form of hydronium ions, occupying K^+-deficient interlayer positions (Loucks 1991; Nieto et al. 2010). In this connection, we should mention that when dry illite is wetted, its layer structure contracts (along the *c*-axis) by up to 0.02 nm due to deprotonation of interlayer hydronium ions (Kühnel et al. 2017).

In contrast to (naturally occurring) micas, illite does not intercalate organic compounds (Sheng et al. 1996; Akçay and Yurdakoç 1999; Chang et al. 2012) although hexadecyltrimethylammonium (HDTMA) ions can apparently displace some interlayer K^+ ions in illite (Jaynes and Boyd 1991b). However, Li et al. (2008, 2020b) reported that both HDTMA and ethidium cations did not penetrate the interlayer space of the illite component, but could freely intercalate into the montmorillonite component, of rectorite. Thus, organic sorption by illite is generally restricted to external particle surfaces and micropores (Morrissey and Grismer 1999).

Even so, illite can be an effective adsorbent of organic compounds as Pura et al. (2014) have found in the case of oleic acid and squalene. Early on, Thompson and Brindley (1969) suggested that the adsorption of purines, pyrimidines, and nucleosides to illite (under acid conditions) took place through a cation exchange mechanism. Indeed, Polubesova and Nir (1999) observed that illite could take up cationic crystal violet (from aqueous solution) in excess of its CEC. Electrostatic (cation exchange) and hydrophobic interactions have also been proposed by Droge and Goss (2013) for the adsorption of various organic cations to clay minerals, including illite. Subsequently, Kang et al. (2016) combined batch experiments and model computation to clarify the mechanism underlying the interaction of kaolinite, illite, and smectite with aflatoxin B_1. They suggested that in the case of illite, complex formation (with AFB_1) involved electron-donor-acceptor interaction. On the other hand, organic acids (salicylic, citric) tend to form monodentate or bidentate complexes with AlOH groups exposed at the edge surface of illite particles, depending on solution pH (Kubicki et al. 1997; Lackovic et al. 2003).

1.5.3 Synthetic Swelling Micas

Synthetic swelling micas (SSM) in which sodium acts as the counterion, instead of potassium (cf. Figure 1.15) show both swelling (and cation exchange) properties although their layer charge density is comparable to that of brittle mica (Franklin and Lee 1996; Kodama and Komarneni 1999; Park et al. 2002; Pavón et al. 2013). Having a trioctahedral character and with fluoride replacing hydroxyl in the layer structure, these minerals may be regarded as synthetic analogues of fluorophlogopite (Alba et al. 2006). We should also mention that SSM are often referred to as Na-*n*-mica where *n* denotes the layer charge per unit cell which is twice the value of the *x* parameter in Table 1.1. Thus, the general unit cell formula may be written as Na_n xH_2O (Mg_6) $(Si_{8-n}Al_n)O_{20}F_4$ giving a basal spacing at 1.21 nm and a CEC of 120 $cmol_c/kg$ (Alba et al. 2006; Leone et al. 2013). Synthetic Na-*n*-mica is marketed under the trade name Somasif ME-100 (and ME-300) by CO-OP Chemical Company, Japan. The thermal stability of Somasif ME-100 intercalated with oligoether- and alkyl-based ammonium salts has been investigated by Souza et al. (2011).

These minerals can intercalate a variety of primary and quaternary ammonium ions as well as stable organic radicals and cationic polyelectrolytes (Yang et al. 2001; Hata et al. 2007; Souza et al. 2011; Zeng et al. 2013; Pazos et al. 2015, 2020; Martín et al. 2019; Orta et al. 2019). In the case of Na-4-mica, Alba et al. (2011) found that intercalated *n*-alkylammonium ions (n = 12, 14, 16, 18) assumed an ordered, paraffin-type bilayer with an all-*trans* conformation (at room temperature) (cf. Figure 4.6). Subsequently, Pesquera et al. (2018) reacted Na-4-mica with hexadecylammonium, hexadecyldimethylammonium, and hexadecyltrimethylammonium ions. Intercalation of primary amine cations gave rise to a fully hydrophobic interlayer. For the complex with the tertiary ammonium cations, the interlayer space contained both inorganic and organic cations. A similar interlayer heterogeneity has also been suggested by Kłapita et al. (2001, 2003) for complexes of Na-fluorotetrasilicic mica and fluorotaenolite (see further) with various quaternary ammonium cations, probably because the layer charge distribution in these synthetic micas is heterogeneous.

Seliem et al. (2011) reported that the interlayer complex of Na-0.5-mica and Na-1-mica with hexadecyltrimethylammonium was a good adsorbent of toluene. Similarly, the complex of Na-4-mica with HDTMA was highly effective in taking up benzene, toluene, and phenol (Pazos et al. 2017), while complexes with octadecylammonium (ODA) and octadecyltrimethylammonium (ODTMA) showed a large capacity for taking up polycyclic aromatic hydrocarbons through a partition mechanism (Satouh et al. 2021). Likewise, Martín et al. (2018, 2019) used ODA-modified Na-4-mica to adsorb ibuprofen together with a range of priority and emerging pollutants, while Orta et al. (2019) prepared a complex with ODTMA for the removal of the fungicide pyrimethanil. Pazos et al. (2020) used octylammonium-intercalated Na-2-mica, functionalized with the *Keggin* Al_{13} cation (Theng 2018) and then grafted with an organosilane, for the simultaneous removal of a phenoxy herbicide and Zn^{2+} from aqueous solutions.

1.5.4 Fluorotetrasilicic Mica

Like SSM, fluorotetrasilicic mica (TSM) is a synthetic swelling layer silicate with the composition of $Na^+(Mg_{2.5}\square_{0.5})(Si_4)O_{10}F_2$, showing the replacement of hydroxyl by fluoride in the octahedral sheet, with sodium counterions (and water) occupying interlayer positions (Kitajima and Daimon 1975; Morikawa 1993). Here the negative layer charge apparently derives from vacancies (defects) in the octahedral sheet, denoted by the open square symbol in the unit cell formula (Okada et al. 2007). The occurrence of similar vacancies in some New Zealand halloysites has also been deduced by Soma and Theng (1998) using X-ray photoelectron spectroscopy. The measured CEC of Na-TSM ranges from 92 to 170 $cmol_c/kg$ which is appreciably lower than the value of 254 $cmol_c/kg$, deduced from its chemical composition. This discrepancy may be ascribed to the non-exchangeability of a proportion of the interlayer sodium ions (Soma et al. 1990; Kłapita et al. 2003). As would be expected, Na-TSM can form interlayer complexes with cationic organic compounds, such as tris(2,2'-bipyridine)ruthenium(II) (Ogawa et al. 2000; Okada et al. 2005) and quaternary ammonium ions (Kłapita et al. 2003). Earlier, Ogawa et al. (1996, 1997) reported on the photochemical isomerization of *p*-aminoazobenzene intercalated into various organoammonium-TSM, and the intercalation by a solid-state reaction of a crown ether (18 crown 6) into a non-swelling K-TSM. Na-TSM is obtainable from Topy Industries, Limited, Japan.

1.5.5 Taenolite

Synthetic sodium taeniolite with the ideal composition of $Na^{+}xH_2O$ $(Mg_2Li)(Si_4)O_{10}F_2$ may be regarded as an analogue of Na-2-mica (Perdigón et al. 2018). Early on, Kitajima et al. (1985) synthesized Na-taenolite with varying layer charge, noting that their expansibility in ethylene glycol and glycerol increased with decreasing layer charge. The sodium- and lithium-exchanged forms, available from Topy Industries Limited, Japan, are known to intercalate various organic species, including acrylamide (Ogawa et al. 1992), trimethylammonium (Tamura and Nakazawa 1996), rhodamine 6G and oxazine 4 (Fujita et al. 1997; Chen et al. 2002), cationic azobenzenes (Iyi et al. 2001), trimethylalkylammonium (Fujita et al. 2003), and protonated polyoxyalkylene monoamines (Kłapita et al. 2011).

1.6 2:1 TYPE LAYER SILICATES WITH LAYER-RIBBON STRUCTURES

1.6.1 Palygorskite and Sepiolite

Palygorskite and sepiolite are hydrous magnesian silicates with a fibrous particle morphology (Figure 1.16). The term 'attapulgite', denoting palygorskite, commonly appears in the early literature (Grim 1968) and is still widely used in publications related to environmental and industrial applications (Miao et al. 2007; Bin et al. 2018; Wang et al. 2019). Palygorskite, however, has priority

FIGURE 1.16 Scanning electron micrograph of sepiolite from New Method Uranium Mine, Amboy, Bristol Mountains, San Bernardino Co., California, USA. (Image reproduced from the 'Images of Clay Archive' of the Mineralogical Society of Great Britain & Ireland and The Clay Minerals Society [www.minersoc.org/images-of-clay.html.])

and should be used in preference to attapulgite (Guggenheim et al. 2006; Brigatti et al. 2013). Palygorskite fibrils, ranging in length from 1000 to 2000 nm, are commonly arranged in bundles or sheaths. By comparison, individual fibrils of sepiolite vary in length from 10 to 5000 nm, in width from 10 to 30 nm, and in thickness from 5 to 10 nm (Jones and Galán 1988; Singer 1989).

Each fibril consists of 2:1 type layers, arranged in ribbons ('chains') along the fibril (*a*-) axis, with the apical oxygens of the silicon tetrahedra in adjacent ribbons pointing in opposite directions. For this reason, palygorskite and sepiolite have been referred to as layer-ribbon or chain-layer silicates (Mackenzie and Mitchell 1966; Sudo et al. 1981; Table 1.1). Being joined at the corners through oxygen ions, the rectangular ribbons alternate with channels (tunnels) of similar size and shape, running the full length of the fibril. Under ambient conditions of humidity and temperature, these channels contain free ('zeolitic') water as well as exchangeable counterions (Brigatti et al. 2013).

Figure 1.17 shows the structure of palygorskite as proposed by Bradley (1940), having the ideal composition of $Mg_5Si_8O_{20}(OH)_2(OH_2)_4.4H_2O$. In reality, however, there is appreciable substitution of aluminium (and iron) for magnesium in octahedral positions, giving rise to a CEC of 5–30 $cmol_c$/kg (Serna and Vanscoyoc 1979; Singer 1989). Four (zeolitic) water molecules are present within the channel structure, and another four are coordinated to Mg ions at the ribbon edge. Each ribbon consists of five octahedral sites, giving rise to a *b*-dimension of 1.8 nm as compared with ca. 2.7 nm in sepiolite. Accordingly, the channel (tunnel) dimension in palygorskite (0.37 nm x 0.64 nm) is appreciably smaller than that in sepiolite (0.37 nm x 1.06 nm).

Figure 1.18 shows the Brauner and Preisinger (1956) structure for sepiolite which is generally preferred to that proposed by Nagy and Bradley (1955). The *b*-dimension of the unit cell is 2.7 nm while the *c*-dimension is 1.34 nm, and the ideal composition may be written as $Mg_8Si_{12}O_{30}(OH)_4(OH_2)_4.8H_2O$. The structure contains four water molecules coordinated (bound) to octahedral magnesium ions, exposed at the ribbon edge together with eight zeolitic (free) water

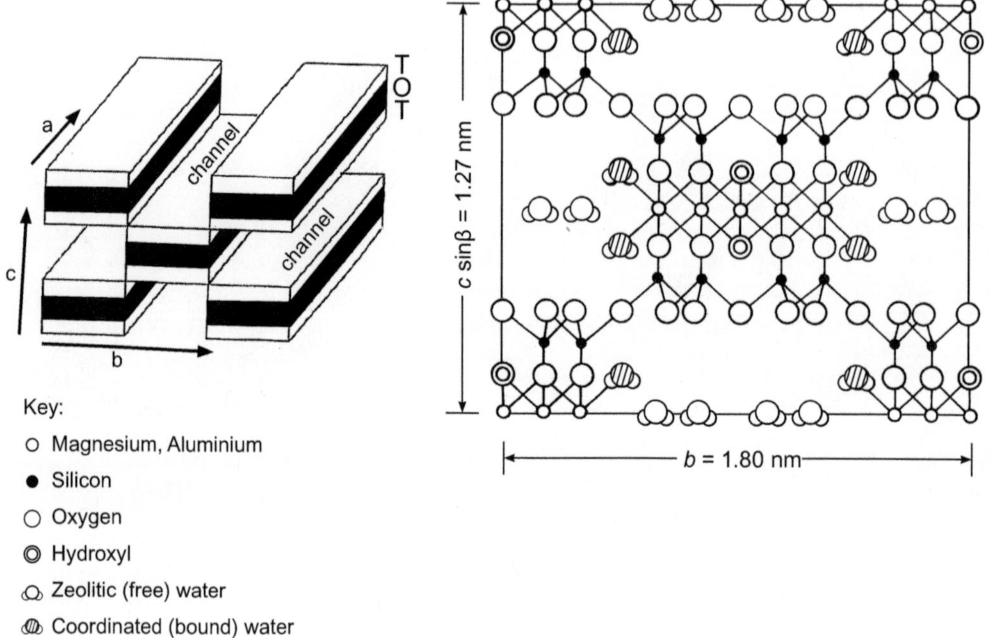

FIGURE 1.17 Right: the structure of palygorskite ('attapulgite') projected on the *bc* plane. The unit cell dimension is indicated by the solid rectangle (from Bradley 1940, Figure 1. Reproduced with permission of the Mineralogical Society of America, publisher of *American Mineralogist*.). Left: diagram showing rectangular channels running parallel to the *a*-axis between opposing 2:1 (T-O-T) ribbons.

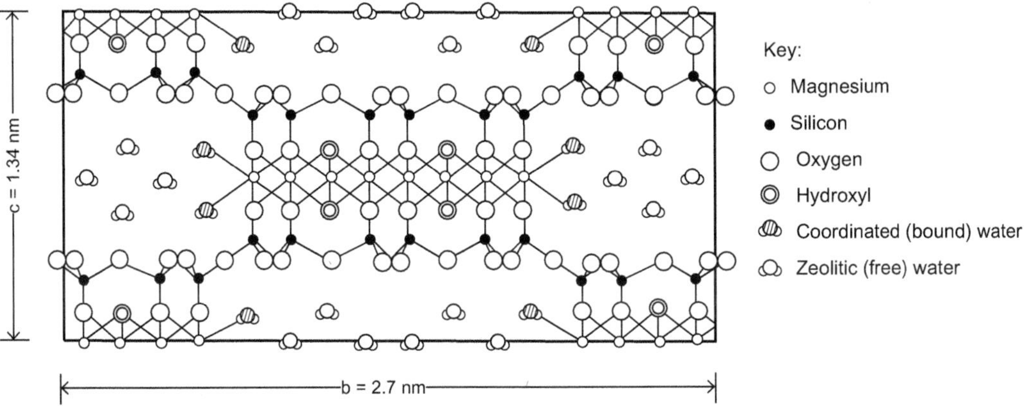

FIGURE 1.18 The structure of sepiolite, viewed along the fibril *a*-axis. The unit cell dimension is indicated by the solid rectangle. (From Brauner and Preisinger 1956, Figure 1. Reused with permission from *Clay Mineral Catalysis of Organic Reactions*, by B.K.G. Theng, Figure 1.20, p. 46, CRC Press, 2018.)

molecules occupying the channel space. In practice, however, not all eight octahedral sites need be occupied. Furthermore, there is always some substitution of Fe^{3+} and Fe^{2+} for Mg^{2+} in the octahedral sheet, and of Al^{3+} and Fe^{3+} for Si^{4+} in tetrahedral positions. As a result, the structure carries a net negative charge which is usually balanced by sorption of extraneous Ca^{2+} ions, giving rise to a CEC of 20–45 $cmol_c$/kg (Hénin and Caillère 1975; Singer 1989). Induced isomorphous substitution of Al^{3+} for Si^{4+} in the tetrahedral sheet increases both the thermal stability and CEC of sepiolite (Sun et al. 1995).

There is a stepwise loss of weight when palygorskite and sepiolite are heated at increasing temperatures (Nagata et al. 1974; Singer 1989; Brigatti et al. 2013). Zeolitic water in sepiolite can be removed by mild heating (< 200°C in air) following which the coordinated water molecules interact with neighbouring hydroxyls or oxygen ions through hydrogen bonding. Further heating (~ 300°C in air) leads to disruption of the hydrogen-bond network, allowing the ribbons to rotate and the structure to fold. At this stage of dehydration, the original structure can be restored by allowing the mineral to take up water at ambient temperature and humidity. Dehydration between 380 and 680°C in air removes the other half of coordinated water, yielding the 'anhydride'. This form can only be rehydrated by exposing the mineral to water vapour at elevated temperatures and pressures, and then often incompletely (Hayashi et al. 1969; Nagata et al. 1974; Serna et al. 1975; Van Scoyoc et al. 1979). Heating at 680 to 900°C causes dehydroxylation and structural collapse (Hayashi et al. 1969; Lokanatha et al. 1985). Palygorskite behaves similarly on heating but the loss of coordinated water, causing structural folding, occurs at a lower temperature than in sepiolite (Van Scoyoc et al. 1979; Gerstl and Yaron 1981). Furthermore, the high-temperature dehydration step in palygorskite may overlap with structural dehydroxylation. Fourier transform infrared spectroscopic analysis by Yan et al. (2012) indicates that most of the hydroxyl groups in palygorskite are removed by heating at about 450°C, and complete structural dehydroxylation occurs at ~ 700°C accompanied by the formation of SiO_4 tetrahedral sheets. We might also add that the tunnels in the anhydride form are no longer accessible to organic compounds although heating at ca. 600°C may increase the volume of pores between individual fibrils (Hénin and Caillère 1975).

In a recent review on the surface properties of palygorskite, Suárez et al. (2022) have pointed out that even very pure samples of the mineral show wide variability in such surface properties as specific surface area, micropore area and volume, and mean equivalent pore diameter. Furthermore, structural folding can occur under soft outgassing conditions, limiting accessibility of the channels to N_2 molecules.

1.7 SILICATES WITH SHORT-RANGE ORDER STRUCTURES

1.7.1 Allophane and Imogolite

Allophane may be defined as 'a group of clay-size minerals with short-range order which contain silica, alumina, and water in chemical combination' (Parfitt 1990). Like illite, allophane should therefore be regarded as a series of minerals rather than as a mineral species. Allophane occurs in the clay fraction of many soils, especially those that derive from volcanic ash and weathered pumice ('Andisols') where it plays a key role in accumulating and stabilizing organic matter (Andreux and Theng 1990; Parfitt et al. 1997; Dahlgren et al. 2004; Goh 2004; Theng and Yuan 2008; Parfitt 2009; Calabi-Floody et al. 2011; McDaniel et al. 2012; Yuan and Wada 2012; Matus et al. 2014). Not all allophanes, however, are associated with soil environments. A prime example is the type that occurs as a deposit on a stream bed near Silica Springs, New Zealand (Wells et al. 1977).

Because of the diffuse nature of its XRD pattern, allophane has often been described as amorphous. High-resolution transmission electron microscopy (HRTEM), however, has consistently shown that the unit particle of allophane consists of a hollow nano-size spherule ('nanoball'). For this reason, the term 'short-range order' (van Olphen 1971) is more appropriate than, and preferable to, 'amorphous' (Wada and Harward 1974). In aqueous suspensions, the spherules tend to form globular micro-aggregates (clusters) of varied size and shape (Figure 1.19) that can be disrupted by shearing (Wells and Theng 1985; Churchman and Tate 1987).

Irrespective of composition, allophane spherules have an outer diameter of 3.5–5.0 nm and a wall thickness of 0.7–1.0 nm (Kitagawa 1971; Henmi and Wada 1976; Wada and Wada 1977; Hall et al. 1985; Abidin et al. 2007a; Calabi-Floody et al. 2009). The spherule wall is essentially made up of a 1:1 type layer with numerous vacancies. Clusters of vacant sites give rise to perforations of ~ 0.3 nm in diameter through which water can freely diffuse into, and out of, the intra-spherule void space (Figure 1.20)

Two extreme compositional forms of allophane have been recognized, namely, $SiO_2Al_2O_3.2.5H_2O$ with an Al/Si ratio of 2 and $2SiO_2Al_2O_3.3H_2O$ giving an Al/Si of 1. It seems likely that natural specimens of allophane with an Al/Si ratio between 1 and 2 represent physical mixtures of the two end members, rather than distinct species within a series (Parfitt et al. 1980). On the other hand, the composition of naturally occurring imogolite is close to the ideal of $SiO_2Al_2O_3.2H_2O$. Because of the similarity to imogolite in terms of composition, and the presence of a well-defined band near 346 cm^{-1} in the infrared spectrum (Farmer et al. 1977), allophane with an Al/Si ratio of $\approx$ 2 has been referred to as 'imogolite-like' allophane (Parfitt and Henmi 1980).

It is generally accepted that the gibbsitic sheet in imogolite-like allophane is on the outer (convex) side of the spherule wall (Figure 1.20). Whether the same arrangement applies to the other end member of allophane (Al/Si = 1), sometimes referred to as 'halloysite-like' allophane, is still open to question (Wada and Wada 1977; Parfitt et al. 1980; Barron et al. 1982; van der Gaast et al. 1985). In this type of allophane about two-thirds of the possible octahedral sites are apparently vacant, and the tetrahedral sheet may form the outer spherule wall as in Silica Springs allophane (Childs et al. 1990, 1997).

In both imogolite-like and halloysite-like allophanes, there is appreciable substitution of Al^{3+} for Si^{4+} in tetrahedral positions (Ildefonse et al. 1994). The proportion of four-fold coordinated aluminium (Al^{IV}) tends to rise as the material becomes more siliceous, making up as high as 50% of the total aluminium for allophane with an Al/Si ratio of $\approx$ 1. The resultant positive charge deficiency, however, is not reflected in the corresponding CEC values (at pH 7) which are always much smaller than would be expected from the Al^{IV} contents. In explanation, Henmi and Wada (1976) have suggested that the (permanent) negative charge from isomorphous substitution is internally compensated by proton adsorption. In addition, coatings of positively charged, amorphous hydrous oxides of aluminium and iron together with 'allophane-like constituents' (Wada 1980) may contribute towards balancing the negative layer charge. A recent Fe K-edge X-ray absorption fine structure (XAFS) study by Baker et al. (2014) indicates that iron in natural allophane and imogolite can both

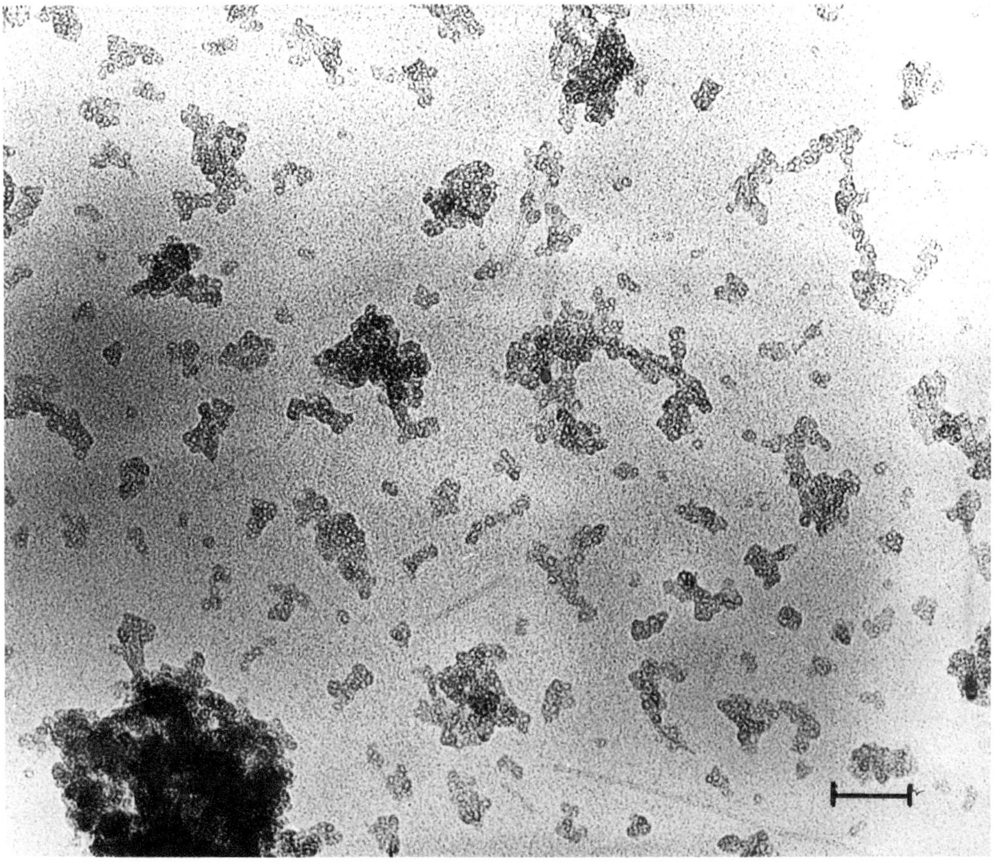

FIGURE 1.19 High-resolution transmission electron micrograph of allophane separated from the Kitakami pumice, Japan, showing the hollow spherule morphology of individual particles. NB: bar = 50 nm. (Courtesy: S.-I. Wada, Kyushu University, Japan.)

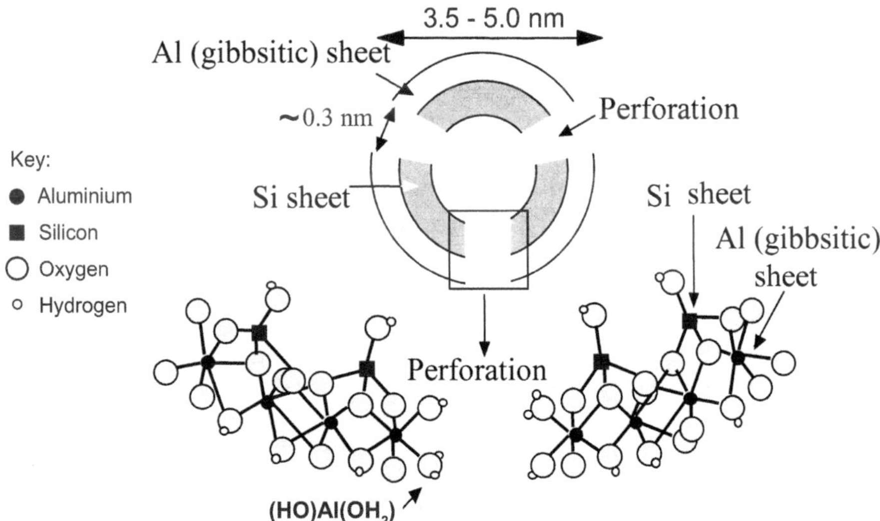

FIGURE 1.20 Diagram of an allophane 'nanoball' showing the hollow spherule morphology with a diameter of ca. 5 nm. The spherule wall (~ 0.7 nm in thickness) consists of an outer gibbsitic sheet and an inner silica sheet with many perforations (~ 0.3 nm in diameter). (From Hashizume and Theng 2007, Figure 1. Reproduced with kind permission of The Clay Minerals Society, publisher of *Clays and Clay Minerals.*)

substitute for aluminium in octahedral positions and occur as a surface-adsorbed species. The presence of octahedrally coordinated Fe^{3+} in allophane has also been indicated by electron paramagnetic resonance spectroscopy (Cervini-Silva et al. 2015).

As might be expected, the charge characteristics of allophane are largely pH-dependent, arising from the dissociation and protonation of $(OH)Al(OH_2)$ groups, exposed at perforations (defect sites) in the spherule wall (Figure 1.20). As described later, these groups are actively involved in, and have a controlling influence on, the interactions of allophane with anionic organic species. In this connection, we would also mention that the PZC of allophanes, derived from ion-retention measurements, ranges from pH 4.5 to 6.5 (Wada 1978; Theng et al. 1982; Clark and McBride 1984). Electrokinetic (Horikawa 1975; Escudey and Galindo 1983) and rheological (Wells and Theng 1985) methods also yield sensible values. Interestingly, prior dehydration or heating often leads to a substantial (30–50%) increase in CEC, suggesting a change in the coordination of some surface aluminium ions (Wada 1978).

The external surface area of allophane, calculated on the basis of 4.0–5.5 nm for the (outer) diameter of a spherule, and 2.6 cm^3/g for the density, is 419–577 m^2/g (Kitagawa 1971). Retention studies, using ethylene glycol (EG) or ethylene glycol monoethyl ether (EGME), and assuming monolayer coverage and molecular areas applicable to planar surfaces, usually yield larger specific surface areas. On the other hand, the corresponding BET-nitrogen areas are generally 30–50% of the values derived from the retention of small, polar organic molecules (Egashira and Aomine 1974; Paterson 1977; Vandickelen et al. 1980; Hall et al. 1985; Hashizume and Theng 2007). This observation suggests that such molecules can penetrate the intra-spherule pore space, at least partially, whereas nitrogen is apparently incapable of doing so. The variation in surface area values, for any given method of measurement, reflects differences among specimens as to composition, origin, and pretreatment.

Pore size distribution studies, using nitrogen as adsorbate, indicate that the porosity of (dry) allophanes is largely contained in pores with a mean radius (r) of < 1.0 nm (Rousseaux and Warkentin 1976; Paterson 1977; Vandickelen et al. 1980). Occupying a volume of 0.12–0.15 cm^3/g, these micropores often show a bimodal distribution, one group of pores (class 1) having r = 0.3–0.6 nm and another (class 2) having r = 0.6–0.9 nm. Both classes are presumably associated with interspherule (intra-aggregate) pore spaces although class 1 pores may include wall perforations and defect structures. A third class of pores with r = 2.0–10 nm, making up ~ 30% of the total porosity (Iyoda et al. 2012), may be identified with inter-aggregate pores. This interpretation of the pore size distribution data is consistent with transmission electron micrographs of allophane (Figure 1.19) and with the results of adsorption studies using *n*-alkylammmonium chlorides (Theng 1972).

Heating to 250°C does not appear to affect the texture and porosity of (synthetic) allophane. Further calcination, however, brings about Si–OH dehydroxylation, Al–OH dissociation, and particle aggregation together with a gradual decrease in specific surface area and porosity (Du et al. 2018). Interestingly, treatment with dilute NaOH can increase the diameter of the perforations in the wall of allophane spherules, allowing methyl- and ethyl-ammonium cations to enter the intra-spherule pores (Abidin et al. 2004, 2007b). Likewise, mild acid (pH 3–4) leaching of (synthetic) allophane can cause an enlargement of wall perforations. This effect is more pronounced under strongly acidic (pH < 3) conditions (Wang et al. 2020).

The occurrence of imogolite in gel films from some Andisols of Japan was discovered by Yoshinaga and Aomine (1962). When examined by HRTEM, imogolite appears as slender tubules forming 10–30 nm thick bundles of several micrometres in length (Figure 1.21). In dilute aqueous suspensions, the bundles of imogolite tubules form a gel structure that can be ruptured by shearing (Wells et al. 1980). The X-ray and electron diffraction patterns further indicate that the mineral has long-range order along the tubule length.

Figure 1.22 gives a diagram of the hollow tubular structure of imogolite, based on the proposal by Cradwick et al. (1972) as elaborated on by Farmer et al. (1983). The imogolite 'nanotube' has an outer diameter of ~2.3 nm, and an inner diameter of 1.0 nm. These values are 10–15%

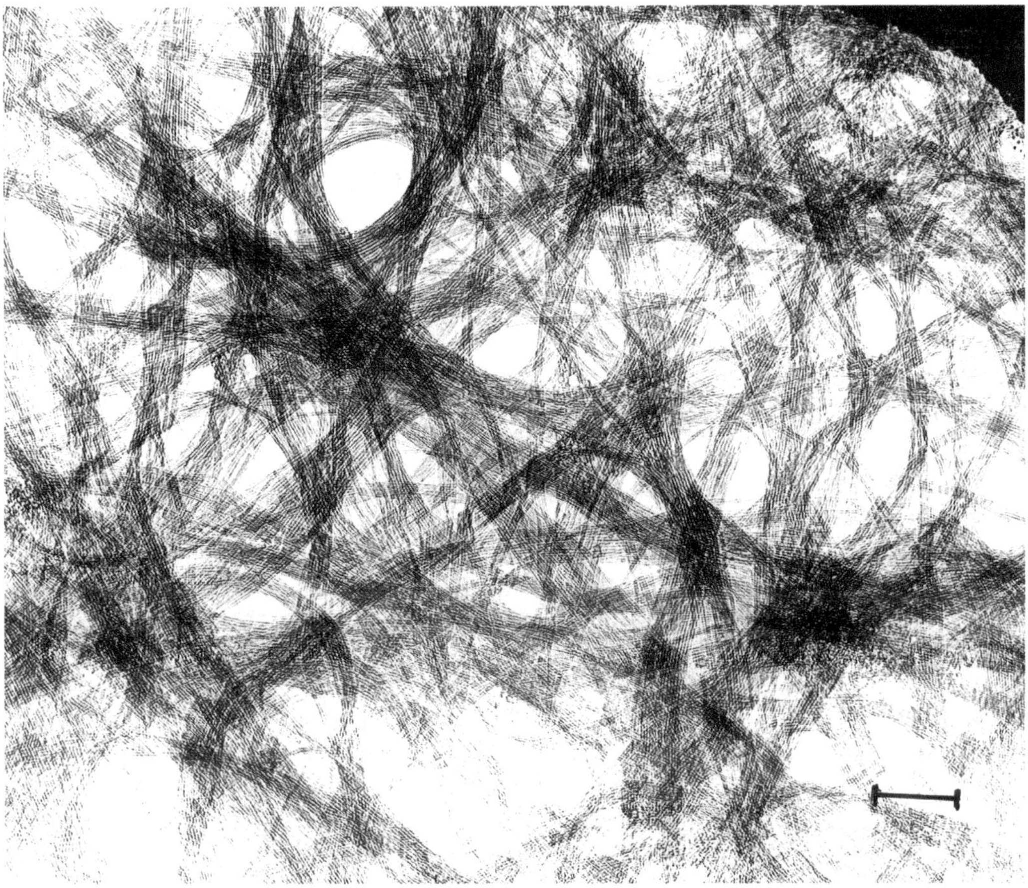

FIGURE 1.21 High-resolution transmission electron micrograph of imogolite separated from a gel film of the Kitakami pumice bed, Japan. NB: bar = 50 nm. (Courtesy: N. Yoshinaga, Ehime University, Japan.)

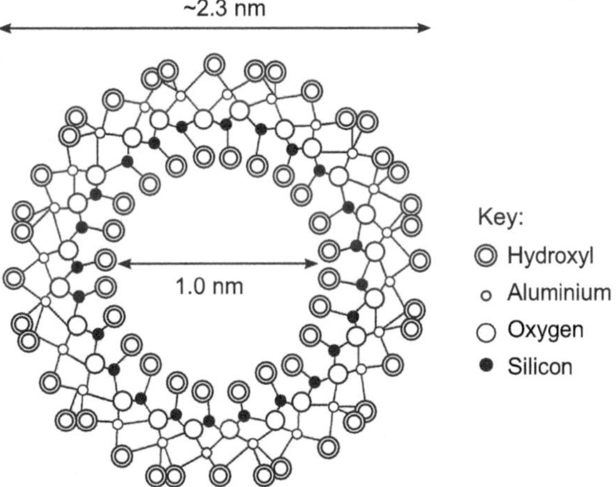

FIGURE 1.22 The structure of a hollow imogolite tubule with an outer diameter of ~2.3 nm and an inner diameter of ~1 nm, having an aluminol outer surface and a silanol inner surface. (From Farmer et al. 1983, Figure 1. Reproduced with kind permission of The Mineralogical Society of the UK and Ireland, publisher of *Clay Minerals*.)

larger for synthetic imogolite. The tubule wall consists of a curved outer gibbsitic sheet to which (O_3SiOH) groups are attached on the inside. The unit formula composition is commonly given as $(OH)_3Al_2O_3SiOH$ to indicate the sequence of ions from the periphery to the centre of the tubule where the orthosilicate group shares three oxygens with aluminium.

There seems to be little substitution of Al^{3+} for Si^{4+} in the imogolite structure and most, if not all, of the aluminium is in six-fold coordination (Henmi and Wada 1976; Wilson et al. 1984) although Fe^{3+} can substitute for Al^{3+} in octahedral sites of both imogolite and allophane (Baker et al. 2014). Imogolite can also develop pH-dependent negative and positive charges due to the dissociation and protonation, respectively, of hydroxyl groups attached to aluminium at the tubule ends. The point of zero charge (PZC), derived from ion-retention measurements at different pH values and acid-base potentiometric titration, occurs at pH 6–6.5 (Theng et al. 1982; Clark and McBride 1984; Yuan et al. 2000; Tsuchida et al. 2005). Gustafsson (2001) has suggested the occurrence of doubly coordinated $-Al_2OH$ groups along the tubule walls carrying a weak positive charge. At the same time, the $-SiOH$ groups on the tubule interior surface can acquire a negative charge capable of adsorbing Na^+ ions as Harsh et al. (1992) has proposed. More importantly, the surface hydroxyls of imogolite, like those in allophane, can take up organic anions by ligand exchange to form inner-sphere complexes (Equation 1.2).

The total surface area of imogolite, based on tubule dimension and a density of 2.65 g/cm^3, amounts to 1400–1500 m^2/g. By comparison, the specific surface area, derived from the retention of water, ethylene glycol (EG), or ethylene glycol monoethylether (EGME) ranges from 700 to 1100 m^2/g (Egashira and Aomine 1974; Wada and Harward 1974). Like water, EG and EGME can presumably enter the intra-tubule pores. Access to inter-tubule surfaces, however, may be restricted because the tubules tend to be aligned in parallel fashion to form bundles (Figure 1.21). Further, strong hydrogen bonding and van der Waals interactions can be established at points where adjacent tubules come into close contact, inhibiting adsorbate entry into these regions. Similar interactions would occur between individual bundles of tubules where they overlap and intertwine. Such a network structure lies behind the pronounced gel-like character, shear-thinning flow behaviour, and high yield value, of dilute aqueous suspensions of imogolite (Wells et al. 1980). Retention studies, using quaternary ammonium chlorides and water, indicate that slightly over half of the total volume of imogolite is contained in pores. Approximately 50% of these pores are inter-tubular, while 25% are intra-tubular with a volume of 0.13 cm^3/g, and the remainder of the porosity is associated with inter-bundle pores (Wada and Henmi 1972). Both the inter- and intra-tubule pores appear to be accessible to benzene (Wilson et al. 2002). For more details regarding the formation, occurrence, properties, and applications of allophane and imogolite, the reader is referred to the reviews written or edited by Fieldes and Claridge (1975), Wada (1978, 1989), Parfitt (1990, 2009), Harsh (2012), Yuan et al. (2016), and Paineau (2018).

REFERENCES

Abdullayev, E. and Y. Lvov. 2015. Halloysite tubule nanoreactors in industrial and agricultural applications. In *Natural Mineral Nanotubes*, (Eds.) P. Pasbakhsh and G.J. Churchman, pp. 363–382. Oakville, ON: Apple Academic Press.

Abidin, Z., N. Matsue, and T. Henmi. 2004. Dissolution mechanism of nano-ball allophane with dilute alkali solution. *Clay Science* 12: 213–222.

Abidin, Z., N. Matsue, and T. Henmi. 2007a. Differential formation of allophane and imogolite: Experimental and molecular orbital study. *Journal of Computer-Aided Materials Design* 14: 5–18.

Abidin, Z., N. Matsue, and T. Henmi. 2007b. Nanometer-scale chemical modification of nano-ball allophane. *Clays and Clay Minerals* 55: 443–449.

Adams, J.M. 1983. Hydrogen atom positions in kaolinite by neutron profile refinement. *Clays and Clay Minerals* 31: 352–356.

Adams, J.M. 1987. Synthetic organic chemistry using pillared, cation-exchanged and acid-treated montmorillonite catalysts – a review. *Applied Clay Science* 2: 309–342.

Akçay, G. and M.K. Yurdakoç. 1999. Nonyl- and dodecyl-amines intercalated bentonite and illite from Turkey. *Turkish Journal of Chemistry* 23: 105–113.

Alba, M.D., M.A. Castro, M. Naranjo, and E. Pavón. 2006. Hydrothermal reactivity of Na-n-micas (n = 2, 3, 4). *Chemistry of Materials* 18: 2867–2872.

Alba, M.D., M.A. Castro, M.M. Orta, E. Pavón, M.C Pazos, and J.S.V. Rios. 2011. Formation of organo-highly charged mica. *Langmuir* 27: 9711–9718.

Aldushin, K., G. Jordan, E. Aldushina, and W.W. Schmahl. 2007. On the kinetics of ion exchange in phlogopite. *Clays and Clay Minerals* 55: 339–347.

Alkan, M., Ö. Demirbaş, and M. Doğan. 2005a. Electrokinetic properties of kaolinite in mono- and multivalent electrolyte solutions. *Microporous and Mesoporous Materials* 83: 51–59.

Allen, F.J., C.L. Truscott, R.J.L. Welbourn, and S.M. Clarke. 2018. Anionic surfactant induced desorption of a cationic surfactant from mica. *Applied Clay Science* 160: 276–281.

Andreux, F. and B.K.G. Theng. 1990. Organic sources of nitrogen and phosphorus nutrients in humus-rich soils with variable charge. *Transactions 14th International Congress of Soil Science, Kyoto, Japan* II: 180–185.

Angel, B.R. and P.L. Hall. 1973. Electron spin resonance studies of kaolins. In *Proceedings of the International Clay Conference 1972*, (Ed.) J.M. Serratosa, pp. 47–60. Madrid, Spain: Division de Ciencias, C.S.I.C.

Arndt, D.S., M. Mattei, C.A. Heist, and M.M. McGuire. 2017. Measurement of swelling of individual smectite tactoids *in situ* using atomic force microscopy. *Clays and Clay Minerals* 65: 92–193.

Atluri, V., J. Jin, K. Shrimali, L. Dang, X. Wang, and J.D. Miller. 2019. The hydrophobic surface state of talc as influenced by aluminum substitution in the tetrahedral layer. *Journal of Colloid and Interface Science* 536: 737–748.

Auerbach, S.M., K.A. Carrado, and P.K. Dutta. 2004. *Handbook of Layered Materials*. New York: Marcel Dekker.

Aylmore, L.A.G. and J.P. Quirk. 1971. Domains and quasi-crystalline regions in clay systems. *Soil Science Society of America Proceedings* 35: 652–654.

Bailey, S.W. 1963. Polymorphism of the kaolin minerals. *American Mineralogist* 48: 1196–1209.

Bailey, S.W. 1980. Structures of layer silicates. In *Crystal Structures of Clay Minerals and Their X-ray Identification*, (Eds.) G.W. Brindley and G. Brown, pp. 1–123. London, UK: Mineralogical Society.

Bailey, S.W. 1990. Halloysite – a critical assessment. In *Proceedings of the 9th International Clay Conference, 1989*, (Eds.) V.C. Farmer and Y. Tardy, pp. 89–98. Strasbourg, France: Sciences Géologiques, Mémoire 86.

Bailey, S.W. and B.E. Brown. 1962. Chlorite polytypism: I. Regular and semi-random one-layer structures. *American Mineralogist* 47: 819–850.

Baker, L.L., R.D. Nickerson, and D.G. Strawn. 2014. XAFS study of Fe-substituted allophane and imogolite. *Clays and Clay Minerals* 62: 20–34.

Balan, E., G. Calas, and D.L. Bish. 2014. Kaolin-group minerals: From hydrogen-bonded layers to environmental recorders. *Elements* 10: 183–188.

Ballantine, J.A., J.H. Purnell, and J.M. Thomas. 1983. Organic reactions in a clay micro-environment. *Clay Minerals* 18: 347–356.

Balogh, M. and P. Laszlo. 1993. *Organic Chemistry Using Clays*. Berlin, Germany: Springer-Verlag.

Barnhisel, R.I. and P.M. Bertsch. 1989. Chlorites and hydroxy interlayered vermiculite and smectite. In *Minerals in Soil Environments*, 2nd edition, (Eds.) J.B. Dixon and S.B. Weed, pp. 729–788. Madison, WI: Soil Science Society of America.

Barrer, R.M. 1978. *Zeolites and Clay Minerals as Sorbents and Molecular Sieves*. London, UK: Academic Press.

Barrer, R.M. and D.L. Jones. 1970. Chemistry of soil minerals. Part VIII. Synthesis and properties of fluorhectorites. *Journal of the Chemical Society A* 1531–1537.

Barron, P.F., M.A. Wilson, A.S. Campbell, and R.L. Frost. 1982. Detection of imogolite in soils using solid state ^{29}Si NMR. *Nature* 299: 616–618.

Bates, T.F. 1959. Morphology and crystal chemistry of 1:1 layer lattice silicates. *American Mineralogist* 44: 78–114.

Bates, T.F., L.B. Sand, and J.F. Mink. 1950. Tubular crystals of chrysotile asbestos. *Science* 3: 512–513.

Bergaya, F. and G. Lagaly (Eds.). 2013. *Handbook of Clay Science*, 2nd edition. Developments in Clay Science, Vol. 5B. Amsterdam, The Netherlands: Elsevier.

Bergaya, F., B.K.G. Theng, and G. Lagaly (Eds.). 2006. *Handbook of Clay Science*. Developments in Clay Science, Vol. 1. Amsterdam, The Netherlands: Elsevier.

Bernstein, D., J. Dunnigan, T. Hesterberg et al. 2013. Health risk of chrysotile revisited. *Critical Reviews in Toxicology* 43: 154–183.
Beyer, J. and H. Graf von Reichenbach. 2002. An extended revision of the interlayer structures of one- and two-layer hydrates of Na-vermiculite. *Clay Minerals* 37: 157–168.
Bin, L.-D., Y. Li, X.-H. Feng et al. 2018. Attapulgite clay and its application in radionuclide treatment. *Science of Advanced Materials* 10: 1529–1542.
Biswas, B., L.N. Warr, E.F. Hilder et al. 2019. Biocompatible functionalisation of nanoclays for improved environmental remediation. *Chemical Society Reviews* 48: 3740–3770.
Blockhaus, F., J.-M. Sequaris, H.D. Narres, and M.J. Schwuger. 1997. Adsorption-desorption behavior of acrylic-maleic acid copolymer at clay minerals. *Journal of Colloid and Interface Science* 186: 234–247.
Bloise, A., E. Belluso, E. Barrese, D. Miriello, and C. Apollaro. 2009. Synthesis of Fe-doped chrysotile and characterization of the resulting chrysotile fibres. *Crystal Research and Technology* 44: 590–596.
Bolland, M.D.A., A.M. Posner, and J.P. Quirk. 1976. Surface charge on kaolinites in aqueous suspension. *Australian Journal of Soil Research* 14: 197–216.
Bolland, M.D.A., A.M. Posner, and J.P. Quirk. 1980. pH-independent and pH-dependent surface charges on kaolinites. *Clays and Clay Minerals* 28: 412–418.
Bookin, A.S., V.A. Drits, A. Plançon, and C. Tchoubar. 1989. Stacking faults in kaolin-group minerals in the light of real structural features. *Clays and Clay Minerals* 37: 297–307.
Bradley, W.F. 1940. The structural scheme of attapulgite. *American Mineralogist* 25: 405–410.
Brady, P.V., R.T. Cygan, and K.L. Nagy. 1996. Molecular controls on kaolinite surface charge. *Journal of Colloid and Interface Science* 183: 356–364.
Braganza, L.F., R.J. Crawford, M.V. Smalley, and R.K. Thomas. 1990. Swelling of n-butylammonium vermiculite in water. *Clays and Clay Minerals* 38: 90–96.
Brauner, K. and A. Preisinger. 1956. Struktur und Entstehung des Sepioliths. *Tschermaks mineralogische und petrographische Mitteilungen* 6: 120–140.
Breu, J., W. Seidl, A.J. Stoll, K.G. Lange, and T.U. Probst. 2001. Charge homogeneity in synthetic fluorohectorite. *Chemistry of Materials* 13: 4213–4220.
Brigatti, M.F., E. Galan, and B.K.G. Theng. 2013. Structure and mineralogy of clay minerals. In *Handbook of Clay Science*, 2nd edition. Developments in Clay Science, Vol. 5A, (Eds.) F. Bergaya and G. Lagaly, pp. 21–81. Amsterdam, The Netherlands: Elsevier.
Brindley, G.W. 1980. Order-disorder in clay mineral structures. In *Crystal Structures of Clay Minerals and Their X-ray Identification*, (Eds.) G.W. Brindley and G. Brown, pp. 125–195. London, UK: Mineralogical Society.
Brindley, G.W. and G. Brown (Eds.). 1980. *Crystal Structures of Clay Minerals and Their X-ray Identification*. London, UK: Mineralogical Society.
Brindley, G.W. and M. Nakahira. 1958. Further considerations of the crystal structure of kaolinite. *Mineralogical Magazine* 31: 781–786.
Brindley, G.W., P.E. Zalba, and C.M. Bethke. 1983. Hydrobiotite, a regular 1:1 interstratification of biotite and vermiculite layers. *American Mineralogist* 68: 420–425.
Brovelli, D., W.R. Caseri, and G. Hähner. 1999. Self-assembled monolayers of alkylammonium ions on mica: Direct determination of the orientation of the alkyl chains. *Journal of Colloid and Interface Science* 216: 418–423.
Brown, G. 1980. Associated minerals. In *Crystal Structures of Clay Minerals and Their X-ray Identification*, (Eds.) G.W. Brindley and G. Brown, pp. 361–410. London, UK: Mineralogical Society.
Brown, G., A.C.D. Newman, J.H. Rayner, and A.H. Weir. 1978. The structures and chemistry of soil clay minerals. In *The Chemistry of Soil Constituents*, (Eds.) D.J. Greenland and M.H.B. Hayes, pp. 29–178. Chichester, UK: John Wiley & Sons.
Browning, K.L., L.R. Griffin, P. Gutfreund et al. 2014. Specular neutron reflection at the mica/water interface – irreversible adsorption of a cationic dichain surfactant. *Journal of Applied Crystallography* 47: 1638–1646.
Brunauer, S., P.H. Emmett, and E. Teller. 1938. Adsorption of gases in multimolecular layers. *Journal of the American Chemical Society* 60: 309–319.
Calabi-Floody, M., J.S. Bendall, A.A. Jara et al. 2011. Nanoclays from an Andisol: Extraction, properties and carbon stabilization. *Geoderma* 161: 159–167.
Calabi-Floody, M., B.K.G. Theng, P. Reyes, and M.L. Mora. 2009. Natural nanoclays: Application and future trends – a Chilean perspective. *Clay Minerals* 44: 161–176.
Call, J.J., S. Rakshit, and M.E. Essington. 2019. The adsorption of tylosin by montmorillonite and vermiculite: Exchange selectivity and intercalation. *Soil Science Society of America Journal* 83: 584–596.

Carr, R.M. and H. Chih. 1971. Complexes of halloysite with organic compounds. *Clay Minerals* 9: 153–166.

Carrado, K.A. 2000. Synthetic organo- and polymer-clays: Preparation, characterization, and materials applications. *Applied Clay Science* 17: 1–23.

Carrado, K.A., P. Thiyagarajan, and K. Song. 1997. A study of organo-hectorite clay crystallization. *Clay Minerals* 32: 29–40.

Carretero, M.I. and M. Pozo. 2007. *Mineralogía Aplicada. Salud y Medio Ambiente*. Madrid, Spain: Thomson.

Cavalcanti, L.P., G.N. Kalantzopoulos, J. Eckert, K.D. Knudsen, and J.O. Fossum. 2018. A nano-silicate material with exceptional capacity for CO_2 capture and storage at room temperature. *Scientific Reports* 8: 11827.

Cervini-Silva, J., A. Nieto-Camacho, V. Gómez-Vidales, S. Kaufhold, and B.K.G. Theng. 2015. The anti-inflammatory activity of natural allophane. *Applied Clay Science* 105–106: 48–51.

Cervini-Silva, J., A. Nieto-Camacho, E. Palacios et al. 2016. Anti-inflammatory, antibacterial, and cytotoxic activity by natural matrices of nano-iron(hydr)oxide/halloysite. *Applied Clay Science* 120: 101–110.

Chang, P.-H., Z. Li, J.-S. Jean, W.-T. Jiang, C.-J. Wang, and K.-H. Lin. 2012. Adsorption of tetracycline on 2:1 layered non-swelling clay mineral illite. *Applied Clay Science* 67–68: 158–163.

Chang, J., B. Liu, J.S. Grundy et al. 2021. Probing specific adsorption of electrolytes at kaolinite-aqueous interfaces by atomic force microscopy. *Journal of Physical Chemistry Letters* 12: 2406–2412.

Chen, G., N. Iyi, R. Sasai, T. Fujita, and K. Kitamura. 2002. Intercalation of rhodamine 6G and oxazine 4 into oriented clay films and their alignment. *Journal of Materials Research* 17: 1035–1040.

Childs, C.W., K. Inoue, H. Seyama, M. Soma, B.K.G. Theng, and G. Yuan. 1997. X-ray photoelectron spectroscopic characterization of Silica Springs allophane. *Clay Minerals* 32: 565–572.

Childs, C.W., R.L. Parfitt, and R.H. Newman. 1990. Structural studies of Silica Springs allophane. *Clay Minerals* 25: 329–341.

Christidis, G. and A.C. Dunham. 1993. Compositional variations in smectites: Part I. Alteration of intermediate volcanic rocks. A case study from Milos Island, Greece. *Clay Minerals* 28: 255–273.

Christidis, G.E. and D.D. Eberl. 2003. Determination of layer charge characteristics of smectites. *Clays and Clay Minerals* 51: 644–655.

Churchman, G.J. 2015. The identification and nomenclature of halloysite (A historical perspective). In *Natural Mineral Nanotubes*, (Eds.) P. Pasbakhsh and G.J. Churchman, pp. 51–68. Oakville, ON: Apple Academic Press.

Churchman, G.J. and R.M. Carr. 1972. Stability fields of hydration states of an halloysite. *American Mineralogist* 57: 914–933.

Churchman, G.J. and R.M. Carr. 1975. The definition and nomenclature of halloysites. *Clays and Clay Minerals* 23: 382–388.

Churchman, G.J., P. Pasbakhsh, D.J. Lowe, and B.K.G. Theng. 2016. Unique but diverse: Observations on the formation, structure and morphology of halloysite. *Clay Minerals* 51: 395–416.

Churchman, G.J. and K.R. Tate. 1987. Stability of aggregates of different size grades in allophanic soils from volcanic ash in New Zealand. *Journal of Soil Science* 38: 19–27.

Churchman, G.J. and B.K.G. Theng. 1984. Interactions of halloysites with amides: Mineralogical factors affecting complex formation. *Clay Minerals* 19: 161–175.

Churchman, G.J., J.S. Whitton, G.G.C. Claridge, and B.K.G. Theng. 1984. Intercalation method using formamide for differentiating halloysite from kaolinite. *Clays and Clay Minerals* 32: 241–248.

Clark, C.J. and M.B. McBride. 1984. Cation and anion retention by natural and synthetic allophane and imogolite. *Clays and Clay Minerals* 32: 291–299.

Cradwick, P.D.G., V.C. Farmer, J.D. Russell, C.R. Masson, K. Wada, and N. Yoshinaga. 1972. Imogolite, a hydrated aluminium silicate of tubular structure. *Nature Physical Science* 240: 187–189.

Czímerová, A., J. Bujdák, and R. Dohrmann. 2006. Traditional and novel methods for estimating the layer charge of smectites. *Applied Clay Science* 34: 2–13.

da Fonseca, M.G. and C. Airoldi. 2001. New amino-inorganic hybrids from talc silylation and copper adsorption properties. *Materials Research Bulletin* 36: 277–287.

Dahlgren, R.A., M. Saigusa, and F.C. Ugolini. 2004. The nature, properties and management of volcanic soils. *Advances in Agronomy* 82: 113–182.

Delineau, T., T. Allard, J.-P. Muller, O. Barres, J. Yvon, and J-M. Cases. 1994. FTIR reflectance *vs.* EPR studies of structural iron in kaolinites. *Clays and Clay Minerals* 42: 308–320.

de Paiva, L.B., A.R. Morales, and F.R. Valenzuela Díaz. 2008. Organoclays: Properties, preparation and applications. *Applied Clay Science* 42: 8–24.

Detellier, C. and R.A. Schoonheydt. 2014. From platy kaolinite to nanorolls. *Elements* 10: 201–206.

Dixon, J.B. 1989. Kaolinite and serpentine group minerals. In *Minerals in Soil Environments*, 2nd edition, (Eds.) J.B. Dixon and S.B. Weed, pp. 467–525. Madison, WI: Soil Science Society of America.

Dixon, J.B. and D.G. Schulze (Eds.). 2002. *Soil Mineralogy with Environmental Applications*. Madison, WI: Soil Science Society of America.

Dixon, J.B. and S.B. Weed (Eds.). 1989. *Minerals in Soil Environments*, 2nd edition. Madison, WI: Soil Science Society of America.

Douglas, L.A. 1989. Vermiculites. In *Minerals in Soil Environments*, 2nd edition, (Eds.) J.B. Dixon and S.B. Weed, pp. 635–674. Madison, WI: Soil Science Society of America.

Droge, S.T.J. and K.-U. Goss. 2013. Sorption of organic cations to phyllosilicate clay minerals: CEC-normalization, salt dependency, and the role of electrostatic and hydrophobic effects. *Environmental Science & Technology* 47: 14224–14232.

Du, M., B. Guo, and D. Jia. 2010. New emerging applications of halloysite nanotubes: A review. *Polymer International* 59: 574–582.

Du, P., P. Yuan, D. Liu, S. Wang, H. Song, and H. Guo. 2018. Calcination-induced changes in structure, morphology, and porosity of allophane. *Applied Clay Science* 158: 211–218.

Du, P., P. Yuan, J. Liu, and B. Ye. 2023. Clay minerals on Mars: An up-to-date review with future perspectives. *Earth Science Reviews* 243: 104491.

Egashira, K. and S. Aomine. 1974. Effects of drying and heating on the surface area of allophane and imogolite. *Clay Science* 4: 231–242.

Ertem, G., A. Steudel, K. Emmerich, G. Lagaly, and R. Schuhmann. 2010. Correlation between the extent of catalytic activity and charge density of montmorillonites. *Astrobiology* 10: 743–749.

Escudey, M. and G. Galindo. 1983. Effect of iron oxide coatings on electrophoretic mobility and dispersion of allophane. *Journal of Colloid and Interface Science* 93: 78–83.

Falini, G., E. Foresti, M. Gazzano et al. 2004. Tubular-shaped stoichiometric chrysotile nanocrytals. *Chemistry – A European Journal* 10: 3043–3049.

Falini, G., E. Foresti, I.G. Lesci, B. Lunelli, P. Sabatino, and N. Roveri. 2006. Interaction of bovine serum albumin with chrysotile: Spectroscopic and morphological studies. *Chemistry – A European Journal* 12: 1968–1974.

Fanning, D.S., V.Z. Keramidas, and M.A. El-Desoky. 1989. Micas. In *Minerals in Soil Environments*, 2nd edition, (Eds.) J.B. Dixon and S.B. Weed, pp. 551–634. Madison, WI: Soil Science Society of America.

Farmer, V.C. 1974. *The Infrared Spectra of Minerals*. London, UK: Mineralogical Society.

Farmer, V.C., M.J. Adams, A.R. Fraser, and F. Palmieri. 1983. Synthetic imogolite: Properties, synthesis, and possible applications. *Clay Minerals* 18: 459–472.

Farmer, V.C., A.R. Fraser, J.D. Russell, and N. Yoshinaga. 1977. Recognition of imogolite structures in allophanic clays by infrared spectroscopy. *Clay Minerals* 12: 55–57.

Ferris, A.P. and W.B. Jepson. 1975. The exchange capacity of kaolinite and the preparation of homoionic clays. *Journal of Colloid and Interface Science* 51: 245–259.

Ferrow, E.A., B.E. Kalinowski, D.R. Veblen, and P. Schweda. 1999. Alteration products of experimentally weathered biotite studies by high-resolution TEM and Mössbauer spectroscopy. *European Journal of Mineralogy* 11: 999–1010.

Fieldes, M. and G.G.C. Claridge. 1975. Allophane. In *Soil Components*. Inorganic Components, Vol. 2, (Ed.) J.E. Gieseking, pp. 351–393. Berlin, Germany: Springer-Verlag.

Fiore, S., J. Cuadros, and F.J. Huertas (Eds.). 2013. *Interstratified Clay Minerals. Origin, Characterization & Geochemical Significance*. Bari, Italy: Digilabs.

Fiore, S. and F.J. Huertas. 2015. The identification and nomenclature of chrysotile (A historical perspective). In *Natural Mineral Nanotubes*, (Eds.) P. Pasbakhsh and G.J. Churchman, pp. 85–92. Oakville, ON: Apple Academic Press.

Foresti, E., E. Fornero, I.G. Lesci, C. Rinaudo, T. Zuccheri, and N. Roveri. 2009. Asbestos health hazard: A spectroscopic study of synthetic geoinspired Fe-doped chrysotile. *Journal of Hazardous Materials* 167: 1070–1079.

Foresti, E., M.F. Hochella, Jr., H. Kornishi et al. 2005. Morphological and chemical/physical characterization of Fe-doped synthetic chrysotile nanotubes. *Advanced Functional Materials* 15: 1009–1016.

Franklin, K.R. and E. Lee. 1996. Synthesis and ion-exchange properties of Na-4-mica. *Journal of Materials Chemistry* 6: 109–115.

Fripiat, J.J. (Ed.). 1981. *Advanced Techniques for Clay Mineral Analysis*. Amsterdam, The Netherlands: Elsevier.

Fripiat, J.J. and M. Della Faille. 1967. Surface properties and texture of chrysotiles. *Clays and Clay Minerals* 15: 305–320.

Frost, R.L., J. Kristof, E. Horvath, and J.T. Kloprogge. 2000. Vibrational spectroscopy of formamide-intercalated kaolinites. *Spectrochimica Acta Part A* 56: 1191–1204.

Fujita, T., N. Iyi, Z. Klapyta, K. Fujii, Y. Kaneko, and K. Kitamura. 2003. Photomechanical response of azobenzene/organophilic mica complexes. *Materials Research Bulletin* 38: 2009–2017.

Fujita, T., N. Iyi, T. Kosugi, A. Ando, T. Deguchi, and T. Sota. 1997. Intercalation characteristics of rhodamine 6G in fluor-taeniolite: Orientation in the gallery. *Clays and Clay Minerals* 45: 77–84.

Fysh, S.A., J.D. Cashion, and P.E. Clark. 1983. Mössbauer effect studies of iron in kaolin. I. Structural iron. *Clays and Clay Minerals* 31: 285–292.

García-García, S., S. Wold, and M. Jonsson. 2009. Effects of temperature on the stability of colloidal montmorillonite particles at different pH and ionic strength. *Applied Clay Science* 43: 21–26.

Gard, J.A. (Ed.). 1971. *The Electron-Optical Investigation of Clays*. London, UK: Mineralogical Society.

Gardolinski, J.E.F.C. and G. Lagaly. 2005. Grafted organic derivatives of kaolinite. II. Intercalation of primary alkylamines and delamination. *Clay Minerals* 40: 547–556.

Garrett, W.G. and G.F. Walker. 1962. Swelling of some vermiculite-organic complexes in water. *Clays and Clay Minerals* 9: 557–567.

Garrett, W.G. and G.F. Walker. 1967. Chemical exfoliation of vermiculite and the production of colloidal dispersions. *Science* 156: 385–387.

Gates, W.P., J.T. Kloprogge, J. Madejová, and F. Bergaya (Eds.). 2017. *Infrared and Raman Spectroscopies of Clay Minerals*. Developments in Clay Science, Vol. 8. Amsterdam, The Netherlands: Elsevier.

Gazzano, E., E. Foresti, I.G. Lesci, M. Tomatis, and C. Riganti. 2005. Different cellular responses evoked by natural and stoichiometric synthetic chrysotile asbestos. *Toxicology and Applied Pharmacology* 206: 356–364.

Gerstl, Z. and B. Yaron. 1981. Stability of parathion on attapulgite as affected by structural and hydration changes. *Clays and Clay Minerals* 29: 53–59.

Giese, R.F. 1975. The effect of F/OH substitution on some layer silicate minerals. *Zeitschrift für Kristallographie* 141: 138–144.

Giese, R.F. 1978. The electrostatic interlayer forces of layer structure minerals. *Clays and Clay Minerals* 26: 51–57.

Giese, R.F. 1988. Kaolin minerals: Structures and stabilities. In *Hydrous Phyllosilicates (Exclusive of Micas)*. Reviews in Mineralogy, Vol. 19, (Ed.) S.W. Bailey, pp. 29–66. Washington, DC: Mineralogical Society of America.

Giese, R.F. and P. Datta. 1973. Hydroxyl orientations in kaolinite, dickite, and nacrite. *American Mineralogist* 58: 471–479.

Goh, K.M. 2004. Carbon sequestration and stabilization in soils: Implications for soil productivity and climate change. *Soil Science and Plant Nutrition* 50: 467–476.

Granquist, W.T. and S.S. Pollack. 1960. A study of the synthesis of hectorite. *Clays and Clay Minerals* 8: 150–169.

Gray, N., D.G. Lumsdon, and S. Hillier. 2016. Effect of pH on the cation exchange capacity of some halloysite nanotubes. *Clay Minerals* 51: 373–383.

Greenland, D.J. and M.H.B. Hayes (Eds.). 1978. *The Chemistry of Soil Constituents*. Chichester, UK: John Wiley & Sons.

Grim, R.E. 1968. *Clay Mineralogy*, 2nd edition. New York: McGraw-Hill.

Grim, R.E., R.H. Bray, and W.F. Bradley. 1937. The mica in argillaceous sediments. *American Mineralogist* 22: 813–829.

Gualtieri, A.F., A. Moen, and D.G. Nicholson. 2000. XANES study of the local environment of iron in natural kaolinites. *European Journal of Mineralogy* 12: 17–23.

Guggenheim, S. 2015. Phyllosilicates used as nanotube substrates in engineered materials: Structures, chemistries, and textures. In *Natural Mineral Nanotubes*, (Eds.) P. Pasbakhsh and G.J. Churchman, pp. 3–50. Oakville, ON: Apple Academic Press.

Guggenheim, S., J.M. Adams, D.C. Bain et al. 2006. Summary of recommendations of nomenclature committees relevant to clay mineralogy: Report of the Association Internationale pour l'Étude des Argiles (AIPEA) nomenclature committee for 2006. *Clay Minerals* 41: 863–877 and *Clays and Clay Minerals* 54: 761–772.

Guggenheim, S. and R.T. Martin. 1995. Definition of clay and clay mineral: Joint report of the AIPEA nomenclature and CMS nomenclature committees. *Clay Minerals* 30: 257–259 and *Clays and Clay Minerals* 43: 255–256.

Gustafsson, J.P. 2001. The surface chemistry of imogolite. *Clays and Clay Minerals* 49: 73–80.

Hall, P.L., G.J. Churchman, and B.K.G. Theng. 1985. Size distribution of allophane unit particles in aqueous suspensions. *Clays and Clay Minerals* 33: 345–349.

Hanif, M., F. Jabbar, S. Sharif, G. Abbas, A. Forooq, and M. Aziz. 2016. Halloysite nanotubes as a new drug-delivery system: A review. *Clay Minerals* 51: 469–477.

Harsh, J.B. 2012. Poorly crystalline aluminosilicate minerals. In *Handbook of Soil Sciences. Properties and Processes*, 2nd edition, (Eds.) P.M. Huang, Y. Li, and M.E. Sumner, pp. 23-1–23-13. Boca Raton, FL: CRC Press.

Harsh, J.B., S.J. Traina, J. Boyle, and Y. Yang. 1992. Adsorption of cations on imogolite and their effect on surface charge characteristics. *Clays and Clay Minerals* 40: 700–706.

Hashizume, H. 2009. Adsorption of aromatic compounds in water by talc. *Clay Science* 14: 61–64.

Hashizume, H. 2013. Removal of some aromatic hydrocarbons from water by pyrophyllite. *Clay Science* 17: 49–55.

Hashizume, H. and B.K.G. Theng. 2007. Adenine, adenosine, ribose, and 5'-AMP adsorption to allophane. *Clays and Clay Minerals* 55: 599–605.

Hata, H., Y. Kobayashi, and T.E. Mallouk. 2007. Encapsulation of anionic dye molecules by a swelling fluoromica through intercalation of cationic polyelectrolytes. *Chemistry of Materials* 19: 79–87.

Hayashi, H., R. Otsuka, and N. Imai. 1969. I.R. study of sepiolite and palygorskite on heating. *American Mineralogist* 53: 1613–1624.

He, H., L. Ma, J. Zhu, R.L. Frost, B.K.G. Theng, and F. Bergaya. 2014. Synthesis of organoclays: A critical review and some unresolved issues. *Applied Clay Science* 100: 22–28.

He, H., Q. Tao, J. Zhu, P. Yuan, W. Shen, and S. Yang. 2013. Silylation of clay mineral surfaces. *Applied Clay Science* 71: 15–20.

Heinz, H., H.J. Castelijns, and U.W. Suter. 2003. Structure and phase transitions of alkyl chains on mica. *Journal of the American Chemical Society* 125: 9500–9510.

Hendricks, S.B. 1942. Lattice structure of clay minerals and some properties of clays. *Journal of Geology* 50: 276–290.

Hendricks, S.B. and M.E. Jefferson. 1938. Structures of kaolin and talc-pyrophyllite hydrates and their bearing on water sorption of the clays. *American Mineralogist* 23: 863–875.

Hénin, S. and S. Caillère. 1975. Fibrous minerals. In *Soil Components*. Inorganic Components, Vol. 2, (Ed.) J.E. Gieseking, pp. 335–349. Berlin, Germany: Springer-Verlag.

Henmi, T. and K. Wada. 1976. Morphology and composition of allophane. *American Mineralogist* 61: 379–390.

Herrington, T.M., A.Q. Clarke, and J.C. Watts. 1992. The surface charge of kaolin. *Colloids and Surfaces* 68: 161–169.

Hillier, S. and P.C. Ryan. 2002. Identification of halloysite (7 Å) by ethylene glycol solvation: The 'MacEwan effect'. *Clay Minerals* 37: 487–496.

Hingston, F.J., R.J. Atkinson, A.M. Posner, and J.P. Quirk. 1967. Specific adsorption of anions. *Nature* 215: 1459–1461.

Hingston, F.J., A.M. Posner, and J.P. Quirk. 1972. Anion adsorption by goethite and gibbsite. I. The role of the proton in determining adsorption envelopes. *Journal of Soil Science* 23: 177–192.

Hofmann, U. 1968. On the chemistry of clays. *Angewandte Chemie International Edition* 7: 681–692.

Hofmann, U., K. Endell, and D. Wilm. 1933. Kristallstruktur und Quellung von Montmorillonit. *Zeitschrift für Kristallographie* 86: 340–348.

Horikawa, Y. 1975. Electrokinetic phenomena of aqueous suspensions of allophane and imogolite. *Clay Science* 4: 255–263.

Huo, X., L. Wu, L. Liao, Z. Xia, and L. Wang. 2012. The effect of interlayer cations on the expansion of vermiculite. *Powder Technology* 224: 241–246.

Ildefonse, P., R.J. Kirkpatrick, B. Montez, G. Calas, A.M. Flank, and P. Lagarde. 1994. ^{27}Al MAS NMR and aluminum X-ray absorption near edge structure study of imogolite and allophanes. *Clays and Clay Minerals* 42: 276–287.

Iyi, N., T. Fujita, C.V. Yelamaggad, and F.L. Arbeloa. 2001. Intercalation of cationic azobenzene derivatives in a synthetic mica and their photoresponse. *Applied Clay Science* 19: 47–58.

Iyoda, F., S. Hayashi, S. Arakawa et al. 2012. Synthesis and adsorption characteristics of hollow spherical allophane nano-particles. *Applied Clay Science* 56: 77–83.

Janek, M., P. Komadel, and G. Lagaly. 1997. Effect of autotransformation on the layer charge of smectites determined by the alkylammonium method. *Clay Minerals* 32: 623–632.

Jaynes, W.F. and S.A. Boyd. 1991a. Hydrophobicity of siloxane surfaces in smectites as revealed by aromatic hydrocarbon adsorption from water. *Clays and Clay Minerals* 39: 428–436.

Jaynes, W.F. and S.A. Boyd. 1991b. Clay mineral type and organic compound sorption by hexadecyltrimethylammonium-exchanged clays. *Soil Science Society of America Journal* 55: 43–48.
Jepson, W.B. 1984. Kaolins: Their properties and uses. *Philosophical Transactions of the Royal Society of London A* 311: 411–432.
Johns, W.D. and P.K. Sen Gupta. 1967. Vermiculite-alkylammonium complexes. *American Mineralogist* 52: 1706–1724.
Johns, C.A., R. Martos-Villa, S. Guggenheim, and C.I. Sainz-Diaz. 2013. Structure determination of trimethylsulfoxonium-exchanged vermiculite. *Clays and Clay Minerals* 61: 218–230.
Johnston, C.T. 2010. Probing the nanoscale architecture of clay minerals. *Clay Minerals* 45: 245–279.
Johnston, C.T., S.A. Boyd, B.J. Teppen, and G. Sheng. 2004. Sorption of nitroaromatic compounds on clay surfaces. In *Handbook of Layered Materials*, (Eds.) S.M. Auerbach, K.A. Carrado, and P.K. Dutta, pp. 155–189. New York: Marcel Dekker.
Jones, B.F. and E. Galán. 1988. Sepiolite and palygorskite. In *Hydrous Phyllosilicates (Exclusive of Micas).* Reviews in Mineralogy, Vol. 19, (Ed.) S.W. Bailey, pp. 631–674. Washington, DC: Mineralogical Society of America.
Joussein, E., S. Petit, J. Churchman, B. Theng, D. Righi, and B. Delvaux. 2005. Halloysite clay minerals – a review. *Clay Minerals* 40: 383–426.
Joussein, E., S. Petit, and B. Delvaux. 2007. Behavior of halloysite clay under formamide treatment. *Applied Clay Science* 35: 17–24.
Kalo, H., M.W. Möller, M. Ziadeh, D. Dolejš, and J. Breu. 2010. Large scale melt synthesis in an open crucible of Na-fluorohectorite with superb charge homogeneity and particle size. *Applied Clay Science* 48: 39–45.
Kang, F., Y. Ge, X. Hu et al. 2016. Understanding the sorption mechanisms of aflatoxin B1 to kaolinite, illite, and smectite clays via a comparative computational study. *Journal of Hazardous Materials* 320: 80–87.
Kaufhold, S., R. Dohrmann, M. Klinkenberg, S. Siegesmund, and K. Ufer. 2010. N_2-BET specific surface area of bentonites. *Journal of Colloid and Interface Science* 349: 275–282.
Kaufhold, S., H. Stanjek, D. Penner, and R. Dohrmann. 2011. The acidity of surface groups of dioctahedral smectites. *Clay Minerals* 46: 583–592.
Keeling, J.L. 2015. The mineralogy, geology and occurrences of halloysite. In *Natural Mineral Nanotubes*, (Eds.) P. Pasbakhsh and G.J. Churchman, pp. 95–115. Oakville, ON: Apple Academic Press.
Keeling, J.L. and P. Pasbakhsh. 2015. Halloysite mineral nanotubes – geology, properties and applied research. *MESA Journal* 77: 20–26.
Kirkman, J.H. 1977. Possible structure of halloysite disks and cylinders observed in some New Zealand rhyolitic tephras. *Clay Minerals* 12: 199–216.
Kirkman, J.H. and W.A. Pullar. 1978. Halloysite in late pleistocene rhyolitic tephra beds near Opotiki, coastal Bay of Plenty, North Island, New Zealand. *Australian Journal of Soil Research* 16: 1–8.
Kitagawa, Y. 1971. The 'unit particle' of allophane. *American Mineralogist* 56: 465–475.
Kitajima, K. and N. Daimon. 1975. Na-fluor-tetrasilicic mica [$Na_2Mg_{2.5}Si_2O_{10}F_2$] and its swelling characteristics. *Nippon Kagaku Kaishi* 6: 991–995.
Kitajima, K., F. Koyama, and N. Takusagawa. 1985. Synthesis and swelling properties of fluorine micas with variable layer charges. *Bulletin Chemical Society, Japan* 58: 1325–1326.
Kłapita, Z., T. Fujita, and N. Iyi. 2001. Adsorption of dodecyl- and octadecyl-trimethylammonium ions on a smectite and synthetic micas. *Applied Clay Science* 19: 5–10.
Kłapita, Z., A. Gaweł, T. Fujita, and N. Iyi. 2003. Structural heterogeneity of alkylammonium-exchanged, synthetic fluorotetrasilicic mica. *Clay Minerals* 38: 151–160.
Kłapita, Z., A. Gaweł, T. Fujita, and N. Iyi. 2011. Intercalation of protonated polyoxyalkylene monoamines into a synthetic Li-fluorotaenolite. *Applied Clay Science* 52: 133–139.
Klebow, B. and A. Meleshyn. 2011. Aggregation of alkyltrimethylammonium ions at the cleaved muscovite mica-water interface: A Monte Carlo study. *Langmuir* 27: 12968–12976.
Kloprogge, J.T. 2017. Raman and infrared spectroscopies of intercalated kaolinite groups minerals. In *Infrared and Raman Spectroscopies of Clay Minerals*. Developments in Clay Science, Vol. 8, (Eds.) W.P. Gates, J.T. Kloprogge, J. Madejová, and F. Bergaya, pp. 343–410. Amsterdam, The Netherlands: Elsevier.
Kloprogge, J.T., S. Komarneni, and J.E. Amonette. 1999. Synthesis of smectite clay minerals: A critical review. *Clays and Clay Minerals* 47: 529–554.
Kodama, T. and S. Komarneni. 1999. Na-4-mica: Cd^{2+}, Ni^{2+}, Co^{2+}, Mn^{2+}, and Zn^{2+} ion exchange. *Journal of Materials Chemistry* 9: 533–539.
Kogure, T. and A. Inoue. 2005a. Determination of defect structures in kaolin minerals by high-resolution transmission electron microscopy (HRTEM). *American Mineralogist* 90: 85–89.

Kogure, T. and A. Inoue. 2005b. Stacking defects and long-period polytypes in kaolin minerals from a hydrothermal deposit. *European Journal of Mineralogy* 17: 465–473.

Kogure, T., K. Morimoto, K. Tamura, H. Sato, and A. Yamagishi. 2012. XRD and HRTEM evidence for fixation of cesium ions in vermiculite clay. *Chemistry Letters* 41: 380–382.

Komusiński, J., L. Stoch, and S.M. Dubiel. 1981. Application of electron paramagnetic resonance and Mössbauer spectroscopy in the investigation of kaolinite-group minerals. *Clays and Clay Minerals* 29: 23–30.

Konta, J. 1995. Clay and man: Clay raw materials in the service of man. *Applied Clay Science* 10: 275–335.

Kubicki, J.D., M.J. Itoh, L.M. Schroeter, and S.E. Apitz. 1997. Bonding mechanisms of salicylic acid adsorbed onto illite clay: An ATR-FTIR and molecular orbital study. *Environmental Science & Technology* 31: 1151–1156.

Kühnel, R.A., S.J. Van der Gaast, M.A.T.M. Broekmans, and B.K.G. Theng. 2017. Wetting-induced layer contraction in illite and mica-family relatives. *Applied Clay Science* 135: 226–233.

Kuroda, Y., K. Ito, K. Itabashi, and K. Kuroda. 2011. One-step exfoliation of kaolinites and their transformation into nanoscrolls. *Langmuir* 27: 2028–2035.

Lackovic, K., B.B. Johnson, M.J. Angove, and J.D. Wells. 2003. Modeling the adsorption of citric acid onto Muloorina illite and related clay minerals. *Journal of Colloid and Interface Science* 267: 49–59.

Lagaly, G. 1981. Characterization of clays by organic compounds. *Clay Minerals* 16: 1–21.

Lagaly, G. 1982. Layer charge heterogeneity in vermiculites. *Clays and Clay Minerals* 30: 215–222.

Lagaly, G., M. Ogawa, and I. Dékány. 2006. Clay mineral organic interactions. In *Handbook of Clay Science*. Developments in Clay Science, Vol. 1, (Eds.) F. Bergaya, B.K.G. Theng, and G. Lagaly, pp. 309–377. Amsterdam, The Netherlands: Elsevier.

Lagaly, G. and A. Weiss. 1970. Inhomogeneous charge distributions in mica-type layer silicates. In *Reunion Hispano-Belga de Minerales de la Arcilla*, (Ed.) J.M. Serratosa, pp. 179–187. Madrid, Spain: C.S.I.C.

Lagaly, G. and A. Weiss. 1976. The layer charge of smectitic layer silicates. In *Proceedings of the International Clay Conference 1975*, (Ed.) S.W. Bailey, pp. 157–172. Wilmette, IL: Applied Publishing.

Laird, D.A. 2006. Influence of layer charge on swelling of smectites. *Applied Clay Science* 34: 74–87.

Ledoux, R.L. and J.L. White. 1964. Infrared studies of selective deuteration of kaolinite and halloysite at room temperature. *Science* 145: 47–49.

Ledoux, R.L. and J.L. White. 1966. Infrared studies of hydrogen bonding of organic compounds on oxygen and hydroxyl surfaces of layer lattice silicates. In *Proceedings of the International Clay Conference, Jerusalem* 1: 361–374.

Lee, L.T., R. Rahbari, J. Lecourtier, and G. Chauveteau. 1991. Adsorption of polyacrylamide on the different faces of kaolinites. *Journal of Colloid and Interface Science* 147: 351–357.

Lee, S.Y., M.L. Jackson, and J.L. Brown. 1975. Micaceous occlusions in kaolinite observed by ultramicrotomy and high-resolution electron microscopy. *Clays and Clay Minerals* 23: 125–129.

Lee, S.Y. and S.J. Kim. 2002. Adsorption of naphthalene by HDTMA modified kaolinite and halloysite. *Applied Clay Science* 22: 55–63.

Leone, G., U. Giovanella, F. Bertini et al. 2013. Poly(styrene)-graft-/rhodamine 6G-fluormica hybrids: Synthesis, characterization and photophysical properties. *Journal of Materials Chemistry C* 1: 1450–1460.

Li, X., N. Liu, L. Tang, and J. Zhang. 2020a. Specific elevated adsorption and stability of cations in the interlayer compared with at the external surface of clay minerals. *Applied Clay Science* 198: 105814.

Li, Y., G. Tian, G. Dong et al. 2018. Research progress on the raw and modified montmorillonites as adsorbents for mycotoxins: A review. *Applied Clay Science* 163: 299–311.

Li, Z., P.-H. Chang, W.-T. Jiang, and Y. Liu. 2020b. Enhanced removal of ethidium bromide (EtBr) from aqueous solution using rectorite. *Journal of Hazardous Materials* 384: 121254.

Li, Z., W.-T. Jiang, and H. Hong. 2008. An FTIR investigation of hexadecyltrimethylammonium intercalation into rectorite. *Spectrochimica Acta Part A: Molecular and Biomolecular Spectroscopy* 71: 1525–1534.

Liétard, O., J. Yvon, J.F. Delon, R. Mercier, and J.-M. Cases. 1980. Determination of the basal and lateral surfaces of kaolins: Variation with types of crystalline defects. In *Fine Particles Processing*, Vol. 1, (Ed.) P. Somasundaran, pp. 558–582. New York: American Institute of Mining, Metallurgical and Petroleum Engineers.

Lim, C.H., M.L. Jackson, R.D. Koons, and P.A. Helmke. 1980. Kaolins: Sources of differences in cation-exchange capacities and cesium retention. *Clays and Clay Minerals* 28: 223–229.

Liu, M., R. He, J. Yang et al. 2016. Polysaccharide-halloysite nanotube composites for biomedical applications: A review. *Clay Minerals* 51: 457–467.

Liu, X., X. Lu, M. Sprik, J. Cheng, E.J. Meijer, and R. Wang. 2013. Acidity of edge surface sites of montmorillonite and kaolinite. *Geochimica et Cosmochimica Acta* 117: 180–190.

Lokanatha, S., B.K. Mathur, B.K. Samantaray, and S. Bhattacherjee. 1985. Dehydration and phase transformation in sepiolite. A radial distribution analysis study. *Zeitschrift für Kristallographie* 171: 69–79.

Loucks, R.R. 1991. The bound interlayer H_2O content of potassic white mica: Muscovite-hydromuscovite-hydropyrophyllite solutions. *American Mineralogist* 76: 1563–1579.

Lv, G., L. Liu, Z. Li, L. Liao, and M. Liu. 2012. Probing the interactions between chlorpheniramine and 2:1 phyllosilicates. *Journal of Colloid and Interface Science* 374: 218–225.

Lvov, Y.M. and E. Abdullayev. 2013 Functional polymer-clay nanotube composites with sustained release of chemical agents. *Progress in Polymer Science* 38: 1690–1719.

Ma, C. and R.A. Eggleton. 1999. Cation exchange capacity of kaolinite. *Clays and Clay Minerals* 47: 174–180.

MacEwan, D.M.C. 1948. Complexes of clays with organic compounds. I. Complex formation between montmorillonite and halloysite and certain organic liquids. *Transactions of the Faraday Society* 44: 349–387.

MacEwan, D.M.C. 1961. Montmorillonite minerals. In *The X-ray Identification and Crystal Structures of Clay Minerals*, (Ed.) G. Brown, pp. 143–207. London, UK: Mineralogical Society.

MacEwan, D.M.C. and A. Ruiz-Amil. 1975. Interstratified clay minerals. In *Soil Components*. Inorganic Components, Vol. 2, (Ed.) J.E. Gieseking, pp. 265–334. Berlin, Germany: Springer-Verlag.

MacEwan, D.M.C., A. Ruiz-Amil, and G. Brown. 1961. Interstratified clay minerals. In *The X-ray Identification and Crystal Structures of Clay Minerals*, 2nd revised and extended edition, (Ed.) G. Brown, pp. 393–445. London, UK: Mineralogical Society.

MacEwan, D.M.C. and M.J. Wilson. 1980. Interlayer and intercalation complexes of clay minerals. In *Crystal Structures of Clay Minerals and Their X-ray Identification*, (Eds.) G.W. Brindley and G. Brown, pp. 197–248. London, UK: Mineralogical Society.

Macht, F., K. Eusterhues, G.J. Pronk, and K.U. Totsche. 2011. Specific surface area of clay minerals: Comparison between atomic force microscopy measurements and bulk-gas (N_2) and liquid (EGME) adsorption methods. *Applied Clay Science* 53: 20–26.

Mackenzie, R.C. and S. Caillère. 1975. The thermal characteristics of soils minerals and the use of these characteristics in the qualitative and quantitative determination of clay minerals in soils. In *Soil Components*. Inorganic Components, Vol. 2, (Ed.) J.E. Gieseking, pp. 529–571. Berlin, Germany: Springer-Verlag.

Mackenzie, R.C. and B.D. Mitchell. 1966. Clay mineralogy. *Earth-Science Reviews* 2: 47–91.

Mackintosh, E.E., D.G. Lewis, and D.J. Greenland. 1971. Dodecylammonium-mica complexes – I. Factors affecting the exchange reaction. *Clays and Clay Minerals* 19: 209–218.

Madejová, J. 2003. FTIR techniques in clay mineral studies. *Vibrational Spectroscopy* 31: 1–10.

Maegdefrau, E. and U. Hofmann. 1937. Die Kristalstruktur des Montmorillonits. *Zeitschrift für Kristallographie* 98: 299–323.

Malden, P.J. and R.E. Meads. 1967. Substitution by iron in kaolinite. *Nature* 215: 844–846.

Malek, N.A.N.N. and N.I. Ramli. 2015. Characterization and antibacterial activity of cetylpyridinium bromide (CPB) immobilized on kaolinite with different CPB loadings. *Applied Clay Science* 109–110: 8–14.

Marshall, C.E. 1935. Layer lattices and base-exchanged clays. *Zeitschrift für Kristallographie* 91: 433–449.

Martín, J., M.D.M. Orta, S. Medina-Carrasco, J.L. Santos, I. Aparicio, and E. Alonso. 2018. Removal of priority and emerging pollutants from aqueous media by adsorption onto synthetic organo-functionalized high-charge swelling micas. *Environmental Research* 164: 488–494.

Martín, J., M.D.M. Orta, S. Medina-Carrasco, J.L. Santos, I. Aparicio, and E. Alonso. 2019. Evaluation of a modified mica and montmorillonite for the adsorption of ibuprofen from aqueous media. *Applied Clay Science* 171: 29–37.

Martin, R.T., S.W. Bailey, D.D. Eberl et al. 1991. Report on the clay minerals society nomenclature committee: Revised classification of clay minerals. *Clays and Clay Minerals* 39: 333–335.

Mathieson, A.M.L. and G.F. Walker. 1954. Crystal structure of magnesium-vermiculite. *American Mineralogist* 39: 231–255.

Matus, F., C. Rumpel, R. Neculman, M. Panichini, and M.L. Mora. 2014. Soil carbon storage and stabilization in andic soils: A review. *Catena* 120: 102–110.

Mbey, J.A., F. Thomas, A. Razafitianamah, and F. Villiéras. 2019. A comparative study of some kaolinites surface properties. *Applied Clay Science* 172: 135–145.

McBride, M.B. 1976. Origin and position of exchange sites in kaolinite: An ESR study. *Clays and Clay Minerals* 24: 88–92.

McBride, M.B. 1994. *Environmental Chemistry of Soils*. New York: Oxford University Press.

McCarney, J. and M.V. Smalley. 1995. Electron microscopy study of *n*-butylammonium vermiculite swelling. *Clay Minerals* 30: 187–194.

McDaniel, P.A., D.J. Lowe, O. Arnalds, and C.-L. Ping. 2012. Andisols. In *Handbook of Soil Sciences: Resource Management and Environmental Impacts*, 2nd edition, (Eds.) P.M. Huang, Y. Li, and M.E. Sumner, pp. 33-29–33-48. Boca Raton, FL: CRC Press.

Meunier, A. and B. Velde. 2004. *Illite. Origins, Evolution and Metamorphism.* Berlin, Germany: Springer-Verlag.

Miao, S., Z. Liu, Z. Zhang et al. 2007. Ionic liquid-assisted immobilization of Rh on attapulgite and its application in cyclohexene hydrogenation. *Journal of Physical Chemistry C* 111: 2185–2190.

Moore, D.M. and R.C. Reynolds, Jr. 1997. *X-ray Diffraction and the Identification and Analysis of Clay Minerals*, 2nd edition. Oxford, UK: Oxford University Press.

Morikawa, Y. 1993. Catalysis by metal ions intercalated in layer lattice silicates. In *Advances in Catalysis*, Vol. 39, (Eds.) D.D. Eley, H. Pines, and P.B. Weisz, pp. 302–327. New York: Academic Press.

Moro, D., G. Ulian, and G. Valdrè. 2015. Single molecule investigations of glycine-chlorite interaction by cross-correlated scanning probe microscopy and quantum mechanics simulations. *Langmuir* 31: 4453–4463.

Moro, D., G. Ulian, and G. Valdrè. 2016. Nanoscale cross-correlated AFM, Kelvin probe, elastic modulus and quantum mechanics investigation of clay mineral surfaces: The case of chlorite. *Applied Clay Science* 131: 175–181.

Moro, D., G. Ulian, and G. Valdrè. 2019. Amino acids-clay interaction at the nano-atomic scale: The L-alanine-chlorite system. *Applied Clay Science* 172: 28–39.

Moro, D., G. Ulian, and G. Valdrè. 2020. Nanoscale oligopeptide adsorption behaviour on chlorite as revealed by scanning probe microscopy and density functional simulations. *Applied Clay Science* 197: 105777.

Morrissey, F.A. and M.E. Grismer. 1999. Kinetics of volatile organic compound sorption/desorption on clay minerals. *Journal of Contaminant Hydrology* 36: 291–312.

Mortland, M.M. 1970. Clay-organic complexes and interactions. *Advances in Agronomy* 22: 75–117.

Mouzon, J., I.U. Bhuiyan, and J. Hedlund. 2016. The structure of montmorillonite gels revealed by sequential cryo-XHR-SEM imaging. *Journal of Colloid and Interface Science* 465: 58–66.

Murray, H.H. 2007. *Applied Clay Mineralogy*. Developments in Clay Science, Vol. 2. Amsterdam, The Netherlands: Elsevier.

Nagata, M., S. Shimoda, and T. Sudo. 1974. On dehydration of bound water of sepiolite. *Clays and Clay Minerals* 22: 285–293.

Nagy, B. and W.F. Bradley. 1955. The structural scheme of sepiolite. *American Mineralogist* 40: 885–892.

Nanzyo, M. and H. Kanno. 2018. *Inorganic Constituents in Soil.* Singapore: SingaporeOpen. https://doi.org/10.1007/978-981-13-1214-4_1.

Neumann, B.S. and K.G. Sansom. 1970a. The formation of stable sols from laponite, a synthetic hectorite-like clay. *Clay Minerals* 8: 389–404.

Neumann, B.S. and K.G. Sansom. 1970b. *Laponite* clay – a synthetic inorganic gelling agent for aqueous solutions of polar organic compounds. *Journal of the Society of Cosmetic Chemists* 21: 237–258.

Newman, A.C.D. 1969. Cation exchange properties of micas. I. The relation between mica composition and potassium exchange in solutions of different pH. *Journal of Soil Science* 20: 357–373.

Newman, A.C.D. (Ed.). 1987. *Chemistry of Clays and Clay Minerals.* Monograph 6. London, UK: Mineralogical Society.

Newman, R.H., C.W. Childs, and G.J. Churchman. 1994. Aluminium coordination and structural disorder in halloysite and kaolinite by ^{27}Al NMR spectroscopy. *Clay Minerals* 29: 305–312.

Newnham, R.E. 1961. A refinement of the dickite structure and some remarks on polymorphism in kaolin minerals. *Mineralogical Magazine* 32: 683–704.

Nieto, F., M. Mellini, and I. Abad. 2010. The role of H_3O^+ in the crystal structure of illite. *Clays and Clay Minerals* 58: 238–246.

Niu, J. 2016. Formation mechanisms of tubular structure of halloysite. In *Nanosized Tubular Clay Minerals.* Developments in Clay Science, Vol. 7, (Eds.) P. Yuan, A. Thill, and F. Bergaya, pp. 387–407. Amsterdam, The Netherlands: Elsevier.

Noll, W. and H. Kircher. 1951. Über die Morphologie von Asbesten und ihren Zusammenhang mit Kristalstruktur. *Neues Jahrbuch für Mineralogie Monatshefte* 10: 219–240.

Noll, W., H. Kircher, and W. Sybertz. 1958. Adsorptionsvermögen und spezifische Oberfläche von Silikaten mit röhrenförmig gebauten Primärkristallen. *Kolloid Zeitschrift* 157: 1–11.

Norrish, K. 1973. Factors in the weathering of mica to vermiculite. In *Proceedings of the International Clay Conference 1972*, (Ed.) J.M. Serratosa, pp. 417–432. Madrid, Spain: Division de Ciencias, C.S.I.C.

Ogawa, M., H. Kadomoto, C. Kato, and K. Kuroda. 1997. Intercalation of a crown-ether into layered silicates. *Clay Science* 10: 185–194.

Ogawa, M., H. Kimura, K. Kuroda, and C. Kato. 1996. Intercalation and photochromism of azo dye in the hydrophobic interlayer spaces of organoammonium-fluor-tetrasilicic mica. *Clay Science* 10: 57–65.

Ogawa, M., Y. Nagafusa, K. Kuroda, and C. Kato. 1992. Solid-state intercalation of acrylamide into smectites and Na-taeniolite. *Applied Clay Science* 7: 291–302.

Ogawa, M., T. Matsutomo, and T. Okada. 2008. Preparation of hectorite-like swelling silicate with controlled layer charge density. *Journal of the Ceramic Society of Japan* 116: 1309–1313.

Ogawa, M., M. Tsujimura, and K. Kuroda. 2000. Incorporation of tris (2,2'-bipyridine)ruthenium(II) in a synthetic swelling mica with poly(vinylpyrrolidone). *Langmuir* 16: 4202–4206.

Okada, T., Y. Ehara, and M. Ogawa. 2007. Adsorption of Eu^{3+} to smectites and fluoro-tetrasilicic mica. *Clays and Clay Minerals* 55: 348–353.

Okada, T., T. Morita, and M. Ogawa. 2005. Tris(2,2'-bipyridine)ruthenium(II)-clays as adsorbents for phenol and chlorinated phenols from aqueous solution. *Applied Clay Science* 29: 45–53.

Okada, T., Y. Seki, and M. Ogawa. 2014. Designed nanostructures of clay for controlled adsorption of organic compounds. *Journal of Nanoscience and Nanotechnology* 14: 2121–2134.

Olejnik, S., A.M. Posner, and J.P. Quirk. 1970. The intercalation of polar organic compounds into kaolinite. *Clay Minerals* 8: 421–434.

Olson, B.G., J.J. Decker, S. Nazarenko et al. 2008. Aggregation of synthetic chrysotile nanotubes in the bulk and in solution probed by nitrogen adsorption and viscosity measurements. *Journal of Physical Chemistry C* 112: 12943–12950.

Orta, M.D.M., F.M. Flores, C.F. Morantes, G. Curutchet, and R.M.T. Sánchez. 2019. Interrelations of structure, electric surface charge, and hydrophobicity of organo-mica and – montmorillonite, tailored with quaternary or primary amine cations: Preliminary study of pyrimethanil adsorption. *Materials Chemistry and Physics* 223: 325–335.

Paineau, E. 2018. Imogolite nanotubes: A flexible nanoplatform with multipurpose applications. *Applied Sciences* 8: 1921.

Parfitt, R.L. 1990. Allophane in New Zealand – a review. *Australian Journal of Soil Research* 28: 343–360.

Parfitt, R.L. 2009. Allophane and imogolite: Role in soil biogeochemical processes. *Clay Minerals* 44: 135–155.

Parfitt, R.L., R.J. Furkert, and T. Henmi. 1980. Identification and structure of two types of allophane from volcanic ash soils and tephra. *Clays and Clay Minerals* 28: 328–334.

Parfitt, R.L. and T. Henmi. 1980. Structure of some allophanes from New Zealand. *Clays and Clay Minerals* 28: 285–294.

Parfitt, R.L., B.K.G. Theng, J.S. Whitton, and T.G. Shepherd. 1997. Effects of clay minerals and land use on organic matter pools. *Geoderma* 75: 1–12.

Park, M., D.H. Lee, C.L. Choi, S.S. Kim, K.S. Kim, and J. Choi. 2002. Pure Na-4-mica: Synthesis and characterization. *Chemistry of Materials* 14: 2582–2589.

Park, Y., G.A. Ayoko, and R.L. Frost. 2011. Application of organoclays for the adsorption of recalcitrant organic molecules from aqueous media. *Journal of Colloid and Interface Science* 354: 292–305.

Pasbakhsh, P. and G.J. Churchman (Eds.). 2015. *Natural Mineral Nanotubes*. Oakville, ON: Apple Academic Press.

Pasbakhsh, P., G.J. Churchman, and J.L. Keeling. 2013. Characterisation of properties of various halloysites relevant to their use as nanotubes and microfibre fillers. *Applied Clay Science* 74: 47–57.

Pasbakhsh, P., R. de Silva, V. Vahedi, and G.J. Churchman. 2016. Halloysite nanotubes: Prospects and challenges of their use as additives and carriers – a focused review. *Clay Minerals* 51: 479–487.

Paterson, E. 1977. Specific surface area and pore structure of allophanic soil clays. *Clay Minerals* 12: 1–9.

Pavón, E., M.A. Castro, M. Naranjo, M.M. Orta, M.C Pazos, and M.D. Alba. 2013. Hydration properties of synthetic high-charge micas saturated with different cations: An experimental approach. *American Mineralogist* 98: 394–400.

Pazos, M.C., L.R. Bravo, S.E. Ramos, F.J. Osuna, E. Pavón, and M.D. Alba. 2020. Multiple pollutants removal by functionalized heterostructures based on Na-2-Mica. *Applied Clay Science* 196: 105749.

Pazos, M.C., M.A. Castro, A. Cota, F.J. Osuna, E. Pavón, and M.D. Alba. 2017. New insights into surface-functionalized swelling high charged micas: Their adsorption performance for non-ionic organic pollutants. *Journal of Industrial and Engineering Chemistry* 52: 179–186.

Pazos, M.C., A. Cota, F.J. Osuna, E. Pavón, and M.D. Alba. 2015. Self-assembling of tetradecylammonium chain on swelling high charge micas (Na-mica-3 and Na-mica-2): Effect of alkylammonium concentration and mica layer charge. *Langmuir* 31: 4394–4401.

Perdigón, A.C., C. Pesquera, A. Cota, F.J. Osuna, E. Pavón, and M.D. Alba. 2018. Heteroatom framework distribution and layer charge of sodium tainiolite. *Applied Clay Science* 158: 246–251.

Pérez-Rodríguez, J.L. and C. Maqueda. 2002. Interactions of vermiculites with organic compounds. In *Organo-Clay Complexes and Interactions*, (Eds.) S. Yariv and H. Cross, pp. 113–173. New York: Marcel Dekker.

Pesquera, C., F. Aguado, F. González, C. Blanco, L. Rodríguez, and A.C. Perdigón. 2018. Tunable interlayer hydrophobicity in a nanostructured high charge organo-mica. *Microporous and Mesoporous Materials* 263: 77–85.

Piperno, S., I. Kaplan-Ashiri, S.R. Cohen et al. 2007. Characterization of geoinspired and synthetic chrysotile nanotubes by atomic force microscopy and transmission electron microscopy. *Advanced Functional Materials* 17: 3332–3338.

Plançon, A., R.F. Giese, R. Snyder, V.A. Drits, and A.S. Bookin. 1989. Stacking faults in the kaolin group minerals: Defect structures of kaolinite. *Clays and Clay Minerals* 37: 203–210.

Polubesova, T. and S. Nir. 1999. Modeling of organic and inorganic cation sorption by illite. *Clays and Clay Minerals* 47: 366–374.

Presti, D., A. Pedone, G. Mancini, C. Duce, M.R. Tiné, and V. Barone. 2016. Insights into structural and dynamical features of water at halloysite interfaces probed by DFT and classical molecular dynamics simulations. *Physical Chemistry Chemical Physics* 18: 2164–2174.

Pura, A., I. Dusenkova, and J. Malers. 2014. Adsorption of organic compounds found in human sebum on Latvian illitic, kaolinitic, and chloritic phyllosilicates. *Clays and Clay Minerals* 62: 500–507.

Radoslovich, E.W. 1963. The cell dimension and symmetry of layer-lattice silicates. VI. Serpentine and kaolin morphology. *American Mineralogist* 48: 368–378.

Ramos, M.E. and F.J. Huertas. 2014. Adsorption of lactate and citrate on montmorillonite in aqueous solutions. *Applied Clay Science* 90: 27–34.

Rand, B. and I.E. Melton. 1975. Isoelectric point of the edge surface of kaolinite. *Nature* 257: 214–216.

Ras, R.H.A., Y. Umemura, C.T. Johnston, A. Yamagishi, and R.A. Schoonheydt. 2007. Ultrathin hybrid films of clay minerals. *Physical Chemistry Chemical Physics* 9: 918–932.

Rausell-Colom, J.A. 1964. Small-angle X-ray diffraction study of the swelling of butylammonium vermiculite. *Transactions of the Faraday Society* 60: 190–201.

Rausell-Colom, J.A. and J.M. Serratosa. 1987. Reactions of clays with organic substances. In *Chemistry of Clays and Clay Minerals*. Monograph 6, (Ed.) A.C.D. Newman, pp. 371–422. London, UK: Mineralogical Society.

Rausell-Colom, J.A., T.R. Sweatman, C.B. Wells, and K. Norrish. 1965. Studies in the artificial weathering of mica. In *Experimental Pedology. Proceedings of the 11th Easter School of Agricultural Science, University of Nottingham*, pp. 40–72. London, UK: Butterworth.

Rawtani, D. and Y.K. Agrawal. 2012. Multifarious applications of halloysite nanotubes: A review. *Reviews on Advanced Materials Science* 30: 282–295.

Rayner, J.H. 1974. The crystal structure of phlogopite by neutron diffraction. *Mineralogical Magazine* 13: 73–84.

Rengasamy, P., G.S.R. Krishna Murti, and V.A.K. Sarma. 1975. Isomorphous substitution of iron for aluminium in some soil kaolinites. *Clays and Clay Minerals* 23: 211–214.

Reynolds, R.C. 1980. Interstratified clay minerals. In *Crystal Structures of Clay Minerals and Their X-ray Identification*, (Eds.) G.W. Brindley and G. Brown, pp. 249–303. London, UK: Mineralogical Society.

Rothbauer, R. 1971. Untersuchung eines $2M_1$-Muskovits mit Neutronenstrahlen. *Neues Jahrbuch für Mineralogie-Monatshefte* 143–154.

Rousseaux, J.M. and B.P. Warkentin. 1976. Surface properties and forces holding water in allophane soils. *Soil Science Society of America Journal* 40: 446–451.

Rouxhet, P.G., N. Samudacheata, H. Jacobs, and O. Anton. 1977. Attribution of the OH stretching bands of kaolinite. *Clay Minerals* 12: 171–179.

Roveri, N., G. Falini, E. Foresti, G. Fracasso, I.G. Lesci, and P. Sabatino. 2006. Geoinspired synthetic chrysotile nanotubes. *Journal of Materials Research* 21: 2711–2725.

Roy, D.M. and R. Roy. 1954. An experimental study of the formation and properties of synthetic serpentines and related layer silicate minerals. *American Mineralogist* 39: 957–975.

Rozalén, M., P.V. Brady, and F.J. Huertas. 2009. Surface chemistry of K-montmorillonite: Ionic strength, temperature dependence and dissolution kinetics. *Journal of Colloid and Interface Science* 333: 474–484.

Satouh, S., J. Martín, M. del Mar Orta et al. 2021. Adsorption of polycyclic aromatic hydrocarbons by natural, synthetic and modified clays. *Environments* 8: 124.

Satterberg, J., T.S. Arnarson, E.J. Lessard, and R.G. Keil. 2003. Sorption of organic matter from four phytoplankton species to montmorillonite, chlorite and kaolinite in seawater. *Marine Chemistry* 81: 11–18.

Sawhney, B.L. 1989. Interstratification in layer silicates. In *Minerals in Soil Environments*, 2nd edition, (Eds.) J.B. Dixon and S.B. Weed, pp. 789–828. Madison, WI: Soil Science Society of America.

Schofield, R.K. and H.R. Samson. 1953. The deflocculation of kaolinite suspensions and the accompanying change-over from positive to negative chloride adsorption. *Clay Minerals Bulletin* 2: 45–51.

Schoonheydt, R.A. and C.T. Johnston. 2013. Surface and interface chemistry of clay minerals. In *Handbook of Clay Science*, 2nd edition. Developments in Clay Science, Vol. 5A, (Eds.) F. Bergaya and G. Lagaly, pp. 139–172. Amsterdam, The Netherlands: Elsevier.
Schwertmann, U. 1979. Non-crystalline and accessory minerals. In *International Clay Conference 1978*. Developments in Sedimentology, Vol. 27, (Eds.) M.M. Mortland and V.C. Farmer, pp. 491–499. Amsterdam, The Netherlands: Elsevier.
Seliem, M.K., S. Komarneni, Y. Cho et al. 2011. Organosilicas and organo-clay minerals as sorbents for toluene. *Applied Clay Science* 52: 184–189.
Senkayi, A.L., J.B. Dixon, L.R. Hossner, and L.A. Kippenberger. 1985. Layer charge evaluation of expandable soil clays by an alkylammonium method. *Soil Science Society of America Journal* 49: 1054–1060.
Serna, C.J., J.L. Ahlrichs, and J.M. Serratosa. 1975. Folding in sepiolite crystals. *Clays and Clay Minerals* 23: 452–457.
Serna, C.J. and G.E. Vanscoyoc. 1979. Infrared study of sepiolite and palygorskite surfaces. In *International Clay Conference 1978*. Developments in Sedimentology, Vol. 27, (Eds.) M.M. Mortland and V.C. Farmer, pp. 197–206. Amsterdam, The Netherlands: Elsevier.
Serratosa, J.M., W.D. Johns, and A. Shimoyama. 1970. I.R. study of alkyl-ammonium vermiculite complexes. *Clays and Clay Minerals* 18: 107–113.
Sheng, G., S. Xu, and S.A. Boyd. 1996. Mechanism(s) controlling sorption of neutral organic contaminants by surfactant-derived and natural organic matter. *Environmental Science & Technology* 30: 1553–1557.
Sheng, J., Y. Xie, and Y. Zhou. 2009. Adsorption of methylene blue from aqueous solution on pyrophyllite. *Applied Clay Science* 46: 422–424.
Shirozu, H. and S.W. Bailey. 1966. Crystal structure of a two-layer Mg-vermiculite. *American Mineralogist* 51: 1124–1143.
Singer, A. 1989. Palygorskite and sepiolite group minerals. In *Minerals in Soil Environments*, 2nd edition, (Eds.) J.B. Dixon and S.B. Weed, pp. 829–872. Madison, WI: Soil Science Society of America.
Singh, B. 1996. Why does halloysite roll? – A new model. *Clays and Clay Minerals* 44: 191–196.
Singh, B. and I.D.R. Mackinnon. 1996. Experimental transformation of kaolinite to halloysite. *Clays and Clay Minerals* 44: 825–834.
Slade, P.G., C. Dean, P.K. Schultz, and P.G. Self. 1987. Crystal structure of a vermiculite-anilinium intercalate. *Clays and Clay Minerals* 35: 177–188.
Slade, P.G. and W.P. Gates. 2004. The ordering of HDTMA in the interlayers of vermiculite and the influence of solvents. *Clays and Clay Minerals* 52: 204–210.
Slade, P.G., M. Raupach, and W.W. Emerson. 1978. The ordering of cetylpyridinium bromide on vermiculite. *Clays and Clay Minerals* 26: 125–134.
Solomon, D.H. 1968. Clay minerals as electron acceptors and/or donors in organic reactions. *Clays and Clay Minerals* 16: 31–39.
Solomon, D.H. and D.G. Hawthorne. 1983. *Chemistry of Pigments and Fillers*. New York: John Wiley & Sons.
Soma, M., G.J. Churchman, and B.K.G. Theng. 1992. X-ray photoelectron spectroscopic analysis of halloysites with different composition and particle morphology. *Clay Minerals* 27: 413–421.
Soma, M., A. Tanaka, H. Seyama, S. Hayashi, and K. Hayamizu. 1990. Bonding states of sodium in tetrasilicic sodium fluor mica. *Clay Science* 8: 1–8.
Soma, M. and B.K.G. Theng. 1998. Surface and interlayer chemical compositions, and structure of clay minerals. In *The Latest Frontiers of the Clay Chemistry*, (Eds.) A. Yamagishi, A. Aramata, and M. Taniguchi, pp. 134–143. Sendai, Japan: The Smectite Forum of Japan.
Souza, D.H.S., K. Dahmouche, C.T. Andrade, and M.L. Dias. 2011. Structure, morphology and thermal stability of synthetic fluorine mica and its organic derivatives. *Applied Clay Science* 54: 226–234.
Sposito, G. 1984. *The Surface Chemistry of Soils*. New York: Oxford University Press.
Sprynskyy, M., J. Niedojadło, and B. Buszewski. 2011. Structural features of natural and acids modified chrysotile nanotubes. *Journal of Physics and Chemistry of Solids* 72: 1015–1026.
Środoń, J. and D.D. Eberl. 1984. Illite. In *Micas. Reviews in Mineralogy*, Vol. 13, (Ed.). S.W. Bailey, pp. 495–544. Washington, DC: Mineralogical Society of America.
Steinfink, H., E.J. Weiss, D.J. Haase, and R.A. Rowland. 1963. An X-ray diffraction study of a hexamethylenediaminevermiculite complex. *Proceedings International Clay Conference, Stockholm* 1: 343–348.
Strese, H. and U. Hofmann. 1941. Synthese von Magnesiumsilikat-gelen mit zweidimensional regelmäβiger Struktur. *Zeitschrift für anorganische und allgemeine Chemie* 247: 65–95.
Stucki, J.W. and W.L. Banwart (Eds.). 1980. *Advanced Chemical Methods for Soil and Clay Minerals Research*. Dordrecht, The Netherlands: D. Reidel.

Stul, M.S. and W.J. Mortier. 1974. The heterogeneity of the charge density in montmorillonites. *Clays and Clay Minerals* 22: 391–396.

Stul, M.S. and L. Van Leemput. 1982. Particle-size distribution, cation exchange capacity and charge density of deferrated montmorillonites. *Clay Minerals* 17: 209–215.

Suárez, M., J. García-Rivas, J. Morales, A. Lorenzo, A. García-Vicente, and E. García-Romero. 2022. Review and new data on the surface properties of palygorskite: A comparative study. *Applied Clay Science* 216: 106311.

Sudo, T., S. Shimoda, H. Yotsumoto, and S. Aita. 1981. *Electron Micrographs of Clay Minerals*. Developments in Sedimentology, Vol. 31. Tokyo, Japan: Kodansha and Amsterdam, The Netherlands: Elsevier.

Suitch, P.R. and R.A. Young. 1983. Atom positions in highly ordered kaolinite. *Clays and Clay Minerals* 31: 357–366.

Sun, A., J.-B. d'Espinose de la Caillerie, and J.J. Fripiat. 1995. A new microporous material: Aluminated sepiolite. *Microporous Materials* 5: 135–142.

Susa, K., H. Steinfink, and W.R. Bradley. 1967. The crystal structure of a pyridine-vermiculite complex. *Clay Minerals* 7: 145–153.

Suzuki, T. and Y. Ono. 1984. The conversion of 2-propanol over chrysotile. *Applied Catalysis* 10: 361–368.

Swartzen-Allen, S.L. and E. Matijević. 1974. Surface and colloid chemistry of clays. *Chemical Reviews* 74: 385–400.

Swift, H.E. 1977. Catalytic properties of synthetic layered silicates and aluminosilicates. In *Advanced Materials in Catalysis*, (Eds.) J.J. Burton and R.L. Garten, pp. 209–233. New York: Academic Press.

Takahashi, T., R.A. Dahlgren, B.K.G. Theng, J.S. Whitton, and M. Soma. 2001. Potassium-selective, halloysite-rich soils formed in volcanic materials from Northern California. *Soil Science Society of America Journal* 65: 516–526.

Tamura, K. and H. Nakazawa. 1996. Intercalation of N-alkyltrimethylammonium into swelling fluoro-mica. *Clays and Clay Minerals* 44: 501–505.

Tan, D., P. Yuan, F. Annabi-Bergaya et al. 2014. Loading and *in vitro* release of ibuprofen in tubular halloysite. *Applied Clay Science* 96: 50–55.

Tao, Q., Y. Fang, T. Li et al. 2016. Silylation of saponite with 3-aminopropyltriethoxysilane. *Applied Clay Science* 132–133: 133–139.

Theng, B.K.G. 1971. Mechanisms of formation of colored clay-organic complexes: A review. *Clays and Clay Minerals* 19: 383–390.

Theng, B.K.G. 1972. Adsorption of ammonium and some primary *n*-alkylammonium cations by soil allophane. *Nature* 238: 150–151.

Theng, B.K.G. 1995. On measuring the specific surface area of clays and soils by adsorption of *para*-nitrophenol: Use and limitations. In *Clays: Conrolling the Environment. Proceedings 10th International Clay Conference, Adelaide, Australia, 1993*, (Eds.) G.J. Churchman, R.W. Fitzpatrick, and R.A. Eggleton, pp. 304–310. Melbourne, Australia: CSIRO Publishing.

Theng, B.K.G. 2008. Clay-organic interactions. In *Encyclopedia of Soil Science*, (Ed.) W. Chesworth, pp. 144–150. Dordrecht, The Netherlands: Springer.

Theng, B.K.G. 2012. *Formation and Properties of Clay-Polymer Complexes*, 2nd edition. Amsterdam, The Netherlands: Elsevier.

Theng, B.K.G. 2018. *Clay Mineral Catalysis of Organic Reactions*. Boca Raton, FL: CRC Press.

Theng, B.K.G., G.J. Churchman, J.S. Whitton, and G.G.C. Claridge. 1984. Comparison of intercalation methods for differentiating halloysite from kaolinite. *Clays and Clay Minerals* 32: 249–258.

Theng, B.K.G., D.J. Greenland, and J.P. Quirk. 1967. Adsorption of alkylammonium cations by montmorillonite. *Clay Minerals* 7: 1–17.

Theng, B.K.G., M. Russell, G.J. Churchman, and R.L. Parfitt. 1982. Surface properties of allophane, halloysite, and imogolite. *Clays and Clay Minerals* 30: 143–149.

Theng, B.K.G. and N. Wells. 1995. The flow characteristics of halloysite suspensions. *Clay Minerals* 30: 99–106.

Theng, B.K.G. and G. Yuan. 2008. Nanoparticles in the soil environment. *Elements* 4: 395–399.

Thomas, J.M. 1984. New ways of characterizing layered silicates and their intercalates. *Philosophical Transactions of the Royal Society of London A* 311: 271–285.

Thompson, T.D. and G.W. Brindley. 1969. Absorption of pyrimidines, purines and nucleosides by Na-, Mg-, and Cu(II)-illite. (Clay-organic studies. XVI). *American Mineralogist* 54: 858–868.

Tiller, K.G. and L.H. Smith. 1990. Limitations of EGME retention to estimate the surface area of soils. *Australian Journal of Soil Research* 28: 1–26.

Tombácz, E. and M. Szekeres. 2004. Colloidal behavior of aqueous montmorillonite suspensions: The specific role of pH in the presence of indifferent electrolytes. *Applied Clay Science* 27: 75–94.
Tombácz, E. and M. Szekeres. 2006. Surface charge heterogeneity of kaolinite in aqueous suspension in comparison with montmorillonite. *Applied Clay Science* 34: 105–124.
Tournassat, C., E. Ferrage, C. Poinsignon, and L. Charlet. 2004. The titration of clay minerals. II. Structure-based model and implications for clay reactivity. *Journal of Colloid and Interface Science* 273: 234–246.
Tsagkaropoulou, G., F.J. Allen, S.M. Clarke, and P.J. Camp. 2019. Self-assembly and adsorption of cetyltrimethylammonium bromide and didodecyldimethylammonium bromide surfactants at the mica-water interface. *Soft Matter* 15: 8402–8411.
Tsuchida, H., S. Ooi, K. Nakaishi, and Y. Adachi. 2005. Effects of pH and ionic strength on electrokinetic properties of imogolite. *Colloids and Surfaces A: Physicochemical and Engineering Aspects* 265: 131–134.
Ulian, G., D. Moro, and G. Valdrè. 2020. Infrared and Raman spectroscopic features of clinochlore $Mg_6Si_4O_{10}(OH)_8$: A density functional theory contribution. *Applied Clay Science* 197: 105779.
Uno, H., K. Tamura, H. Yamada, K. Umeyama, T. Hatta, and Y. Moriyoshi. 2009. Preparation and mechanical properties of exfoliated mica-polyamide 6 nanocomposites using sericite mica. *Applied Clay Science* 46: 81–87.
Utracki, L.A. 2007. Interphase between nanoparticles and molten polymeric matrix: Pressure-volume-temperature measurements. *Composite Interfaces* 14: 229–242.
Vahedi-Faridi, A. and S. Guggenheim. 1997. Crystal structure of tetramethylammonium-exchanged vermiculite. *Clays and Clay Minerals* 45: 859–866.
Vahedi-Faridi, A. and S. Guggenheim. 1999a. Structural study of tetrametylphosphonium-exchanged vermiculite. *Clays and Clay Minerals* 47: 219–225.
Vahedi-Faridi, A. and S. Guggenheim. 1999b. Structural study of monomethylammonium and dimethylammonium-exchanged vermiculites. *Clays and Clay Minerals* 47: 338–347.
Valdrè, G. 2007. Natural nanoscale surface potential of clinochlore and its ability to align nucleotides and drive DNA conformational change. *European Journal of Mineralogy* 19: 309–319.
van der Gaast, S.J., K. Wada, S-I. Wada, and Y. Kakuto. 1985. Small-angle X-ray powder diffraction, morphology, and structure of allophane and imogolite. *Clays and Clay Minerals* 33: 237–243.
Vandickelen, R., G. de Roy, and E.F. Vansant. 1980. New Zealand allophanes: A structural study. *Journal of the Chemical Society, Faraday Transactions* 76: 2542–2551.
van Olphen, H. 1971. Amorphous clay materials. *Science* 171: 90–91.
van Olphen, H. 1977. *An Introduction to Clay Colloid Chemistry*, 2nd edition. New York: John Wiley & Sons.
van Olphen, H. and J.J. Fripiat (Eds.). 1979. *Data Handbook for Clay Materials and Other Non-Metallic Minerals*. Oxford, UK: Pergamon Press.
van Scoyoc, G.E., J. Serna, and J.L. Ahlrichs. 1979. Structural changes in palygorskite during dehydration and dehydroxylation. *American Mineralogist* 64: 215–222.
Varghese, S. and J. Karger-Kocsis. 2004. Melt-compounded natural rubber nanocomposites with pristine and organophilic layered silicates of natural and synthetic origin. *Journal of Applied Polymer Science* 91: 813–819.
Vinci, D., B. Dazas, E., Ferrage et al. 2020. Influence of layer charge on hydration properties of synthetic octahedrally-charged Na-saturated trioctahedral swelling phyllosilicates. *Applied Clay Science* 184: 105404.
Vogels, R.J.M.J., J.T. Kloprogge, and J.W. Geus. 2005. Catalytic activity of synthetic saponite clays: Effects of tetrahedral and octahedral composition. *Journal of Catalysis* 231: 443–452.
Wada, K. 1978. Allophane and imogolite. In *Clays and Clay Minerals of Japan*. Developments in Sedimentology, Vol. 26, (Eds.) T. Sudo and S. Shimoda, pp. 147–187. Tokyo: Kodansha and Amsterdam: Elsevier.
Wada, K. 1980. Mineralogical characteristics of Andisols. In *Soils with Variable Charge*, Ed. B.K.G. Theng, pp. 87–107. Lower Hutt: New Zealand Society of Soil Science.
Wada, K. 1989. Allophane and imogolite. In *Minerals in Soil Environments*, 2nd edition, (Eds.) J.B. Dixon and S.B. Weed, pp. 1051–1087. Madison, WI: Soil Science Society of America.
Wada, K. and M.E. Harward. 1974. Amorphous clay constituents. *Advances in Agronomy* 26: 211–260.
Wada, K. and T. Henmi. 1972. Characterization of micropores of imogolite by measuring retention of quaternary ammonium chlorides and water. *Clay Science* 4: 127–136.
Wada, K. and S.-I. Wada. 1977. Density and structure of allophane. *Clay Minerals* 12: 289–298.
Walker, G.F. 1960. Macroscopic swelling of vermiculite crystals in water. *Nature* 187: 312–313.
Walker, G.F. 1967. Interactions of *n*-alkylammonium ions with mica-type layer lattices. *Clay Minerals* 7: 129–143.

Walker, G.F. 1975. Vermiculites. In *Soil Components*. Inorganic Components, Vol. 2, (Ed.) J.E. Gieseking, pp. 155–189. Berlin, Germany: Springer-Verlag.

Wan, J. and T.K. Tokunaga. 2002. Partitioning of clay colloids at air-water interfaces. *Journal of Colloid and Interface Science* 247: 54–61.

Wan, H., A. Yan, H. Xiong et al. 2020. Montmorillonite: A structural evolution from bulk through unilaminar nanolayers to nanotubes. *Applied Clay Science* 194: 105695.

Wang, L., R. Liu, Y. Hu, and W. Sun. 2016. pH effects on adsorption behavior and self-aggregation of dodecylamine at muscovite/aqueous interfaces. *Journal of Molecular Graphics and Modelling*. 67: 62–68.

Wang, S., P. Du, P. Yuan et al. 2020. Structural alterations of synthetic allophane under acidic conditions for understanding the acidification of allophanic Andosols. *Geoderma* 376: 114561.

Wang, W. and A. Wang. 2019. Vermiculite nanomaterials: Structure, properties, and potential applications. In *Nanomaterials from Clay Minerals*, (Eds.) A. Wang and W. Wang, pp. 415–484. Amsterdam, The Netherlands: Elsevier.

Wang, Y., Y. Feng, J. Jiang, and J. Yao. 2019. Designing of recyclable attapulgite for wastewater treatments: A review. *ACS Sustainable Chemical Engineering* 7: 1855–1869.

Weaver, C.E. and L.D. Pollard. 1973. *The Chemistry of Clay Minerals*. Developments in Sedimentology, Vol. 15. Amsterdam, The Netherlands: Elsevier.

Weiss, A. 1963. Mica-type layer silicates with alkylammonium ions. *Clays and Clay Minerals* 10: 191–224.

Weiss, A., A. Mehler, and U. Hofmann. 1956. Kationenaustausch und innerkristallines Quellungsvermögen bei den Mineralen der Glimmergruppe. *Zeitschrift für Naturforschung*, 116: 435–438.

Weiss, A. and J. Russow. 1963. Über die Lage der austauschbaren Kationen bei Kaolinit. In *International Clay Conference 1963*, Vol. 1, (Eds.) I.T. Rosenqvist and P. Graff-Petersen, pp. 203–213. Oxford, UK: Pergamon Press.

Wells, N., C.W. Childs, and C.J. Downes. 1977. Silica Springs, Tongariro National Park, New Zealand – analysis of the spring water and characterization of the alumino-silicate deposit. *Geochimica et Cosmochimica Acta* 41: 1497–1506.

Wells, N. and B.K.G. Theng. 1985. Factors affecting the flow behavior of soil allophane suspensions under low shear rates. *Journal of Colloid and Interface Science* 104: 398–408.

Wells, N., B.K.G. Theng, and G.D. Walker. 1980. Behaviour of imogolite gels under shear. *Clay Science* 5: 257–265.

Whalley, W.R. and C.E. Mullins. 1991. Effect of saturating cation on tactoid size distribution in bentonite suspensions. *Clay Minerals* 26: 11–17.

Wicks, F.J. and E.J.W. Whittaker. 1975. A reappraisal of the structures of the serpentine minerals. *Canadian Mineralogist* 13: 227–243.

Williams-Daryn, S. and R.K. Thomas. 2002. The intercalation of a vermiculite by cationic surfactants and its subsequent swelling with organic solvents. *Journal of Colloid and Interface Science* 255: 303–311.

Wilson, M.A., P.F. Barron, and A.S. Campbell. 1984. Detection of aluminium coordination in soils and clay fractions using ^{27}A1 magic angle spinning N.M.R. *Journal of Soil Science* 35: 201–207.

Wilson, M.A., G.S.H. Lee, and R.C. Taylor. 2002. Benzene displacement on imogolite. *Clays and Clay Minerals* 50: 348–351.

Wilson, M.J. (Ed.). 1994. *Clay Mineralogy: Spectroscopic and Chemical Determinative Methods*. London, UK: Chapman & Hall.

Xu, Y., Y.-L. Liu, S. Gao, Z.-W. Jiang, D. Su, and G.-S. Liu. 2014. Monolayer adsorption of dodecylamine surfactant at the mica/water interface. *Chemical Engineering Science* 114: 58–69.

Xu, Y., Y.-L. Liu, and G.-S. Liu. 2015. Molecular dynamics simulation of primary ammonium ions with different alkyl chains on the muscovite (001) surface. *International Journal of Mineral Processing* 145: 48–56.

Yada, K. 1967. Study of chrysotile asbestos by a high resolution electron microscope. *Acta Crystallographica* 23: 704–707.

Yada, K. 1971. Study of microstructure of chrysotile asbestos by high resolution electron microscopy. *Acta Crystallographica A* 27: 659–664.

Yan, W., D. Liu, D. Tan, P. Yuan, and M. Chen. 2012. FTIR spectroscopy study of the structure changes of palygorskite under heating. *Spectrochimica Acta Part A: Molecular and Biomolecular Spectroscopy* 97: 1052–1057.

Yang, J.-H., Y.-S. Han, J.-H. Choy, and H. Tateyama. 2001. Intercalation of alkylammonium cations into expandable fluorine mica and its application for the evaluation of heterogeneous charge distribution. *Journal of Materials Chemistry* 11: 1305–1312.

Yariv, S. 1992. The effect of tetrahedral substitution of Si by Al on the surface acidity of the oxygen plane of clay minerals. *International Reviews in Physical Chemistry* 11: 345–375.

Yariv, S. and H. Cross (Eds.). 2002. *Organo-Clay Complexes and Interactions*. New York: Marcel Dekker.

Yoshinaga, N. and S. Aomine. 1962. Imogolite in some Ando soils. *Soil Science and Plant Nutrition* 8: 22–29.

Yuan, G.D. and B.K.G. Theng. 2012. Clay-organic interactions in soil environments. In *Handbook of Soil Sciences: Resource Management and Environmental Impacts*, 2nd edition, (Eds.) P.M. Huang, Y. Li, and M.E. Sumner, pp. 2-1–2-20. Boca Raton, FL: CRC Press.

Yuan, G.D., B.K.G. Theng, G.J. Churchman, and W.P. Gates. 2013a. Clays and clay minerals for pollution control. In *Handbook of Clay Science*, 2nd edition. Developments in Clay Science, Vol. 5A, (Eds.) F. Bergaya and G. Lagaly, pp. 587–644. Amsterdam, The Netherlands: Elsevier.

Yuan, G.D., B.K.G. Theng, R.L. Parfitt, and H.J. Percival. 2000. Interactions of allophane with humic acid and cations. *European Journal of Soil Science* 51: 35–41.

Yuan, G.D. and S.-I. Wada. 2012. Allophane and imogolite nanoparticles in soil and their environmental applications. In *Nature's Nanostructures*, (Eds.) A.S. Barnard and H. Guo, pp. 485–508. Singapore: Pan Stanford Publishing.

Yuan, P., D. Tan, and F. Annabi-Bergaya. 2015. Properties and applications of halloysite nanotubes: Recent research advances and future prospects. *Applied Clay Science* 112–113: 75–93.

Yuan, P., D. Tan, F. Annabi-Bergaya, W. Yan, D. Liu, and Z. Liu. 2013b. From platy kaolinite to aluminosilicate nanoroll via one-step delamination of kaolinite: Effect of the temperature of intercalation. *Applied Clay Science* 83–84: 68–76.

Yuan, P., A. Thill, and F. Bergaya (Eds.). 2016. *Nanosized Tubular Clay Minerals: Halloysite and Imogolite*. Developments in Clay Science, Vol. 7. Amsterdam, The Netherlands: Elsevier.

Zeng, Z., D. Matushek, A. Studer, C. Schwickert, R. Pöttgen, and H. Eckert. 2013. Synthesis and characterization of inorganic-organic hybrid materials based on the intercalation of stable organic radicals into a fluoromica clay. *Dalton Transactions* 42: 8585–8596.

Zhang, C., H. He., Q. Tao et al. 2017. Metal occupancy and its influence on thermal stability of synthetic saponites. *Applied Clay Science* 135: 282–288.

Zhang, H., J. Wu, C. Zhang, and Y. Dong. 2023. Water adsorption on kaolinite basal and edge surfaces. *Langmuir* 39: 7539–7547. https://doi.org/10.1021/acs.langmiur.2c03282.

Zhang, J., C.H. Zhou, S. Petit, and H. Zhang. 2019. Hectorite: Synthesis, modification, assembly and applications. *Applied Clay Science* 177: 114–138.

Zhang, Y., A. Tang, H. Yang, and J. Ouyang. 2016. Applications and interfaces of halloysite nanocomposites. *Applied Clay Science* 119: 8–17.

Zhou, C.H., Q. Zhou, Q.Q. Wu et al. 2019. Modification, hybridization and applications of saponite: An overview. *Applied Clay Science* 168: 136–154.

Zhu, R., Q. Chen, Q. Zhou, Y. Xi, J. Zhu, and H. He. 2016a. Adsorbents based on montmorillonite for contaminant removal from water: A review. *Applied Clay Science* 123: 239–258.

Zhu, R., L. Zhu, J. Zhu, and L. Xu. 2008. Structure of cetyltrimethylammonium intercalated hydrobiotite. *Applied Clay Science* 42: 224–231.

Zhu, X., Z. Zhu, X. Lei, and C. Yan. 2016b. Defects in structure as the sources of the surface charges of kaolinite. *Applied Clay Science* 124–125: 127–136.

2 Interactions of Clay Minerals with Uncharged Organic Compounds

2.1 INTRODUCTION

The term *clay minerals* in the title refers to the 2:1 type layer silicates (cf. Table 1.1) among which the smectite group stands out in terms of its affinity for, and reactivity toward, organic molecules. The six representative species in this group are montmorillonite, beidellite, and nontronite which are dioctahedral together with trioctahedral saponite, hectorite, and sauconite. Montmorillonite has held the limelight because of its propensity for adsorbing, intercalating, and transforming a wide range and variety of small and polymeric organic molecules (Mortland 1970; Kowalska et al. 1994; Yariv and Cross 2002; Lagaly et al. 2006; Theng 2012, 2018). In addition, the interlayer space of montmorillonite provides a peculiar micro-environment as regards surface geometry, charge density, and the retention of hydrated counterions (Ballantine et al. 1983; Li et al. 2020).

This chapter describes the interactions of organic compounds with smectite, especially montmorillonite, and to a lesser extent with vermiculite. The limited information on organic complex formation involving other 2:1 type layer silicates, such as pyrophyllite, talc, illite, and mica (natural and synthetic), has been summarized in the relevant sections of Chapter 1.

2.2 ROLE OF LAYER/SURFACE CHARGE DENSITY

In terms of surface reactivity, the site (origin) of isomorphous substitution is as important as, if not more so than, its extent. Substitution of Mg^{2+} for Al^{3+} in the octahedral sheet, as in montmorillonite (cf. Figure 1.11), gives rise to a negative layer charge that is delocalized (over 8–10 oxygen atoms) on each side of the layer structure. On the other hand, the negative charge arising from substitution of Al^{3+} for Si^{4+} in the tetrahedral sheet is largely confined to one side of the layer structure (Farmer 1978; Schoonheydt and Johnston 2013). Depending on the extent of isomorphous substitution, the surface charge density (σ) of the smectite group of phyllosilicates ranges from 0.117 to 0.175 C/m^2, corresponding to a negative charge separation of 0.9–1.35 nm (Środoń and McCarty 2008). The effective charge density, however, is twice as large because the silicate layers are stacked one on top of another, while the negative charge separation is reduced to 0.65–0.84 nm (Johnston 2010; Schoonheydt and Johnston 2013).

The importance of negative layer charge separation to the clay-organic interaction may be illustrated by the competitive adsorption of two divalent cationic herbicides, diquat and paraquat (Figure 2.1), by 2:1 type layer silicates with varying surface charge density (Weed and Weber 1968; Philen et al. 1970, 1971). In the case of montmorillonite, paraquat is preferred to diquat because the negative intercharge distance in the mineral was close to the positive charge separation of 0.7–0.8 nm in the herbicide molecule. As the surface density of charge increases in going from montmorillonite through vermiculite ($\sigma = 0.289$ C/m^2) to mica ($\sigma = 0.321$–0.332 C/m^2), the distance separating pairs of negative charges decreases. Now diquat with a positive charge separation of 0.3–0.4 nm is preferred (to paraquat) because it can readily fit into the negatively charged sites on the interlayer surface of vermiculite as well as on the external surface of mica. The interactions of cationic

DOI: 10.1201/9781003080244-2

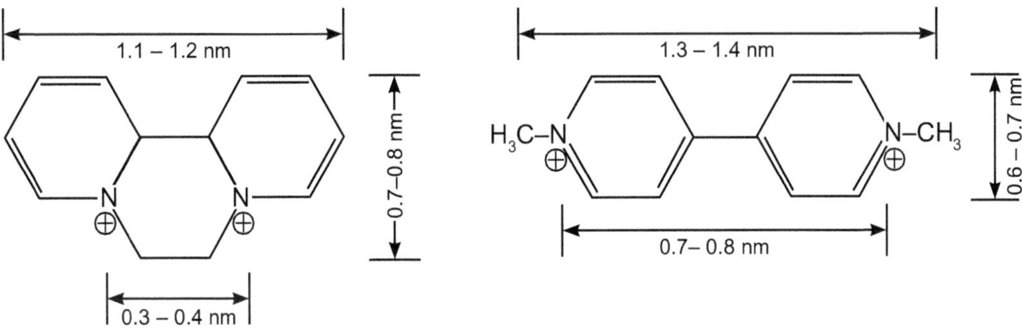

FIGURE 2.1 Molecular dimensions of the diquat (1,1'-ethylene-2,2'-dipyridilium) cation (*left*), and of the paraquat (1,1'-dimethyl-4,4'-dipyridilium) cation (*right*). (From Weed and Weber 1968, Figure 1. Reproduced with permission of the Mineralogical Society of America publisher of *American Mineralogist.*)

porphyrins with synthetic saponite provide a more recent illustration of this 'size-matching' effect/rule (Takagi et al. 2002; Eguchi et al. 2004). These observations support, and are consistent with, the concept that the negatively charged sites on smectite are discrete rather than smeared out over the layer surface (Edwards et al. 1965).

Surface charge density can also influence the orientation of organic molecules in the interlayer space of 2:1 layer silicates. Using infrared spectroscopy, Serratosa (1966) has provided compelling evidence for such an effect based on the marked dichroism (sensitivity to orientation) of certain absorption bands. Thus, the band intensity changes when the angle that the incident beam makes with the normal is varied. For the interlayer complex of montmorillonite with pyridinium, the bands at 677 and 748 cm^{-1} (assigned to out-of-plane vibrations) intensify when the angle of incidence is changed (from 0 to 40°). This observation indicates that the pyridinium ion adopts an orientation with the plane of the ring parallel to the silicate layer, giving a basal spacing of 1.25 nm (Figure 2.2a). On the other hand, in the spectrum of the interlayer vermiculite-pyridinium complex it is the 1491 cm^{-1} band (due to in-plane ring stretching vibration) that intensifies as the incidence angle is increased. Here the pyridinium ion is oriented with the plane of the ring perpendicular to the silicate layers, giving a basal spacing of 1.38 nm (Figure 2.2b).

The reason why intercalated pyridinium cations adopt a 'flat' configuration in montmorillonite and an 'upright' posture in vermiculite may be sought in the difference in surface charge density between the two mineral species which, in turn, determines the space requirement of intercalated species. In the flat configuration, all the pyridinium ions can be accommodated in the interlayer space of montmorillonite at full exchange (Greene-Kelly 1955b; Serratosa 1966). However, if the pyridinium ions in vermiculite were to adopt a flat orientation, only 60–70% of the total amount can be accommodated since the projected area of flat-lying pyridinium (~0.4 nm^2) is substantially larger than that of its upright counterpart (~0.24 nm^2). Thus, by adopting an upright orientation a full complement of the organic cations can be intercalated into vermiculite. Interestingly, the orientation of pyridinium ions in montmorillonite changed from flat to upright following intercalation of chlorobenzene (Serratosa 1968a). This observation may also be explained in terms of the space requirement on the part of the introduced aromatic molecule. Likewise, surface charge density and space requirement have a determining effect on the conformation of intercalated *n*-butylammonium ions. In montmorillonite the alkyl chain adopts a flat orientation, while in vermiculite the chain axis is inclined 55°, with respect to the silicate layers (Martin-Rubi et al. 1974). A similar interlayer chain conformation has been observed for a vermiculite complex with cetylpyridinium bromide (Slade et al. 1978). For a given (single) clay mineral species, the orientation of intercalated organic molecules (e.g., quinoline) may also vary with solute-surface coverage (Helmy et al. 1983).

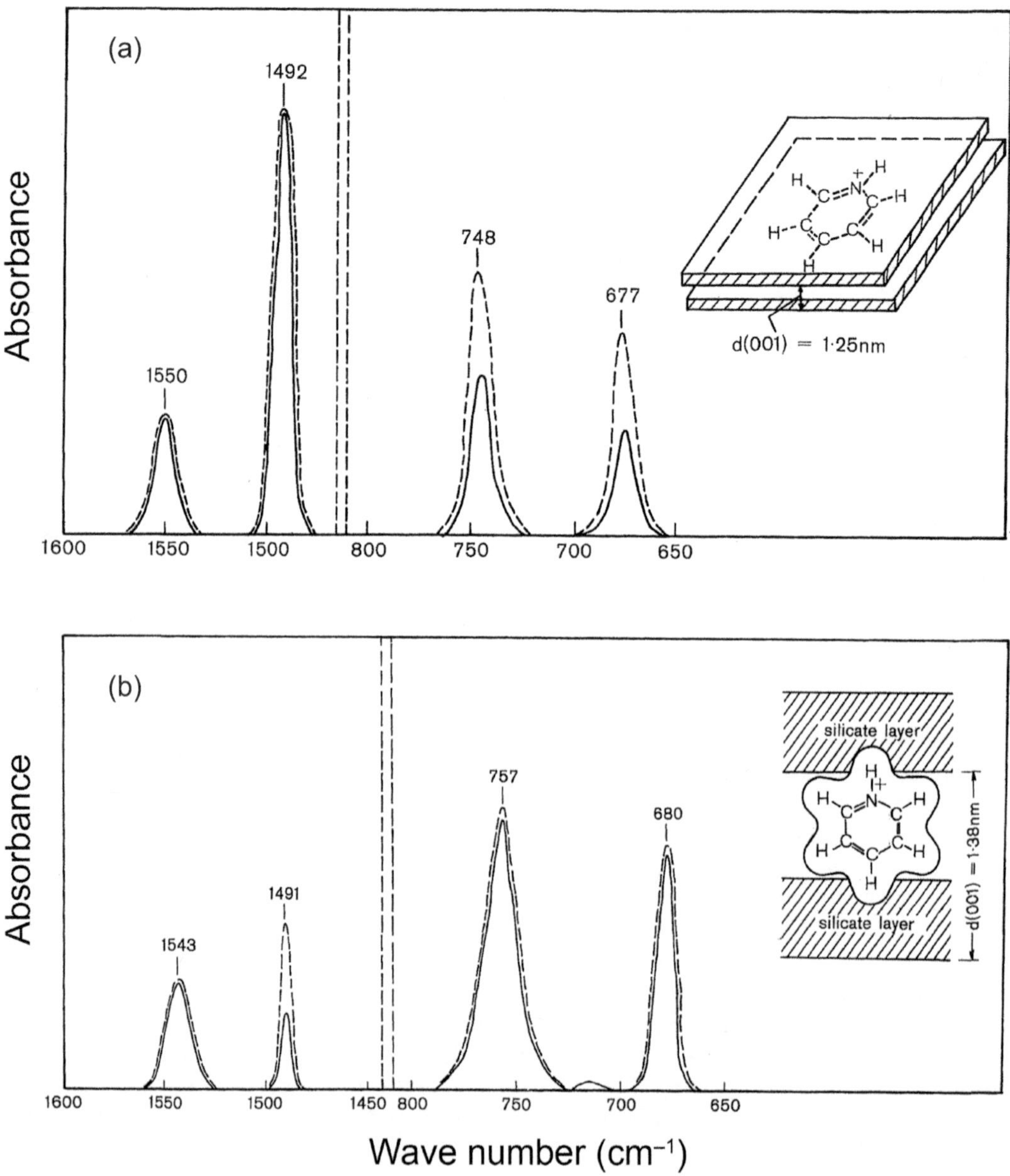

FIGURE 2.2 (a) infrared spectrum of interlayer montmorillonite-pyridinium complex with a basal spacing of 1.25 nm; and (b) spectrum of the corresponding vermiculite complex with a basal spacing of 1.38 nm. The spectra were recorded at two angles of incidence, namely, zero degree (—) and 40° (– – –). (From Serratosa 1966, Figures 3 and 4. Reproduced with kind permission of The Clay Minerals Society, publisher of *Clays and Clay Minerals*.)

A basal spacing of 1.38 nm measured for the fully exchanged complex with vermiculite is consistent with an orientation in which the NH^+ and *para*-CH groups of intercalated pyridinium ions penetrate ('key') into the ditrigonal cavity (~0.26 nm equivalent spherical diameter) in the siloxane surface (Figure 2.2b), causing an apparent shortening of contact distances. Similarly, a basal spacing of 1.37 nm for the interlayer complex of vermiculite with 1,4-diazabicyclo[2,2,2]octane (DABCO) indicates the 'keying' of one amino group of DABCO, and the 'riding' of the other amino group on the opposite siloxane surface (Slade et al. 1989). In a vermiculite-aniline intercalate, however, the nitrogen atoms of the organic cation lie over, and are keyed into, the ditrigonal cavities (Slade and Stone 1983). The keying of intercalated alkylammonium cations in montmorillonite has also been suggested by Theng et al. (1967) (cf. Table 4.1). A rather extreme example of keying is the snug fitting of K^+ into the space between opposing ditrigonal cavities in the mica layer structure (cf. Figure 1.15) as the size of this counterion is comparable with the cavity diameter (Sposito 1984).

The keying effect is generally observed with respect to the intercalation of uncharged polar aromatic and aliphatic compounds into smectite (Greene-Kelly 1955a; Greenland 1956; Brindley and Hoffmann 1962). As a result, the observed interlayer distance ('Δ-value') for the clay-organic complex is less than the normal van der Waals diameter of the intercalated molecule. For interlayer complexes of montmorillonite with aromatic compounds, Greene-Kelly (1955a) suggested that ~0.1 nm be added to the observed basal spacings to give values within a few hundredths of a nanometre of the expected spacing if normal van der Waals contact were to occur.

Both Bradley (1945a) and MacEwan (1948a) have suggested that the methylene group of adsorbed polar organic molecules can form C–H···O hydrogen bonds with surface oxygens of montmorillonite. This process could account for, or at least contribute to, the apparent shortening of contact distances. However, Greene-Kelly (1955a) found that the extent of shortening (~0.05 nm per contact) for small, strongly polar molecules was similar to that for large, weakly polar compounds. This behaviour would not be expected if C–H···O bonding was operative. Subsequent X-ray diffraction and infrared spectroscopy measurements (Greenland 1956; Mortland 1970; Farmer 1971) have also cast doubt on the importance of C–H···O bonding in smectite-organic systems. Nevertheless, density functional theory indicates that this type of interaction could occur with adsorbed quaternary ammonium cations (Scholtzová et al. 2013; Peng et al. 2018).

The orientation and keying of organic molecules are well illustrated by the intercalation into smectite of monovalent tetrathiafulvalene (TTF) through electron transfer to structural (intralayer) Fe^{3+} (Van Damme et al. 1984). At full cation exchange, the interlayer complex with glauconite has a basal [*d*(001)] spacing of 1.675 nm, while a slightly larger value was measured by Soma and Soma (1988a) for the complex with montmorillonite. By taking 0.96 nm as the (van der Waals) thickness of an individual silicate (T-O-T) layer (Figure 2.3), a Δ-value of 0.715 nm is derived. Combined with infrared dichroic measurements, described earlier, this value indicates that intercalated TTF is keyed and adopts an orientation with its short axis perpendicular to the smectite layer. Other examples of organic molecule keying are described as follows.

In 2:1 type layer silicates with little, if any, isomorphous substitution (e.g., pyrophyllite, talc), the basal oxygens of the tetrahedral sheet, making up the 'siloxane' surface, behave as very weak electron donors or Lewis bases (Farmer 1978; Sposito and Prost 1982; Sposito 1984). Being neutral, non-polar, and hydrophobic the siloxane surface of such minerals interacts only weakly with water

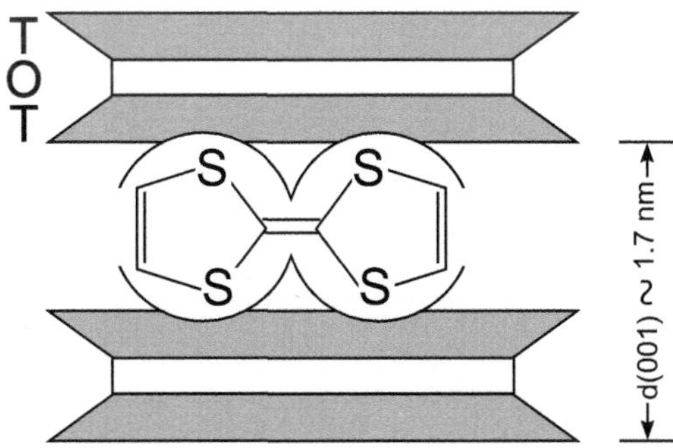

FIGURE 2.3 Diagram showing the orientation and keying of monovalent tetrathiafulvalene cation in the interlayer space of smectite, giving a basal spacing of 1.675 nm and an interlayer separation of 0.715 nm. (After Van Damme, H. et al., *Nouveau J. Chim.*, 8, 681–683, 1984.)

and uncharged polar organic molecules (Schrader and Yariv 1990; Malandrini et al. 1997; Atluri et al. 2019). By comparison, in octahedrally substituted montmorillonite the Lewis base character of the basal surface oxygens is enhanced, allowing the counterions to form stable complexes with both water and polar organic molecules.

More importantly, the 'neutral siloxane surface' exposed between negatively charged sites can bind uncharged organic molecules through a variety of mechanisms, including electron-acceptor complexation, H-bonding together with hydrophobic, and van der Waals interactions (Lee et al. 1990; Jaynes and Boyd 1991a; Weissmahr et al. 1997; Lawrence et al. 1998; Laird and Fleming 1999; Chun et al. 2003; Johnston et al. 2004; Boyd et al. 2011; Zhang et al. 2010, 2011; Zhu et al. 2016). Figure 2.4 illustrates this point for smectites of varying charge density and where the negatively charged sites are occupied by trimethylphenylammonium (TMPA) ions. Being oriented with their phenyl rings perpendicular (or nearly so) to the interlayer siloxane surface, these and similarly compact quaternary ammonium counterions could act as 'pillars', allowing polar and non-polar gases (Barrer and MacLeod 1954, 1955; Önal and Sarikaya 2008) as well as aliphatic and aromatic organic molecules (Diamond and Kinter 1963; Kinter and Diamond 1963; Stevens and Anderson 1996b; Ruan et al. 2008; Zhu et al. 2012, 2014, 2015) to occupy the 'interpillar' space. For example, phenol can adsorb to the exposed siloxane surface in smectite exchanged with tetramethylphosphonium (Lawrence et al. 1998), benzyltrimethylammonium (Shen 2004), hexamethonium (Xu and Zhu 2007), and even cetyltrimethylammonium ions (Zhou et al. 2015).

Thus, the capacity of smectite for adsorbing uncharged organic solutes, such as atrazine (Laird et al. 1992), 4,6-dinitro-*o*-cresol (DNOC) (Sheng et al. 2002), and aflatoxin (Deng et al. 2012;

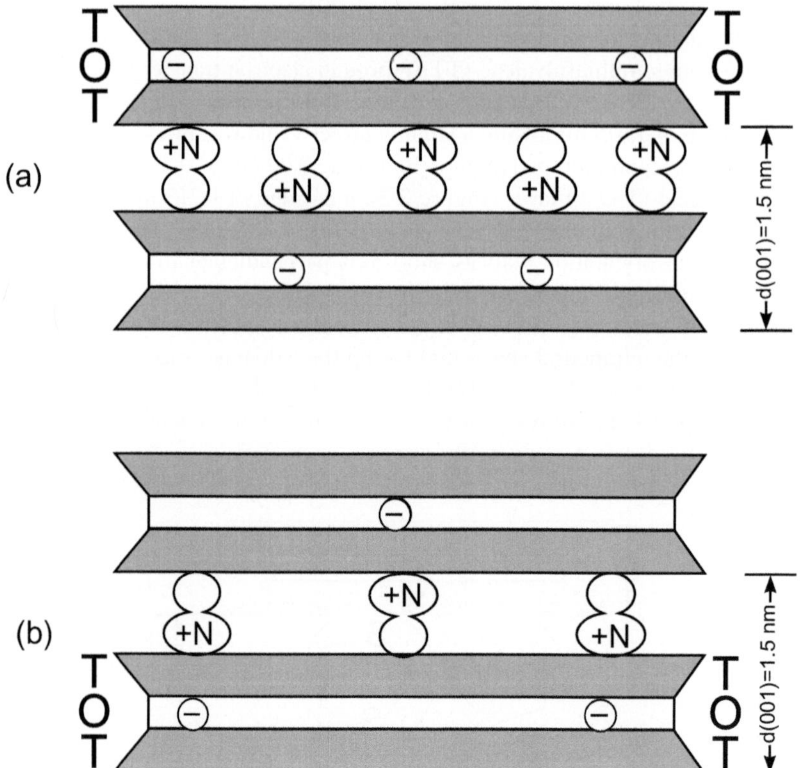

FIGURE 2.4 Diagram showing the intercalation of compact quaternary ammonium cations, such as trimethylphenylammonium (TMPA) into (a) high-charged smectites (SWa, SAz) and (b) low-charged montmorillonite (SAC). In both instances, the basal spacing is 1.5 nm, but the distance separating individual TMPA cations is markedly different. (From Jaynes and Boyd 1991, Figure 2. Reproduced with kind permission of The Clay Minerals Society, publisher of *Clays and Clay Minerals*.)

Barrientos-Velázquez et al. 2016), was inversely related to layer charge density in that the higher the charge density the smaller the extent of the neutral siloxane surface. Using molecular dynamics simulations, Willemsen et al. (2019) also found that the adsorption of phthalate esters to Ca^{2+}-montmorillonite decreased with an increase in surface charge density. The effect of charge density was also apparent during the uptake of phenols by saponite, fluorotetrasilicic mica, and montmorillonite intercalated with tris(2,2'-bipyridine)ruthenium(II) cations (Okada et al. 2005). This effect is also well illustrated by the interaction of biphenyl with reduced-charged montmorillonites that have been exchanged with tetramethylammonium and hexadecyltrimethylammonium ions (Pálková et al. 2009).

As might be expected, steric factors also come into play. For example, appreciably more DNOC can be accommodated (into K^+-montmorillonite) than the bulkier 4,6-dinitro-*sec*-butylphenol (Sheng et al. 2002). Furthermore, DNOC can apparently penetrate the interlayers of a low-charge (SWy-2) Ca^{2+}-montmorillonite quasi-crystal but not those of its high-charge (SAz-1) counterpart because of steric restrictions (Pereira et al. 2008; Figure 2.4). A shape-selective adsorption mechanism has also been observed by Lawrence et al. (1998) for the adsorption of phenol and mono-chlorinated phenols by tetramethylphosphonium-exchanged montmorillonite and by Liu et al. (2015) for the interaction of coplanar and non-planar polychlorinated biphenyls with montmorillonite.

In tetrahedrally substituted saponite where the negative layer charge is highly localized, the exposed (lateral) neutral siloxane surface is relatively large, especially when the small and weakly hydrated Cs^+ serves as the counterion. Thus, Cs^+-saponite shows a high capacity for adsorbing uncharged organic species; even such non-polar molecules as dibenzo-*p*-dioxin and dibenzofuran (Liu et al. 2009; Rana et al. 2009). Similarly, the high uptake of trichloroethylene by Cs^+-saponite, relative to the Ca^{2+}-exchanged counterpart, may be ascribed to the low hydration energy and small hydrated radius of the cesium ion (Table 2.1), and the accompanying increase in the 'lateral' adsorption area, exposed between interlayer Cs^+ ions (Aggarwal et al. 2006).

TABLE 2.1
Ionic Radius, Ionic Potential (z/r), Hydration Enthalpy ($-\Delta H_{hyd}$), and Probable Number of Coordinated Water Molecules (N) for Some Cations

Cation	Radius[1] (pm)	z/r (x 100)[2]	$-\Delta H_{hyd}$ (kJ/mol)[3]	N	$-\Delta H_{hyd}/N$ (kJ)
Li^+	74	1.35	515	4	129
Na^+	102	0.98	405	4	101
K^+	138	0.72	321	4	80
Rb^+	149	0.67	296	4	74
Cs^+	170	0.59	263	4	66
Cu^{2+}	73	2.74	2100	6	350
Mg^{2+}	72	2.78	1922	6	320
Fe^{2+}	78	2.56	1920	6	320
Ca^{2+}	100	2.0	1592	6	265
Sr^{2+}	116	1.72	1445	6	241
Ba^{2+}	136	1.47	1304	6	217
Al^{3+}	53	5.66	4660	6	777
Fe^{3+}	65	4.62	4376	6	729

[1] Taken from Atkins (1982), based on 6-coordination; 1 pm = 0.001 nm
[2] z = cation valency; r = cation radius
[3] Taken from Burgess (1978)
(Reused with permission from *Clay Mineral Catalysis of Organic Reactions*, by B.K.G. Theng, Table 2.7, p. 98, CRC Press, 2018.)

2.3 ROLE OF INTERLAYER WATER

The smectite-organic interaction is also highly influenced by the presence and activity (a_w) of interlayer water. Depending on the nature of the counterions up to four sheets of water molecules may be intercalated over a range of a_w from zero to 0.98 (Prost et al. 1998). At the same time, the basal spacing increases stepwise from ~0.96 nm (for the fully dehydrated mineral) to ~1.9 nm (Ferrage 2016; cf. Table 1.2). The associated interlayer expansion is referred to as restricted ('crystalline') swelling (Slade et al. 1991; Quirk 1994; Laird and Shang 1997; Laird 2006). The extent of interlayer hydration can profoundly affect the ability of smectites to take up polar organic molecules, such as atrazine (Chappell et al. 2005).

We might add here that Li^+- and Na^+-montmorillonites can swell in water to give interlayer separations of some tens of nanometres (Norrish 1954). Under optimum conditions, the process can lead to delamination of the particle ('quasi-crystal') into individual silicate layers. This type of expansion is referred to as extensive ('osmotic') swelling (cf. Table 1.2). Interestingly, Na^+-montmorillonite can also show osmotic swelling in formamide and N-methyl formamide, while Cs^+-montmorillonite can do the same in formamide (Olejnik et al. 1974).

Yotsuji et al. (2021) used molecular dynamics (MD) simulation to assess the (stepwise) crystalline swelling of Na^+-, K^+-, Cs^+-, Ca^{2+}-, and Sr^{2+}-exchanged montmorillonites. The results confirm that the free energy of hydration of the counterion controls interlayer expansion. In the case of NH_4^+-montmorillonite, MD simulation indicates that at a low water content the NH_4^+counterion forms an inner-sphere complex with water as well as N–H···O hydrogen bonds to the siloxane surface (Peng et al. 2020). Using MD simulation, Shen et al. (2018) also found that charge density has a marked influence on the hydration characteristics of montmorillonite saturated with tetramethylammonium (TMA) ions. As charge density increases the water molecules are more attracted to the siloxane surface than to the TMA counterions.

Infrared (IR) spectroscopy has served as the single, most useful technique for probing the nature and properties of surface-adsorbed and interlayer water in clay minerals. The large volume of literature on this topic has been summarized by Farmer (1978) and Johnston (2010, 2017), while the application of different instrumental and computational techniques to characterizing the clay-water interaction has been reviewed by Wang et al. (2021).

Thus, IR spectroscopy has provided strong evidence to show that, like water, uncharged polar organic molecules are either directly coordinated to the exchangeable cation or indirectly linked to the counterion by means of (bridging) water molecules. Many examples of each type of cation-dipole interactions are described in the relevant sections that follow. What is important to appreciate here is that polar organic compounds must successfully compete with water for essentially the same ligand positions or sites around the exchangeable cation for adsorption to occur. All things being equal, the water in the outer coordination spheres (type II) of the cation is less strongly held than that in the primary hydration shell (type I). Many polar molecules such as pyridine, nitrobenzene, and benzoic acid can therefore displace type II water in montmorillonite at ambient humidity. If the saturating cation is weakly polarizing, such as Na^+ and Ba^{2+}, type I water may similarly be displaced.

The tendency of interlayer water molecules to coordinate to, and form hydration shells around, the counterions may be ascribed to the high enthalpies of ion hydration, ranging from −263 to −515 kJ/mol (for alkali metal ions) and from −1304 to −1922 kJ/mol (for alkaline earth metal ions) (Burgess 1978; Table 2.1). By comparison, the enthalpy of hydration for small organic cations (e.g., TMA) is much lower (Johnston et al. 2004). Thus, it is the presence of hydrated counterions that makes the surface of naturally occurring smectites hydrophilic (Yamamoto et al. 2018) and hence ineffective in taking up small uncharged organic molecules from water (Figure 2.6). Infrared and NMR spectroscopy studies by Pálková et al. (2015) have confirmed that TMA- and hexadecyltrimethylammonium (HDTMA)-exchanged montmorillonite are more hydrophobic than the Li-saturated counterpart.

Figure 2.5 shows the structures of one- and two-sheet hydrates of Cu^{2+}-hectorite as deduced from XRD analysis and (polarized) electron spin resonance (ESR) spectroscopy (Clementz et al. 1973;

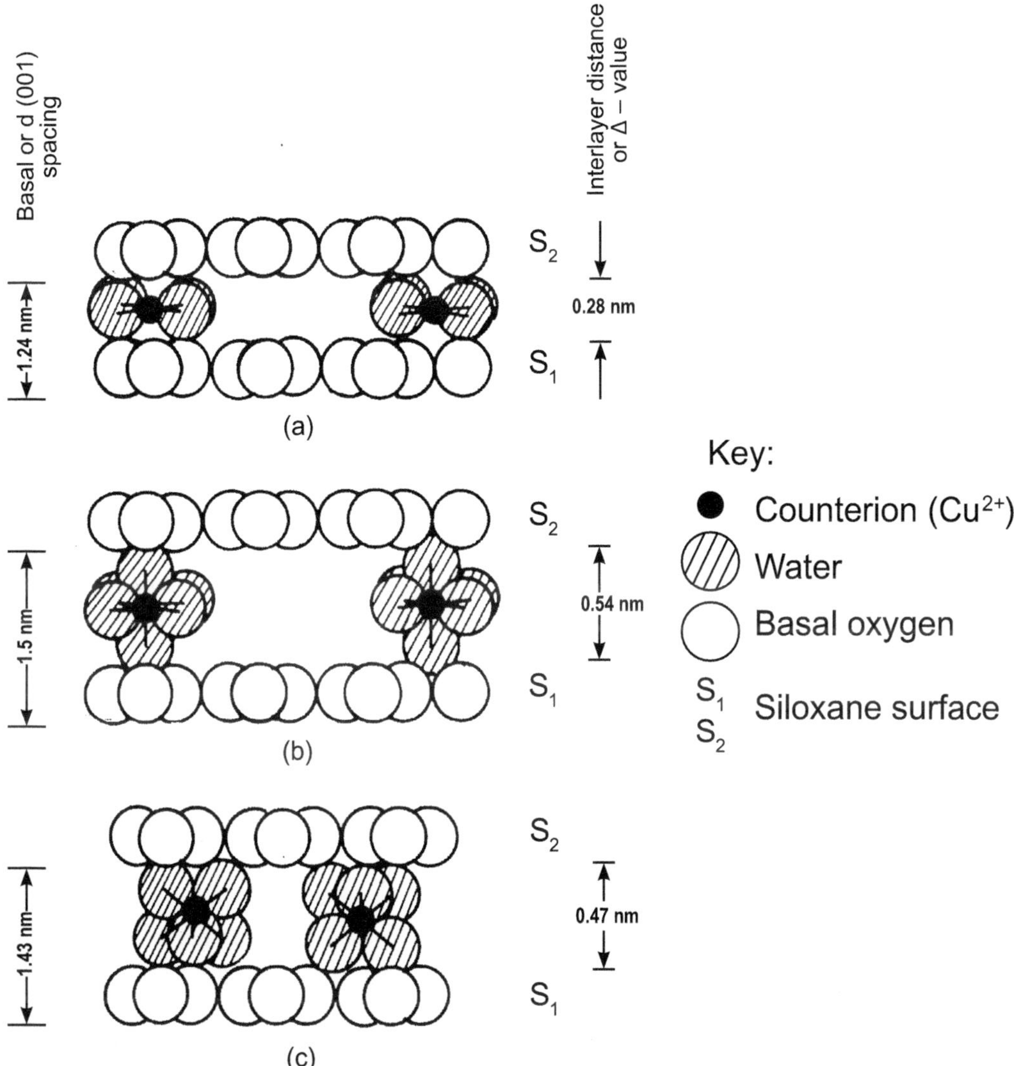

FIGURE 2.5 Diagram showing the arrangement of: (a) interlayer $Cu(H_2O)_4^{2+}$ counterions in a one-sheet hydrate of hectorite; (b) interlayer $Cu(H_2O)_6^{2+}$ counterions in a two-sheet hydrate of hectorite; (c) interlayer $Cu(H_2O)_6^{2+}$ counterions in a two-sheet hydrate of vermiculite. Interlayer distance ('Δ-value') is obtained by subtracting 0.96 nm from the corresponding basal spacing value. (Sources: Clementz, D.M. et al., *J. Phys. Chem.*, 77, 196–200, 1973; McBride, M.B. et al., *J. Phys. Chem.*, 79, 2430–2435, 1975)

McBride et al. 1975). In the one-sheet hydrate, each Cu^{2+} ion is coordinated to four water molecules in a square planar configuration, forming an *inner-sphere* (coordination) complex with a basal spacing of 1.24 nm and a Δ-value of 0.28 nm (Figure 2.5a). Here the $Cu(H_2O)_4^{2+}$ plane lies parallel to the silicate layers and the copper ion is directly bonded to the siloxane surface (without the intervention of water molecules).

ESR spectroscopy indicated that pyridine, glycine, and β-alanine similarly formed planar or axially elongated tetragonal complexes on intercalation into Cu(II)-montmorillonite (Nagai et al. 1974). Interestingly, the divalent tetragonally distorted complexes of Cu(II) with diethylenetriamine and tetraethylenepentamine in hectorite could partially transform into their corresponding planar configurations (Schoonheydt et al. 1979). The formation of a square planar inner-sphere complex of Cu(II) with ehylenediamine in montmorillonite, showing a basal spacing of ~1.27 nm, was reported by Burba and McAtee (1977). By comparison, a chelation complex of Cu(II) with 1,10-phenanthroline in hectorite gave a basal spacing of 1.74 nm (Berkheiser and Mortland 1977). The formation of a square planar four-coordinate Cu(II) complex with methanol (and ethanol) has also been reported by Brown and Kevan (1988). Bissada et al. (1967) have proposed similarly for a Ca^{2+}-ethanol complex in montmorillonite as did Annabi-Bergaya et al. (1981) for a complex with methanol.

On the other hand, in a two-sheet hydrate (Figure 2.5b) each interlayer Cu^{2+} ion is octahedrally coordinated to six water molecules. In this instance, the cation-water octahedron is oriented with its long symmetry axis perpendicular to the silicate layers, giving a basal spacing of 1.50 nm and a Δ-value of 0.54 nm. Here the water molecules serve both as ligands to the counterions and as dielectric bridges between cation and the siloxane surface. Since the counterions now interact with the surface through a *water bridge*, such a configuration is referred to as an *outer-sphere* complex. A similar silicate layer-counterion arrangement applies to the two-sheet hydrate of Ca^{2+}-montmorillonite (Sposito 1984). Molecular dynamics simulation studies also indicate that Na^+, Ca^{2+}, and Sr^{2+} counterions give rise to outer-sphere complexes while inner-sphere complex formation is favoured with K^+ and Cs^+ (Yotsuji et al. 2021).

With rising isomorphous substitution in the tetrahedral sheet, as in vermiculite, the distance separating the negatively charged sites decreases while the layer charge is increasingly localized to those surface oxygens near the sites of substitution. As a result, the electron-donating character (Lewis base strength) of the basal siloxane surface is highly enhanced (Yariv 1992). In a two-sheet hydrate of Cu^{2+}-vermiculite, for example, this effect induces the $Cu(H_2O)_6^{2+}$ complex to adopt an orientation in which its ligand axis is inclined at a high angle (ca. 45°) to the silicate layers, giving a basal spacing of 1.43 nm and a Δ-value of 0.47 nm (Clementz et al. 1974). Figure 2.5c shows that all six water ligands can be hydrogen-bonded to surface oxygens, allowing the central Cu^{2+} ion to get closer to the negatively charged site than would otherwise be possible.

We might interpolate here that the water molecules surrounding interlayer counterions become increasingly polarized when smectites are dehydrated to a water content below ~5% w/w and much of the water of hydration is removed. Infrared spectroscopy has indicated that the cation-coordinated water molecules are then less H-bonded than in bulk water (Xu et al. 2000). Under these conditions, the forces of polarization act on the remaining water molecules in the inner coordination sphere (hydration shell) of the cation, causing them to dissociate. Indeed, the degree of dissociation of cation-coordinated water can be several orders of magnitude greater than that of bulk water (Fripiat et al. 1965; Touillaux et al. 1968; Liu et al. 2011).

The enhanced dissociation of these 'residual' water molecules, $(H_2O)_n$, may be represented by Equation 2.1,

$$[M(H_2O)_n]^{z+} \rightleftharpoons [M(H_2O)_{n-1}(OH)]^{(z-1)+} + H^+ \qquad \text{(Equation 2.1)}$$

where M represents a metal counterion of valency z, and n is an integer. Thus, the higher the valency (z) and the smaller the radius (r) of the counterion, the greater is its polarizing effect (Mortland 1968a; Mortland and Raman 1968; Russell et al. 1968). Similarly, Liu et al. (2013a) reported that the proton-donating capacity (Brønsted acidity) of cation-exchanged smectites was positively correlated with the ionic potential (z/r ratio) of the counterion (Table 2.1). The protons – in reality hydronium ions – derived in this manner can protonate a variety of adsorbed organic bases (B) as indicated by Equation 2.2. More importantly, such protons can catalyze the conversion and transformation of various organic species (Theng 2018). The number and variety of adsorbed and intercalated organic bases that can be protonated in this manner are listed in Table 2.2.

$$H^+ + B \rightleftharpoons BH^+ \qquad \text{(Equation 2.2)}$$

TABLE 2.2
Protonation of some Organic Species (Bases) adsorbed on Montmorillonite Surfaces

Organic species (in alphabetical order)	Exchangeable cation (counterion)	Remark	Reference
Alkenes	Al^{3+}, Cr^{3+}, Cu^{2+}, Fe^{3+}	Formation of carbocations, yielding either ethers/alcohols or dimers/oligomers	Adams et al. (1979, 1981); Adams and Clapp (1985)
Amides	H^+ (H^+/Al^{3+})[1]	Protonation commonly occurs on the oxygen atom, and influenced by substituent group	Tahoun and Mortland (1966a); Cloos (1972); Stutzmann and Siffert (1977)
Amines	various	With some amines, protonation may yield colored surface complexes	Swoboda and Kunze (1968); Yariv et al. (1968); Santos et al. (1970); Yariv and Heller (1970); Laura and Cloos (1975a, 1975b); Pusino et al. (1989); Šnircová et al. (2009)
Aminopyridine	Ca^{2+}, K^+		Ikhsan et al. (2005)
3-Aminotriazole	various	Extent of protonation increases with polarizing power of cation	Russell et al. (1968); Morillo et al. (1991)
Asulam[1]	H^+ (H^+/Al^{3+})[1]		Fusi et al. (1980)
Atrazine	Fe^{3+}		Herwig et al. (2001)
3-Butylpyridine	Ca^{2+}		Laird and Fleming (1999)
2,6-Dichlorothio-benzamide	Al^{3+}	No protonation with Ca^{2+}-montmorillonite	Ristori et al. (1987)
Fenarimol[2]	various	Decomposed when heated at 100 °C	Fusi et al. (1983)
Fluazifop[3]	Cu^{2+}, Fe^{3+}, Al^{3+}	Protonation of pyridine-nitrogen	Micera etal. (1988)
Histidine	various	Concomitant formation of cation hydroxide	Heller-Kallai et al. (1973)
Porphyrins and metalloporphyrins	various	Metalloporphyrins may also be demetallated	Cady and Pinnavaia (1978); Van Damme et al. (1978); Abdo et al. (1980)
Pyridine	Mg^{2+}	Concomitant formation of $Mg(OH)_2$	Farmer and Mortland (1966); Swoboda and Kunze (1968)
Thiabendazole	Na^+		Gamba et al. (2017)
s-Triazines	various	Hydrolysis often follows protonation with loss of herbicidal activity	Cruz et al. (1968); Russell et al. (1968); Brown and White (1969); White (1976)
Triphenylcarbinol	Ca^{2+}, Mg^{2+}	Formation of triphenylcarbonium ion	Fripiat et al. (1964); Tarasevich and Doroshenko (1983)
Urea	H^+ (H^+/Al^{3+})[4]		Mortland (1966)

Note: The active protons either derive from dissociation of water molecules, coordinated to exchangeable cations, or introduced into exchange sites by acid washing/activation

[1] 4-amino benzensulphonyl-methylcarbamate

[2] α-(2-chlorophenyl)-α-(4-chlorophenyl)-5-pyrimidinemethanol

[3] (RS)-2-[4-[[5-trinuoromethyl)-2-pyridin-y]oxy]phenoxy]propanoic acid

[4] exchange complex of acid-washed/activated montmorillonite is occupied by both protons and aluminium ions

The important conclusion that has emerged from studies on the clay-water interaction is that the properties of interlayer water are affected more by the counterions than the silicate surface. Like water, uncharged organic molecules can either be directly coordinated to the cation or indirectly linked to the cation through a water bridge (Yariv et al. 1966; Mortland 1970; Johnston 2010). The respective mechanisms are referred to as *ion-dipole interactions* and *water-bridging*.

For example, nitrobenzene cannot easily displace the more polar water molecule that is directly coordinated to Mg^{2+} counterions in montmorillonite. Nitrobenzene, however, can readily form a water bridge complex with Mg^{2+}-montmorillonite as does benzoic acid (Yariv et al. 1966), pyridine (Farmer and Mortland 1966), and benzonitrile (Serratosa 1968b). The interaction of montmorillonite with benzonitrile is further described.

On the other hand, in montmorillonite saturated with weakly hydrated counterions (NH_4^+, K^+, Cs^+), the polar nitro group of nitrobenzene can coordinate directly to the counterion. The same applies to 1,3- and 1,4-dinitrobenzenes (Qu et al. 2011). Ion-dipole interactions can also occur with montmorillonite that has previously been dehydrated, irrespective of the nature of the counterion (Yariv et al. 1966; Farmer 1978; Li et al. 2004). Similar findings have been reported for ethyl *N,N*-di-*n*-propylthiolcarbamate (Mortland and Meggitt 1966), pyridine *N*-oxide (Olejnik et al. 1973), dimethyl sulfoxide (Garwood and Condrate 1978), 3-aminotriazole (Morillo et al. 1991), dinitrophenols (Johnston et al. 2002; Sheng et al. 2002), nitroaromatics (Haderlein et al. 1996; Johnston et al. 2001, 2004; Chen et al. 2008), di- and tri-nitrobenzenes (Boyd et al. 2001; Chen and Huang 2009; Qu et al. 2011; Zhang et al. 2011), and caffeine (Yamamoto et al. 2018).

We can therefore say that the interactions of smectite with uncharged polar organic molecules are influenced by the valency and size (radius) of the counterion and the hydration status of the clay mineral (and exchangeable cation) as well as by the basicity and orientation (packing) of the intercalated organic species (Farmer and Mortland 1966; Kellomäki et al. 1987; Nguyen et al. 1987; Stevens and Anderson 1996a; Johnston et al. 2004; Liu et al. 2012). Prior dehydration of the clay mineral, and introducing the organic compound as a vapour, or in an organic solvent, would therefore be conducive to inner-sphere complex formation through ion-dipole interactions (Yariv et al. 1966; Serratosa 1968b; Yamanaka et al. 1974; Fusi et al. 1982). By the same token, the solid-solid reaction, in which dry smectite is blended with the solid form of the organic compound, is an effective and facile means of forming interlayer clay-organic complexes (Khaorapapong and Ogawa 2011). Molecular dynamic simulation, however, would indicate that hydration can assist the intercalation of even non-polar uncharged organic molecules, such as benzene (Liu et al. 2013b). Similarly, Ca^{2+}-montmorillonite (previously heated at 120°C) can take up more benzene than its Na^+-exchanged counterpart because the former material has a relatively large interlayer space/volume (Deng et al. 2017). On the other hand, Morozov et al. (2014) experimentally found that hydration tended to inhibit hydrocarbon adsorption. Hatta et al. (2011) also reported that the basal spacings of complexes with sugar alcohols (erythritol, xylitol, sorbitol), formed in the presence of water ('wet intercalation') were about 0.4 nm larger than their counterparts obtained by the 'dry' solid-solid reaction.

2.4 ADSORPTION FROM AQUEOUS SOLUTIONS

In keeping with the concept that uncharged organic molecules compete with water for ligand positions around the exchangeable cation, little or no adsorption of short-chain compounds by montmorillonite occurs from dilute (< 0.5 mol/l) aqueous solutions. For a range of non-ionic aliphatic compounds, intercalating into Ca^{2+}-montmorillonite, Hoffmann and Brindley (1960) found that a chain length of at least six carbon atoms (units) was required for appreciable adsorption to occur (Figure 2.6). Dékány et al. (1996b) have observed similarly with respect to the adsorption of nitrobenzene and *n*-pentanol. The effect of chain length extends to about ten units, beyond which it becomes less pronounced. As we will see later, however, there are exceptions to the simple chain length 'rule'. Nevertheless, adsorption may be promoted by raising the solute concentration, eliminating water from the system (MacEwan 1948b; Brindley and Rustom 1958; Hoffmann and Brindley

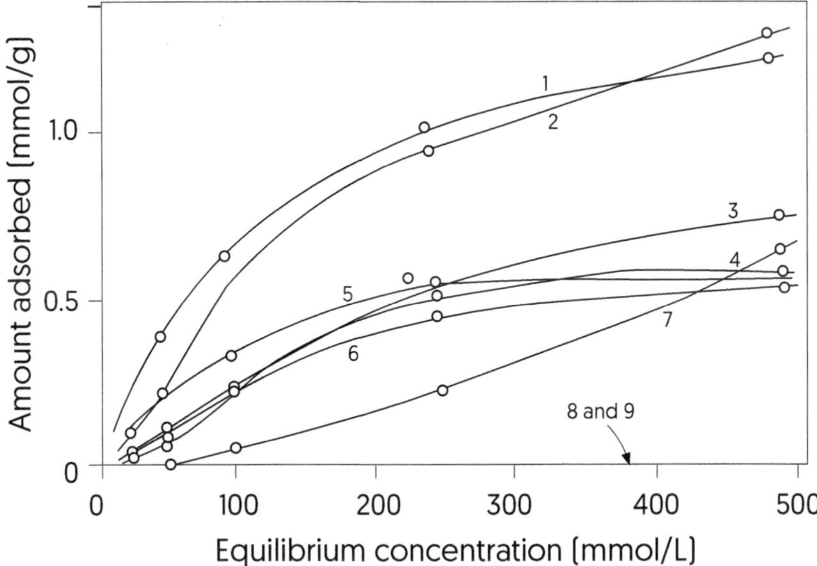

FIGURE 2.6 Diagram showing the influence of chain length on the adsorption of some polar organic molecules from aqueous solution to montmorillonite. (1) 2,5,8-nonanetrione; (2) bis(2-ethoxyethyl) ether; (3) 2,5-hexanedione; (4) bis(2-methoxyethyl) ether; (5) ethylene glycol diglycidyl ether; (6) triethylene glycol; (7) 1,6-hexanediol; (8) diethylene glycol; (9) 1,5-pentanediol. (From Hoffmann, R.W. and Brindley, G.W., *Geochim. Cosmochim. Acta* 20, 15–29, 1960.)

1960; Svensson and Hansen 2010), or introducing the organic compound as a liquid or vapour (MacEwan 1948a; Dowdy and Mortland 1967; German and Harding 1971; Breen et al. 1987b; Chiou and Rutherford 1997; Onikata et al. 1999; Mosser-Ruck et al. 2005). Thus, ethylene glycol, glycerol, and pentanediol-1:5, none of which shows preferential uptake over water, can form stable interlayer complexes with montmorillonite by evaporating the aqueous clay suspension containing these substances (Brindley and Rustom 1958; Hoffmann and Brindley 1961). Removal of the more volatile water, leaving behind the clay and the organic compound, may be likened to immersing the montmorillonite sample in the corresponding organic liquid (MacEwan 1948b).

On the other hand, aliphatic and aromatic molecules with more than five units may be taken up to an appreciable extent in the presence of excess water (Bradley 1945a; Doehler and Young 1962; Bower 1963; German and Harding 1969) by displacing the water molecules associated with the exchangeable cations. The increase in affinity with chain length (molecular size) may be ascribed to the enhanced contribution of van der Waals forces to the overall adsorption energy (Hendricks 1941; Theng et al. 1967). As the size of the solute molecule increases, these forces become important because they are essentially additive, orienting the molecule to maximize the number of solute-surface contacts (De Boer and Heller 1937; Hamaker 1937).

Since the adsorption of a single organic molecule is often accompanied by the desorption of several (cation-coordinated) water molecules, an appreciable amount of entropy is gained by the system. Indeed, the adsorption of uncharged linear polymers to montmorillonite is largely an 'entropy-driven' process (Theng 1982, 2012). Besides chain length (molecular size) the chemical character of the organic molecule influences adsorption behaviour. If the molecule is sufficiently large to be adsorbed from dilute aqueous solution the adsorption process is determined by its 'character'; that is, the presence or absence of certain structural groupings.

For many aliphatic chain molecules, a useful index of character is their *CH activity* denoting the activation of methylene groups by neighbouring electron-withdrawing structures, such as C=O and C≡N (Hoffmann and Brindley 1960). Thus, molecules with many carbonyl or nitrile groups

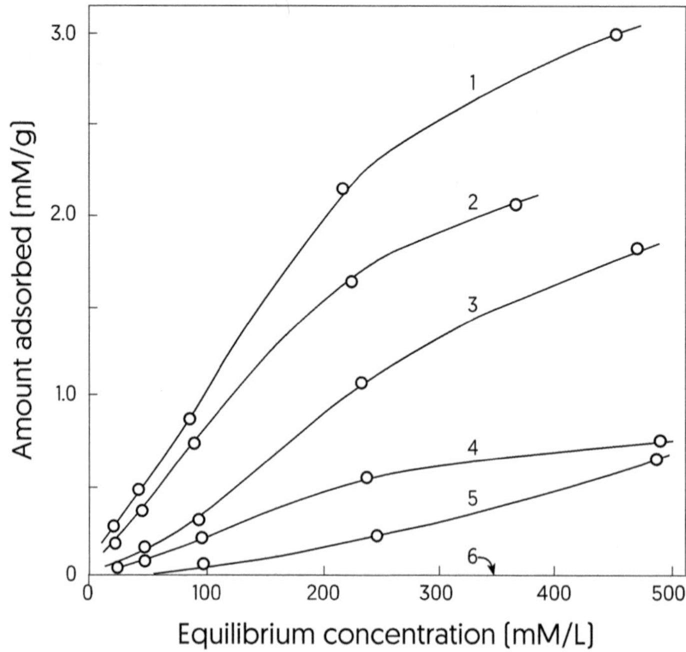

FIGURE 2.7 Diagram showing the influence of chemical character on the adsorption to montmorillonite of some polar organic molecules with a chain length of 6–7 units from aqueous solution. (1) α-methoxy acetylacetone; (2) ethyl acetoacetate; (3) β-ethoxypropionitrile; (4) 2,5-hexanedione; (5) 1,6-hexanediol; (6) 2,4-hexadiynediol-1,6. (From Hoffmann, R.W. and Brindley, G.W. *Geochim. Cosmochim. Acta*, 20, 15–29, 1960.)

adjacent to methylene groups would have a high CH activity, and hence be more strongly adsorbed than compounds of comparable length but where such groupings are few or absent (Figure 2.7). Although CH activity is related to the polarity by which it is induced, the polarity of the molecule *per se*, as measured by its respective dipole moment, does not appear to be closely and consistently correlated with adsorption behaviour.

Hoffmann and Brindley (1960) further noted that chain length and CH activity appeared to exert their influence independently. Thus pentanediol-1:5 ($OHCH_2–CH_2–CH_2–CH_2–CH_2OH$) was not measurably adsorbed by Ca^{2+}-montmorillonite from dilute aqueous solution (Figure 2.6) whereas acetylacetone ($CH_3–CO–CH_2–CO–CH_3$) with high CH activity was taken up under similar conditions. By the same token, low CH activity may be overcome by an increase in chain length. Thus diethyleneglycol [$HOCH_2CH_2(OCH_2CH_2)OH$], like pentanediol-1:5, showed no preferential uptake over water but triethyleneglycol [$HOCH_2CH_2(OCH_2CH_2)_2OH$] with six units to the molecule was appreciably adsorbed.

Tensmeyer et al. (1960) have compared the infrared spectra of 2,5,8-nonanetrione [$CH_3CO(CH_2)_2CO(CH_2)_2COCH_3$] and 2,5-hexanedione [$CH_3CO(CH_2)_2COCH_3$], dissolved in carbon disulphide and carbon tetrachloride, with the spectra of the compounds in the solid state and when adsorbed to Ca^{2+}-montmorillonite. Since each methyl/methylene group was 'activated' by an adjacent carbonyl group, both compounds were strongly adsorbed by the clay mineral (Figure 2.7). No decrease in C–H stretching frequency, however, was observed; rather, the spectrum of intercalated 2,5,8-nonanetrione closely resembled that of its solid-state counterpart. Thus, intermolecular interactions are apparently more important than those operating between the organic molecule and the clay surface, and CH activity is related more to the organic-organic than the clay-organic interaction. Similar conclusions have been made by Mortland (1966) and Farmer and Ahlrichs (1969) for the montmorillonite-urea interaction.

2.5 MOLECULAR ORIENTATION

The orientation and spatial arrangement of polar organic molecules in the interlayer space of montmorillonite and vermiculite can ideally be deduced from X-ray diffraction data using one-dimensional electron density projections (Fourier synthesis) based on the *00l* reflections of the respective complexes. Thus, the arrangement of ethylene glycol in the interlayers of these minerals as well as the way the molecules are related to one another and to the silicate surface, have been reasonably well determined (Walker 1958; Brindley and Hoffmann 1962; Bradley et al. 1963; Reynolds 1965). Similar Fourier 'sketches' have been drawn for complexes of montmorillonite with some aromatic compounds, such as benzidine (Bradley 1945b; Greene-Kelly 1955a). Likewise, Steinfink and co-workers (Haase et al. 1963; Steinfink et al. 1963) have used two- and three-dimensional Fourier syntheses to locate the position of hexamethylene diamine molecules and pyridinium ions in the interlayers of vermiculite. Their results emphasize the limitations of X-ray diffraction methods in studies of this kind. The vermiculite-organic interaction is inherently amenable to being characterized by such methods because macrocrystalline forms of this clay mineral are available. On the other hand, the small size of montmorillonite quasi-crystals seldom permits sufficient resolution for structural analysis to be carried out. Hence, the orientation and arrangement of interlayer organic molecules in montmorillonite has mostly been deduced from basal spacing measurements (Greene-Kelly 1955a; MacEwan 1961; Brindley and Hoffmann 1962; Greenland 1965). In this respect, dichroic/polarised infrared spectroscopy has been of great value as illustrated in Figure 2.2 for the interlayer complex with pyridinium ions.

Using polarised IR attenuated total reflectance, Raupach and Janik (1976) reported that ornithine adopted an almost flat orientation, forming two adjacent layers, in the interlayer space of vermiculite. On the other hand, 6-aminohexanoic acid formed a single-layer complex with the plane of the carbon chain tilted at an angle of 34° to the silicate surface. As would be expected, the orientation of organic molecules, such as acrylonitrile, is also influenced by the amount of adsorbed water and molecule-surface interactions (Yamanaka et al. 1975). We might add here that the orientation of dye molecules in smectite interlayers has received particular attention because the associated clay-dye hybrid materials can have superior non-linear optical properties (Suzuki et al. 2012).

From an analysis of basal spacing data on single-layer montmorillonite-organic complexes, Brindley and Hoffmann (1962) have concluded that aliphatic chain molecules, adsorbed with their shortest axis perpendicular to the silicate surface, may adopt two kinds of orientation. The first type, designated α_I, refers to an interlayer arrangement in which the carbon zig-zag plane is perpendicular to the silicate layer. Molecules that adopt an α_I orientation give rise to complexes with a basal spacing between 1.325 and 1.365 nm. In the second type of orientation, referred to as α_{II}, the carbon zig-zag plane is parallel to the silicate surface, yielding complexes with basal spacings of 1.30–1.31 nm. Uncharged organic compounds having strongly polar groups, such as ketones and esters, tend to adopt an α_{II} orientation with the dipole parallel to the surface. On the other hand, aliphatic chain molecules without such groups (e.g., alcohols) tend to favour the α_I arrangement.

An analogous situation applies to aromatic ring compounds. An orientation in which the plane of the ring is perpendicular to the silicate layer may be regarded as α_I, while the arrangement in which this plane is parallel to the siloxane surface may be likened to an α_{II} orientation. Saturated ring molecules with a puckered or (non-planar) zig-zag structure, such as piperidine, are more difficult to classify in this way, and a 'long-hand' description of their interlayer orientation may be required. Saturated ring compounds often adopt the α_I arrangement (Greene-Kelly 1956). On the other hand, unsaturated ring molecules tend to adopt an α_{II} configuration (Brindley and Hoffmann 1962), giving basal spacings that are 0.08–0.10 nm less than those measured for complexes with an α_I arrangement.

The type of orientation adopted by many aromatic compounds is also dependent on their concentration. When the concentration is low the α_{II} arrangement is favoured. Rearrangement of the molecules to give an α_I orientation, as indicated by an increase in basal spacing, may occur at high concentrations (Greene-Kelly 1955b). Thus, montmorillonite saturated with nitrobenzene and pyridine gives a basal spacing of 1.52 and 1.48 nm, respectively, which on heating to 100°C (to evaporate a part of the intercalated compound) reduces to 1.25 nm. The low-spacing complex with pyridine is also produced when the molecule is introduced as the cation (Figure 2.2a). Sometimes a simple change of solvent polarity can alter the orientation of adsorbed organic compounds as Eguchi et al. (2006) have observed for dicationic porphyrins intercalated in (synthetic) saponite, using polarized absorption spectroscopy.

2.6 COMPLEX FORMATION WITH VARIOUS CLASSES OF UNCHARGED ORGANIC COMPOUNDS

2.6.1 Complexes with Primary *n*-Alcohols

The adsorption of primary *n*-alcohols (C_2–C_9) from dilute aqueous 'solutions' (< 0.6 mol/l) by calcium- and sodium-exchanged montmorillonite has been investigated by German and Harding (1969). The finding that ethanol, *n*-propanol, and *n*-butanol were adsorbed to an appreciable extent (~ 3.5–5 mol/kg) seemed to be at variance with the earlier observations (Hoffmann and Brindley 1960) that polar organic compounds with chain lengths less than six units were not measurably adsorbed (Figure 2.6).

German and Harding (1969) did not explain this apparent departure from the simple chain-length requirement 'rule'. Infrared spectroscopy, however, has shown that ethanol, at least, is capable of displacing water in the primary hydration shell around exchangeable calcium, copper(II), and aluminium cations (Dowdy and Mortland 1967). This ability of ethanol, and presumably of *n*-propanol and *n*-butanol, to compete successfully for ligand sites around the counterion may partly account for its positive adsorption from an aqueous solution.

MacEwan (1948b) was the first to investigate complex formation between NH_4^+-montmorillonite and a homologous series of primary *n*-alcohols. Except for methanol and ethanol which gave a double-layer complex with a *d*(001) spacing of ~1.7 nm, primary aliphatic alcohols formed single-layer complexes (*d*(001) = 1.35–1.40 nm) with the alkyl chain lying parallel to the silicate layer. Subsequently, Barshad (1952) reported similarly for Ca^{2+}-montmorillonite but also observed that *n*-nonanol (C_9) and *n*-decanol (C_{10}) gave 'long-spacing' complexes with *d*(001) > 2.5 nm (Figure 2.8).

Brindley and Ray (1964) reported that complex formation between Ca^{2+}-montmorillonite and even-numbered *n*-alcohols (C_2 to C_{18}) gave rise to four series of basal spacings. The two series of short-spacing complexes corresponded to intercalation of a single (*d*(001) ~1.4 nm) and a double layer (*d*(001) ~1.7 nm) of alcohol molecules, oriented with their alkyl chain parallel to the silicate surface, as MacEwan (1948b) had reported earlier. In addition, two series of long-spacing complexes were obtained for C_6 to C_{18}, one above and one below the melting point of the respective pure alcohols (Figure 2.9). The temperature at which the transition from one to the other type of complex occurred coincided within 2–4°C with the melting point of the alcohol used.

We might interpolate that the intercalation of long-chain molecules, including *n*-alcohols, often proceeds by a stepwise replacement process. For example, Brindley and Ray (1964) used the interlayer complex of Ca^{2+}-montmorillonite to obtain the hexanol complex which, in turn, served as a base for intercalating longer chain alcohols up to *n*-octadecanol.

The basal spacing values suggested that the alcohol molecules here were intercalated with their alkyl chain oriented away from, and making a high angle with, the silicate surface. Similarly, Weiss (1969) obtained long-spacing complexes by reacting an identical set of *n*-alcohols (C_6–C_{18}) with four smectite samples but failed to observe the formation of (short-spacing) single- and double-layer complexes.

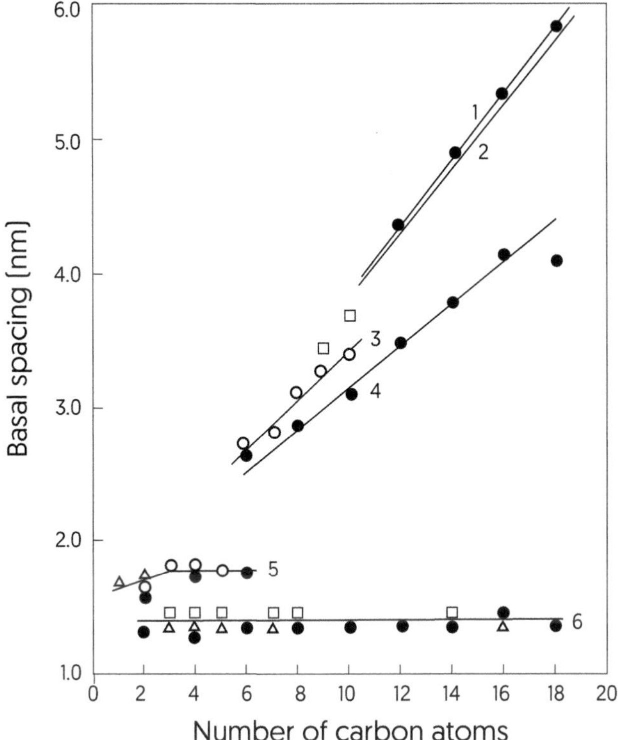

FIGURE 2.8 Relationship between the basal spacing of interlayer Ca^{2+}-montmorillonite complexes with primary *n*-alcohols and number of carbon atoms in the alcohol molecules. Lines 1, 2, 3, and 4 are 'long-spacing' complexes, while lines 5 and 6 are referred to as 'short-spacing' complexes. (Sources: [•], Brindley, G.W. and Ray, S., *Am. Mineral.*, 49, 106–115, 1964; [○], German, W.L. and Harding, D.A., *Clay Miner.*, 9, 167–175, 1971; [Δ], MacEwan, D.M.C. *Trans. Faraday Soc.*, 44, 349–367, 1948; [□], Barshad, I., *Soil Sci. Soc. Am. Proc.*, 16, 176–182, 1952.)

By adopting suitable models for the arrangement of the intercalated organic molecules and making certain assumptions, an equation to fit the observed basal spacing data can be derived. As an example, the *d*(001) spacing for the 'below-melting-point' series of long spacings increases by increments of 0.248 nm per carbon atom in the complex whereas the corresponding value in an alkyl chain is 0.127 nm. This discrepancy may be ascribed to intercalation of a double layer of molecules with their alkyl chain inclined at a high angle (77°) to the silicate surface. Assuming further that the terminal methyl groups of one alcohol layer make van der Waals contact with those of the opposite organic layer, and taking the generally accepted values of bond length and bond angle, the basal spacing may be derived from the Equation 2.3 (Brindley and Ray 1964),

$$d(001) = 1.916 + (n-2)0.248 \text{ nm} \qquad \text{(Equation 2.3)}$$

where *n* is the number of carbon atoms in the molecule.

Equation 2.3 corresponds to line 1 in Figure 2.8. Similar equations have been developed to fit lines 2, 3, and 4 by making certain assumptions as to the position of the carbon zig-zag plane in relation to the silicate surface, the inclination angle of the alkyl chain, the possibility of H-bonding between the hydroxyl group of the alcohol and siloxane surface oxygens, and bond rotation around (terminal) OC_I (Emerson 1957; German and Harding 1971). On the other hand, short-chain alcohols

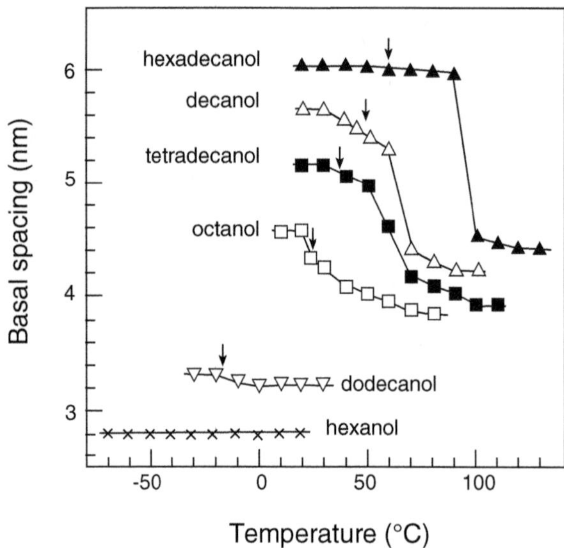

FIGURE 2.9 Basal spacings of interlayer Ba^{2+}-nontronite complexes with some long-chain *n*-alkanols as a function of temperature. Arrows indicate the melting point of the respective pure compounds. (From Pfirrmann et al. 1973, Figure 1a. Reproduced with kind permission of The Clay Minerals Society, publisher of *Clays and Clay Minerals.*)

(C_6 and below) tend to form a double layer of flatly intercalated molecules (Figure 2.8; lines 5 and 6) presumably because in this orientation maximum molecule-to-molecule and molecule-to-surface interactions can be achieved. By the same token, single-layer complexes with short-chain alcohols are more difficult to prepare than the corresponding double-layer complexes (German and Harding 1971).

Following Brindley and Ray (1964), Pfirrman et al. (1973) measured the basal spacings of interlayer complexes between long-chain alcohols and nontronite, saturated with Li^+, K^+, Mg^{2+}, Ca^{2+}, Sr^{2+}, and Ba^{2+}, at temperatures from –70°C to 130°C (Figure 2.9). In the low temperature range (~20°C below the melting point of the alcohols), the intercalated molecules formed a double layer with their chain axes perpendicular to the silicate layers but with some sharp bends ('kinks'), giving *d*(001) spacings between ~3 and ~6 nm (Figure 2.8). The transition from a low to a high-temperature form involved only a small change in basal spacing (~0.12 nm) and probably accompanied by the increased formation of gentle bends ('jogs') in the chain.

To account for the decrease in basal spacing of the complexes as the temperature is raised from below to above the melting point of the corresponding pure alcohols, Brindley and Ray (1964) have offered the following explanation. While the alcohol molecules may still be attached to the silicate surface by O–H···O bonds, a rise in temperature would impart a greater mobility to the adsorbed compounds owing to an increase in thermal energy and the weakening of intermolecular van der Waals attraction. This, coupled with the tendency of the silicate layers to contract owing to cation-surface attraction, would result in bending of the molecules and, hence, a decrease in inclination angle. Alternatively, Pfirrmann et al. (1973) have proposed that chain rearrangement was caused by a cooperative transition of structures with low 'kinks' into conformations dominated by 'jogs'.

At this point, we wish to make the following remarks. Firstly, the interlayer cation exerts a profound influence on complex formation. Indeed, for short-chain *n*-alcohols (e.g., ethanol), the nature of the counterion determines the amount adsorbed and the type of complex formed (either a single or a double layer of flatly oriented molecules). The saturating cation is less important in determining the adsorption of long-chain (> C_6) *n*-alcohols. In forming double layers with steeply inclined hydrocarbon chains in the interlayer space, van der Waals attraction between adjacent alkyl chains are more important than cation-dipole or H-bonding interactions (Laby and Walker 1970; Lagaly et al. 2006) although Beall and Goss (2004) have argued for the importance of ion-dipole interactions. Thus, the adsorption of *n*-hexanol and higher *n*-alcohols from a non-aqueous medium tends to yield long-spacing complexes (Brindley and Ray 1964; Weiss 1969; German and Harding 1971; Pfirrmann et al. 1973) whereas intercalation from water usually give rise to short-spacing double-layer complexes with the alkyl chains lying parallel to the silicate surface (German and Harding 1969).

We might also mention that the rate of entry into the interlayer region, particularly of long-chain alcohol molecules (C_7–C_{10}), is influenced by whether the solute is presented as a vapour or a liquid (German and Harding 1971). This is another manifestation of the concentration effect in that the frequency of solute-mineral collisions is less in the clay-alcohol vapour system than in the corresponding liquid system. Furthermore, the vapour pressure of the alcohol at ambient temperature decreases with an increase in molecular weight.

The second remark concerns the formation of O–H···O–Si hydrogen bonds between adsorbed alcohol molecules and siloxane surface oxygens as Emerson (1957) has suggested. Although such bonding is clearly possible, there is good evidence to indicate that, at least for short-chain *n*-alcohols, cation-dipole interactions are more important than solute-surface H-bonding in the formation of interlayer montmorillonite-alcohol complexes (Dowdy and Mortland 1967; Bissada et al. 1967).

Thirdly, differences in basal spacing for the above-melting-point series of complexes between the values reported by Brindley and Ray (1964) and those obtained by German and Harding (1971) (Figure 2.8, lines 3 and 4) are too large to ascribe to general uncertainties in the van der Waals radii of the constituent atoms. The cause of such discrepancies in these and similar systems, such as complexes with long-chain amines (described later) is yet to be clarified. Subtle variations in the amount and site of isomorphous substitution between montmorillonite samples (Stul and Uytterhoeven 1975) as well as in methods of complex preparation may conceivably lie at the source of such small, but real, differences in the data obtained by various workers.

Figure 2.8, line 5 shows that the basal spacing of the two-layer (flat orientation) complexes with methanol (*d*(001) ~1.65 nm) and ethanol (*d*(001) ~1.68 nm) is smaller than that obtained with *n*-propanol to *n*-hexanol (~1.78 nm). Using an assumed model, Emerson (1957) calculated that the basal spacing of a two-layer methanol complex would be 1.60 nm (the corresponding value of a single-layer complex is 1.27 nm). On this basis, the observed *d*(001) spacing is greater than the theoretical value and this is contrary to the general rule that contact distance is shorter than the normal van der Waals thickness of the adsorbed molecule (cf. Chapter 2). This analysis provides further evidence against O–H···O bonding.

The importance of cation-dipole interactions in interlayer complexes of montmorillonite with small polar molecules is further supported by the data of Bissada et al. (1967) on the adsorption of ethanol and acetone (from *n*-dodecane) by oven-dry (225°C) montmorillonite, exchanged with Na^+, K^+, Ca^{2+}, and Ba^{2+} cations. The number of organic molecules, associated with each counterion, at maximum adsorption together with the X-ray diffraction data are summarized in Table 2.3 where the smaller (2–3) and the larger (8–10) of these numbers correspond to single- and double-layer complexes, respectively.

Bissada et al. (1967) have proposed that ethanol is intercalated with its plane of symmetry perpendicular to the silicate surface while acetone has its double bond parallel, and the plane of symmetry containing both methyl groups, perpendicular to the surface. In this orientation the van der Waals cross-sectional thickness is ~0.50 nm for ethanol and ~0.61 nm for acetone. Since the measured Δ-value is 0.42 and 0.41nm, respectively, while O–H···O and C–H···O interactions seem unlikely, both molecules are apparently keyed into the ditrigonal depressions of the silicate surface.

It is noteworthy that despite the difference in dipole moment (μ_p) between ethanol (1.69 D) and acetone (2.88 D) as well as in molecular structure (the possibility of O–H···O–Si hydrogen bonding, for example, does not exist for acetone), the number of organic molecules associated with a particular counterion is essentially the same for both compounds. Ethanol and acetone, like water, can thus

TABLE 2.3
Isotherm and Basal Spacing Data for the Intercalation of Ethanol and Acetone into Montmorillonite Exchanged with Different Inorganic Cations.

	Ethanol				Acetone			
			Amount adsorbed				Amount adsorbed	
Exchangeable cation (counterion)	Basal spacing (nm)	Number of molecular layers	g/g clay	molecules/cation	Basal spacing (nm)	Number of molecular layers	g/g clay	molecules/cation
K^+	1.36	single	0.088	2	1.34	single	0.096	2
Na^+	1.35	single	0.125	3	1.32	single	0.174	3
Ba^{2+}	1.72	double	0.160	8	1.73	double	0.198	8
Ca^{2+}	1.73	double	0.215	10	1.73	double	0.203	8

Source: Bissada, K.K. et al. *Clay Miner.*, 7, 155–166, 1967

be said to 'solvate' the interlayer counterion. The solvation energy, the coordination number, and hence the stability of the complex, increase in the order $K^+ < Na^+ < Ba^{2+} < Ca^{2+}$. These observations strongly indicate that cation-dipole interactions have a determining influence on complex formation.

MacEwan's (1948b) failure to obtain short- and long-spacing double-layer complexes of C_3–C_7 *n*-alcohols with NH_4^+-montmorillonite may thus be ascribed to weak cation-dipole interactions. As Dowdy and Mortland (1967) have shown by infrared spectroscopy, the behaviour of NH_4^+-montmorillonite towards ethanol is very similar to that of the Na^+-exchanged clay. Nevertheless, highly polar organic compounds, such as cyclic organic carbonates, can form strong ion-dipole interactions even with Na^+ counterions in montmorillonite (Gates et al. 2016).

Figure 2.8, line 5 shows that C_3–C_6 *n*-alcohols, like methanol and ethanol, give two-layer complexes with Ca^{2+}-montmorillonite. The basal spacing of these complexes (1.72–1.78 nm) however, is about 0.1 nm larger than that of the complex with methanol or ethanol. For an orientation, as described for ethanol, and assuming normal van der Waals contact, the expected basal spacing is ~1.85 nm. The amount of contraction (0.03–0.05 nm per clay/organic contact) is consistent with what is generally observed for complexes with polar organic molecules. Thus, geometrical packing (keying) rather than chemical bonding lies behind the shortening of contact distances (Brindley and Hoffmann 1962).

German and Harding (1971) have suggested that the increase in basal spacing of Ca^{2+}-montmorillonite complexes with C_3–C_5 *n*-alcohols is due to the increased contribution of van der Waals interactions between alcohol molecules and the silicate surface. Steric factors, however, must also influence the extent of intercalation. The interlayer space may possibly just fit two layers of five-fold planar groups of *n*-propanol around calcium ions. This coordination (solvation) number, however, cannot be realized for *n*-butanol and *n*-pentanol without a considerable amount of steric overlap if the model of Bissada et al. (1967) for ethanol is adopted. Interestingly, butanol, hexanol, and octanol can apparently self-associate when they intercalate into alkylammonium-exchanged montmorillonite (Stul et al. 1979), while vermiculite exchanged with octadecylammonium ions can expand to a basal spacing of 3.87 nm in dilute solutions of *n*-pentanol (Dékány et al. 1996a).

The formation of single-layer complexes between Ca^{2+}-montmorillonite and C_2–C_{18} *n*-alcohols with the alkyl chain lying parallel to the silicate surface (Figure 2.8, line 6) has also been reported by many workers (MacEwan 1948a; Barshad 1952; Glaeser 1954; Brindley and Ray 1964; German and Harding 1971). Since some difficulty attaches to their formation, double-layer complexes where the alcohol molecules adopt an orientation with their hydrocarbon chain either parallel or steeply inclined to the silicate layer, are preferentially formed. Brindley and Ray (1964) were able to obtain single-layer complexes with even-numbered C_2–C_{18} *n*-alcohols by removing part of the initially adsorbed alcohol. Similarly, German and Harding (1971) had to evaporate excess liquid from the corresponding double-layer complexes with C_4–C_6 *n*-alcohols although they did not observe an integral series of basal reflections in the X-ray diffractograms.

Within the limits of uncertainty regarding basal spacing values, the *d*(001) spacing of one-layer complexes is about 1.40 nm although a value as low as 1.32 nm has been given for the methanol complex (Brindley and Ray 1964). If the van der Waals dimension of a hydrocarbon chain in an α_I and α_{II} orientation is taken as 0.45 and 0.42 nm, respectively, and the thickness of an individual silicate layer as 0.95 nm, the basal spacing corresponding to the respective orientation is 1.40 and 1.35 nm. An α_I arrangement is therefore indicated for these complexes in accord with Emerson's (1957) model but without having to invoke O–H···O hydrogen bonding.

Using Ca^{2+}-montmorillonite, Barshad (1952) measured basal spacing values for C_3 to C_8 *n*-alcohol complexes that were consistent with the presence of a single layer of flatly oriented alcohol molecules in the interlayer space (Figure 2.8, line 6). The high cation exchange capacity (120 $cmol_c$/kg) of the montmorillonite (Goldfield) sample used was probably responsible for the limited interlayer expansion. Thus, the basal spacings of the C_3 and C_4 complex with the Ca^{2+}-exchanged Goldfield material was 1.44 nm whereas the values for similar complexes with Wyoming montmorillonite (CEC ~95 $cmol_c$/kg), were 1.74 and 1.66 nm, respectively.

It could be argued that basal spacing measurements, by themselves, cannot provide unequivocal evidence for the type of orientation adopted by intercalated organic compounds. Thus, a basal spacing of ~1.78 nm, measured for a montmorillonite complex with *n*-propanol, may indicate formation of a single-layer complex in which the alcohol molecules adopt an α_I orientation, rather than the intercalation of two layers of molecules in an α_{II} orientation. Similarly, Radul and Ovcharenko (1971) have proposed that *n*-propanol, intercalated into Na^+-montmorillonite, adopted an α_I conformation based on infrared spectroscopy and X-ray diffraction measurements.

Here we will briefly describe the clay-phenol interaction although phenol is not an alcohol. The interaction of clay minerals and organoclays with phenol and substituted phenols have attracted a great deal of attention in relation to the removal of these highly toxic compounds from industrial effluents and wastewater (Ko et al. 2007; Lee and Tiwari 2012; Nafees and Waseem 2014; Ouardi et al. 2019). Early on, Saltzman and Yariv (1975) reported that montmorillonite-adsorbed *p*-nitrophenol could either be protonated or directly coordinated to the exchangeable cation, depending on the hydration status of the mineral. Subsequently, Isaacson and Sawhney (1983) found that the adsorption of phenols by montmorillonite was influenced by the nature of the interlayer cation. For 2,6-dimethylphenol adsorption decreased in the order $Fe^{3+} > Al^{3+} > Cu^{2+} > Ca^{2+}$. The results are strongly indicative of ion-dipole interactions since the polarising power (ionic potential) of the counterions (Table 2.1) decreases in the same sequence. Cation-dipole interactions have also been proposed by Guo et al. (2022) as the mechanism underlying the adsorption of bisphenols by Ca^{2+}-montmorillonite. The observation by Yu et al. (2004) that adsorption by Ca^{2+}-montmorillonite increases in the sequence phenol < 2-chlorophenol < 2,4-dichlorophenol is also in line with the proposed mechanism since the hydrophobicity and polarity of these compounds increase in the same order (Mortland et al. 1986). The same or similar sequence was observed by Dentel et al. (1995), Juang et al. (2002), and Rawajfih and Nsour (2006) for the adsorption of phenol and related compounds to montmorillonite modified by long-chain quaternary ammonium ions.

Using reduced-charge montmorillonites, modified with hexamethylene bispyridinium (HMBP), Luo et al. (2015) noted that uptake increased in the order *p*-nitrophenol < *p*-methylphenol < *p*-chlorophenol < phenol. They suggested that π-π interactions between the pyridine ring of HMBP and the benzene ring of phenols played an important role in adsorption. Wibulswas et al. (1999) also found that phenol adsorption to Al-pillared (Al-PILC) montmorillonite, incorporating hexadecyltrimethylammonium (HDTMA) ions, increased in the order phenol < 3-chlorophenol < 3,5-dichlorophenol. Bentonite, modified with 'mixed' Al-PILC/HDTMA, is a better adsorbent of phenol than either the Al-PILC- or HDTMA-modified material (Al-Asheh et al. 2003). Similarly, montmorillonite modified with one or two cationic surfactants is a useful adsorbent of phenol (Wang et al. 2017). Interlayer complexes of montmorillonite with adamantylammonium ions are effective and selective in taking up chlorophenols from water (Spyrou et al. 2014) as did complexes with neostigmine and octadecyltrimethylammonium with respect to 2-phenylphenol (Seki et al. 2015). Likewise, bentonite complexes with octadecyltrimethylammonium and benzyldimethyltetradecylammonium show a strong affinity for 2,4,5- and 2,4,6-trichlorophenol, respectively (Zaghouane-Boudiaf et al. 2014; Ghezali et al. 2018), while vermiculite modified with gemini surfactants has a large capacity for taking up *p*-nitrophenol (Yu et al. 2018).

As might be expected, phenol adsorption from aqueous solution to montmorillonite/bentonite increases with solute concentration and decreases with a rise in solution pH (Viraraghavan and Alfaro 1998; Banat et al. 2000; Ahmadi and Igwegbe 2018). More importantly, the capacity of montmorillonite/bentonite for adsorbing phenol (and its derivatives) can be greatly enhanced by replacing the inorganic counterions by cationic organic compounds and surfactants, notably compact and long-chain quaternary ammonium ions. The adsorption process often follows pseudo-second order kinetics and obeys the Freundlich or Langmuir model (Roberts et al. 1964; Mortland et al. 1986; Stockmeyer and Kruse 1991; Huh et al. 2000; Koh and Dixon 2001; Jiang et al. 2002; Yılmaz and Yapar 2004; Yıldız et al. 2005; Richards and Bouazza 2007; Okada et al. 2008; Shakir et al. 2008; Alkaram et al. 2009; Froehner et al. 2009; Senturk et al. 2009; Praus et al. 2011; Su et al. 2011; Mirmohamadsadeghi et al. 2012; Nguyen et al. 2013; Xue et al. 2013; Zheng et al. 2013; Anirudhan and Ramachandran 2014; Cao et al. 2014; Liu et al. 2014; Zhu et al. 2015; Ma et al. 2016; Shah et al. 2018; Shen and Gao 2019).

Majdan et al. (2009) have reported that the isotherm for the adsorption of phenol by phenyltrimethylammonium-bentonite is linear, indicative of solute partitioning between the aqueous solution and the organoclay. On the other hand, the isotherm for the benzyltrimethylammonium-exchanged clay is convex, consistent with an adsorptive mechanism, such as π-π interactions between the benzene ring of phenol and BTMA and phenol-water H-bonding (Shen 2002). Charge-transfer interactions are also involved in the adsorption of 2,4-dichlorophenol to an interlayer complex of saponite with 1,1'-dimethyl-4,4'-bipyridinium ions (Okada and Ogawa 2003), and in the interactions between chlorophenols and montmorillonite exchanged with transition metal complexes (Boufatit et al. 2007). Montmorillonite modified with quaternary ammonium cations (Witthuhn et al. 2005; Andini et al. 2006; Khenifi et al. 2009; Yang et al. 2017) or sodium dodecyl sulphate (Shah et al. 2017) is also an efficient sorbent of chlorophenols from aqueous solution.

Xu et al. (2018) have suggested that both partitioning and adsorption are involved in the interaction of phenols with montmorillonite that has been modified with a multi amine-containing gemini surfactant. Gemini surfactant-modified montmorillonites are also effective in removing *p*-nitrophenol, 2-naphthol, and bisphenol A from aqueous solutions (Xue et al. 2013; Yang et al. 2015, 2016). Furthermore, intercalated gemini surfactants do not easily leach out in the process of removing phenol (Luo et al. 2019). The review by Shen and Gao (2019) has more details on the synthesis and applications of gemini surfactant-modified clay minerals.

The partitioning of phenol and related compounds into surfactant-modified smectites has been reported by many workers (Boyd et al. 1988a; Dentel et al. 1995; Zhu and Zhu 2008; Díaz-Nava et al. 2012; Park et al. 2013; Yang et al. 2015; Martinez-Costa and Leyva-Ramos 2017; Li et al. 2018). There is evidence to indicate that chemical interaction is dominant at low solute concentrations while partitioning becomes important at high concentrations (Zhu et al. 1998, 2000; Gonen and Rytwo 2006; Rawajfih and Nsour 2006; Moshe and Rytwo 2018). The reviews by Yuan et al. (2013), Nafees and Waseem (2014), and Zhu et al. (2016) should be consulted for more details.

2.6.2 Complexes with Polyhydric Alcohols

Among the polyhydric alcohols, ethylene glycol (1,2-ethanediol, or simply glycol) and glycerol (1,2,3-propanetriol) have received the greatest attention as complexing agents of expanding 2:1 type layer silicates (Brindley 1966). Only the salient features of the formation, properties, and applications of glycol and glycerol complexes with smectite and vermiculite are discussed here. The use of ethylene glycol in characterizing high charge smectite and vermiculite by X-ray diffraction (XRD) has been summarized by Mosser-Ruck et al. (2005).

Bradley (1945b) and MacEwan (1948a) were the first to report that ethylene glycol (EG) and glycerol formed a double-layer complex with smectite, giving basal spacings of 1.7 and 1.77 nm, respectively. Since then both compounds have been routinely used, in conjunction with XRD analysis, to indicate the presence of smectite in soils and sediments (Whitton and Churchman 1987), and to distinguish (finely divided) 'clay vermiculite' from montmorillonite. All things being equal, vermiculite tends to take up a single layer, while montmorillonite commonly intercalates a double layer of EG and glycerol.

Differentiating smectite from vermiculite is inherently difficult because the former grades into the latter with increasing charge, and both groups of layer silicates cover a range of charge per formula unit (value of x in Table 1.1). In other words, low-charge vermiculites may show interlayer expansion characteristics that closely resemble those displayed by high-charge montmorillonites. Furthermore, vermiculites can form either single- or double-layer complexes depending on such factors as the nature of the interlayer cation and the magnitude of layer charge (Walker 1957, 1958).

A widely used technique for differentiating montmorillonite from vermiculite is to saturate the clay sample with Mg^{2+} ions prior to adding EG or glycerol. Montmorillonite forms a double-layer complex with glycerol, giving a basal or $d(001)$ spacing of ~1.78 nm whereas vermiculite generally intercalates only a single layer of molecules, giving a $d(001)$ spacing of ~1.43 nm (Brindley 1966; Kowalik et al. 2021). A complementary approach is to replace the counterions, initially present, with K^+ ions. The resultant inhibition of interlayer expansion, when EG or glycerol is added to the clay mineral, has been referred to as 'potassium contraction' (Weaver 1958). We might add here that EG treatment of K^+- (and NH_4^+-) montmorillonite may give rise to interstratified layer structures, consisting of non-expanded, partially expanded, and fully expanded layers (Čičel and Machajdík 1981). Interlayer expansion of K^+-montmorillonite or Mg^{2+}-vermiculite is also influenced by the method of sample saturation by exposure to either EG vapour or liquid EG (Mosser-Ruck et al. 2005).

Interlayer expansion may also be inhibited by heating the minerals to high temperatures (> 400°C) prior to treatment with EG or glycerol (Grim and Kulbicki 1961). As might be expected, the temperature to which montmorillonite samples must be taken to prevent interlayer swelling in both organic liquids is dependent on the nature of the counterion and layer charge density (Greene-Kelly 1953). Molecular dynamics simulation studies by Szczerba et al. (2014) confirm that EG and water compete for the same ligand positions around the interlayer counterion in montmorillonite (Svensson and Hansen 2010), while the interlayer packing of these molecules is controlled by the density and location of layer charge. Curiously, however, Eltantawy and Arnold (1974) found that the amount of EG required to give a complete monolayer surface coverage was not influenced by the nature of the counterion.

To answer the question whether total charge or layer charge location exerts the stronger influence on the intercalation of EG and glycerol, Harward and Brindley (1965) examined the expansion characteristics of (synthetic) montmorillonites and beidellites. When the total charge was comparable, both groups of minerals behaved similarly towards ethylene glycol. Montmorillonites, where most of the negative layer charge arises from isomorphous substitution in octahedral positions, expanded to 1.67–1.77 nm. On the other hand, tetrahedrally substituted beidellites expanded to a basal spacing of 1.42–1.46 nm. This observation would indicate that beidellite is intermediate between montmorillonite and vermiculite with respect to solvation by polyhydric alcohols.

Tettenhorst et al. (1962) also noted that EG and glycerol expanded Ca^{2+}- and Na^{+}-montmorillonite samples to a similar extent ($d(001)$ = 1.70–1.78 nm). By contrast, polyhydric alcohols in which the methylene group was not directly linked to an oxygen atom (e.g., 1,3-butanediol and 1,5-pentanediol), gave rise to complexes with different basal spacings depending on the amount and site of isomorphous substitution in the structure.

Evidently, both total charge and source of charge are important in determining the interlayer expansion of 2:1 layer silicates when treated with polar organic liquids or vapours. To emphasize the former or the latter factor in this regard would give an oversimplified picture of the nature of the intercalation process. We may also add that ethylene glycol, glycerol, and other small polyhydric alcohols interact more with (strongly polarizing) counterions than with the silicate surface (Dowdy and Mortland 1968).

The orientation and packing of ethylene glycol in a two-layer complex with Ca^{2+}-montmorillonite ($d(001)$ = 1.69 nm), deduced by Reynolds (1965) from X-ray diffraction data, are shown in Figure 2.10. The glycol molecules are apparently oriented with their symmetry plane parallel to the *b*-*c* plane, that is, normal to the silicate surface. The second glycol layer is slightly displaced in the *b* direction to the extent of about half an oxygen diameter between the two opposing siloxane surface oxygens. The calcium and (residual) water molecules are situated between the two ethylene glycol layers, just above and below the *a*-*b* plane that bisects the organic layers. The cation-glycol complex depicted is clearly of the octahedral type and weak interactions between one of the hydroxyl groups of each glycol molecule (through its oxygen atom) and the calcium ion is possible in accord with the infrared results of Dowdy and Mortland (1968). The proposed structure also allows for the methylene groups of EG to be keyed into the ditrigonal holes of the silicate surface, giving the observed basal spacing. Fujimine et al. (1995) have proposed that in a double-layer complex with Cu^{2+}-montmorillonite, the interlayer cation is coordinated to six EG molecules with the symmetry axis inclined at an angle of ~45° to the silicate surface.

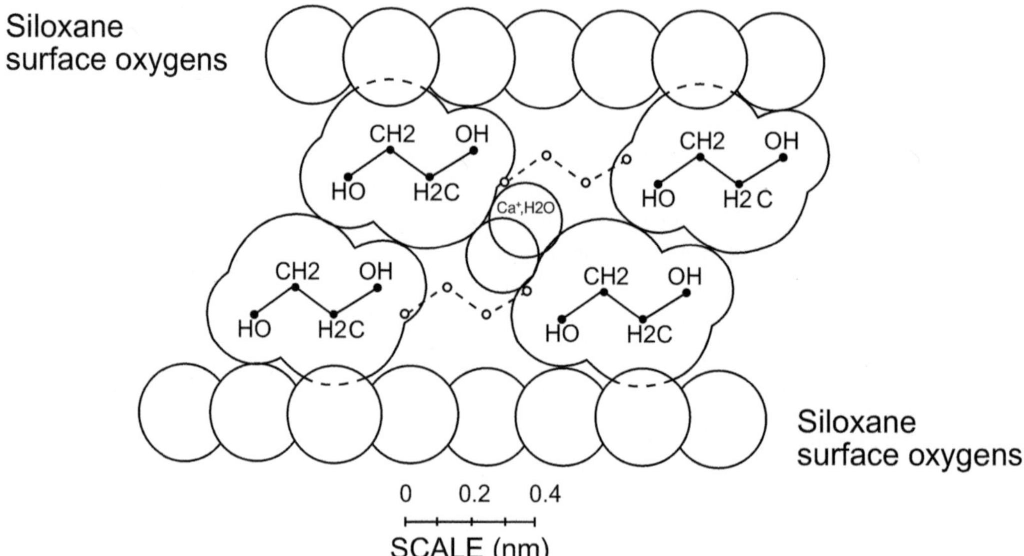

FIGURE 2.10 Diagram showing the orientation and packing of ethylene glycol molecules in the interlayer space of Ca^{2+}-montmorillonite. The basal spacing of this double-layer complex is 1.69 nm. (From Reynolds 1969, Figure 1. Reproduced with permission of the Mineralogical Society of America publisher of *American Mineralogist.*)

Complex formation with ethylene glycol and glycerol has been widely used to estimate the (total) surface area of expanding layer silicates. Dyal and Hendricks (1950) were the first to determine the specific surface area of montmorillonites, based on the amount of ethylene glycol retained in a monolayer on all planar surfaces. If interlayer penetration by EG is prevented (e.g., by preheating the sample to 600°C), the external particle area can be derived, while the internal (interlayer) area may be obtained by difference. Modifications and refinements of the Dyal and Hendricks (1950) method have subsequently been proposed with respect to both ethylene glycol (Bower and Gschwend 1952; Martin 1955; Bower and Goertzen 1959; Sor and Kemper 1959; Chassin 1976) and glycerol (Diamond and Kinter 1956; Kinter and Diamond 1958; Mehra and Jackson 1959; Milford and Jackson 1962; Moore and Dixon 1970).

In this regard, the use of ethylene glycol monoethyl ether (EGME) has received much attention because this polar molecule is more volatile than EG and, hence, requires less time for equilibrium to be reached (Carter et al. 1965; Eltantawy and Arnold 1973; Cihacek and Bremner 1979; Tiller and Smith 1990; Churchman et al. 1991; Quirk and Murray 1999; Cerato and Lutenegger 2002). From a one-dimensional Fourier synthesis, based on 12 orders of reflections, Reynolds (1969) deduced that like ethylene glycol (Figure 2.10), EGME was intercalated as a double layer with the plane of symmetry of the aliphatic chain perpendicular to the silicate surface, giving a basal spacing of 1.60 nm. Subsequent FTIR spectroscopy, however, would indicate that the plane of the molecular skeleton is tilted at an angle of 69° to the interlayer surface of Ca^{2+}-montmorillonite (Nguyen et al. 1987).

The principle behind these techniques is that the rate of desorption of polar organic liquids levels off when all the 'free' liquid has been removed by evacuation. At this point, the quantity retained by the clay mineral is assumed to correspond to that required for a monolayer coverage over the total surface. Since this quantity is proportional to the surface area, the latter parameter can be estimated if the effective area occupied by the organic probe molecule, a_s, is known. For montmorillonite Brindley (1966) has suggested a figure of 30 m^2/g for each 1% EG retained as probably the best compromise in calculating specific surface areas from *retention* measurements.

Besides being dependent on surface/interlayer orientation, the value of a_s is apparently influenced by clay mineral type (Heister 2014). In the case of montmorillonite, Dyal and Hendricks (1950) calculated $a_s = 0.33$ nm^2 for EG, while Brindley (1966) suggested $a_s = 0.269$ nm^2 for glycerol, and Carter et al. (1965) used $a_s = 0.52$ nm^2 for EGME. The values for EG and EGME are appreciably higher than those derived by Helmy et al. (1999) from *adsorption* measurements, probably because the amount (of EG or EGME) retained is dependent on the nature of the counterion (McNeal 1964; Ugochukwu 2017). A similar dependence attaches to the measurement of surface areas of clays and soils using water as adsorbate (Quirk 1955; Newman 1983). Chiou and Rutherford (1997) have also pointed out that both EGME and water, when added as a vapour, can condense into the small pores of the clay particle assemblage.

Thus, procedures based on the retention of polar organic liquids, notably EG, glycerol, and EGME, are more suited for comparative than absolute determinations of surface areas (Brindley 1966; Churchman et al. 1991). Nevertheless, such methods have wide appeal partly because of the ease and simplicity of operation, requiring no elaborate and expensive equipment, and partly because they seem 'to work' reasonably well (Brindley 1966; Cerato and Lutenegger 2002).

The surface area of clay minerals (and soils) can also be estimated from the (monolayer) adsorption of selected organic compounds, such as *o*-phenanthroline (Bower 1963). Thus, Lawrie (1961) was able to calculate surface areas by assuming that the molecule, lying flat on the clay mineral surface, occupied an area (a_s) of 0.6 nm^2. Likewise, the adsorption of *para*-nitrophenol (*p*NP) provides a simple method of measuring the specific surface area (SSA) of clays and soils (Ristori et al. 1985, 1989; Theng et al. 1999). The isotherms for the adsorption of *p*NP by non-expanding clay minerals and metal oxides are of the Langmuir type, giving a distinct plateau above a certain equilibrium concentration (Figure 2.11). The SSA values derived from the respective plateau adsorption, using $a_s = 0.424$ nm^2, are in excellent agreement with the corresponding BET-N_2 surface areas as shown in Figure 2.12 (Theng 1995).

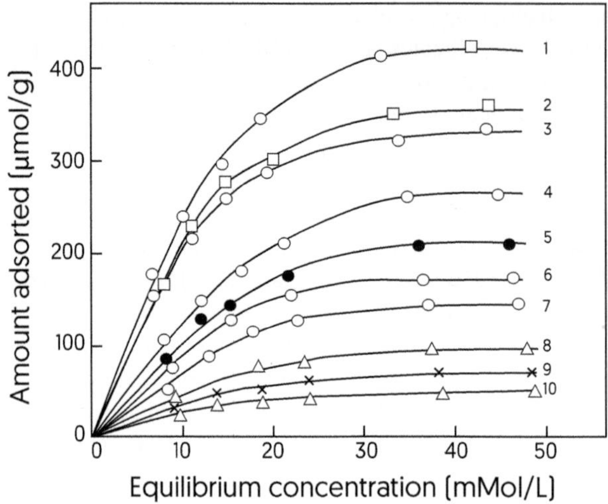

FIGURE 2.11 Isotherms for the adsorption of *p*-nitrophenol (from xylene) by some non-expanding clay minerals and a sample of silica. The halloysite samples are from New Zealand (NZ). (1) halloysite (Hamilton, NZ); (2) illite (Fithian); (3) halloysite (Opotiki, NZ); (4) halloysite (Te Akatea, NZ); (5) Ca^{2+}-halloysite (Te Puke, NZ); (6) Na^{+}-halloysite (Te Puke, NZ); (7) Matauri Bay, NZ); (8) kaolinite (KGa-2); (9) silica; (10) kaolinite (KGa-1). (From Theng, B.K.G., *Clays: Controlling the Environment. Proc. 10th Int. Clay Conf., 1993, Adelaide*, pp. 304–310, 1995.)

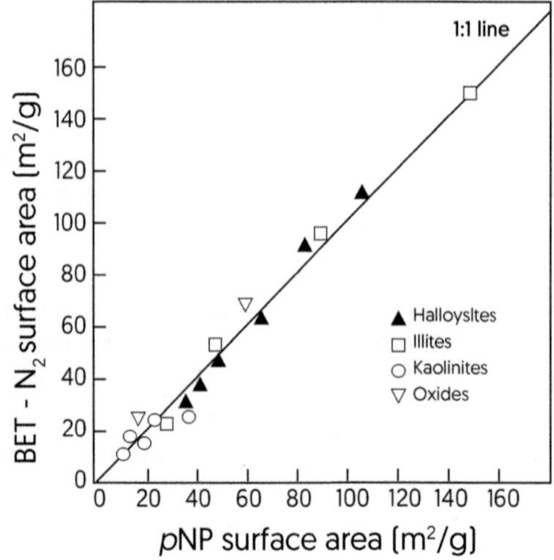

FIGURE 2.12 Relationship between the surface areas, derived from the plateau adsorption of *p*-nitrophenol, and the corresponding BET-nitrogen surface areas for some halloysites, illites, kaolinites, and metal oxides. (From Theng, B.K.G., *Clays: Controlling the Environment. Proc. 10th Int. Clay Conf., 1993, Adelaide*, pp. 304–310, 1995.)

A similarly good relationship between the water content of air-dry topsoils and BET-H_2O areas was reported by Parfitt et al. (2001). More recently, Kirschbaum et al. (2020) have proposed that the SSA of soils can be reliably estimated from water adsorption by air-dry samples, while allowing the contribution of soil organic carbon to be assessed. We might mention here that the SSA of clays (and soils) can also be estimated by adsorption of cationic organic molecules, such as cetylpyridinium bromide (Greenland and Quirk 1964) and methylene blue (Pham and Brindley 1970). The interactions of layer silicates with positively charged organic species are described in Chapter 4.

The assumptions and limitations of deriving the SSA of clay minerals from retention of polar organic molecules have been discussed by Brindley (1966), Guyot (1969), van Olphen (1969), Tiller and Smith (1990), Quirk and Murray (1999), Michot and Villiéras (2006), and Heister (2014).

2.6.3 Complexes with Amines

Complex formation of 2:1 type layer silicates with aliphatic and aromatic amines has received much attention. In common with alcohols, alkylamines (RNH_2) from R = 1 to R = 18 are available or may be synthesized. An advantage of using a homologous series of a given compound is that the properties of the corresponding complexes might be expected to vary regularly with molecular size. Moreover, amines often exist, or are added to the clay mineral, as the corresponding alkylammonium ions (Weiss et al. 1956; Sutherland and MacEwan 1961; Wolfe et al. 1985). Indeed, the 'amine' part in the clay-amine interaction is commonly identifiable with the corresponding cationic (salt) form (Wang and Lee 1993; Pérez-Santano et al. 2005; Ahmad et al. 2009; Xie et al. 2017; Peng et al. 2018; Slaný et al. 2019). The interactions of layer silicates with positively charged organic compounds are described in Chapter 4.

Early on, Bradley (1945a) reported the formation of an interlayer montmorillonite-methylamine complex with a basal spacing of ~1.27 nm. Subsequently, Haxaire and Bloch (1956) reacted Na^+-montmorillonite with methylamine and *n*-heptylamine, yielding complexes with a closely similar basal spacing. In this instance, the amines were added to the clay sample as the respective aqueous salt solution. Both compounds would therefore be intercalated as the corresponding alkylammonium ions, releasing a stoichiometric amount of sodium ions. A basal spacing of ~1.27 nm for the complex with methylammonium is compatible with an orientation in which the organic cation is keyed to the extent of 0.07 nm (Rowland and Weiss 1963). A similar interlayer arrangement may be assumed for methylamine. Complexes of vermiculite with ethyl-, propyl-, butyl-, and hexyl-diamines have basal spacings between 1.28 and 1.47 nm (da Fonseca et al. 2006).

Primary *n*-amines with less than 5 carbon atoms in the molecule apparently prefer an orientation in which the hydrocarbon chain lies parallel to the silicate layer. In this arrangement the plane of the carbon zig-zag may either be perpendicular (α_I) or parallel (α_{II}) to the clay surface. Fripiat and co-workers (Fripiat et al. 1962; Léonard et al. 1962; Servais et al. 1962) have proposed that either type of orientation is possible, depending on the amount of amine present. Adsorbed amines may also be retained in their cationic form when a ready source of protons is available at the silicate surface (Laura and Cloos 1975a, 1975b). An example of a surface protonation reaction is the formation of the respective alkylammonium and alkyldiammonium ions when *n*-propylamine, *n*-butylamine, ethylenediamine, and propylenediamine are intercalated by acid-treated montmorillonite. Similarly, octadecylamine and oleylamine are apparently protonated when they intercalate into a natural acidic clay (Hayakawa et al. 2016). Protonation (of cyclohexylamine and *n*-butylamine) is also indicated with Ni^{2+}- and Co^{2+}-montmorillonite (Breen 1991) but does not occur with Na^+- and Ca^{2+}-montmorillonite where adsorption is apparently determined by ion-dipole interactions. Thus, at a given solute concentration, more amine is adsorbed by the calcium-saturated material than the sodium-saturated material (Servais et al. 1962). Interestingly, the diethylammonium-montmorillonite complex, prepared by Boufatit and Ait-Amar (2007), can remove N,N-dimethylaniline from water.

Sieskind and Wey (1958a, 1958b) reported that the adsorption of some primary *n*-amines by H^+-montmorillonite was influenced by the pH of the system and the size (chain length) of the organic molecule. At pH 11 the amine entered the interlayer space as an uncharged species capable of accepting a proton, while at pH 3 the molecule was intercalated as the corresponding *n*-alkylammonium ion. These workers further observed that below pH 5 adsorption of monoamines increased with chain length. Indeed, *n*-decylamine was adsorbed in excess of the cation exchange capacity of the clay mineral. Cowan and White (1958) have reported similarly for the adsorption of some *n*-alkylammonium ions by Na^+-montmorillonite, the excess being present as the free amine. This behaviour is generally attributed to the increased contribution of van der Waals forces to the adsorption energy (Hendricks 1941; Theng et al. 1967; Fripiat et al. 1969) although H-bonding (Farmer and Mortland 1965) and ion hydration effects (Vansant and Uytterhoeven 1972; Vansant and Peeters 1978) may also be important.

The kinetics for the adsorption of monomethylamine (Me_1), dimethylamine (Me_2), trimethylamine (Me_3), and monoethylamine (Et_1) from the vapour phase by (a previously degassed) Na^+-montmorillonite, have been examined by Palmer and Bauer (1961). The results indicated that the adsorption process was diffusion-controlled. The decrease in the apparent rate constant with molecular size was ascribed to steric hindrance in that the individual silicate layers must be forced apart before intercalation occurred. Similarly, Loduhová et al. (2015) obtained interlayer complexes by exposing dry Hg^{2+}-montmorillonite to Me_1, Me_2, and Me_3 vapours. Alternatively, such complexes may be prepared by titrating the clay mineral with the amine. For example, Bodenheimer et al. (1962) titrated Cu^{2+}-montmorillonite with Me_1, Et_1, diethylamine, triethylamine, and *n*-hexylamine (Hx_1) in an aqueous medium. X-ray diffraction and differential thermal analysis indicated the formation of interlayer $[Cu (amine)]^{2+}$ and $[Cu (amine)_2]^{2+}$ complexes whose stability increased in going from the primary (e.g., Et_1) to the tertiary (e.g., Et_3) amine and with increasing chain length (from Me_1 to Hx_1). We might add that titration with *n*-butylamine (in an organic solvent) is widely used to assess the surface acidity of clay mineral catalysts (Theng 2018).

Using infrared spectroscopy, Farmer and Mortland (1965) have confirmed that in the absence of water, ethylamine can coordinate directly to interlayer copper ions in montmorillonite, forming a (blue) square planar coordination complex, $[Cu(Et_1)_4]^{2+}$, with the plane of the molecule lying parallel to the silicate surface, giving a basal spacing of ~1.29 nm. Bruque et al. (1982) have observed similarly for complexes of Ln^{3+}-montmorillonite (Ln = La, Sm, Gd, Er, Y) with *n*-butylamine, sec-butylamine, and iso-butylamine. By reacting Fe^{2+}-exchanged montmorillonite with *o*-phenanthroline (L) in the solid state, Kabadagi et al. (2018) obtained an interlayer clay-$[FeL_3]^{2+}$ complex. Likewise, Ogawa et al. (1991) and Khaorapapong et al. (2001) were able to prepare complexes of Co^{2+}-, Ni^{2+}-, and Cu^{2+}-montmorillonite with 4,4'-bipyridine and 1,2-di(4-pyridine) ethylene by solid-state intercalation. The formation of fluorescent interlayer complexes of Li^+-, Zn^{2+}-, and Mn^{2+}-exchanged montmorillonite with 8-hydroxyquinoline has been reported by Khaorapapong and Ogawa (2007). Interestingly, the mixed-ligand complexes of Zn^{2+}-montmorillonite with 8-hydroxyquinoline and 4,4'-bipyridine show intense photoluminescence (Pimchan et al. 2014).

Interlayer complexes of this type between transition metal counterions and diamines in montmorillonite and vermiculite have also been described by Bodenheimer et al. (1962, 1963a, 1963b, 1963c), Laura and Cloos (1970), Santos et al. (1969, 1970), and Burba and McAtee (1977). The formation and stability of such complexes are related to the propensity of transition metal cations, having unfilled *d* orbitals, for accepting electrons from electron-donating groups in the organic compounds, such as amines. We might interpolate here that stable cationic complexes of Cu(II) with ethylene diamine, triethylene tetramine, and triethylene pentamine have a high affinity for negatively charged materials, and hence are useful for assessing the cation exchange capacity of clay minerals (Bergaya and Vayer 1997; Meier and Kahr 1999; Ammann et al. 2005).

We would also mention the possible formation of 'hemisalt' complexes when the amount of adsorbed base, B, exceeds that of protons available for protonating B. The protonated species, BH^+, then shares its proton with the neutral (excess) base to yield a cation of the type $(B...H...B)^+$. Thus, when ethylamine was brought into contact with ethylammonium-montmorillonite (Et^+-M), the corresponding interlayer ethylammonium-ethylamine hemisalt complex was formed (Farmer and Mortland 1965). This complex (Scheme 2.1) was stable against evacuation, but on exposure to air at ambient humidity ethylamine was lost and the infrared spectrum of the (exposed) complex then resembled that of the (parent) Et^+-M material. The formation of hemisalt structures has also been observed in complexes of montmorillonite with ethylenediammonium-ethylenediamine (Fripiat

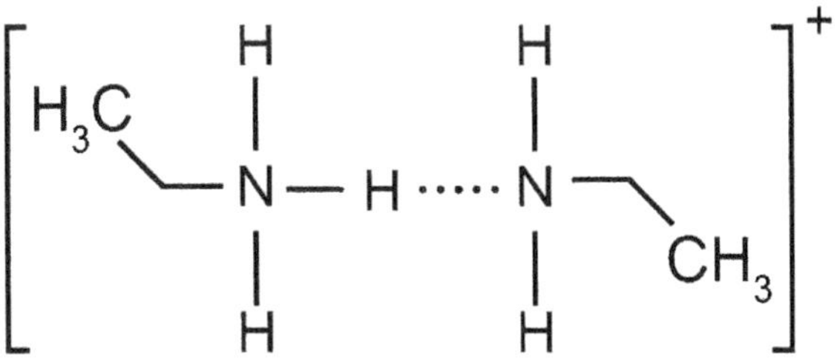

SCHEME 2.1 Formation of a 'hemisalt' complex through intercalation of ethylamine into an ethylammonium-exchanged montmorillonite.

et al. 1962), pyridinium-pyridine (Farmer and Mortland 1966), urea (Mortland 1966), amides (Tahoun and Mortland 1966a), anilinium-aniline (Yariv et al. 1968; Homenauth and McBride 1994), and isopropylammonium-isopropylamine (Shoval and Yariv 1979). Both protonation reactions and hemisalt formation have been reported by Santos et al. (1970) for complexes of Co^{2+}-, Ni^{2+}-, Cu^{2+}, Mg^{2+}-montmorillonites, and Cu^{2+}-vermiculite, with *n*-propylamine, *n*-pentylamine, ethylenediamine, 1,2-propylene diamine and 1,2-butylamine.

A related process, described by Raman and Mortland (1969), is the transfer of a proton from a protonated base (BH^+) to another adsorbed organic species, B* (Equation 2.4),

$$BH^+ + B^* \rightleftharpoons B^*H^+ + B \qquad \text{(Equation 2.4)}$$

In which direction this equilibrium will move would clearly depend on the relative basicity of B and B* as well as on the concentration of the reactants and products. Using infrared spectroscopy, Russell et al. (1968) have shown that treatment of NH_4^+-montmorillonite with 3-aminotriazole yielded 3-aminotriazolium cations to the extent of the cation exchange capacity of the parent clay.

By analogy with primary *n*-alcohols, long-chain primary *n*-amines containing five or more carbon atoms per molecule, tend to intercalate into montmorillonite (and vermiculite) as a double-layer with the alkyl chain inclined at a high angle (~65°) to the silicate surface, giving rise to 'β-type' interlayer complexes. Aragon et al. (1959) reported that the basal spacing of such complexes between Na^+-montmorillonite and primary *n*-amines (from C_4 to C_{18}) increased linearly with chain length (Figure 2.13). Ogawa et al. (1990) have found similarly for *n*-tetradecyl-, *n*-hexadecyl-, and *n*-octadecylamines. The slope of the line corresponded to ~0.26 nm per carbon atom. Such a relationship was also observed by Sutherland and MacEwan (1961) with two vermiculites (Irish and West Chester), but the slope of the line relating the number of C atoms to basal spacing was slightly less (~0.23 nm per C atom). Inclination angles of 76–95° were reported by Weiss et al. (2003) for a two-layer magnesium-rich vermiculite complex with octadecylamine.

By analysing the data afresh and comparing them with those obtained for complexes of montmorillonite with *n*-alcohols (Figure 2.8), Brindley (1965) has proposed an alternative model that could account for the incremental increase of ~0.23 nm per C atom. The results of this analysis are shown in Figure 2.14. In this model the orientation and packing of *n*-amines in β-type complexes offer scope for hydrogen bonding between the terminal amino groups and surface oxygen ions as well as for the keying of these groups into the hexagonal – actually ditrigonal – holes of the silicate surface. The model further indicates that β-type complexes would not obtain with short-chain amines ($n < 6$). Being adsorbed by ion-dipole interactions, the molecules tend to adopt an orientation with the alkyl chain lying parallel to the silicate surface. On the other hand, the adsorption of amines with $n > 8$ is largely determined by van der Waals attractive forces between adjacent alkyl chains, resulting in a tilted (extended) chain conformation. We might also add that the adsorption and interlayer orientation of amines are sensitive to variations in suspension pH (Peng et al. 2017).

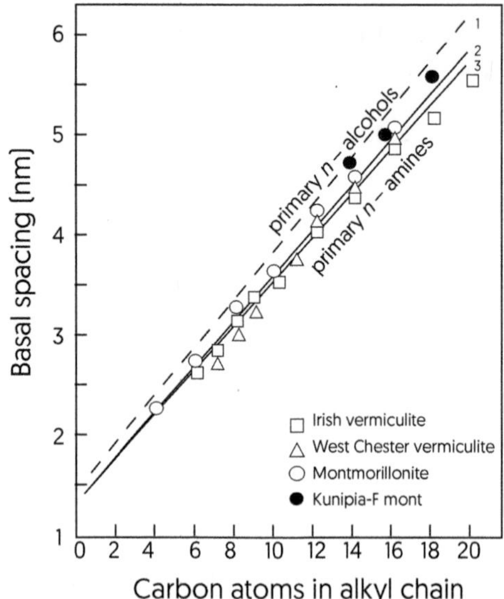

FIGURE 2.13 Variation in basal spacing with number of carbon atoms in the alkyl chain for complexes of montmorillonite and vermiculite with primary *n*-alcohols and *n*-amines. Line 1 (broken) refers to complexes of montmorillonite with *n*-alcohols. (Brindley, G.W. and Ray, S., *Am. Mineral.*, 49, 106–115, 1964). Lines 2 and 3 (solid) refer to complexes of montmorillonite with *n*-amines (Aragon, F. et al., *Nature* 183, 740–741, 1959; Ogawa et al. *Clay Sci.*, 8, 31–36, 1990) and vermiculite with *n*-amines (Sutherland, H.H. and MacEwan, D.M.C., *Clay Min. Bull.* 4, 229–233, 1961).

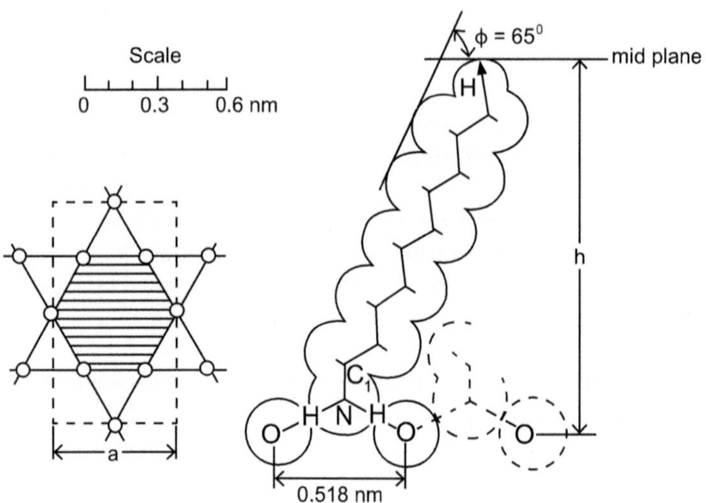

FIGURE 2.14 Model for the arrangement of straight-chain aliphatic amines in a double-layer complex with montmorillonite and vermiculite. Left: idealized structure of siloxane surface oxygens, showing the hexagonal 'hole' (hatched). Right: orientation of amine molecule with two NH–O contacts to surface oxygens and an inclination angle of 65°. The basal spacing is given by 2h + 0.66 nm. The broken line indicates the possible position of an adjacent molecule. (From Brindley 1965, Figure 1. Reproduced with kind permission of The Mineralogical Society of the UK and Ireland publisher of *Clay Minerals*.)

Yariv and Heller (1970) prepared complexes of montmorillonite with cyclohexylamine by immersing thin, self-supporting films of the clay in the liquid amine. Except for the Na^+- and Cs^+-exchanged montmorillonite, protonation occurred immediately after immersion. Prolonged exposure to the liquid led to the intercalation of (uncharged) cyclohexylamine. The total (cationic and uncharged amine) amount taken up after 24-hour exposure was insensitive to, although the rate of sorption was much influenced by, the nature of the exchangeable cations. X-ray diffraction showed a basal spacing of ~2.05 nm, analogous to the 2.3 nm pyridine complex of Farmer and Russell (1967). The amino nitrogen of cyclohexylamine was apparently hydrogen-bonded to water coordinated to the interlayer counterion, while the hydrogens were similarly bonded to surface oxygens.

Complex formation between montmorillonite and aniline has attracted considerable attention as this amine serves as a model of how aromatic organic bases interact with expanding layer silicates, and many amine-based pesticides consist of substituted anilines.

Being a weak base (pK_a ~4.6), aniline is less reactive toward clay minerals than aliphatic amines under comparable conditions of pH and concentration. For example, Haxaire and Bloch (1956) reported that Na^+-montmorillonite adsorbed less aniline and other aromatic amines (N, N-dimethylaniline, diphenylamine) than their aliphatic counterparts. Earlier, Hendricks (1941) observed that organic bases such as benzidine, *p*-aminodimethylaniline, *p*-phenylenediamine and α-naphthylamine could neutralise most of the exchangeable hydrogen ions in montmorillonite. On the other hand, *o*- and *m*-nitroanilines failed to form salts with the clay mineral because these compounds were very weak bases.

Using Na^+- and H^+/Al^{3+}-montmorillonite, Bailey and White (1964) have observed that the amount of aniline adsorbed is greater for the acid-treated clay than for the Na^+-exchanged material. This observation was ascribed to the ability of H^+/Al^{3+}-montmorillonite to protonate the intercalated organic molecule which is then retained as the anilinium ion. In the sodium system, on the other hand, the neutral amine molecules must compete with water for sorption sites around the interlayer cation (Praus et al. 2011). That aniline is protonated when brought into contact with acid-treated montmorillonite has been demonstrated by Harter and Ahlrichs (1969) who noted that the infrared spectrum of H^+/Al^{3+}-montmorillonite containing aniline showed features identical with those of the complex obtained by adding a solution of aniline hydrochloride to the clay (at pH 2.5).

The interaction of aniline with montmorillonite containing different interlayer cations has been studied in some detail by Yariv et al. (1968), using infrared spectroscopy, X-ray diffraction, and differential thermal analysis. H^+/Al^{3+}-montmorillonite formed an anilinium-aniline hemisalt complex when an excess of aniline is present, while the NH_4^+-exchanged clay gave rise to ammonium-aniline ions. On the other hand, Li^+-, Na^+-, K^+-, Mg^{2+}-, and Ca^{2+}-montmorillonites adsorbed aniline principally through a water-bridging mechanism, while aniline could displace all the interlayer water from Cs^+-montmorillonite. A similar range of adsorption mechanisms was proposed by Cloos et al. (1979) and Janíková et al. (2016) for the interaction of aniline with H^+-, Al^{3+}-, Cu^{2+}-, Ca^{2+}-, and Fe^{3+}-montmorillonites.

Heller and Yariv (1969) have described the interaction of aniline and some of its derivatives with montmorillonite containing interlayer transition metal cations. Aniline could directly coordinate to the cation or indirectly through a water bridge, giving rise to type I or type II complexes, respectively. For a given compound, the tendency to form a type I complex decreased in the order Cd = Cu > Zn > Ni > Co > Mn. Both types of complexes were formed with aniline, *m*- and *p*-toluidine and *m*-chloroaniline, while only type II obtained with other aniline derivatives. Interestingly, N,N-dimethylaniline and aniline, adsorbed to Cu^{2+}-montmorillonite from the vapour phase, can dimerise to yield N,N,N',N'-tetramethylbenzidine and benzidine dications (Soma and Soma 1988b). The polymerization of aniline in the interlayer space of Cu^{2+}-montmorillonite has been reported by Ilic et al. (2000).

Essington (1994) reported that the adsorption of aniline and *o*-, *m*-, and *p*-toluidine by Ca^{2+}- and K^+-montmorillonite was at a maximum when the solution pH was close to the pK_a (4.45–5.08) of the organic bases. Maximum uptake increased with a decrease in solution ionic strength. Homenauth and McBride (1994) also found that aniline adsorption was greater at acid than neutral pH. These observations indicated that the amines were largely retained by a cation exchange mechanism. Likewise, Vansant and Uytterhoeven (1973) observed that the cationic forms of aniline, *p*-toluidine, and pyridine (pK_a 4.63–5.23) were adsorbed in excess of the cation exchange capacity (CEC) of the montmorillonite sample, the excess amount being present as the respective salt. On the other hand, benzylammonium, piperidinium, and cyclohexylammonium (pK_a 9.37–11.21) showed maximum adsorption close to the CEC of the clay sample, indicative of a stoichiometric cation exchange process that was fully reversible. Likewise, the adsorption of tyramine and histamine (as the hydrochloride salts) to Ca^{2+}-montmorillonite proceeds through a cation exchange mechanism (Malferrari et al. 2017). Adsorption of tyramine (to SAz-2 montmorillonite) conforms to the Langmuir type, yielding complexes with a basal spacing of 1.5 nm (Chang et al. 2018). We would also mention that aminoazobenzene (and azobenzene), intercalated into alkylammonium-exchanged smectites, can isomerize by irradiating the complex with UV light, causing a shift in basal spacing (Koteja et al. 2017).

Heller-Kallai and Yariv (1981) have shown that formation of interlayer coordination complexes of ethylene diamine with transition metal ions can inhibit the interlayer swelling of montmorillonite. Complexes of Na^+-bentonite with 1,5-pentanediamine, 3-(2-aminoethyl)pentane-1,5-diamine, and 3,3,-bis(2-aminoethyl)-1,5-pentanediamine can do likewise, their inhibitory effect increasing with the number of primary amine groups in the molecule (Xie et al. 2017; Ahmed et al. 2019). In this respect, polyamines and polyether amines as well as amine-based non-ionic surfactants have attracted considerable attention because of their application in water-based drilling fluids to inhibit clay swelling and improve shale stability (Anderson et al. 2010; Cui and van Duijneveldt 2010; Wang et al. 2011; Peng et al. 2013; Silva et al. 2014; Zhang et al. 2015, 2016; Xie et al. 2020; Mao et al. 2021), while polyethylene glycol is perhaps the classic and most widely known example (Theng 1982; Villabona-Estupiñán et al. 2017). The clay-polymer interaction has been comprehensively reviewed by Theng (2012).

A well-known example of the smectite-diamine interaction is the oxidation of adsorbed benzidine (Bz) to its blue monovalent radical cation ($Bz^{\bullet +}$). This reaction (Scheme 2.2) can occur in an aqueous medium, even under nitrogen (Dodd and Ray 1960; Solomon et al. 1968; Hakusui et al. 1970; Lahav and Raziel 1971; Furukawa and Brindley 1973; Lahav and Anderson 1973; McBride 1979; do Nascimento et al. 2006). Prior reduction of the clay mineral with hydrazine, or 'masking'

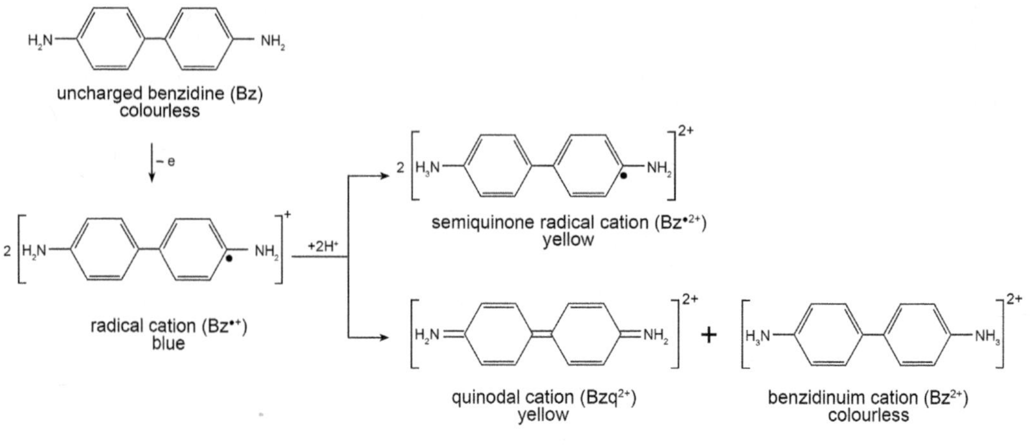

SCHEME 2.2 Transformation of adsorbed benzidine to the corresponding blue radical cation through electron transfer from the uncharged, colourless diamine to structural ferric ions or edge surface aluminium of montmorillonite. In turn, the radical cation can disproportionate or transform into several cationic species under low pH conditions.

the particle edge surface with sodium hexametaphosphate, inhibits or reduces colour development. This finding has been rationalized in terms of electron transfer from the uncharged colourless diamine (Scheme 2.2) to ferric ions in the montmorillonite layer structure and/or aluminium atoms exposed at particle edges (Solomon et al. 1968; Theng 1971; Tennakoon et al. 1974a, 1974b). The blue complex turns yellow after drying/dehydration or when the suspension pH is ~2 (Scheme 2.2). The blue-to-yellow transformation results from the protonation of $Bz^{\bullet+}$ to yield the divalent semiquinone radical cation ($Bz^{\bullet 2+}$) and quinodal cation (Bzq^{2+}), both of which are yellow, plus the colourless divalent benzidinium cation (Bz^{2+}). In the case of 3,3',5,5'-tetramethylbenzidine, adsorbed to hectorite, oxidation of the diamine is mediated more by dissolved O_2, diffusing to the clay surface, than through electron transfer since hectorite has a low content of structural Fe^{3+} (Furukawa and Brindley 1973; McBride 1985)

2.6.4 Complexes with Amides

In acidic media, amides may *a priori* accept a proton on either the oxygen or the nitrogen atom, but the weight of evidence (Fraenkel and Franconi 1960; Janssen 1961; Smith and Yates 1972) is in favour of the former alternative. Tahoun and Mortland (1966a) have deduced that O-protonation predominates from the occurrence of two bands near 1700 and 1500 cm^{-1} in the IR spectra of H^+-montmorillonite complexes with acetamide. Similar conclusions may be drawn from the spectra of N-ethylacetamide and N,N-diethylacetamide adsorbed by H^+- and Al^{3+}-montmorillonite. Being stronger bases, these compounds are protonated to a greater extent than acetamide. In addition, hemisalt formation occurs when excess amide is present.

IR spectroscopy further indicates that adsorbed acetamide, N-ethylacetamide and N,N-diethylacetamide coordinate to the exchangeable cation and adopt a flat orientation in the interlayer space, at least for low amounts adsorbed (Tahoun and Mortland 1966b). Coordination apparently occurs through the oxygen atom of the amide and is strongest with transition metal ions, weakest with alkali ions, and intermediate with alkaline earth ions. For a given counterion, bond strength falls in the order tertiary > secondary > primary amide. There is evidence to show that N,N'-diethylacetamide is linked to the counterion either directly or through a water bridge. A similar mechanism has been proposed for methylcarbamate (Fusi et al. 1989b) and 5-chlorouracil (Akyuz and Akyuz 2018). Interestingly, coordination of oxamide to Cu^{2+} counterions in montmorillonite promotes its decomposition to NH_3 and CO_2 (Ristori et al. 1991). Similarly, thioacetamide intercalated into Cd^{2+}-montmorillonite (by a solid-solid reaction), was partially decomposed to CdS when the interlayer complex was stored in air for several months (Khaorapapong et al. 2002). Earlier, Ruiz-Conde et al. (1997) made the curious observation that treatment of Mg^{2+}-vermiculite with aqueous solutions of formamide, acetamide and propionamide gave rise to NH_4^+-vermiculite through a series of interstratified Mg^{2+}- and NH_4^+-rich phases.

IR spectroscopy was also used by Doner and Mortland (1969a) to assess the interactions of trimethylammonium (Me_3NH^+)-exchanged montmorillonite with a series of dialkylamides, including N,N-dimethylformamide (DMF), N,N-diethylformamide (DEF), N,N-dimethylacetamide (DMA), N,N-diethylacetamide (DEA), and N,N-dipropylacetamide (DPA). The results indicated that dialkylamides were adsorbed by H-bonding between the oxygen of the carbonyl group and the hydrogen of Me_3NH^+ ions as shown in Scheme 2.3. The basal spacing of 1.29 nm for Me_3NH^+-montmorillonite, however, did not measurably increase after intercalation of the dialkylamides except for DEA and DPA. More recently, Zhou et al. (2019) reported that the bulky erucamide (cis-13-docosenamide) and oleamide (cis-9-octadecenamide) by themselves failed to intercalate into Na^+-montmorillonite but did so in the presence of octadecyltrimethylammonium ions.

H-bonding has also been proposed by Mortland (1968b) for the interaction of ethyl N,N-di-*n*-propylthiolcarbamate (EPTC) with pyridinium-exchanged montmorillonite (Scheme 2.4) on the basis of the changes in the NH and CO vibrational bands observed in the IR spectrum of the corresponding complex. Beall and Goss (2004) measured a basal spacing of ~3.72 nm for a 3:1

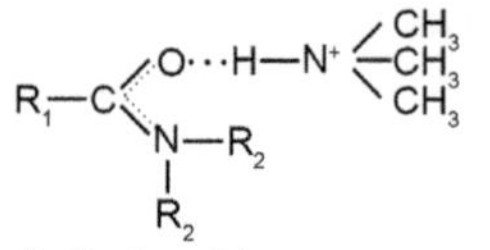

trimethylammonium

dialkylamide

SCHEME 2.3 H-bonding interaction between dialkylamide and trimethylammonium ions occupying exchange sites in montmorillonite.

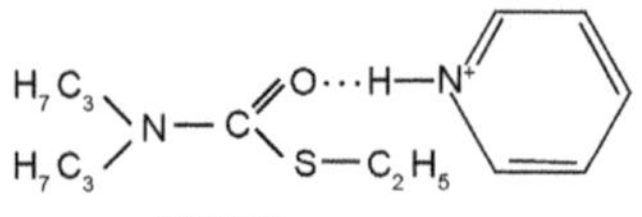

pyridinium

EPTC

SCHEME 2.4 Proposed mode of interaction between EPTC and pyridinium cations occupying exchange sites in the interlayer space of montmorillonite.

dodecylpyrrolidone:Na^+ counterion complex, indicative of 'self-assembly' of organic molecules in the interlayer space of Na^+-montmorillonite.

Because of their polar nature and high dielectric constant, liquid amides lend themselves to being used as a swelling agent of clay minerals (Olejnik et al. 1974). In this regard, Weismiller (1970) has reported on the interlayer expansion of montmorillonite and vermiculite in N-ethylacetamide. Irrespective of the nature of the exchangeable cation, montmorillonite expands to a basal spacing of ~2.05 nm, indicating intercalation of three layers of N-ethylacetamide molecules with the plane of the carbon atoms and the C=O group parallel to the silicate surface as Tahoun and Mortland (1966b) have suggested. Breen et al. (2000), however, reported that N-methylformamide only forms a two-layer complex with Ca^{2+}-, Mg^{2+}-, and Na^+-exchanged montmorillonite.

Interlayer swelling is less pronounced with vermiculite because the individual silicate layers making up a particle are more strongly held together as compared with montmorillonite. Some vermiculite samples, however, may expand to a basal spacing of 6.3 nm after prolonged immersion in liquid N-ethylacetamide. This type of expansion is also shown by complexes of vermiculite with *n*-butylammonium ions in water (Garrett and Walker 1962). Extensive or 'osmotic' swelling also occurs with vermiculite on immersion in concentrated solutions of some amino acids (Barshad 1952; Walker and Garrett 1961). Osmotic swelling has also been reported for montmorillonite-formamide complexes immersed in various polar organic liquids, including water, acetone, tetrahydrofuran, and acetonitrile (Onikata et al. 1999).

Among the amides, urea occupies a special position because of its importance as a soil nitrogen fertilizer while some urea derivatives possess herbicidal properties (Davies and Jabeen 2002; Park et al. 2004). Furthermore, intercalation into montmorillonite allows urea to be slowly released into soil (Wanyika 2014; Golbashy et al. 2017). Like acrylamide and methacrylamide, urea can intercalate into smectite by a solid-solid reaction (Ogawa et al. 1989). Here we would mention the massive intercalation of urea by mixing dehydrated montmorillonite with a molten $Ca(urea)_6^{2+}$complex (Kim et al. 2011), while co-intercalation of metal cations and urea into (Palabora) vermiculite significantly lowers the onset temperature of clay particle exfoliation (Muiambo et al. 2015).

In common with formamide and acetamide, urea can be protonated on intercalation into acidic (H^+/Al^{3+}) montmorillonite or be coordinated to the counterion when adsorbed to metal ion-exchanged clay minerals (Mortland 1966). The IR spectra of Mg^{2+}- Ca^{2+}-, Li^+-, Na^+-, and K^+-montmorillonite complexes indicate coordination of urea to the exchangeable cation through its nitrogen atom as Penland et al. (1957) have reported for Pd(II)- and Pt(II)-urea complexes. On the other hand, in complexes with Cu^{2+}-, Mn^{2+}-, and Ni^{2+}-montmorillonites urea appears to be coordinated to the transition metal ion through its oxygen atom as in the Cu $[OC(NH_2)_2]Cl_2$ complex of Penland et al. (1957). In accord with Mortland's (1966) proposal, deuterated urea and urea derivatives interact with Ca^{2+}-, Ni^{2+}-, and Al^{3+}-montmorillonite by coordination to the exchangeable cation/counterion through the oxygen atom of the carbonyl group (Farmer and Ahlrichs 1969). The IR spectra further indicate that the strength of coordination increases in the order $Ca^{2+} < Ni^{2+} < Al^{3+}$.

As might be expected, the interlayer swelling of montmorillonite immersed in aqueous urea solutions is influenced by the nature of the saturating counterion. Using a range of concentrated urea solutions (1–10 mol/L), Shiga (1961) found that Na^+-, Mg^{2+}-, and Ba^{2+}-exchanged montmorillonite expanded to a basal spacing of 1.7–2.1 nm, whereas the K^+- and NH_4^+-saturated materials failed to swell. Similar results were reported by Libor et al. (1970, 1971) using X-ray diffractometry, differential thermal, and thermogravimetric analyses, while Khaorapapong (2010) recorded a basal spacing of ~1.7 nm for an interlayer complex of Cu^{2+}-montmorillonite with thiourea. We might mention that intercalation of a small amount of urea into Mg^{2+}-, Al^{3+}-, and Fe^{3+}-vermiculite can appreciably lower the exfoliation onset temperature of the mineral (Muiambo et al. 2015). Earlier, Brigatti et al. (2005) found that intercalation of $[Al(urea)_6]^{3+}$ into vermiculite promoted layer stacking order, while the opposite effect was observed following interlayer adsorption of $[Cr(urea)_6]^{3+}$. In this respect, we would mention the wide use of a silver-thiourea complex $[Ag(TU)_2]^+$ (Pleysier and Cremers 1975) to measure the cation exchange capacity of clay minerals and soils (Pleysier and Juo 1980; Searle 1986; Dohrmann 2006).

The clay-urea interaction has found practical application in the manufacture of nitrogenous fertilizers based on urea. Since the amount intercalated by, and the strength of bonding to, clay minerals are influenced by the exchangeable cation, it is possible to vary the rate of urea release into the soil solution by using montmorillonite containing different interlayer cations (Libor et al. 1970; Wanyika 2014; Golbashy et al. 2017). It is clearly important that the nitrogen requirement of crops is supplied in small amounts over an extended period of its growth rather than all at once when much may be lost by leaching (Pereira et al. 2012). A related objective is to control the release, and reduce the leaching potential, of applied herbicides. In this respect, we would mention that intercalation of the phenylurea herbicide, diuron, into an organoclay is an effective way of meeting this objective (Trigo et al. 2009).

Similarly, the mode of bonding has a large effect on the bioactivity and persistence of herbicidal urea derivatives added to soil. Van Bladel and Moreale (1974) reported that adsorption of fenuron (3-phenyl-1,1-dimethylurea) and monuron [3-(*p*-chlorophenyl)-1,1-dimethylurea] to montmorillonite increased with the ionic potential or polarizing power (Table 2.1) of the exchangeable cation. Fusi et al. (1989a) have observed similarly for the herbicide isoxaben. Interestingly, Farmer (1967) failed to observe measurable intercalation of phenylurea, fenuron, and monuron by montmorillonite. Likewise, isoproturon [3-(4-isopropylphenyl)-1,1 dimethylurea)] does not form an interlayer complex with montmorillonite (Pantani et al. 1997). On the other hand, thiazafluron can penetrate the interlayer space of montmorillonite where the carbonyl group of the herbicide interacts with the exchangeable cation through a water bridge (Cox et al. 1995). A similar mechanism has been proposed by Davies and Jabeen (2002) for the interaction of *N,N*-dimethylurea and isoproturon with montmorillonite. The formation of a chelate complex between rimsulfuron and Cu^{2+} counterions in montmorillonite stabilizes this sulfonylurea herbicide against chemical degradation (Calamai et al. 1997).

The failure of these herbicides to enter the interlayer space of montmorillonite may be ascribed to concentration and steric factors. Thus, Kim (1970) observed the formation of mono- and double-layer intercalation complexes of fenuron, OMU (3-cyclo octyl-1,1-dimethylurea), herban [3-(hexahydro-4,7-methano indan-5-yl)-1,1-dimethylurea], and monuron with Na^+-, Ca^{2+}-, and Al^{3+}-montmorillonite, using high application rates.

IR spectroscopy indicated that urea-based herbicides were adsorbed by coordination to the exchangeable cations through a water bridge involving the C=O group of the molecules, while hydrophobic bonding would be important with alkylammonium-exchanged montmorillonite. There was little, if any, evidence for coordination involving the NH group of the herbicides although some N–H···O–Si bonding could not be excluded. Direct cation-dipole interactions only occurred with clay samples that had been dehydrated (Fusi et al. 1989a; Pantani et al. 1997; Aguer et al. 2000). The interactions of clay minerals with pesticides are described more fully in Chapter 3.

2.6.5 Complexes with Other Classes of Uncharged Organic Compounds: Ketones, Aldehydes, Ethers, Nitriles, Hydrocarbons

Glaeser (1948) and MacEwan (1948a) did much of the initial investigation on the formation of interlayer complexes between clay minerals and various uncharged organic compounds. For example, MacEwan (1948b) prepared two-layer complexes of Ca^{2+}-montmorillonite with acetone and nitrobenzene by treating the clay mineral with a large excess of the respective organic liquid (with or without boiling). Under the same conditions, acetonitrile and nitromethane gave rise to three-layer complexes. These observations indicate that the number of intercalated layers is dependent on the ratio of dipole moment (μ) to molecular size (volume) of the organic compound. Subsequently, Hoffmann and Brindley (1960) suggested that the extent of intercalation – at least from aqueous solution – was controlled more by the CH activity of the organic molecule than by its polarity. Thus, if polarity *per se* were important, the nitriles with a dipole moment (μ) of 3.5–4.0 D would show higher molar adsorption than the 1,3-diketones (μ = 3.0 D) of comparable chain length and molecular weight. Similarly, bis-(2-ethoxyethyl)-ether (μ = 2.0 D) would be adsorbed to a lesser extent than triethylene glycol (μ ~2.5 D). In both cases, however, the opposite behaviour was observed (Figure 2.15).

As would be expected, complex formation between clay minerals and ketones is profoundly affected by the nature of the counterion and the water content (hydration status) of the clay mineral. Early on, Glaeser (1948) reported that dehydrated Na^+- and Ca^{2+}-montmorillonites gave one-layer complexes (*d*(001) ~1.32 nm) when treated with acetone vapour. On the other hand, two-layer complexes (*d*(001) = 1.75 nm) were formed in the presence of saturated acetone and water vapour. Moreover, dehydrated Ca^{2+}-montmorillonite invariably yielded a two-layer complex, while the Na^+-exchanged clay gave either a single- or a double-layer complex. This observation accords with that by Bissada et al. (1967) and may be ascribed to the greater solvation energy of the calcium ion relative to that of the sodium ion.

The interaction of acetone, 2,4-pentanedione (acetylacetone), and 2,5-hexanedione with montmorillonite containing different interlayer cations has been examined by Parfitt and Mortland (1968), using X-ray diffraction and infrared spectroscopic techniques. The results indicate that adsorption to Na^+- montmorillonite occurs through (direct) cation-dipole interaction between the counterion and the carbonyl group of the ketone, giving rise to inner-sphere complexes. This type of bonding can also occur with montmorillonite containing dehydrated polyvalent counterions as Dios-Cancela et al. (2000) have postulated for acetone, acetonitrile and dimethylsulphoxide, and Sohn and Kim (2000) for methyl methacrylate. With Ca^{2+}- and Mg^{2+}-montmorillonites, on the other hand, the carbonyl group is linked to the counterion through a water bridge, forming outer-sphere complexes. Earlier, Tensmeyer et al. (1960) have proposed similarly for the interaction of 2,5-hexanedione and 2,5,8-nonanetrione with Ca^{2+}-montmorillonite. In addition, electrovalent bonding through proton donation to the clay is indicated for acetylacetone adsorbed to Al^{3+}-and Cu^{2+}-montmorillonites, resulting in the formation of the corresponding metal-acetylacetonates. More recently, Charaabi et al. (2021) have suggested that the adsorption of benzophenone-3 (from ethanol) to laponite and montmorillonite is consistent with a pseudo-second order kinetic model, involving surface diffusion followed by π-π interactions between the aromatic ketone and the clay mineral.

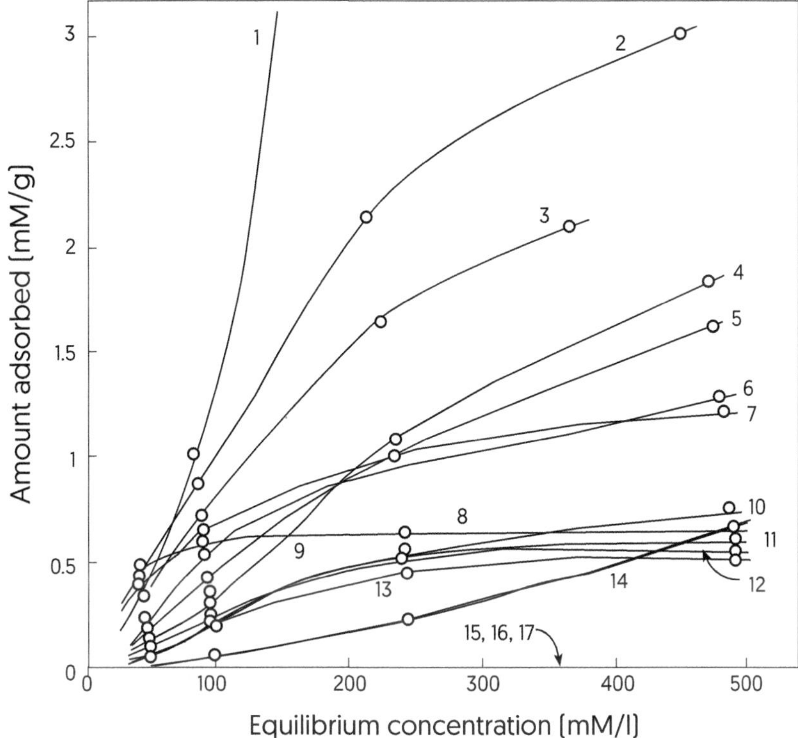

FIGURE 2.15 Isotherms for the apparent adsorption to Ca^{2+}-montmorillonite of some uncharged aliphatic compounds from aqueous solution. (1) acetylacetone; (2) α-methoxy acetylacetone; (3) acetoacetic ethylester; (4) β-ethoxypropionitrile; (5) β:β'-oxydipropionitrile; (6) bis-(2-ethoxyethyl)ether; (7) 2,5,8-nonantrione; (8) triethylene glycoldiacetate; (9) diethylene glycoldiacetate; (10) 2,5-hexanedione; (11) bis-(2-methoxyethyl) ether; (12) ethyleneglycol diglycidether; (13) triethyleneglycol; (14) 1,6-hexanediol; (15) 1,5-pentanediol; (16) 2,4-hexadynedol-1,6; (17) diethyleneglycol. (From Hoffmann, R.W. and Brindley, G.W. *Geochim. Cosmochim. Acta* 20, 15–29, 1960.)

MacEwan (1948a) was able to obtain a one-layer complex of halloysite with acetaldehyde by treating the mineral with a large excess of the (cold) liquid. Using a similar technique, followed by heating the clay-organic liquid mixture at 60°C, Larson and Sherman (1964) prepared complexes of bentonite (montmorillonite) with several aldehydes (aldol, acrolein, benzaldehyde, crotonaldehyde). The infrared spectra of the complexes indicated that the carbonyl group of the aldehyde was hydrogen-bonded to the silicate layer. Like ketones, however, aldehydes are likely to interact with the exchangeable cations of the clay mineral, either directly or through a bridging water molecule. Since neither the basal spacing of the complexes, nor the amount adsorbed was measured, the occurrence of intercalation could not be ascertained.

Following up this work, Sohn et al. (2005) observed an appreciable downward shift of the carbonyl stretching band in the infrared spectra of acetaldehyde, acrolein, and crotonaldehyde adsorbed to dehydrated Ni^{2+}-, Ca^{2+}-, and Al^{3+}-montmorillonites. This finding strongly indicated that complex formation here occurred by ion-dipole interaction involving the carbonyl group of the alkyl aldehydes. Koyuncu (2008) later reported that the kinetics for the adsorption of 3-hydroxybenzaldehyde (from ethanol) to bentonite could be fitted to a pseudo second-order model. Glaeser and Méring (1952) further suggested that acetaldehyde could convert to acetic acid when adsorbed to Ca^{2+}-montmorillonite, but the underlying mechanism was not specified. That clay minerals can catalyze the oxidation of adsorbed organic molecules, however, is now well established (Theng 2018).

The formation of (two-layer) complexes of montmorillonite with ethers and polyethers was first reported by Bradley (1945a). Subsequently, MacEwan (1948a) prepared (one-layer) complexes of halloysite with some alcohol ethers. The montmorillonite complex with 1,4-dioxane had a basal spacing of 1.47–1.50 nm. This value was too low for a double-layer of intercalated molecules but comparable to that shown by a single-layer pyridine complex with the plane of the ring lying normal to the silicate surface (Figure 2.2b).

The adsorption of some ethers, ether-alcohols, and ether-esters from dilute aqueous solutions by Ca^{2+}-montmorillonite has been investigated by Hoffmann and Brindley (1960). Both single- and double-layer complexes were formed giving basal spacings of 1.31–1.34 and 1.57–1.76 nm, respectively. The shortening of contact distances and the probable interlayer orientation of the organic molecules in single-layer complexes, deduced from the respective Δ-values, are summarized in Table 2.4.

TABLE 2.4
Basal Spacings of Single- and Double-Layer Complexes Formed Between Ca^{2+}-Montmorillonite and Various Polar Organic Compounds. Also Indicated Are the Δ-Value of Single-Layer Complexes and the Calculated Thickness and Probable Orientation of the Intercalated Compounds.

	Basal spacing (nm)			Calculated molecular thickness[b] (nm)		Shortening of contact distance (nm)		
Compound	Single-layer	Double-layer	Δ-value[a] (nm)	α_I	α_{II}	α_I	α_{II}	Probable orientation
Bis-(2-ethoxyethyl) ether	1.34	1.67	0.38	0.45	0.41	0.07	0.03	Perpendicular
Bis-(2-methoxyethyl) ether	1.32	1.70	0.36	0.45	0.41	0.07	0.05	Perpendicular
Ethyleneglycol diglycidyl ether	1.34	1.76	0.38	n.d.	n.d.	n.d.	n.d.	n.d.
Triethylene glycol	1.33	1.73	0.37	0.45	0.41	0.08	0.04	Perpendicular
Diethylene glycol	1.33	1.57	0.37	0.45	0.41	0.08	0.04	Perpendicular
Ethoxy trimethylene glycol	n.a.	1.67	n.a.	n.a.	n.a.	n.a.	n.a.	n.a.
1,4-Dioxane	1.47–1.50	n.a.	0.51–0.54	0.59	n.a.	0.04	n.a.	Perpendicular (?)
Carbitol	n.a.	1.69	n.a.	n.a.	n.a.	n.a.	n.a.	n.a.
Methoxy triethylene glycol	n.a.	1.74	n.a.	n.a.	n.a.	n.a.	n.a.	n.a.
Dimethylether of ethylene glycol	n.a.	1.70	n.a.	n.a.	n.a.	n.a.	n.a.	n.a.
Dimethylether of triethylene glycol	n.a.	1.75	n.a.	n.a.	n.a.	n.a.	n.a.	n.a.
Dimethylether of tetraethylene glycol	n.a.	1.77	n.a.	n.a.	n.a.	n.a.	n.a.	n.a.
Triethylene glycol diacetate	1.33	1.65	0.37	0.45	0.41	0.08	0.04	Perpendicular
Diethylene glycol diacetate	1.31	1.65		0.45	0.41	1.0	0.06	Parallel (?)
β:β'-Oxypropionitrile	1.31	1.57	0.35	0.48	0.41	1.3	0.06	Parallel
β-Ethoxypropionitrile	1.32	n.a.	0.38	0.48	0.41	1.2	0.05	Parallel

Sources: Hoffmann, R.W. and Brindley, G.W., *Geochim. Cosmochim Acta* 20, 15–29, 1960; Bradley, W.F., *J. Am. Chem. Soc.*, 67, 975–981, 1945; MacEwan, D.M.C., *Trans. Faraday Soc.*, 44, 349–367, 1948.

[a] single-layer spacing minus 0.96 nm (silicate layer thickness)

[b] (α_I), plane of aliphatic chain perpendicular to the silicate surface; (α_{II}), plane of aliphatic chain parallel to silicate surface
n.a. = not applicable; n.d. = not determined

Brindley et al. (1963) have compared the adsorption of acetoacetic ethylester, β:β'-oxydipropionitrile, 2,5-hexanedione, and triethylene glycoldiacetate from aqueous solutions by gibbsite, kaolinite, and montmorillonite. Because of its relatively large surface area, montmorillonite had the largest adsorption capacity in terms of moles per unit weight of mineral. However, gibbsite came top when the results were expressed per unit surface or as an equivalent number of packed organic layers. Being devoid of exchangeable cations, gibbsite could not form complexes through ion-dipole interactions. Interactions by O···H–O bonding, however, are possible for both gibbsite and kaolinite.

Ion-dipole interactions have been proposed by Brindley and Tsunashima (1972) in the formation of montmorillonite complexes with dioxane, morpholine, and piperidine. The latter two compounds can exist as cations under acid conditions and be taken up by a cation-exchange mechanism. Breen et al. (1987a) have also alluded to the possibility of dioxane forming a bidentate coordination complex with aluminium ions exposed at the edge surface of montmorillonite particles. Desorption studies by Breen (1994) also indicate that cyclic ethers, such as tetrahydropyran, tetrahydrofuran, and dioxan are more strongly adsorbed than water by various cation-exchanged montmorillonites. More recently, Abbas et al. (2017) have found that montmorillonite, exchanged with hexadecyltrimethylammonium ions, is an effective adsorbent of methyl *tert*-butyl ether. The adsorption of anisole and related aromatic ethers to transition metal-exchanged smectites (under dehydrating conditions) is described later.

The interactions of 2:1 type layer silicates with crown ethers and their three-dimensional analogues, cryptands, have been the subject of many studies (Ruiz-Hitzky and Casal 1978; Guignard and Pezerat 1979; Casal and Ruiz-Hitzky 1986; Yao et al. 2002; Chrayet et al. 2017). These macrocyclic compounds (Scheme 2.5) intercalate into montmorillonite by displacing the water molecules around, and acting as ligands of, the interlayer counterions. X-ray diffraction and dichroic infrared spectroscopy indicated the formation of one- and two-layer complexes with the plane of the organo-cation complex lying flat or tilted to the interlayer surface (Casal et al. 1994). The intercalation of crown ethers (15-crown-5 and 18-crown-6) into NH_4^+-montmorillonite markedly altered the infrared spectrum of the counterion (Casal et al. 1984). This finding was ascribed to the lowering of the T_d symmetry of the ammonium counterion following its encapsulation by the macrocylic compound. More recently, Liu et al. (2006) prepared complexes of Na^+-bentonite with 4'-substituted monobenzo-15 crown-5 ethers. X-ray diffraction indicated that these compounds adopted a flat-lying or tilted conformation in the interlayer space. Furthermore, the complexes were good adsorbents of naphthalene and tetrachloroethane. Qu et al. (2011) reported that complexation of K^+ counterions by 18-crown-6 ether dramatically reduced 1,3-dinitrobenzene adsorption by montmorillonite, while Chrayet et al. (2017) used an interlayer smectite-crown ether complex to remove various heavy metals from aqueous solution.

Guégan et al. (2009) reported the synthesis of a double-layer complex between Ca^{2+}-montmorillonite and triethylene glycol mono-*n*-decyl ether ($C_{10}E_3$). Above the critical micelle

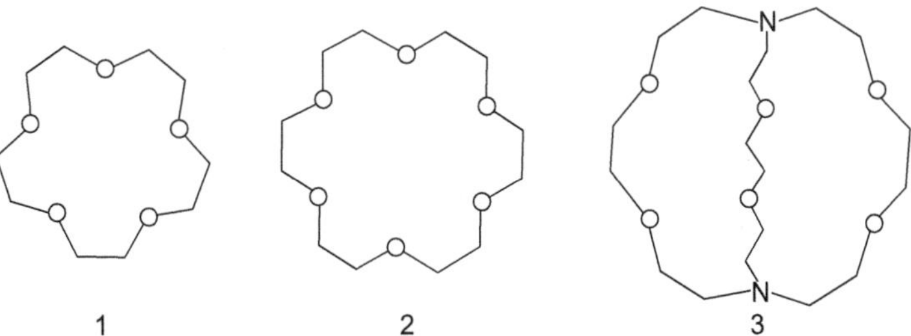

SCHEME 2.5 Structures of some crown ethers. These macrocyclic molecules can intercalate into montmorillonite. (1) 15-crown-5; (2) 18-crown-6; (3) 2,2,2-cryptand.

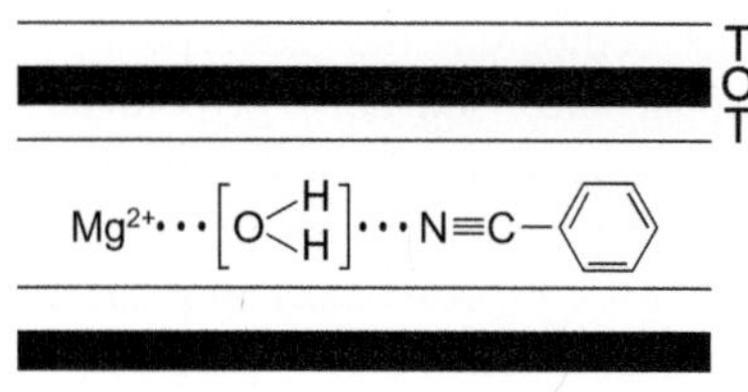

FIGURE 2.16 Proposed mode of interaction between benzonitrile and Mg^{2+}-montmorillonite showing H-bonding of the nitrile group to interlayer Mg^{2+} ion by means of a water bridge. (From Serratosa 1968b, Figure 1. Reproduced with permission of the Mineralogical Society of America publisher of *American Mineralogist.*)

concentration (cmc), this uncharged (non-ionic) surfactant can self-assemble into a lamellar phase. Accordingly, $C_{10}E_3$ can form different intercalation complexes with Na^+-montmorillonite (Guégan 2010, 2013). Recent molecular dynamic simulation studies by Mao et al. (2021) indicate that (uncharged) polyether amines, intercalated into Na^+-montmorillonite, adopt a crown-like structure enclosing the sodium counterions. Similarly, Malinová et al. (2016) were able to intercalate lactim methyl ethers of varying ring size, in the form of the respective hydrochloride salts, into Na^+-montmorillonite. The basal spacing of the complexes increased regularly with the number of carbon atoms in the ring, from 1.28 to 1.55 nm.

The intercalation of benzonitrile into montmorillonite containing divalent counterions, has been investigated by Serratosa (1968b) using infrared spectroscopy. In the interlayer complex with Mg^{2+}-montmorillonite, benzonitrile was hydrogen-bonded through its –C≡N group to the counterion by way of a water bridge as shown in Figure 2.16).

When some of the Mg^{2+}-coordinated water was removed (by evacuation), the organic molecule could interact directly with the counterion. Prolonged evacuation (at ~80°C), however, failed to reduce the water content below that corresponding to the presence of two water molecules (and four organic molecules) per Mg^{2+} ion. Under these conditions, benzonitrile was intercalated with the aromatic ring inclined at a high angle, and the –C≡N bond parallel to the silicate surface. At the same time, the cation-coordinated water molecules had their molecular axes at a high angle to the surface while their oxygens pointed inwards towards the interlayer Mg^{2+} counterion. The basal spacing of this complex was 1.51 nm, giving a Δ-value of ~0.56 nm, indicating slight keying of the organic molecule into the ditrigonal depression in the surface oxygen plane (Figure 2.17). Ca^{2+}- and Ba^{2+}-montmorillonite yielded similar complexes with benzonitrile, but the cation-coordinated water molecules here could be removed by evacuation.

Like benzonitrile, acetonitrile interacted with the interlayer counterion of montmorillonite through its –C≡N group (Tarasevich et al. 1968). Little, if any, interaction occurred between acetonitrile and interlayer K^+ since the C≡N stretching frequency, $v_{(C\equiv N)}$, in the infrared spectrum of adsorbed acetonitrile was virtually coincident with that of the pure liquid (2254 cm^1). On the other hand, $v_{(C\equiv N)}$ in the spectra of complexes with Na^+-, Ca^{2+}-, and Cu^{2+}-montmorillonite (exposed to acetonitrile vapour at $P/P_0 = 0.1$) showed an appreciable increase, in the order Na^+ (13 cm^{-1}) < Ca^{2+} (24 cm^{-1}) < Co^{2+} (64 cm^{-1}) < Cu^{2+} (76 cm^{-1}). In the case of Cu^{2+}-montmorillonite the magnitude of the frequency shift (~80 cm^{-1}) is comparable to that observed for the interaction of acetonitrile with $AlCl_3$ and $SnCl_4$ which are strong electron acceptors (Lewis acids). This finding strongly indicates that the nitrile group of the molecule is directly coordinated to interlayer Cu^{2+} ion.

Entirely parallel observations were reported by Yariv et al. (1966) for complexes of montmorillonite with nitrobenzene and benzoic acid in that adsorption to K^+- and NH_4^+-montmorillonite occurred by cation-dipole interactions, and by water-bridging to montmorillonite exchanged with Cu^{2+}, Mg^{2+}, and Ca^{2+} ions. Infrared spectroscopy and X-ray diffraction indicated that in freshly prepared complexes, the nitrobenzene molecule was intercalated in an orientation like that shown by

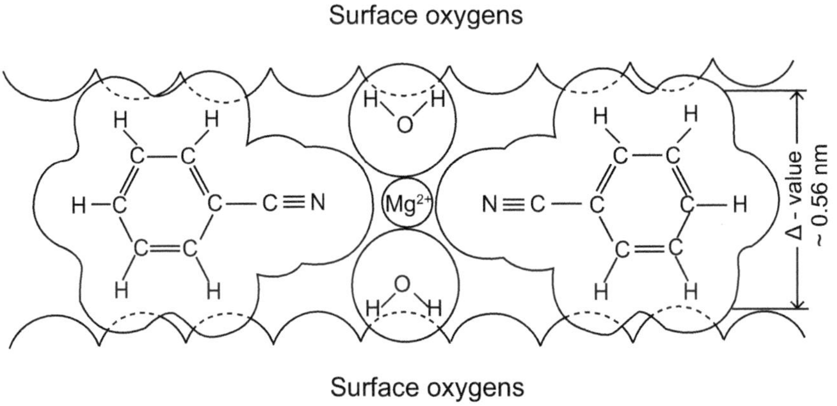

FIGURE 2.17 Diagram showing the direct interaction of benzonitrile with Mg^{2+} counterion in (partially) dehydrated montmorillonite. The aromatic ring of benzonitrile is inclined at a high angle, and the –C≡N bond is parallel to the silicate layer giving a interlayer separation (Δ-value) of ~ 0.56 nm. (From Serratosa 1968b, Figure 2. Reproduced with permission of the Mineralogical Society of America publisher of *American Mineralogist.*)

benzonitrile (Figure 2.17) with the plane of the aromatic ring inclined at a high angle to the silicate surface. Exposure of the complex to water vapour caused a change in molecular conformation from an upright orientation to one in which the benzene ring lay parallel to the silicate surface.

Similarly, benzoic acid interacted (through its carboxyl group) either directly with the interlayer cation, or indirectly by means of a water bridge, depending on the polarity of the cation and the hydration status of the clay sample. Basal spacings of 1.24–1.28 nm were recorded for complexes of montmorillonite exchanged with copper and monovalent cations, indicating a flat orientation of the intercalated molecule. The complex with Ca^{2+}-montmorillonite, however, gave a basal spacing of 1.49 nm indicative of an upright orientation.

Hydrocarbons are non-polar or, at best, only very weakly polar. For this reason, cation-dipole interactions would not be expected to play an important part in their adsorption to clay minerals. Rather, these compounds interact with smectites primarily through cation-π bonding or electron donor-acceptor complexation, supplemented by H-bonding and van der Waals forces (Clementz and Mortland 1972; Warren et al. 1986; Siantar et al. 1994; Zhu et al. 2004; Keiluweit and Kleber 2009; Morozov et al. 2014; Jia et al. 2015; Zhao et al. 2017, 2022).

As already remarked on, in dehydrated or calcined systems where the silicate layers of smectite are fully collapsed, intercalation of non-polar organic liquids and gases is either absent or proceeds only with difficulty. Adsorption then takes place predominantly on external particle surfaces (Barrer 1978). In partly dehydrated or air-dry systems where the silicate layers are slightly separated, intercalation can occur but is limited by the inability of non-polar molecules to replace interlayer water or establish links to the exchangeable cation through water bridges although the hydrated siloxane surface can provide adsorption sites (Liu et al. 2013b).

Between these two states of hydration, optimum conditions may be found under which interlayer sorption can occur although the results are not always self-consistent. The early X-ray diffraction measurements by Barshad (1952), for example, indicated that liquid benzene and *n*-hexane could enter the interlayer space of montmorillonite and vermiculite that have previously been dried at 20°C but no intercalation was observed after drying the samples at 250°C. On the other hand, MacEwan (1948b) reported that *n*-hexane and *n*-heptane did not intercalate into air-dry Ca^{2+}- or NH_4^+-montmorillonite by boiling the mineral with the organic liquids, but benzene, naphthalene, and tetrahydronaphthalene did so. Indeed, benzene formed a double-layer complex as Bradley

(1945a) had observed earlier. Greene-Kelly (1955a), however, failed to detect interlayer sorption of benzene by montmorillonite that had been dried at 80°C.

Eltantawy and Arnold (1972) reported the formation of single-layer complexes of several cation-exchanged montmorillonites with *n*-hexane and *n*-dodecane by exposing air-dry samples of the mineral to the hydrocarbon vapour for 3 hours at room temperature. The interlayer complex of Ca^{2+}-montmorillonite with *n*-hexane gave a basal spacing of 1.41 nm. Slight variations in basal spacing between complexes were ascribed to differences in the packing and orientation of the hydrocarbon molecules in the interlayer space.

The presence of water apparently affects the rate of intercalation by slowing down the diffusion of the *n*-alkanes into the interlayer space. Furthermore, the intercalated molecules must be distributed and ordered in a certain way, causing a delay in complex formation. Thus, failure on the part of some workers to observe complex formation with non-polar organic molecules may be ascribed, at least partly, to insufficient time being allowed for intercalation to occur. Here we might add that treatment of hectorite with variably wet supercritical methane (under reservoir conditions) can promote the sorption of CH_4 to interlayer and external particle surfaces of the mineral. Maximum intercalation occurs when the mean basal spacing is ~1.15 nm, allowing the methane molecules to occupy interlayer vacancies (Bowers et al. 2018). High pressures (up to 18.0 MPa) facilitate and promote methane adsorption to clay minerals. With kaolinite and illite uptake is confined to external particle surfaces but montmorillonite can intercalate up to ~6 cm^3/g of methane (Liu et al. 2013c).

The influence of surface hydration on complex formation is well illustrated by the interactions of aromatic hydrocarbons with montmorillonite containing transition metal cations. For example, when benzene is presented as a vapour to Cu^{2+}-montmorillonite (under dehydrating conditions), a portion is physically adsorbed while the remainder is coordinated to the Cu^{2+} counterion forming two types of coloured complexes. Which type is formed appears to be critically dependent on the degree of hydration of the clay sample. Infrared spectroscopy indicates that type I forms at a higher water content than type II, both types being interconvertible by simply changing the hydration status of the system within a given, narrow limit (Doner and Mortland 1969b; Mortland and Pinnavaia 1971; Van de Poel et al. 1973). Like benzene, symmetrical arenes (biphenyl, naphthalene, anthracene) (Rupert 1973; Tricker et al. 1975) as well as thiophene (Cloos et al. 1973), and anisole (Fenn et al. 1973) form both types of complexes. On the other hand, methyl-substituted benzenes (toluene, mesitylene, xylene), phenyl ether and benzyl methyl ether form only type I complexes (Pinnavaia and Mortland 1971; Fenn et al. 1973). Type II complexes are also formed with montmorillonite containing Fe^{3+}, Al^{3+}, Ru^{3+}, and VO^{2+} counterions (Pinnavaia et al. 1974; Cloos et al. 1979; Soma et al. 1986; Gu et al. 2011; Jia et al. 2016).

In type I (green) complex, benzene is apparently edge-π-bonded to Cu^{2+} retaining both ring planarity and aromaticity. ESR spectroscopy indicates that in type II (red) complex the benzene ring is distorted, and aromaticity is lost following the transfer of a single electron from benzene to Cu^{2+}, giving rise to either a dication or a radical cation plus a diamagnetic Cu^{+} ion (Rupert 1973) as shown in Figure 2.18.

Using resonance Raman spectroscopy, Soma et al. (1984) have observed that para-substituted benzenes and 4,4'-substituted biphenyls also form radical cations in the interlayer space of transition metal-exchanged montmorillonite. Radical cation formation also occurs with dioxins (Boyd and Mortland 1985), pentachlorophenol (Boyd and Mortland 1986; Gu et al. 2011), 4-chloroanisole (Govindaraj et al. 1987), benzidine and N,N,N',N'-tetramethylbenzidine (Soma and Soma 1988b), chloroethenes (Mortland and Boyd 1989), 1-aminonaphthalene (Ainsworth et al. 1991), *p*-xylene (Johnston et al. 1992), and toluene (Tipton and Gerdom 1992).

Such radical cations can interact mutually to yield dimeric, oligomeric, and polymeric products (Mortland and Halloran 1976; Mortland and Boyd 1989). Thus, benzene and biphenyl yield poly(*p*-phenylene) as shown in Figure 2.18 (Stoessel et al. 1977; Soma et al. 1983, 1986; Walter et al. 1990; Hinedi et al. 1993), while anisole gives rise to 4,4'-dimethoxybiphenyl (Fenn et al. 1973), pyrene to its dimer (Joseph-Ezra et al. 2014), and thiophene and methylthiophene yield the respective polymers

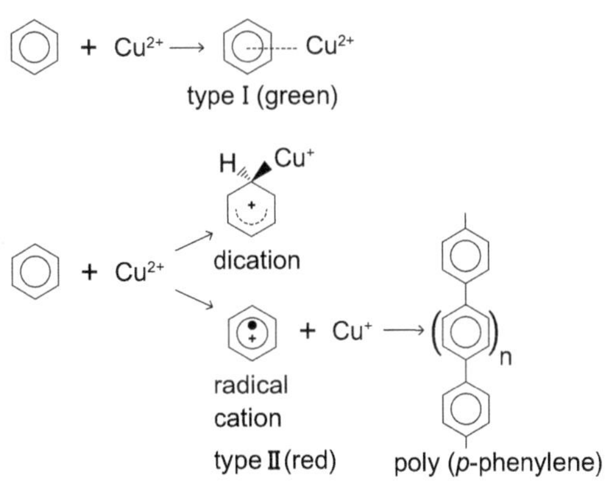

FIGURE 2.18 Formation of type I and type II interlayer complexes between benzene and Cu^{2+}-montmorillonite under dehydrating conditions. Rapid electron interchange may take place between the dication and the radical cation in type II complex. The radical cation may interact mutually to yield poly(*p*-phenylene). (Sources: Doner, H.E. and Mortland, M.M., *Science*, 166, 1406–1407, 1969b; Mortland, M.M. and Pinnavaia, T.J., *Nature Phys. Sci.* 229, 75–77; Rupert, J.P., *J. Phys. Chem.*, 77, 784–790, 1973; Stoessel, F. et al., *Clay Miner.*, 12, 255–259, 1977; Soma, Y. et al., *Chem. Phys. Lett.*, 99, 153–156, 1983.)

(Soma et al. 1987). No significant dimerization or polymerization, however, occurs when the *para* positions on the benzene ring are occupied as in *p*-dimethoxybenzene (Soma et al. 1986; Johnston et al. 1991). The monograph by Theng (2018) has more details on the clay-catalyzed dimerization, oligomerization, and polymerization of hydrocarbons.

Incidentally, both saturated and unsaturated hydrocarbons may be produced in the interlayer space of montmorillonite by the *in-situ* decomposition of intercalated organic compounds. Thus, a mixture of *n*-alkanes and *n*-alkenes were formed when interlayer complexes of montmorillonite with alkylammonium ions were heated (Weiss and Roloff 1963; Chaussidon and Calvet 1965; Durand et al. 1972). Similar results have been reported for montmorillonite complexes with alcohols (Galway 1969, 1970) and fatty acids (Shimoyama and Johns 1971).

As already alluded to, intercalation of non-polar aliphatic and aromatic molecules into smectite is problematic even when the silicate layers are slightly separated by the presence of interlayer water. Indeed, prior dehydration of clay mineral sorbents by heating at 120°C is conducive to benzene intercalation (Deng et al. 2017). For the larger phenanthrene molecules, capillary condensation into the network of micro- and nano-sites of smectite quasi-crystals, rather than intercalation, is highly probable (Hundal et al. 2001).

There is no doubt, however, about the ease and capacity of organically modified montmorillonite in intercalating these molecules, even bulky hydrocarbons, such as β-carotene (Kakegawa and Ogawa 2002). Much of the early research on the sorption and intercalation of non-polar gases and vapours by alkylammonium-exchanged smectites has been summarized by Barrer (1978).

We should also mention that organoclays can be highly selective in taking up aliphatic and aromatic hydrocarbons (Barrer and Reay 1957; Barrer and Perry 1961; Barrer 1989; Lee et al. 1989, 1990), a property which forms the basis of their wide use as gas-chromatographic separation media (Barrer and Hampton 1957; Mortimer and Gent 1964; Taramasso and Veniale 1969; Slabaugh and Vasofsky 1975; Berezkin and Gavrichev 1976; Chukwunenye and McAtee 1987; Eiceman and Lara 1991). In this regard, alkylammonium- and surfactant-modified smectites have attracted a great deal of attention because of their potential for removing organic contaminants and pollutants from the environment (Park et al. 2011; Lee and Tiwari 2012; Sarkar et al. 2012; Yuan et al. 2013; Liu et al. 2014; Zhu et al. 2015). The interactions of clay minerals with positively charged (cationic) organic compounds are further described in Chapter 4.

The capacity and efficiency of organically modified clays in sorbing aromatic hydrocarbons and related compounds are influenced by the size, shape, and packing density of the modifying compounds as well as on the structure (polarity) and hydrophobicity of the contaminant/pollutant molecules. As such, the sorption of uncharged organic molecules to organoclays is controlled by a variety of mechanisms.

Boyd and co-workers (Boyd et al. 1988b; Jaynes and Boyd 1990, 1991b; Roberts et al. 2006) together with Smith et al. (1990) did much of the research on the sorption of benzene and its derivatives by smectite exchanged with either compact quaternary ammonium ions, notably trimethylphenylammonium (TMPA) and tetramethylammonium (TMA), or their long-chain counterparts, such as hexadecyltrimethylammonium (HDTMA). With HDTMA-exchanged clay minerals, solute sorption from water increases with mineral charge density and interlayer separation. Figure 2.19 shows this relationship for ethylbenzene. The sorption of this non-polar aromatic compound to HDTMA-modified clay minerals gives rise to linear isotherms indicating that the underlying mechanism is one of solute partitioning. Such a mechanism is also operative in the uptake of naphthalene and phenanthrene by dodecylpyridinium-exchanged montmorillonite (Changchaivong and Khaodhiar 2009). On the other hand, the isotherms for the uptake of polar aromatics (e.g., phenol and *p*-nitrophenol) by HDTMA-montmorillonite and similar organoclays are commonly non-linear, indicative of surface-solute interaction (Zhu et al. 1998; Ko et al. 2007; Ozola et al. 2019), while anionic and zwitterionic aromatic compounds can adsorb through electrostatic interactions with HDTMA (Xu et al. 2011).

The isotherms for the sorption of non-polar aromatics to TMPA-smectite are generally of the Langmuir form (Jaynes and Boyd 1991a; Jaynes and Vance 1999). Furthermore, sorption increases with a decrease in layer charge density (and quaternary ammonium ion occupancy) as shown in Figure 2.20 for the sorption of benzene by high- and low-charge TMA-smectites (Lee et al. 1990). Here the solute molecules interact with the (hydrophobic) siloxane surface, exposed between the interlayer TMA 'pillars' although a limited amount of benzene can apparently be incorporated into TMA (Chun et al. 2003; Deng et al. 2020). Similar results have been reported by Shen (2004) for the sorption of phenol by benzyltrimethylammonium-exchanged smectite.

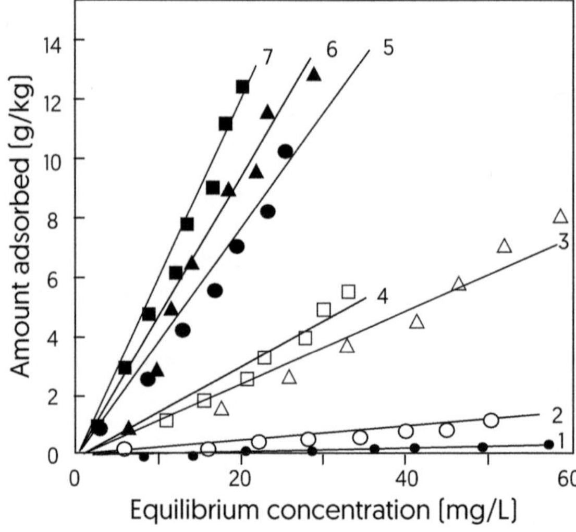

FIGURE 2.19 Isotherms for the adsorption of ethylbenzene (from water) to different clay minerals fully exchanged with hexadecyltrimethylammonium (HDTMA) ions. Line 1: parent Mg^{2+}-montmorillonite (SWy-1, low charge); line 2: kaolinite (KGa-2); line 3: montmorillonite (SWy-1); line 4: illite ((IMt-1); line 5: nontronite (SWa-1); line 6: vermiculite (VSC); line 7: montmorillonite (SAz-1, high-charge). (From Jaynes, W.F and Boyd, S.A., *Soil Sci. Soc. Am. J.* 55, 43–48, 1991b.)

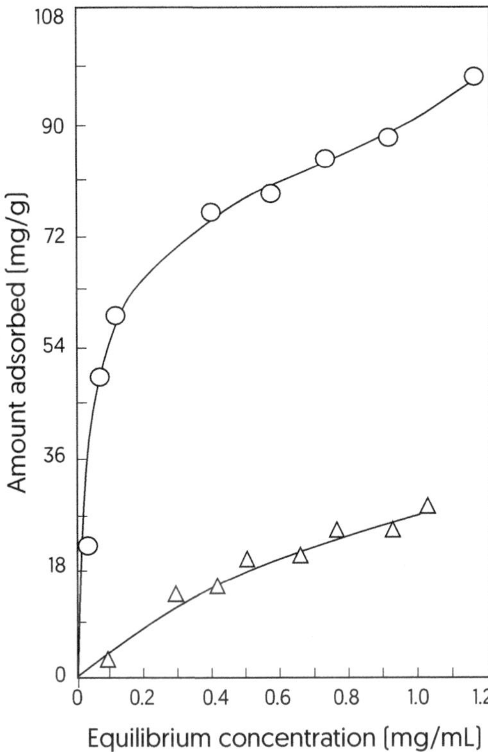

FIGURE 2.20 Isotherms for the adsorption of benzene (from water) to interlayer complexes of montmorillonite with tetramethylammonium (TMA) ions. (Δ), high-charge montmorillonite (SAz-1); (○), low-charge montmorillonite (SWy-1). (From Lee et al. 1990, Figure 1. Reproduced with kind permission of The Clay Minerals Society, publisher of *Clays and Clay Minerals*.)

Kukkadapu and Boyd (1995) also made the interesting observation that TMA- and tetramethylphosphonium (TMP)-exchanged smectite could take up more aromatic and chlorinated hydrocarbons from the vapour phase than from water. They proposed that hydration of TMA and TPA cations reduced the access to, and availability of, the siloxane surface to the solutes. In keeping with this proposal, FTIR spectroscopy indicates that water preferentially hydrates the interlayer TMA (and TMPA) counterions, thereby shielding the interionic siloxane surface (Stevens and Anderson 1996b; Stevens et al. 1996). Thus, the sorption of nitrobenzene to tetraethylammonium-, benzyltrimethylammonium-, and hexadecylammonium-montmorillonite is appreciably enhanced when the organoclays are preheated at 150°C (Borisover et al. 2010a, 2010b).

Sheng and Boyd (2000) have reported that the sorption of dichlorobenzene isomers to HDTMA-smectites is influenced by the layer charge density of the clays as well as by the polarity of the solutes and the basal spacing of the organoclays. The heterogeneity of layer charge distribution within individual smectite particles (cf. Chapter 1) is an added complication (Sheng et al. 1997). ^{13}C NMR spectroscopy further indicates that the partitioning of sorbed phenanthrene into tetradecyltrimethylammonium counterions in montmorillonite changes the interlayer orientation of TDTMA. As a result, the TDTMA chains become disordered and intimately mixed with phenanthrene (Theng et al. 1998) which then becomes physically inaccessible to degrading bacteria and their enzymes (Theng et al. 2001). Interestingly, Ogawa et al. (1993) found that anthracene and pyrene were dispersed among the alkyl chains of dimethyldioctadecylammonium counterions in montmorillonite. A similar change in interlayer microstructure has also been proposed by Zhu et al. (2008) and Ghavami et al. (2017) with respect to the sorption of naphthalene by HDTMA-montmorillonite.

Molecular dynamics simulations further suggest that intercalated naphthalene is H-bonded to the siloxane surface although solute partitioning into HDTMA is the dominant mechanism. The facile intercalation of naphthalene and anthracene into HDTMA-montmorillonite through a solid-state reaction has been reported by Ogawa et al. (1992).

^{1}H-NMR spectroscopy points to the probable involvement of (specific) π-π interactions between the aromatic ring of naphthalene and cetyldimethylbenzylammonium counterion (Xu et al. 2014). This type of intermolecular interaction has also been implied in the adsorption of several herbicides (alachlor, metolachlor, norflurazon) (Nir et al. 2000) and phenanthrene (El-Nahhal and Safi 2004) to benzyltrimethylammonium-exchanged montmorillonite. Similarly, π-π interactions have been invoked by Kameda et al. (2012) for the uptake of various polar aromatic compounds by tetraphenylphosphonium-montmorillonite, and by Pei et al. (2012) for the sorption of polycyclic aromatic hydrocarbons (PAH) to montmorillonite modified with norfloxacin. More recently, Qin et al. (2018) suggested that DNA, coordinated to Na^+-, Ca^{2+}, and Fe^{3+}-montmorillonite, could bind PAH through π-π interactions with the nucleobases.

The enhanced uptake of polycyclic aromatic hydrocarbons (PAHs), relative to chlorobenzenes, by tetra-alkylammonium-exchanged smectites has been ascribed to cation-π interactions between the organic cations and PAHs (Qu et al. 2008). Density functional theory investigation by Hou et al. (2015) also indicates that cation-π interactions play an important role in the intercalation of toluene into (dry) tetramethylammonium-montmorillonite, while intermolecular cation-π interactions have been proposed by Vasudevan et al. (2013) for the adsorption of cationic aromatic amines to montmorillonite.

In closing we should mention the use of organoclays as sorbents of benzene, toluene, ethylbenzene, and xylene (BTEX). This topic has attracted considerable attention since these compounds are classified as priority pollutants by the U.S. Environmental Protection Agency. In comparing the efficacy of montmorillonite and vermiculite exchanged with long-chain quaternary ammonium cations in taking up BTEX compounds from water, Jaynes and Vance (1996) found that sorption increased with the layer charge density of the minerals and basal spacing of the organoclays. More interestingly, sorption promoted further uptake of BTEX, giving rise to curvilinear, concave up isotherms. Subsequently, Jaynes and Vance (1999) measured BTEX sorption by hectorite modified with compact aromatic cations. Not surprisingly, Langmuir-type isotherms were obtained with TPMA-, methylphenylpyridinium- and trimethylammonium indan-modified clays. On the other hand, methylpyridinium-, trimethylammonium biphenyl-, and trimethylammonium fluorene-exchanged hectorite gave rise to linear isotherms, probably because these organic cations were hydrated, preventing BTEX compounds from accessing the intercation siloxane surface.

Sharmasarkar et al. (2000) have also observed that montmorillonite, exchanged with TMPA and trimethylammonium adamantane, can take up BTEX compounds by surface adsorption, while a partition mechanism is operative with respect to the HDTMA-exchanged clay. Carvalho et al. (2012) and Fatimah and Huda (2013), however, used a Langmuir-type isotherm to describe the sorption of BTEX and toluene by HDTMA-montmorillonite. Similarly, Nourmoradi et al. (2013) found that the uptake of BTEX by tetradecyl trimethylammonium-montmorillonite could be described by the Freundlich isotherm, while Moshe and Rytwo (2018) proposed a dual-mode mechanism for the adsorption of phenol by thiamine-modified montmorillonite, namely, a high-affinity isotherm at low solute concentration, and a linear/partitioning one at high concentration.

In comparison with benzene and toluene, the higher molecular weight and hydrophobic alkylbenzenes show a larger partitioning affinity toward organoclays. The capacity of montmorillonite, modified with polyethylene glycol (PEG), in removing BTEX compounds (from water) increases with PEG loading, following the order benzene < toluene < ethylbenzene < xylene (Nourmoradi et al. 2012). A similar order was reported by Gitipour et al. (1997a, 1997b) for the removal of BTEX from oily liquid wastes and gasoline-contaminated soil by a commercial organoclay (Claytone 40). The use of organoclays for cleaning up oil spills has been described by Carmody et al. (2007).

REFERENCES

Abbas, A., A.S. Sallam, A.R.A. Usman, and M.I. Al-Wabel. 2017. Organoclay-based nanoparticles from montmorillonite and natural clay deposits: Synthesis, characteristics, and application for MTBE removal. *Applied Clay Science* 142: 21–29.

Abdo, S., M.I. Cruz, and J.J. Fripiat. 1980. Metallation-demetallation reaction of tin tetra (4-pyridyl)porphyrin in Na-hectorite. *Clays and Clay Minerals* 28: 125–129.

Adams, J.M., J.A. Ballantine, S.H. Graham et al. 1979. Selective chemical conversions using sheet silicate intercalates: Low temperature addition of water to 1-alkenes. *Journal of Catalysis* 58: 238–252.

Adams, J.M., A. Bylina, and S.H. Graham. 1981. Shape selectivity in lowtemperature reactions of C_6-alkenes catalysed by a Cu^{2+}-exchanged montmorillonite. *Clay Minerals* 16: 325–332.

Adams, J.M. and T.V. Clapp. 1985. Infrared studies of 1-hexene adsorbed onto Cr^{3+}-exchanged montmorillonite. *Clays and Clay Minerals* 33: 15–20.

Aggarwal, V., H. Li, S.A. Boyd, and B.J. Teppen. 2006. Enhanced sorption of trichloroethene by smectite clay exchanged with Cs^{+}. *Environmental Science & Technology* 40: 894–899.

Aguer, J.P., M.C. Hermosin, M.J. Calderon, and J. Cornejo. 2000. Fenuron sorption on homoionic natural and modified smectites. *Journal of Environmental Science and Health, Part B* 35: 279–296.

Ahmad, M.B., W.H. Hoidy, N.A.B. Ibrahim, and E.A.J. Al-Mulla. 2009. Modification of montmorillonite by new surfactants. *Journal of Engineering and Applied Sciences* 4: 184–188.

Ahmadi, S. and C.A. Igwegbe. 2018. Adsorptive removal of phenol and aniline by modified bentonite: Adsorption isotherm and kinetics study. *Applied Water Science* 8: 170.

Ahmed, H.M., M.S. Kamal, and M. Al-Harthi. 2019. Polymeric and low molecular weight shale inhibitors: A review. *Fuel* 251: 187–217.

Ainsworth, C.C., B.D. McVeety, S.C. Smith, and J.M. Zachara. 1991. Transformation of 1-aminonaphthalene at the surface of smectite clays. *Clays and Clay Minerals* 39: 416–427.

Akyuz, S. and T. Akyuz. 2018. Investigation of adsorption of 5-chlorouracil onto montmorillonite: An IR and Raman spectroscopic study. *Applied Clay Science* 164: 54–57.

Al-Asheh, S., F. Banat, and L. Abu-Aitah. 2003. Adsorption of phenol using different types of activated bentonites. *Separation and Purification Technology* 33: 1–10.

Alkaram, U.F., A.A. Mukhlis, and A.H. Al-Dujaili. 2009. The removal of phenol from aqueous solutions by adsorption using surfactant-modified bentonite and kaolinite. *Journal of Hazardous Materials* 169: 324–332.

Ammann, L., F. Bergaya, and G. Lagaly. 2005. Determination of the cation exchange capacity of clays with copper complexes revisited. *Clay Minerals* 40: 441–453.

Anderson, R.L., I. Ratcliffe, H.C. Greenwell, P.A. Williams, S. Cliffe, and P.V. Coveney. 2010. Clay swelling – a challenge in the oilfield. *Earth-Science Reviews* 98: 201–216.

Andini, S., R. Cioffi, F. Montagnaro, F. Pisciotta, and L. Santoro. 2006. Simultaneous adsorption of chlorophenol and heavy metal ions on organophilic bentonite. *Applied Clay Science* 31: 126–133.

Anirudhan, T.S. and M. Ramachandran. 2014. Removal of 2,4,6-trichlorophenol from water and petroleum refinery industry effluents by surfactant-modified bentonite. *Journal of Water Process Engineering* 1: 46–53.

Annabi-Bergaya, F., M.I. Cruz, L. Gatineau, and J.J. Fripiat. 1981. Adsorption of alcohols by smectites: IV. Models. *Clay Minerals* 16: 115–122.

Aragon, F., J. Cano Ruiz, and D.M.C. MacEwan. 1959. β-Type interlamellar sorption complexes. *Nature* 183: 740–741.

Atkins, P.W. 1982. *Physical Chemistry*, 2nd edition, p. 750. Oxford, UK: Oxford University Press.

Atluri, V., J. Jin, K. Shrimali, L. Dang, X. Wang, and J.D. Miller. 2019. The hydrophobic surface state of talc as influenced by aluminum substitution in the tetrahedral layer. *Journal of Colloid and Interface Science* 536: 737–748.

Bailey, G.W. and J.L. White. 1964. Soil-pesticide relationships, adsorption and desorption of organic pesticides by soil colloids, with implications concerning pesticide bioactivity. *Journal of Agricultural and Food Chemistry* 12: 324–332.

Ballantine, J.A., J.H. Purnell, and J.M. Thomas. 1983. Organic reactions in a clay micro-environment. *Clay Minerals* 18: 347–356.

Banat, F.A., B. Al-Bashir, S. Al-Asheh, and O. Hayajneh. 2000. Adsorption of phenol by bentonite. *Environmental Pollution* 197: 391–398.

Barrer, R.M. 1978. *Zeolites and Clay Minerals as Sorbents and Molecular Sieves*. London, UK: Academic Press.

Barrer, R.M. 1989. Shape-selective sorbents based on clay minerals: A review. *Clays and Clay Minerals* 37: 385–395.

Barrer, R.M. and M.G. Hampton. 1957. Gas chromatography and mixture isotherms in alkylammonium bentonites. *Transactions of the Faraday Society* 53: 1462–1475.

Barrer, R.M. and D.M. MacLeod. 1954. Intercalation and sorption by montmorillonite. *Transactions of the Faraday Society* 50: 980–989.

Barrer, R.M. and D.M. MacLeod. 1955. Activation of montmorillonite by ion-exchange and sorption complexes of tetra-alkylammonium montmorillonites. *Transactions of the Faraday Society* 51: 1290–1300.

Barrer, R.M. and G.S. Perry. 1961. Sorption of mixtures, and selectivity in alkylammonium montmorillonites. I. Monomethylammonium bentonite. *Journal of the Chemical Society* 842–849.

Barrer, R.M. and J.S.S. Reay. 1957. Sorption and intercalation by methylammonium montmorillonites. *Transactions of the Faraday Society* 53: 1253–1261.

Barrientos-Velázquez, A.L., A.M. Cardona, L. Liu, T. Phillips, and Y. Deng. 2016. Influence of layer charge origin and layer charge density of smectites on their aflatoxin adsorption. *Applied Clay Science* 132–133: 281–289.

Barshad, I. 1952. Factors affecting the interlayer expansion of vermiculite and montmorillonite with organic substances. *Soil Science Society of America Proceedings* 16: 176–182.

Beall, G.W. and M. Goss. 2004. Self-assembly of organic molecules on montmorillonite. *Applied Clay Science* 27: 179–186.

Berezkin, V.G. and V.S. Gavrichev. 1976. Retention of hydrocarbons on gas chromatographic sorbents containing an organo-clay. *Journal of Chromatography A* 116: 9–16.

Bergaya, F. and M. Vayer. 1997. CEC of clays: Measurement by adsorption of a copper ethylenediamine complex. *Applied Clay Science* 12: 327–336.

Berkheiser, V.E. and M.M. Mortland. 1977. Hectorite complexes with Cu(II) and Fe(II)-1,10-phenanthroline chelates. *Clays and Clay Minerals* 25: 105–112.

Bissada, K.K., W.D. Johns, and F.S. Cheng. 1967. Cation-dipole interactions in clay organic complexes. *Clay Minerals* 7: 155–166.

Bodenheimer, W., L. Heller, B. Kirson, and S. Yariv. 1962. Organo-metallic clay complexes. Part II. *Clay Minerals Bulletin* 5: 145–154.

Bodenheimer, W., L. Heller, B. Kirson, and S. Yariv. 1963b. Organo-metallic clay complexes. III. Copper-polyamine-clay complexes. *Proceedings of the International Clay Conference, Stockholm* 1: 351–363.

Bodenheimer, W., L. Heller, B. Kirson, and S. Yariv. 1963c. Organo-metallic clay complexes. IV. Nickel and mercury aliphatic polyamines. *Israel Journal of Chemistry* 1: 391–403.

Bodenheimer, W., B. Kirson, and S. Yariv. 1963a. Organometallic clay complexes. I. *Israel Journal of Chemistry* 1: 69–78.

Borisover, M., N. Bukhanovsky, I. Lapides, and S. Yariv. 2010a. Mild pre-heating of organic cation-exchanged clays enhances their interactions with nitrobenzene in aqueous environment. *Adsorption* 16: 223–232.

Borisover, M., N. Bukhanovsky, I. Lapides, and S. Yariv. 2010b. Thermal treatment of organoclays: Effect on the aqueous sorption of nitrobenzene on *n*-hexadecyltrimethyl ammonium montmorillonite. *Applied Surface Science* 256: 5539–5544.

Boufatit, M. and H. Ait-Amar. 2007. Removal of N,N-dimethylaniline from a dilute aqueous solution by Na^+/K^+ saturated montmorillonite. *Desalination* 206: 300–310.

Boufatit, M., H. Ait-Amar, and W.R. McWhinnie. 2007. Development of an Algerian material montmorillonite clay. Adsorption of phenol, 2-dichlorophenol and 2,4,6-trichlorophenol from aqueous solutions onto montmorillonite exchanged with transition metal complexes. *Desalination* 206: 394–406.

Bower, C.A. 1963. Adsorption of *o*-phenanthroline by clay minerals and soils. *Soil Science* 95: 192–195.

Bower, C.A. and J.O. Goertzen. 1959. Surface area of soils and clays by an equilibrium ethylene glycol method. *Soil Science* 87: 289–292.

Bower, C.A. and F.B. Gschwend. 1952. Ethylene glycol retention by soils as a measure of surface area and interlayer swelling. *Soil Science Society of America Proceedings* 16: 342–345.

Bowers, G.M., J.S. Loring, H.T. Schaef et al. 2018. Interaction of hydrocarbons with clays under reservoir conditions: In situ infrared and nuclear magnetic resonance spectroscopy and X-ray diffraction for expandable clays with variably wet supercritical methane. *ACS Earth Space Chemistry* 2: 640–652.

Boyd, S.A., C.T. Johnston, D.A. Laird, B.J. Teppen, and H. Li. 2011. Comprehensive study of organic contaminant adsorption by clays: Methodologies, mechanisms, and environmental implications. In

Biophysico-Chemical Processes of Anthropogenic Organic Compounds in Environmental Systems, (Eds.) B. Xing, N. Senesi, and P.M. Huang, pp. 51–71. New York: John Wiley & Sons.

Boyd, S.A. and M.M. Mortland. 1985. Dioxin radical formation and polymerization on Cu(II)-smectite. *Nature* 316: 532–535.

Boyd, S.A. and M.M. Mortland. 1986. Radical formation and polymerization of chlorophenols and chloroanisole on Cu(II)-smectite. *Environmental Science & Technology* 20: 1056–1058.

Boyd, S.A., M.M. Mortland, and C.T. Chiou. 1988b. Sorption characteristics of organic compounds on hexadecyltrimethylammonium-smectite. *Soil Science Society of America Journal* 52: 652–657.

Boyd, S.A., G. Sheng, B.J. Teppen, and C.T. Johnston. 2001. Mechanisms for the adsorption of substituted nitrobenzenes by smectite clays. *Environmental Science & Technology* 35: 4227–4234.

Boyd, S.A., S. Sun, J.-F. Lee, and M.M. Mortland. 1988a. Pentachlorophenol sorption by organo-clays. *Clays and Clay Minerals* 36: 125–130.

Bradley, W.F. 1945a. Molecular associations between montmorillonite and some polyfunctional organic liquids. *Journal of the American Chemical Society* 67: 975–981.

Bradley, W.F. 1945b. Diagnostic criteria. *American Mineralogist* 30: 704–713.

Bradley, W.F., E.J. Weiss, and R.A. Rowland. 1963. A glycol-sodium vermiculite complex. *Clays and Clay Minerals* 10: 117–122.

Breen, C. 1991. Thermogravimetric and infrared study of the desorption of butylamine, cyclohexylamine and pyridine from Ni- and Co-exchanged montmorillonite. *Clay Minerals* 26: 487–496.

Breen, C. 1994. Thermogravimetric, infrared and mass spectroscopic analysis of the desorption of tetrahydropyran, tetrahydrofuran and 1,4-dioxan from montmorillonite. *Clay Minerals* 29: 115–121.

Breen, C., F. Clegg, T.L. Hughes, and J. Yarwood. 2000. Thermal and spectroscopic characterization of *N*-methylformamide/Ca-, Mg-, and Na-exchanged montmorillonite intercalates. *Langmuir* 16: 6648–6656.

Breen, C., A.T. Deane, and J.J. Flynn 1987a. Vapor-phase sorption kinetics for tetrahydrofuran, tetrahydropyran, and 1,4-dioxane by Al^{3+}- and Cr^{3+}-exchanged montmorillonite. *Clays and Clay Minerals* 35: 343–346.

Breen, C., A.T. Deane, J.J. Flynn, and D. Reynolds. 1987b. Vapor-phase sorption kinetics for methanol, propan-2-ol, and 2-methylpropan-2-ol on Al^{3+}-, Cr^{3+}-, and Fe^{3+}-exchanged montmorillonite. *Clays and Clay Minerals* 35: 336–342.

Brigatti, M.F., A. Laurora, D. Malferrari, L. Medici, and L. Poppi. 2005. Adsorption of $[Al(Urea)_6]^{3+}$ and $[Cr(Urea)_6]^{3+}$ complexes in the vermiculite interlayer. *Applied Clay Science* 30: 21–32.

Brindley, G.W. 1965. Clay-organic studies. X. Complexes of primary amines with montmorillonite and vermiculite. *Clay Minerals* 6: 91–96.

Brindley, G.W. 1966. Ethylene glycol and glycerol complexes of smectites and vermiculites. *Clay Minerals* 6: 237–259.

Brindley, G.W., R. Bender, and S. Ray. 1963. Sorption of non-ionic aliphatic molecules from aqueous solutions on clay minerals. Clay-organic studies VII. *Geochimica et Cosmochimica Acta* 27: 1129–1137.

Brindley, G.W. and R.W. Hoffmann. 1962. Orientation and packing of aliphatic chain molecules on montmorillonite. Clay-organic studies VI. *Clays and Clay Minerals* 9: 546–556.

Brindley, G.W. and S. Ray. 1964. Complexes of Ca-montmorillonite with primary monohydric alcohols (Clay-organic studies VIII). *American Mineralogist* 49: 106–115.

Brindley, G.W. and M. Rustom. 1958. Adsorption and retention of an organic material by montmorillonite in the presence of water. *American Mineralogist* 43: 627–640.

Brindley, G.W. and A. Tsunashima. 1972. Montmorillonite complexes with dioxane, morpholine, and piperidine: Mechanisms of formation. *Clays and Clay Minerals* 20: 233–240.

Brown, C.B. and J.L. White. 1969. Reactions of 12 *s*-triazines with soil clays. *Soil Science Society of America Proceedings* 33: 863–867.

Brown, D.R. and L. Kevan. 1988. Solvation of exchangeable Cu^{2+} cations by primary alcohols in montmorillonite clay studied by electron spin resonance and electron spin echo modulation spectroscopies. *Journal of Physical Chemistry* 92: 1971–1974.

Bruque, S., L. Moreno-Real, T. Mozas, and A. Rodriquez-Garcia. 1982. Interlayer complexes of lanthanide-montmorillonites with amines. *Clay Minerals* 17: 201–208.

Burba, J.L., III and J.L. McAtee, Jr. 1977. The orientation and interaction of ethylenediamine copper(II) with montmorillonite. *Clays and Clay Minerals* 25: 113–118.

Burgess, J. 1978. *Metal Ions in Solution*. Chichester, UK: Ellis Horwood.

Cady, S.S. and T.J. Pinnavaia. 1978. Porphyrin intercalation in mica-type silicates. *Inorganic Chemistry* 17: 1501–1507.

Calamai, L., O. Pantani, A. Pusino, C. Gessa, and P. Fusi. 1997. Interaction of rimsulfuron with smectites. *Clays and Clay Minerals* 45: 23–27.

Cao, C.-Y., L.-K. Meng, and Y.-H. Zhao. 2014. Adsorption of phenol from wastewater by organo-bentonite. *Desalination and Water Treatment* 52: 3504–3509.

Carmody, O., R. Frost, Y. Xi, and S. Kokot. 2007. Adsorption of hydrocarbons on organo-clays – Implications for oil spill remediation. *Journal of Colloid and Interface Science* 305: 17–24.

Carter, D.L., M.D. Heilman, and C.L. Gonzalez. 1965. Ethylene glycol monoethyl ether for determining surface area of silicate minerals. *Soil Science* 100: 356–360.

Carvalho, M.N., M. da Motta, M. Benachour, D.C.S. Sales, and C.A.M. Abreu. 2012. Evaluation of BTEX and phenol removal from aqueous solution by multi-solute adsorption onto smectite organoclay. *Journal of Hazardous Materials* 239–240: 95–101.

Casal, B., P. Aranda, J. Sanz, and E. Ruiz-Hitzky. 1994. Interlayer adsorption of macrocyclic compounds (crown-ethers and cryptands) in 2:1 phyllosilicates: II. Structural features. *Clay Minerals* 29: 191–203.

Casal, B. and E. Ruiz-Hitzky. 1986. Interlayer adsorption of macrocyclic compounds (crown-ethers and cryptands) in 2:1 phyllosilicates: I. Isotherms and kinetics. *Clay Minerals* 21: 1–7.

Casal, B., E. Ruiz-Hitzky, J.M. Serratosa, and J.J. Fripiat. 1984. Vibrational spectra of ammonium ions in crown-ether-NH_4^+-montmorillonite complexes. *Journal of the Chemical Society, Faraday Transactions* 80: 2255–2232.

Cerato, A.B. and A.J. Lutenegger. 2002. Determination of surface area of fine-grained soils by the ethylene glycol monoethyl ether (EGME) method. *Geotechnical Testing Journal* 25: 1–9.

Chang, P.-H., W.-T. Jiang, and Z. Li. 2018. Mechanism of tyramine adsorption on Ca-montmorillonite. *Science of the Total Environment* 642: 198–207.

Changchaivong, S. and S. Khaodhiar. 2009. Adsorption of naphthalene and phenanthrene on dodecylpyridinium-modified bentonite. *Applied Clay Science* 43: 317–321.

Chappell, M.A., D.A. Laird, M.L. Thompson et al. 2005. Influence of smectite hydration and swelling on atrazine sorption behavior. *Environmental Science & Technology* 39: 3150–3156.

Charaabi, S., R. Absi, A.-M. Pensé-Lhéritier, M. Le Borgne, and S. Issa. 2021. Adsorption studies of benzophenone-3 onto clay minerals and organosilicates: Kinetics and modelling. *Applied Clay Science* 202: 105937.

Chassin, P. 1976. Signification de la mésure des surfaces totals des argiles avec l'éthane 1.2 diol. *Clay Minerals* 11: 23–29.

Chaussidon, J. and R. Calvet. 1965. Evolution of amine cations adsorbed on montmorillonite with dehydration of the mineral. *Journal of Physical Chemistry* 69: 2265–2268.

Chen, B. and W. Huang. 2009. Effect of background electrolytes on the adsorption of nitroaromatic compounds onto bentonite. *Journal of Environmental Sciences* 21: 1044–1052.

Chen, B., W. Huang, J. Mao, and S. Lv. 2008. Enhanced sorption of naphthalene and nitroaromatic compounds to bentonite by potassium and cetyltrimethylammonium cations. *Journal of Hazardous Materials* 158: 116–123.

Chiou, C.T. and D.W. Rutherford. 1997. Effects of exchanged cation and layer charge on the sorption of water and EGME vapors on montmorillonite clays. *Clays and Clay Minerals* 45: 867–880.

Chrayet, B., F. Ammari, A. Ouerghui, and F. Meganem. 2017. Removal of heavy metals from aqueous solutions using Tunisian smectite clay intercalated by functionalized crown ethers. *Journal of the Tunisian Chemical Society* 19: 256–263.

Chukwunenye, C. and J.L. McAtee. 1987. Organo-modified smectite for the gas chromatographic separation of various compounds. *Journal of Chromatography A* 410: 121–128.

Chun, Y., G. Sheng, and S.A. Boyd. 2003. Sorptive characteristics of tetraalkylammonium-exchanged smectite clays. *Clays and Clay Minerals* 51: 415–420.

Churchman, G.J., C.M. Burke, and R.L. Parfitt. 1991. Comparison of various methods for the determination of specific surfaces of subsoils. *Journal of Soil Science* 42: 449–461.

Čičel, B. and D. Machajdík. 1981. Potassium- and ammonium-treated montmorillonites. I. Interstratified structures with ethylene glycol and water. *Clays and Clay Minerals* 29: 40–46.

Cihacek, L.J. and J.M. Bremner. 1979. A simplified ethylene glycol monoethyl ether procedure for assessment of soil surface area. *Soil Science Society of America Journal* 43: 821–822.

Clementz, D.M. and M.M. Mortland. 1972. Interlamellar metal complexes in layer silicates. III. Silver(I)-arene complexes in smectites. *Clays and Clay Minerals* 20: 181–187.

Clementz, D.M., M.M. Mortland, and T.J. Pinnavaia. 1974. Properties of reduced-charge montmorillonites: Hydrated Cu(II) ions as a spectroscopic probe. *Clays and Clay Minerals* 22: 49–57.

Clementz, D.M., T.J. Pinnavaia, and M.M. Mortland. 1973. Stereochemistry of hydrated copper (II) ions on the interlamellar surfaces of layer silicates: An electron spin resonance study. *Journal of Physical Chemistry* 77: 196–200.

Cloos, P. 1972. Interactions entre les pesticides et la fraction minérale du sol. *Pédologie* 22: 148–173.

Cloos, P., A. Moreale, C. Broers, and C. Badot. 1979. Adsorption and oxidation of aniline and *p*-chloroaniline by montmorillonite. *Clay Minerals* 14: 307–321.

Cloos, P., D. Van de Poel, and J.P. Camerlynck. 1973. Thiophene complexes on montmorillonite saturated with different cations. *Nature Physical Science* 243: 54–55.

Cox, L., M.C. Hermosin, and J. Cornejo. 1995. Adsorption mechanisms of thiazafluron in mineral soil clay components. *European Journal of Soil Science* 46: 431–438.

Cowan, C.T. and D. White. 1958. The mechanism of exchange reactions occurring between sodium montmorillonite and various *n*-primary aliphatic amine salts. *Transactions of the Faraday Society* 54: 691–697.

Cruz, M., J.L. White, and J.D. Russell. 1968. Montmorillonite-*s*-triazine interactions. *Israel Journal of Chemistry* 6: 315–323.

Cui, Y. and J.S. van Duijneveldt. 2010. Adsorption of polyetheramines on montmorillonite at high pH. *Langmuir* 26: 17210–17217.

da Fonseca, M.G., A.F. Wanderley, K. Sousa, L.N.H. Arakaki, and J.G.P. Espinola. 2006. Interaction of aliphatic diamines with vermiculite in aqueous solution. *Applied Clay Science* 32: 94–98.

Davies, J.E.D. and N. Jabeen. 2002. The adsorption of herbicides and pesticides on clay minerals and soils. Part 1. Isoproturon. *Journal of Inclusion Phenomena and Macrocyclic Chemistry* 43: 329–336.

De Boer, J.H. and G. Heller. 1937. The anisotropy of van der Waals forces. *Physica* 4: 1045–1057.

Dékány, I., A. Farkas, Z. Kiraly, E. Klumpp, and H.D. Narres. 1996a. Interlamellar adsorption of 1-pentanol from aqueous solution on hydrophobic clay mineral. *Colloids and Surfaces A: Physicochemical and Engineering Aspects* 119: 7–13.

Dékány, I., A. Farkas, I. Regdon, E. Klumpp, H.D. Narres, and M.J. Schwuger. 1996b. Adsorption of nitrobenzene and *n*-pentanol from aqueous solution on hydrophilic and hydrophobic clay minerals. *Colloid and Polymer Science* 274: 981–988.

Deng, L., Y. Liu, G. Zhuang et al. 2020. Dynamic benzene adsorption performance of microporous TMA^+-exchanged montmorillonite: The role of TMA^+ cations. *Microporous and Mesoporous Materials* 296: 109994.

Deng, L., P. Yuan, D. Liu et al. 2017. Effects of microstructure of clay minerals, montmorillonite, kaolinite and halloysite, on their benzene adsorption behaviors. *Applied Clay Science* 143: 184–191.

Deng, Y., L. Liu, A.L.B. Velázquez, and J.B. Dixon. 2012. The determinative role of the exchange cation and layer-charge density of smectite on aflatoxin adsorption. *Clays and Clay Minerals* 60: 374–386.

Dentel, S.K., J.Y. Bottero, K. Khatib, H. Demougeot, J.P. Duguet, and C. Anselme. 1995. Sorption of tannic acid, phenol, and 2,4,5-trichlorophenol on organoclays. *Water Research* 29: 1273–1280.

Diamond, S. and E.B. Kinter. 1956. Surface areas of clay minerals as derived from measurements of glycerol retention. *Clays and Clay Minerals* 5: 334–347.

Diamond, S. and E.B. Kinter. 1963. Characterization of montmorillonite saturated with short chain amine cations: I. Interpretation of basal spacing measurements. *Clays and Clay Minerals* 10: 163–173.

Díaz-Nava, M.C., M.T. Olguín, and M. Solache-Ríos. 2012. Adsorption of phenol onto surfactants modified bentonite. *Journal of Inclusion Phenomena and Macrocyclic Chemistry* 74: 67–75.

Dios-Cancela, G., L. Alfonso-Méndez, F.J. Huertas, E. Romero-Taboada, C.I. Sainz-Díaz, and A. Hernández-Laguna. 2000. Adsorption mechanism and structure of the montmorillonite complexes with (CH3)2XO (X=C, and S), (CH3O)3PO, and CH3-CN molecules. *Journal of Colloid and Interface Science* 222: 125–136.

Dodd, C.G. and S. Ray. 1960. Semiquinone cation adsorption on montmorillonite as a function of surface acidity. *Clays and Clay Minerals* 8: 237–251.

Doehler, R.W. and W.A. Young. 1962. Some conditions affecting the adsorption of quinoline by clay minerals in aqueous suspensions. *Clays and Clay Minerals* 9: 468–483.

Dohrmann, R. 2006. Cation exchange capacity methodology II: A modified silver-thiourea method. *Applied Clay Science* 34: 38–46.

do Nascimento, G.M., P.S.M. Barbosa, V.R.L. Constantino, and M.L.A. Temperini. 2006. Benzidine oxidation on cationic clay surfaces in aqueous suspension monitored by in situ resonance Raman spectroscopy. *Colloids and Surfaces A: Physicochemical and Engineering Aspects* 289: 39–46.

Doner, H.E. and M.M. Mortland. 1969a. Intermolecular interaction in montmorillonites: NH-CO systems. *Clays and Clay Minerals* 17: 265–270.

Doner, H.E. and M.M. Mortland. 1969b. Benzene complexes with Cu(II) montmorillonite. *Science* 166: 1406–1407.

Dowdy, R.H. and M.M. Mortland. 1967. Alcohol-water interactions on montmorillonite surfaces. I. Ethanol. *Clays and Clay Minerals* 15: 259–271.

Dowdy, R.H. and M.M. Mortland. 1968. Alcohol-water interactions on montmorillonite surfaces. II. Ethylene glycol. *Soil Science* 105: 36–43.

Durand, B., R. Pelet, and J.J. Fripiat. 1972. Alkylammonium decomposition on montmorillonite surfaces in an inert atmosphere. *Clays and Clay Minerals* 20: 21–35.

Dyal, R.S. and S.B. Hendricks. 1950. Total surface of clays in polar liquids as a characteristic index. *Soil Science* 69: 421–432.

Edwards, D.G., A.M. Posner, and J.P. Quirk. 1965. Discreteness of charge on clay surfaces. *Nature* 206: 168–170.

Eguchi, M., S. Takagi, and H. Inoue. 2006. The orientation control of dicationic porphyrins on clay surfaces by solvent polarity. *Chemistry Letters* 35: 14–15.

Eguchi, M., S. Takagi, H. Tachibana, and H. Inoue. 2004. The 'size matching rule' in di-, tri-, and tetra-cationic charged porphyrin/synthetic clay complexes: Effect of the inter-charge distance and the number of charged sites. *Journal of Physics and Chemistry of Solids* 65: 403–407.

Eiceman, G.A. and A.S. Lara. 1991. Gas chromatographic properties of organoammonium exchanged montmorillonites. *Journal of Chromatography A* 549: 273–281.

El-Nahhal, Y. and J.M. Safi. 2004. Adsorption of phenanthrene on organoclays from distilled and saline water. *Journal of Colloid and Interface Science* 269: 265–273.

Eltantawy, I.M. and P.W. Arnold. 1972. Adsorption of *n*-alkanes by Wyoming montmorillonite. *Nature Physical Science* 237: 123–125.

Eltantawy, I.M. and P.W. Arnold. 1973. Reappraisal of ethylene glycol mono-ethyl ether (EGME) method for surface area estimations of clays. *Journal of Soil Science* 24: 232–238.

Eltantawy, I.M. and P.W. Arnold. 1974. Ethylene glycol sorption by homoionic montmorillonites. *Journal of Soil Science* 25: 99–110.

Emerson, W.W. 1957. Organo-clay complexes. *Nature* 180: 46–49.

Essington, M.E. 1994. Adsorption of aniline and toluidines on montmorillonite. *Soil Science* 158: 181–188.

Farmer, V.C. 1971. The characterization of adsorption bonds in clays by infrared spectroscopy. *Soil Science* 112: 62–68.

Farmer, V.C. 1978. Water on particle surfaces. In *The Chemistry of Soil Constituents*, (Eds.) D.J. Greenland and M.H.B. Hayes, pp. 405–448. Chichester: John Wiley & Sons.

Farmer, V.C. and M.M. Mortland. 1965. An infrared study of complexes of ethylamine with ethylammonium and copper ions in montmorillonite. *Journal of Physical Chemistry* 69: 683–686.

Farmer, V.C. and M.M. Mortland. 1966. An infrared study of the co-ordination of pyridine and water to exchangeable cations in montmorillonite and saponite. *Journal of the Chemical Society A* 344–351.

Farmer, V.C. and J.D. Russell. 1967. Infrared absorption spectrometry in clay studies. *Clays and Clay Minerals* 15: 121–142.

Farmer, W.J. 1967. Infrared studies of the mechanisms of adsorption of urea and its herbicidal derivatives by montmorillonite. *Dissertation Abstracts* 27: 3749.11.

Farmer, W.J. and J.L. Ahlrichs. 1969. Infrared studies of the mechanism of adsorption of urea-d_4, methylurea-d_3, and 1,1-dimethylurea-d_2 by montmorillonite. *Soil Science Society of America Proceedings* 33: 254–258.

Fatimah, I. and T. Huda. 2013. Preparation of cetyltrimethylammonium intercalated Indonesian montmorillonite for adsorption of toluene. *Applied Clay Science* 74: 115–120.

Fenn, D.B., M.M. Mortland, and T.J. Pinnavaia. 1973. The chemisorption of anisole on Cu(II) hectorite. *Clays and Clay Minerals* 21: 315–322.

Ferrage, E. 2016. Investigation of the interlayer organization of water and ions in smectite from the combined use of diffraction experiments and molecular simulations: A review of methodology, applications, and perspectives. *Clays and Clay Minerals* 64: 348–373.

Fraenkel, A. and C. Franconi. 1960. Protonation of amides. *Journal of the American Chemical Society* 82: 4478–4483.

Fripiat, J.J., J. Helsen, and L. Vielvoye. 1964. Formation des radicaux libres sur la surface des montmorillonites. *Bulletin Groupe Français des Argiles* 15: 3–10.

Fripiat, J.J., A.N. Jelli, G. Poncelet, and J. André. 1965. Thermodynamic properties of adsorbed water molecules and electrical conduction in montmorillonites and silicas. *Journal of Physical Chemistry* 69: 2185–2197.

Fripiat, J.J., M. Pennequin, G. Poncelet, and P. Cloos. 1969. Influence of the van der Waals force on the infrared spectra of short aliphatic alkylammonium cations held on montmorillonite. *Clay Minerals* 8: 119–134.

Fripiat, J.J., A. Servais, and A. Léonard. 1962. Étude de l'adsorption des amines par les montmorillonites. III. La nature de la liaison amine-montmorillonite. *Bulletin de la Société Chimique de France* 635–644.

Froehner, S., R.F. Martins, W. Furukawa, and M.R. Errera. 2009. Water remediation by adsorption of phenol onto hydrophobic modified clay. *Water, Air, and Soil Pollution* 199: 107–113.

Fujimine, Y., A. Hanaki, S. Ikoma, and J.-I. Kobayashi. 1995. Structural study on the intercalation of water and ethylene glycol into copper(II) montmorillonite. *Clay Science* 9: 285–297.

Furukawa, T. and G.W. Brindley. 1973. Adsorption and oxidation of benzidine and aniline by montmorillonite and hectorite. *Clays and Clay Minerals* 21: 279–288.

Fusi, P., M. Franci, and G.G. Ristori. 1989a. Interactions of isoxaben with montmorillonite. *Applied Clay Science* 4: 235–245.

Fusi, P., M. Franci, and G.G. Ristori. 1989b. Adsorption of methylcarbamate by montmorillonite. *Applied Clay Science* 4: 403–409.

Fusi, P., G.G. Ristori, S. Cecconi, and M. Franci. 1983. Adsorption and degradation of fenarimol on montmorillonite. *Clays and Clay Minerals* 31: 312–314.

Fusi, P., G.G. Ristori, and M. Franci. 1982. Adsorption and catalytic decomposition of 4-nitrobenzenesulphonyl methylcarbamate by smectite. *Clays and Clay Minerals* 30: 306–310.

Fusi, P., G.G. Ristori, and A. Malquori. 1980. Montmorillonite-asulam interactions: I. Catalytic decomposition of asulam adsorbed on H- and Al-clay. *Clay Minerals* 15: 147–155.

Galway, A.K. 1969. Reactions of alcohols adsorbed on montmorillonite and the role of minerals in petroleum genesis. *Journal of the Chemical Society D* 577–578.

Galway, A.K. 1970. Reactions of alcohols and of hydrocarbons on montmorillonite surfaces. *Journal of Catalysis* 19: 330–342.

Gamba, M., P. Kovár, M. Pospíšil, and R.M. Torres Sánchez. 2017. Insight into thiabendazole interaction with montmorillonite and organically modified montmorillonites. *Applied Clay Science* 137: 59–68.

Garrett, W.G. and G.F. Walker. 1962. Swelling of some vermiculite-organic complexes in water. *Clays and Clay Minerals* 9: 557–567.

Garwood, G.A., Jr. and R.A. Condrate, Sr. 1978. The I.R. spectra of dimethyl sulfoxide adsorbed on several cation-substituted montmorillonites. *Clays and Clay Minerals* 26: 273–278.

Gates, W.P., U. Shaheen, T.W. Turney, and A.F. Patti. 2016. Cyclic carbonate-sodium smectite intercalates. *Applied Clay Science* 124–125: 94–101.

German, W.L. and D.A. Harding. 1969. The adsorption of aliphatic alcohols by montmorillonite. *Clay Minerals* 8: 213–227.

German, W.L. and D.A. Harding. 1971. Primary aliphatic alcohol-homoionic montmorillonite interactions. *Clay Minerals* 9: 167–175.

Ghavami, M., Q. Zhao, S. Javadi, J.S.D. Jangam, J.B. Jasinski, and N. Saraei. 2017. Change of organobentonite interlayer microstructure induced by sorption of aromatic and petroleum hydrocarbons – a combined study of laboratory characterization and molecular dynamics simulations. *Colloids and Surfaces A: Physicochemical and Engineering Aspects* 520: 324–334.

Ghezali, S., A. Mahdad-Benzerdjeb, M. Ameri, and A.Z. Bouyakoub. 2018. Adsorption of 2,4,6-trichlorophenol on bentonite modified with benzyldimethyltetradecylammonium chloride. *Chemistry International* 4: 24–32.

Gitipour, S., M.T. Bowers, and A. Bodocsi. 1997b. The use of modified bentonite for removal of aromatic organics from contaminated soil. *Journal of Colloid and Interface Science* 196: 191–198.

Gitipour, S., M.T. Bowers, W. Huff, and A. Bodocsi. 1997a. The efficiency of modified bentonite clays for removal of aromatic organics from oily liquid wastes. *Spill Science & Technology Bulletin* 4: 155–164.

Glaeser, R. 1948. On the mechanism of formation of montmorillonite-acetone complexes. *Clay Minerals Bulletin* 1: 88–90.

Glaeser, R. 1954. *Complexes Organo-Argileux et Rôle des Cations Échangeables*. Thèse de doctorat d'état de Université, Paris.

Glaeser, R. and J. Méring. 1952. Les propriétes des associations organo-argileuses. *Comptes Rendus Congress Géologique International Alger* 117–121.

Golbashy, M., H. Sabahi, I. Allahdadi, H. Nazokdast, and M. Hosseini. 2017. Synthesis of highly intercalated urea-clay nanocomposite via domestic montmorillonite as eco-friendly slow-release fertilizer. *Archives of Agronomy and Soil Science* 63: 84–95.

Gonen, Y. and G. Rytwo. 2006. Using the dual-mode model to describe adsorption of organic pollutants onto an organoclay. *Journal of Colloid and Interface Science* 299: 95–101.

Govindaraj, N., M.M. Mortland, and S.A. Boyd. 1987. Single electron transfer mechanism of oxidative dichlorination of 4-chloroanisole on copper(II)-smectite. *Environmental Science & Technology* 21: 1119–1123.

Greene-Kelly, R. 1953. Irreversible dehydration in montmorillonite. II. *Clay Minerals Bulletin* 2: 52–56.

Greene-Kelly, R. 1955a. Sorption of aromatic organic compounds by montmorillonite. Part 1. Orientation studies. *Transactions of the Faraday Society* 51: 412–424.

Greene-Kelly, R. 1955b. Sorption of aromatic organic compounds by montmorillonite. Part 2. Packing studies with pyridine. *Transactions of the Faraday Society* 51: 425–430.

Greene-Kelly, R. 1956. The sorption of saturated organic compounds by montmorillonite. *Transactions of the Faraday Society* 52: 1281–1286.

Greenland, D.J. 1956. The adsorption of sugars by montmorillonite. I. X-ray studies. *Journal of Soil Science* 7: 319–328.

Greenland, D.J. 1965. Interaction between clays and organic compounds in soils. I. Mechanism of interaction between clays and defined organic compounds. *Soils Fertilizers* 28: 415–425.

Greenland, D.J. and J.P. Quirk. 1964. Determination of the total specific surface areas of soils by adsorption of cetylpyridinium bromide. *Journal of Soil Science* 15: 178–191.

Grim, R.E. and G. Kulbicki. 1961. Montmorillonite: High temperature reaction classification. *American Mineralogist* 46: 1329–1369.

Gu, C., C. Liu, C.T. Johnston, B.J. Teppen, H. Li, and S.A. Boyd. 2011. Pentachlorophenol radical cations generated on Fe(III)-montmorillonite initiate octachlorodibenzo-*p*-dioxin formation in clays: Density functional theory and Fourier transform infrared studies. *Environmental Science & Technology* 45: 1399–1406.

Guégan, R. 2010. Intercalation of a nonionic surfactant ($C_{10}E_3$) bilayer into a Na-montmorillonite clay. *Langmuir* 26: 19175–19180.

Guégan, R. 2013. Self-assembly of a non-ionic surfactant onto a clay mineral for the preparation of hybrid layered materials. *Soft Matter* 9: 10913–10920.

Guégan, R., M. Gautier, J.-M. Beny, and F. Muller. 2009. Adsorption of a $C_{10}E_3$ non-ionic surfactant on a Ca-smectite. *Clays and Clay Minerals* 57: 502–509.

Guignard, J. and H. Pezerat. 1979. Étude des interactions entre montmorillonites et agents organiques complexants. *Clay Minerals* 14: 259–265.

Guo, F., D. Li, J.B. Fein et al. 2022. Roles of hydrogen bond and ion bridge in adsorption of two bisphenols onto montmorillonite: An experimental and DFT study. *Applied Clay Science* 217: 106406.

Guyot, J. 1969. Mésure des surfaces spécifiques des argiles par adsorption. *Annales Agronomiques* 20: 333–359.

Haase, D.J., E.J. Weiss, and H. Steinfink. 1963. The crystal structure of a hexamethylenediamine-vermiculite complex. *American Mineralogist* 48: 261–270.

Haderlein, S.B., K.W. Weissmahr, and R.P. Schwarzenbach. 1996. Specific adsorption of nitroaromatic explosives and pesticides to clay minerals. *Environmental Science & Technology* 30: 612–622.

Hakusui, A., Y. Matsunaga, and K. Umehara. 1970. Diffuse reflectance spectra of acid clays colored with benzidine and some other diamines. *Bulletin of the Chemical Society of Japan* 43: 707–712.

Hamaker, H.C. 1937. London-van der Waals attraction between spherical particles. *Physica* 4: 1058–1072.

Harter, R.D. and J.L. Ahlrichs. 1969. Effect of acidity on reactions of organic acids and amines with montmorillonite clay surfaces. *Soil Science Society of America Proceedings* 33: 859–863.

Harward, M.E. and G.W. Brindley. 1965. Swelling properties of synthetic smectites in relation to lattice substitutions. *Clays Clay Minerals* 13: 209–222.

Hatta, T., T. Echhigo, S. Nemoto, K. Tamura, and H. Yamada. 2011. Intercalation of sugar alcohols into the interlayer of montmorillonite by wet and dry processes. *Clay Science* 15: 61–66.

Haxaire, A. and J.M. Bloch. 1956. Sorption de molécules organiques azotées par la montmorillonite. Étude du mécanisme. *Bulletin de la Société française de Minéralogie et de Cristallographie* 79: 464–475.

Hayakawa, T., M. Minase, K.-I. Fujita, and M. Ogawa. 2016. Green synthesis of organophilic clays: Solid-state reaction of acidic clay with organoamine. *Industrial & Engineering Chemistry Research* 55: 6325–6330.

Heister, K. 2014. The measurement of the specific surface area of soils by gas and polar liquid adsorption methods – limitations and potentials. *Geoderma* 216: 75–87.

Heller, L. and S. Yariv. 1969. Sorption of some anilines by Mn-, Co-, Ni-, Cu-, Zn-, and Cd- montmorillonite. *Proceedings of the International Clay Conference, Tokyo* 1: 741–755.

Heller-Kallai, L. and S. Yariv. 1981. Swelling of montmorillonite containing coordination complexes of amines with transition metal cations. *Journal of Colloid and Interface Science* 79: 479–485.

Heller-Kallai, L., S. Yariv, and M. Riemer. 1973. The formation of hydroxy interlayers in smectites under the influence of organic bases. *Clay Minerals* 10: 35–40.

Helmy, A.K., S.G. de Bussetti, and E.A. Ferreiro. 1983. Adsorption of quinoline from aqueous solutions by some clays and oxides. *Clays and Clay Minerals* 31: 29–36.

Helmy, A.K., E.A. Ferreiro, and S.G. de Bussetti. 1999. Surface area evaluation of montmorillonite. *Journal of Colloid and Interface Science* 210: 167–171.

Hendricks, S.B. 1941. Base exchange of the clay mineral montmorillonite for organic cations and its dependence upon adsorption due to van der Waals forces. *Journal of Physical Chemistry* 45: 65–81.

Herwig, U., E. Klumpp, H.-D. Narres, and M.J. Schwuger. 2001. Physicochemical interactions between atrazine and clay minerals. *Applied Clay Science* 18: 211–222.

Hinedi, Z.R., C.T. Johnston, and C. Erickson. 1993. Chemisorption of benzene on Cu-montmorillonite as characterized by FTIR and ^{13}C MAS NMR. *Clays and Clay Minerals* 41: 87–94.

Hoffmann, R.W. and G.W. Brindley. 1960. Adsorption of non-ionic aliphatic molecules from aqueous solutions on montmorillonite. Clay-organic studies II. *Geochimica et Cosmochimica Acta* 20: 15–29.

Hoffmann, R.W. and G.W. Brindley. 1961. Adsorption of ethylene glycol and glycerol by montmorillonite. *American Mineralogist* 46: 450–452.

Homenauth, O.P. and M.B. McBride. 1994. Adsorption of aniline on layer silicate clays and an organic soil. *Soil Science Society of America Journal* 58: 347–354.

Hou, X.-J., H. Li, P. He, S. Li, and Q. Liu. 2015. Molecular-level investigation of the adsorption mechanisms of toluene and aniline on natural and organically modified montmorillonite. *Journal of Physical Chemistry A* 119: 11199–11207.

Huh, J.-K., D.-I. Song, and Y.-W. Jeon. 2000. Sorption of phenol and alkylphenols from aqueous solution onto organically modified montmorillonite and applications of dual-mode sorption. *Separation Science and Technology* 35: 243–259.

Hundal, L.S., M.L. Thompson, D.A. Laird, and A.M. Carmo. 2001. Sorption of phenanthrene by reference smectites. *Environmental Science & Technology* 35: 3456–3461.

Ikhsan, J., M.J. Angove, J.D. Wells, and B.B. Johnson. 2005. Surface complexation modeling of the sorption of 2-, 3-, and 4-aminopyridine by montmorillonite. *Journal of Colloid and Interface Science* 284: 383–392.

Ilic, M., E. Koglin, A. Pohlmeier, H.D. Narres, and M.J. Schwuger. 2000. Adsorption and polymerization of aniline on Cu(II)-montmorillonite: Vibrational spectroscopy and ab initio calculation. *Langmuir* 16: 8946–8951.

Isaacson, P.J. and B.L. Sawhney. 1983. Sorption and transformation of phenols on clay surfaces: Effect of exchangeable cations. *Clay Minerals* 18: 253–265.

Janíková, V., E. Jóna, R. Janík, V. Pavlík, D. Ondrušová, and M. Ďurčeková. 2016. Formation of coordination compounds with aniline in the interlayer space of Ca^{2+}-, Cu^{2+}- and Fe^{3+}-exchanged montmorillonite. *Chemical Papers* 70: 131–134.

Janssen, M.J. 1961. The structure of protonated amides and ureas and their thio analogues. *Spectrochimica Acta* 17: 475–485.

Jaynes, W.F. and S.A. Boyd. 1990. Trimethylammonium-smectite as an effective adsorbent of water-soluble aromatic hydrocarbons. *Journal of the Air & Waste Management Association* 40: 1649–1653.

Jaynes, W.F. and S.A. Boyd. 1991a. Hydrophobicity of siloxane surfaces in smectites as revealed by aromatic hydrocarbon adsorption from water. *Clays and Clay Minerals* 39: 428–436.

Jaynes, W.F. and S.A. Boyd. 1991b. Clay mineral type and organic compound sorption by hexadecyltrimethylammonium-exchanged clays. *Soil Science Society of America Journal* 55: 43–48.

Jaynes, W.F. and G.F. Vance. 1996. BTEX sorption by organo-clays: Cosorptive enhancement and equivalence of interlayer complexes. *Soil Science Society of America Journal* 60: 1742–1749.

Jaynes, W.F. and G.F. Vance. 1999. Sorption of benzene, toluene, ethylbenzene, and xylene (BTEX) compounds by hectorite clays exchanged with aromatic organic cations. *Clays and Clay Minerals* 47: 358–365.

Jia, H., L. Li, H. Chen, Y. Zhao, X. Li, and C. Wang. 2015. Exchangeable cations-mediated photodegradation of polycyclic aromatic hydrocarbons (PAHs) on smectite surface under visible light. *Journal of Hazardous Materials* 287: 16–23.

Jia, H., G. Nulaji, H. Gao, F. Wang, Y. Zhu, and C. Wang. 2016. Formation and stabilization of environmentally persistent free radicals induced by the interaction of anthracene with Fe(III)-modified clays. *Environmental Science & Technology* 50: 6310–6319.

Jiang, J.-Q., C. Cooper, and S. Ouki. 2002. Comparison of modified montmorillonite adsorbents Part I: Preparation, characterization, and phenol adsorption. *Chemosphere* 47: 711–716.

Johnston, C.T. 2010. Probing the nanoscale architecture of clay minerals. *Clay Minerals* 45: 245–279.

Johnston, C.T. 2017. Infrared studies of clay mineral-water interactions. In *Infrared and Raman Spectroscopies of Clay Minerals*. Developments in Clay Science, Vol. 8, (Eds.) W.P. Gates, J.T. Kloprogge, J. Madejová, and F. Bergaya, pp. 288–309. Amsterdam, The Netherlands: Elsevier.

Johnston, C.T., S.A. Boyd, B.J. Teppen, and G. Sheng. 2004. Sorption of nitroaromatic compounds on clay surfaces. In *Handbook of Layered Materials*, (Eds.) S.M. Auerbach, K.A. Carrado, and P.K. Dutta, pp. 155–189. New York: Marcel Dekker.

Johnston, C.T., M.F. de Oliveira, B.J. Teppen, G. Sheng, and S.A. Boyd. 2001. Spectroscopic study of nitroaromatic-smectite sorption mechanisms. *Environmental Science & Technology* 35: 4767–4772.

Johnston, C.T., G. Sheng, B.J. Teppen, S.A. Boyd, and M.F. de Oliveira. 2002. Spectroscopic study of dinitrophenol herbicide sorption on smectite. *Environmental Science & Technology* 36: 5067–5074.

Johnston, C.T., T. Tipton, D.A. Stone, C. Erickson, and S.L. Trabue. 1991. Chemisorption of *p*-dimethoxybenzene on copper-montmorillonite. *Langmuir* 7: 289–296.

Johnston, C.T., T. Tipton, S.L. Trabue, C. Erickson, and D.A. Stone. 1992. Vapor-phase sorption of *p*-xylene on Co- and Cu-exchanged SAz-1 montmorillonite. *Environmental Science & Technology* 26: 382–390.

Joseph-Ezra, H., A. Nasser, and U. Mingelgrin. 2014. Surface interactions of pyrene and phenanthrene on Cu-montmorillonite. *Applied Clay Science* 95: 348–356.

Juang, R.-S., S.-H. Lin, and K.-H. Tsao. 2002. Mechanism of sorption of phenols from aqueous solutions onto surfactant-modified montmorillonite. *Journal of Colloid and Interface Science* 254: 234–241.

Kabadagi, A., S. Chikkamath, J. Manjanna, and S. Kobayashi. 2018. *In-situ* complexation of *o*-phenanthroline in the interlayer of Fe(II)-montmorillonite. *Applied Clay Science* 165: 148–154.

Kakegawa, N. and M. Ogawa. 2002. The intercalation of β-carotene into the organophilic interlayer space of dialkyldimethylammonium-montmorillonites. *Applied Clay Science* 22: 137–144.

Kameda, T., S. Shimamori, and T. Yoshioka. 2012. Specific uptake of aromatic compounds from aqueous solution by montmorillonite modified with tetraphenylphosphonium. *Journal of Physics and Chemistry of Solids* 73: 120–123.

Keiluweit, M. and M. Kleber. 2009. Molecular-level interactions in soils and sediments: The role of aromatic π-systems. *Environmental Science & Technology* 43: 3421–3429.

Kellomäki, A., P. Nieminen, and L. Ritamäki. 1987. Sorption of ethylene glycol monoethylether (EGME) on homoionic montmorillonites. *Clay Minerals* 22: 297–303.

Khaorapapong, N. 2010. *In situ* complexation of thiourea in the interlayer space of copper(II)-montmorillonite. *Applied Clay Science* 50: 414–417.

Khaorapapong, N., K. Kuroda, and M. Ogawa. 2002. Incorporation of thioacetamide into the interlayer space of montmorillonite by solid-solid reactions. *Clay Science* 11: 549–564.

Khaorapapong, N., K. Kuroda, H. Hashizume, and M. Ogawa. 2001. Solid-sate intercalation of 4,4'-bipyridine and 1,2-di(4-pyridine) ethylene into the interlayer spaces of Co(II)-, Ni(II)- and Cu(II)-montmorillonites. *Applied Clay Science* 19: 69–76.

Khaorapapong, N. and M. Ogawa. 2007. Solid-state intercalation of 8-hydroxyquinoline into Li(I)- Zn(II)- and Mn(II)-montmorillonites. *Applied Clay Science* 35: 31–38.

Khaorapapong, N. and M. Ogawa. 2011. Solid-state intercalation of organic and inorganic substances in smectites. *Clay Science* 15: 147–159.

Khenifi, A., B. Zohra, B. Kahina, H. Houari, and D. Zoubir. 2009. Removal of 2,3-DCP from wastewater by CTAB/bentonite using one-step and two-step methods: A comparative study. *Chemical Engineering Journal* 146: 345–354.

Kim, J.T. 1970. The absorption of substituted urea compounds on montmorillonite. *Dissertation Abstracts* 30: 5326-B.

Kim, K.S., M. Park, W.T. Lim, and S. Komarneni. 2011. Massive intercalation of urea in montmorillonite. *Soil Science Society of America Journal* 75: 2361–2366.

Kinter, E.B. and S. Diamond. 1958. Gravimetric determination of monolayer glycerol complexes by clay minerals. *Clays and Clay Minerals* 5: 318–333.

Kinter, E.B. and S. Diamond. 1963. Characterization of montmorillonite saturated with short chain amine cations: II. Interlayer surface coverage by the amine cations. *Clays and Clay Minerals* 10: 174–190.

Kirschbaum, M.U.F., D.L. Giltrap, S.R. McNally et al. 2020. Estimating the mineral surface area of soils by measured water adsorption: Adjusting for the confounding effect of water adsorption by soil organic carbon. *European Journal of Soil Science* 71: 382–391.

Ko, C.H., C. Fan, P.N. Chiang, M.K. Wang, and K.C. Lin. 2007. *p*-Nitrophenol, phenol and aniline sorption by organo-clays. *Journal of Hazardous Materials* 149: 275–282.

Koh, S.-M. and J.B. Dixon. 2001. Preparation and application of organo-minerals as sorbents of phenol, benzene and toluene. *Applied Clay Science* 18: 111–122.

Koteja, A., M. Szczerba, and J. Matusik. 2017. Smectites intercalated with azobenzene and aminoazobenzene: Structure changes at nanoscale induced by UV light. *Journal of Physics and Chemistry of Solids* 111: 294–303.

Kowalik, M., M. Szczerba, B. Barylska, and M. Skiba. 2021. Structure of glycerol-Mg2+-smectites/vermiculites complex based on molecular dynamics and implementation of the model for X-ray diffraction modeling. *Applied Clay Science* 206: 106066.

Kowalska, M., H. Güler, and D.L. Cocke. 1994. Interactions of clay minerals with organic pollutants. *Science of the Total Environment* 141: 223–240.

Koyuncu, H. 2008. Adsorption kinetics of 3-hydroxybenzaldehyde on native and activated bentonite. *Applied Clay Science* 38: 279–287.

Kukkadapu, R.K. and S.A. Boyd. 1995. Tetramethylphosphonium and tetramethylammonium-smectites as adsorbents of aromatic and chlorinated hydrocarbons: Effect of water on adsorption efficiency. *Clays and Clay Minerals* 43: 318–323.

Laby, R.H. and G.F. Walker. 1970. Hydrogen bonding in primary alkylammonium-vermiculite complexes. *Journal of Physical Chemistry* 74: 2369–2373.

Lagaly, G., M. Ogawa, and I. Dékány. 2006. Clay mineral organic interactions. In *Handbook of Clay Science*. Developments in Clay Science, Vol. 1, (Eds.) F. Bergaya, B.K.G. Theng, and G. Lagaly, pp. 309–377. Amsterdam, The Netherlands: Elsevier.

Lahav, N. and D.M. Anderson. 1973. Montmorillonite-benzidine reactions in the frozen and dry states. *Clays and Clay Minerals* 21: 137–139.

Lahav, N. and S. Raziel. 1971. Interaction between montmorillonite and benzidine in aqueous solutions – I. Adsorption of benzidine on montmorillonite. *Israel Journal of Chemistry* 9: 683–689.

Laird, D.A. 2006. Influence of layer charge on swelling of smectites. *Applied Clay Science* 34: 74–87.

Laird, D.A., E. Barriuso, R.H. Dowdy, and W.C. Koskinen. 1992. Adsorption of atrazine on smectites. *Soil Science Society of America Journal* 56: 62–67.

Laird, D.A. and P.D. Fleming. 1999. Mechanism for adsorption of organic bases on hydrated smectite surfaces. *Environmental Toxicology and Chemistry* 18: 1668–1672.

Laird, D.A. and C. Shang. 1997. Relationship between cation exchange selectivity and crystalline swelling in expanding 2:1 phyllosilicates. *Clays and Clay Minerals* 45: 681–689.

Larson, G.O. and L.R. Sherman. 1964. Infrared spectrophotometric analysis of some carbonyl compounds adsorbed on bentonite clay. *Soil Science* 98: 328–331.

Laura, R.D. and P. Cloos. 1970. Adsorption of ethylenediamine (EDA) on montmorillonite saturated with different cations. I. Copper-montmorillonite: Coordination. *Reunion Hispano-Belga do Minerales de la Arcilla, Madrid* 76–86.

Laura, R.D. and P. Cloos. 1975a. Adsorption of ethylene diamine (EDA) on montmorillonite saturated with different cations – III. Na-, K-, and Li-montmorillonite: Ion-exchange, protonation, co-ordination and hydrogen- bonding. *Clays and Clay Minerals* 23: 61–69.

Laura, R.D. and P. Cloos. 1975b. Adsorption of ethylenediamine (EDA) on montmorillonite saturated with different cations – IV. Al-, Ca-, and Mg-montmorillonite: Ion-exchange, protonation, co-ordination and hydrogen- bonding. *Clays and Clay Minerals* 23: 343–348.

Lawrence, M.A.M., R.K. Kukkadapu, and S.A. Boyd. 1998. Adsorption of phenol and chlorinated phenols from aqueous solution by tetramethylammonium- and tetramethylphosphonium-exchanged montmorillonite. *Applied Clay Science* 13: 13–20.

Lawrie, D.C. 1961. A rapid method for the determination of approximate surface areas of clays. *Soil Science* 92: 188–191.

Lee, J.-F., M.M. Mortland, S.A. Boyd, and C.T. Chiou. 1989. Shape-selective adsorption of aromatic molecules from water by tetramethylammonium-smectite. *Journal of the Chemical Society Faraday Transactions* 85: 2953–2962.

Lee, J-F., M.M. Mortland, C.T. Chiou, D.E. Kile, and S.A. Boyd. 1990. Adsorption of benzene, toluene, and xylene by two tetramethylammonium-smectites having different charge density. *Clays and Clay Minerals* 38: 113–120.

Lee, S.M. and D. Tiwari. 2012. Organo and inorgano-organo-modified clays in the remediation of aqueous solutions: An overview. *Applied Clay Science* 59–60: 84–102.

Léonard, A., A. Servais, and J.J. Fripiat. 1962. Étude de l'adsorption des amines par les montmorillonites. II. La structure des complexes. *Bulletin de la Société Chimique de France* 625–635.

Li, H., B.J. Teppen, C.T. Johnston, and S.A. Boyd. 2004. Thermodynamics of nitroaromatic compound adsorption from water by smectite clay. *Environmental Science & Technology* 38: 5433–5442.

Li, Y., X. Hu, X. Liu et al. 2018. Adsorption behavior of phenol by reversible surfactant-modified montmorillonite: Mechanism, thermodynamics, and regeneration. *Chemical Engineering Journal* 334: 1214–1221.

Li, X., N. Liu, L. Tang, and J. Zhang. 2020. Specific elevated adsorption and stability of cations in the interlayer compared with the external surface of clay minerals. *Applied Clay Science* 198: 105814.

Libor, O., L. Graber, and E. Donath. 1970. Montmorillonites treated with urea. I. Montmorillonites treated with urea solutions. *Acta Mineralogica-Petrographica* 20: 97–106.

Libor, O., L. Graber, and E. Donath. 1971. Investigation on montmorillonites contacting urea. II. Thermal analysis of Na- and H-montmorillonites containing urea. *Agrokémia és Talajtan* 19: 293–310.

Liu, C., H. Li, C.T. Johnstone, S.A. Boyd, and B.J. Teppen. 2012. Relating clay structural factors to dioxin adsorption by smectites: Molecular dynamics simulations. *Soil Science Society of America Journal* 76: 110–120.

Liu, C., H. Li, B.J. Teppen, C.T. Johnstone, and S.A. Boyd. 2009. Mechanisms associated with the high adsorption of dibenzo-*p*-dioxin from water by smectite clays. *Environmental Science & Technology* 43: 2777–2783.

Liu, D., P. Yuan, H. Liu et al. 2013c. High-pressure adsorption of methane on montmorillonite, kaolinite and illite. *Applied Clay Science* 85: 25–30.

Liu, H., D. Liu, P. Yuan et al. 2013a. Studies on the solid acidity of heated and cation-exchanged montmorillonite using *n*-butylamine titration in non-aqueous system and diffuse reflectance Fourier transform infrared (DRIFT) spectroscopy. *Physics and Chemistry of Minerals* 40: 479–489.

Liu, X., X. Lu, R. Wang, E.J. Meijer, and H. Zhou. 2011. Acidities of confined water in interlayer space of clay minerals. *Geochimica et Cosmochimica Acta* 75: 4978–4986.

Liu, X., R. Zhu, J. Ma, F. Ge, Y. Xu, and Y. Liu. 2013b. Molecular dynamics simulation study of benzene adsorption to montmorillonite: Influence of the hydration status. *Colloids and Surfaces A: Physicochemical and Engineering Aspects* 434: 200–206.

Liu, Y., M. Gao, Z. Gu, Z. Luo, Y. Ye, and L. Lu. 2014. Comparison between the removal of phenol and catechol by modified montmorillonite with two novel hydroxyl-containing Gemini surfactants. *Journal of Hazardous Materials* 267: 71–80.

Liu, Y., H. Gong, Y. Yu et al. 2006. Stability of 4'-substituted monobenzo-15-crown-5-bentonites and its adsorption properties for tetrachloroethane and naphthalene. *Applied Clay Science* 33: 7–12.

Liu, C., C. Gu, K. Yu et al. 2015. Integrating structural and thermodynamic mechanisms for sorption of PCBs by montmorillonite. *Environmental Science & Technology* 49: 2796–2805.

Loduhová, M., E. Jóna, V. Bizovská et al. 2015. Thermal, spectral and diffraction properties of Hg(II-exchanged montmorillonite with alkylamines. *Journal of Thermal and Analytical Calorimetry* 119: 879–884.

Luo, Z., M. Gao, S. Yang, and Q. Yang. 2015. Adsorption of phenols on reduced-charge montmorillonites modified by bispyridinium dibromides: Mechanism, kinetics and thermodynamics studies. *Colloids and Surfaces A: Physicochemical and Engineering Aspects* 482: 222–230.

Luo, W., J. Ouyang, P. Antwi, M. Wu., Z. Huang, and W. Qin. 2019. Microwave/ultrasound-assisted modification of montmorillonite by conventional and Gemini alkyl quaternary ammonium salts for adsorption of chromate and phenol: Structure-function relationship. *Science of the Total Environment* 655: 1104–1112.

Ma, L., Q. Chen, J. Zhu et al. 2016. Adsorption of phenol and Cu(II) onto cationic and zwitterionic surfactant modified montmorillonite in single and binary systems. *Chemical Engineering Journal* 283: 880–888.

MacEwan, D.M.C. 1948a. Complexes of clays with organic compounds. 1. Complex formation between montmorillonite and halloysite and certain organic liquids. *Transactions of the Faraday Society* 44: 349–367.

MacEwan, D.M.C. 1948b. Clay mineral complexes with organic liquids. *Clay Minerals Bulletin* 1: 44–46.

MacEwan, D.M.C. 1961. Montmorillonite minerals. In *The X-ray Identification and Crystal Structures of Clay Minerals*, (Ed.) G. Brown, pp. 143–207. London, UK: Mineralogical Society.

Majdan, M., M. Bujacka, E. Sabah et al. 2009. Unexpected difference in phenol sorption on PTMA- and BTMA-bentonite. *Journal of Environmental Management* 91: 195–205.

Malandrini, H., F. Clauss, S. Partyka, and J.M. Douillard. 1997. Interactions between talc particles and water and organic solvents. *Journal of Colloid and Interface Science* 194: 183–193.

Malferrari, D., F. Bernini, F. Tavanti, L. Tuccio, and A. Pedone. 2017. Experimental and molecular dynamics investigation proves that montmorillonite traps the biogenic amines histamine and tyramine. *Journal of Physical Chemistry* 121: 27493–27503.

Malinová, L., D. Jaksch, and J. Brožek. 2016. Montmorillonite modified with lactim methyl ethers having different ring sizes. *Applied Clay Science* 129: 20–26.

Mao, H., Y. Huang, J. Luo, and M. Zhang. 2021. Molecular simulation of polyether amines intercalation into Na-montmorillonite interlayer as a clay swelling inhibitor. *Applied Clay Science* 202: 105991.

Martin, R.T. 1955. Ethylene glycol retention by clays. *Soil Science Society of America Proceedings* 19: 160–164.

Martinez-Costa, J.I. and R. Leyva-Ramos. 2017. Effect of surfactant loading and type upon the sorption capacity of organobentonite towards pyrogallol. *Colloids and Surfaces A: Physicochemical and Engineering Aspects* 520: 676–685.

Martin-Rubi, J.A., J.A. Rausell-Colom, and J.M. Serratosa. 1974. Infrared absorption and X-ray diffraction study of butylammonium complexes of phyllosilicates. *Clays and Clay Minerals* 22: 87–90.

McBride, M.B. 1979. Reactivity of adsorbed and structural iron in hectorite as indicated by oxidation of benzidine. *Clays and Clay Minerals* 27: 224–230.

McBride, M.B. 1985. Surface reactions of 3,3'5,5'-tetramethylbenzidine on hectorite. *Clays and Clay Minerals* 33: 510–516.

McBride, M.B., T.J. Pinnavaia, and M.M. Mortland. 1975. Electron spin resonance studies of cation orientation in restricted water layers on phyllosilicate (smectite) surfaces. *Journal of Physical Chemistry* 79: 2430–2435.

McNeal, B.L. 1964. Effect of exchangeable cations on glycol retention by clay minerals. *Soil Science* 97: 96–102.

Mehra, O.P. and M.L. Jackson. 1959. Specific surface determination by duo-interlayer and mono-interlayer glycerol sorption for vermiculite and montmorillonite analysis. *Soil Science Society of America Proceedings* 23: 351–354.

Meier, L.P. and G. Kahr. 1999. Determination of the cation exchange capacity (CEC) of clay minerals using the complexes of copper(II) ion with triethylenetetramine and tetraethylenepentamine. *Clays and Clay Minerals* 47: 386–388.

Micera, G., A. Pusino, C. Gessa, and S. Petretto. 1988. Interaction of fluazifop with Al-, Fe^{3+}-, and Cu^{2+}-saturated montmorillonite. *Clays and Clay Minerals* 36: 354–358.

Michot, L.J. and F. Villiéras. 2006. Surface area and porosity. In *Handbook of Clay Science*. Developments in Clay Science, Vol. 1, (Eds.) F. Bergaya, B.K.G. Theng, and G. Lagaly, pp. 965–978. Amsterdam, The Netherlands: Elsevier.

Milford, M.H. and M.L. Jackson. 1962. Specific surface determination of expansible layer silicates. *Science* 135: 929–930.

Mirmohamadsadeghi, S., T. Kaghazchi, M. Soleimani, and N. Asasian. 2012. An efficient method for clay modification and its application for phenol removal from wastewater. *Applied Clay Science* 59–60: 8–12.

Moore, D.E., and J.B. Dixon. 1970. Glycerol vapor adsorption on clay minerals and montmorillonite soil clays. *Soil Science Society of America Proceedings* 34: 816–822.

Morillo, E., J.L. Pérez-Rodríguez, and C. Maqueda. 1991. Mechanisms of interaction between montmorillonite and 3-aminotriazole. *Clay Minerals* 26: 269–279.

Morozov, G., V. Breus, S. Nekludov, and I. Breus. 2014. Sorption of volatile organic compounds and their mixtures on montmorillonite at different humidity. *Colloids and Surfaces A: Physicochemical and Engineering Aspects* 454: 159–171.

Mortimer, J.V. and P.L. Gent. 1964. The use of organo-clays as gas chromatographic stationary phases. *Analytical Chemistry* 36: 754–756.

Mortland, M.M. 1966. Urea complexes with montmorillonite: An infrared absorption study. *Clay Minerals* 6: 143–156.

Mortland, M.M. 1968a. Protonation of compounds at clay mineral surfaces. *Transactions 9th International Congress of Soil Science, Adelaide* 1: 691–699.

Mortland, M.M. 1968b. Pyridinium-montmorillonite complexes with ethyl N,N-di-*n*-propylthiolcarbamate (EPTC). *Journal of Agricultural and Food Chemistry* 16: 706–707.

Mortland, M.M. 1970. Clay-organic complexes and interactions. *Advances in Agronomy* 22: 75–117.

Mortland, M.M. and S.A. Boyd. 1989. Polymerization and dechlorination of chloroethenes on copper(II)-smectite via radical-cation intermediates. *Environmental Science & Technology* 23: 223–227.

Mortland, M.M. and L.J. Halloran. 1976. Polymerization of aromatic molecules on smectite. *Soil Science Society of America Journal* 40: 367–370.

Mortland, M.M. and W.F. Meggitt. 1966. Interaction of ethyl *N,N*-di-*n*-propylthiolcarbamate (EPTC) with montmorillonite. *Journal of Agricultural and Food Chemistry* 14: 126–129.

Mortland, M.M. and T.J. Pinnavaia. 1971. Formation of copper(II) arene complexes on the interlamellar surfaces of montmorillonite. *Nature Physical Science* 229: 75–77.

Mortland, M.M. and K.V. Raman. 1968. Surface acidity of smectites in relation to hydration, exchangeable cation, and structure. *Clays and Clay Minerals* 16: 393–398.

Mortland, M.M., S. Sun, and S.A. Boyd. 1986. Clay-organic complexes as adsorbents for phenol and chlorophenols. *Clays and Clay Minerals* 34: 581–585.

Moshe, A.B. and G. Rytwo. 2018. Thiamine-based organoclay for phenol removal from water. *Applied Clay Science* 155: 50–56.

Mosser-Ruck, R., K. Devineau, D. Charpentier, and M. Cathelineau. 2005. Effects of ethylene glycol saturation protocols on XRD patterns: A critical review and discussion. *Clays and Clay Minerals* 53: 631–638.

Muiambo, H.F., W.W. Focke, M. Atanasova, and A. Benhamida. 2015. Characterization of urea-modified Palabora vermiculite. *Applied Clay Science* 105–106: 14–20.

Nafees, M. and A. Waseem. 2014. Organoclays as sorbent material for phenolic compounds: A review. *CLEAN – Soil Air Water* 41: 1–9.

Nagai, S., S. Onishi, I. Nitta, A. Tsunashima, F. Kanamaru, and M, Koizumi. 1974. ESR study of Cu(II) ion complexes adsorbed on interlamellar surface of montmorillonite. *Chemical Physics Letters* 26: 517–520.

Newman, A.C.D. 1983. The specific surface of soils determined by water sorption. *Journal of Soil Science* 34: 23–32.

Nguyen, T.T., M. Raupach, and L.J. Janik. 1987. Fourier-transform infrared study of ethylene glycol monoethyl ether adsorbed on montmorillonite: Implications for surface area measurements of clays. *Clays and Clay Minerals* 35: 60–67.

Nguyen, V.N., T.D.C. Nguyen, T.P. Dao, H.T. Tran, D.B. Nguyen, and D.H. Ahn. 2013. Synthesis of organoclays and their application for the adsorption of phenolic compounds from aqueous solution. *Journal of Industrial and Engineering Chemistry* 19: 640–644.

Nir, S., T. Undabeytia, D.Y. Marcovich et al. 2000. Optimization of adsorption of hydrophobic herbicides on montmorillonite preadsorbed by monovalent organic cations: Interaction between phenyl rings. *Environmental Science & Technology* 34: 1269–1274.

Norrish, K. 1954. The swelling of montmorillonite. *Discussions of the Faraday Society* 18: 120–134.

Nourmoradi, H., M. Nikaeen, and M. Khiadani (Hajian). 2012. Removal of benzene, toluene, ethylbenzene and xylene (BTEX) from aqueous solutions by montmorillonite modified with nonionic surfactant: Equilibrium, kinetic and thermodynamic study. *Chemical Engineering Journal* 191: 341–348.

Nourmoradi, H., M. Khiadani, and M. Nikaeen. 2013. Multi-component adsorption of benzene, toluene, ethylbenzene, and xylene from aqueous solutions by montmorillonite modified with tetradecyl trimethylammonium bromide. *Journal of Chemistry* 2013, Article ID 589354. https://doi.org/1155/2013/589354.

Ogawa, M., T. Aono, K. Kuroda, and C. Kato. 1993. Photophysical probe study of alkylammonium-montmorillonites. *Langmuir* 9: 1529–1533.

Ogawa, M., T. Hashizume, K. Kuroda, and C. Kato. 1991. Intercalation of 2,2'-bipyridine and complex formation in the interlayer space of montmorillonite by solid-solid reactions. *Inorganic Chemistry* 30: 584–585.

Ogawa, M., K. Kato, K. Kuroda, and C. Kato. 1990. Preparation of montmorillonite-alkylamine intercalation compounds by solid-solid reactions. *Clay Science* 8: 31–36.

Ogawa, M., K. Kuroda, and C. Kato. 1989. Preparation of montmorillonite-organic intercalation compounds by solid-solid reactions. *Chemistry Letters* 18: 1659–1662.

Ogawa, M., H. Shirai, K. Kuroda, and C. Kato. 1992. Solid-state intercalation of naphthalene and anthracene into alkylammonium montmorillonites. *Clays and Clay Minerals* 40: 485–490.

Okada, T., T. Morita, and M. Ogawa. 2005. Tris(2,2'-bipyridine)ruthenium(II)-clays as adsorbents for phenol and chlorinated phenols from aqueous solution. *Applied Clay Science* 29: 45–53.

Okada, T. and M. Ogawa. 2003. 1,1'-Dimethyl-4,4'-bipyridinium-smectites as a novel adsorbent of phenols from water through charge-transfer interactions. *Chemical Communications* 1378–1379.

Okada, T., H. Sakai, and M. Ogawa. 2008. The effect of the molecular structure of a cationic azo dye on the photoinduced intercalation of phenol in a montmorillonite. *Applied Clay Science* 40: 187–192.

Olejnik, S., A.M. Posner, and J.P. Quirk. 1973. Adsorption of pyridine *N*-oxide onto montmorillonite. *Clays and Clay Minerals* 21: 191–198.

Olejnik, S., A.M. Posner, and J.P. Quirk. 1974. Swelling of montmorillonite in polar organic liquids. *Clays and Clay Minerals* 22: 361–365.

Önal, M. and Y. Sarikaya. 2008. The effect of organic cation content on the interlayer spacing, surface area and porosity of methyltributylammonium-smectite. *Colloids and Surfaces A: Physicochemical and Engineering Aspects* 317: 323–327.

Onikata, M., M. Kondo, and S. Yamanaka. 1999. Swelling of formamide-montmorillonite complexes in polar liquids. *Clays and Clay Minerals* 47: 678–681.

Ouardi, M.E., M. Laabd, H.A. Oualid et al. 2019. Efficient removal of *p*-nitrophenol from water using montmorillonite clay: Insights into the adsorption mechanism, process optimization, and regeneration. *Environmental Science and Pollution Research* 26: 19615–19631.

Ozola, R., A. Krauklis, J. Burlakovs, M. Klavins, Z. Vincevica-Gaile, and W. Hogland. 2019. Surfactant-modified clay sorbents for the removal of *p*-nitrophenol. *Clays and Clay Minerals* 67: 132–142.

Pálková, H., V. Hronsky, V. Bizovská, and J. Madejová. 2015. Spectroscopic study of water adsorption on Li^+, TMA^+ and $HDTMA^+$ exchanged montmorillonite. *Spectrochimica Acta Part A: Molecular and Biomolecular Spectroscopy* 149: 751–761.

Pálková, H., J. Madejová, and P. Komadel. 2009. The effect of layer charge and exchangeable cations on sorption of biphenyl on montmorillonites. *Central European Journal of Chemistry* 7: 494–504.

Palmer, J. and N. Bauer. 1961. Sorption of amines by montmorillonite. *Journal of Physical Chemistry* 65: 894–895.

Pantani, O.L., S. Dousset, M. Schiavon, and P. Fusi. 1997. Adsorption of isoproturon on homoionic clays. *Chemosphere* 35: 2619–2626.

Parfitt, R.L. and M.M. Mortland. 1968. Ketone adsorption on montmorillonite. *Soil Science Society of America Proceedings* 32: 355–363.

Parfitt, R.L., J.S. Whitton, and B.K.G. Theng. 2001. Surface reactivity of A horizons towards polar compounds estimated from water adsorption and water content. *Australian Journal of Soil Research* 39: 1105–1110.

Park, M., C.Y. Kim, D.H. Lee et al. 2004. Intercalation of magnesium-urea complex into swelling clay. *Journal of Physics and Chemistry of Solids* 65: 409–412.

Park, Y., G.A. Ayoko, and R.L. Frost. 2011. Application of organoclays for the adsorption of recalcitrant organic molecules from aqueous media. *Journal of Colloid and Interface Science* 354: 292–305.

Park, Y., G.A. Ayoko, R. Kurdi, E. Horváth, J. Kristóf, and R.L. Frost. 2013. Adsorption of phenolic compounds by organoclays: Implications for the removal of organic pollutants from aqueous media. *Journal of Colloid and Interface Science* 406: 196–208.

Pei, Z., J. Kong, X.-Q. Shan, and B. Wen. 2012. Sorption of aromatic hydrocarbons onto montmorillonite as affected by norfloxacin. *Journal of Hazardous Materials* 203–204: 137–144.

Peng, B., P.-Y. Luo, W.-Y. Guo, and Q. Yuan. 2013. Structure-property relationship of polyetheramines as clay-swelling inhibitors in water-based drilling fluids. *Journal of Applied Polymer Science* 129: 1074–1079.

Peng, C., F. Min, and L. Liu. 2017. Effect of pH on the adsorption of dodecylamine on montmorillonite: Insights from experiments and molecular dynamics simulations. *Applied Surface Science* 425: 996–1005.

Peng, C., Y. Zhong, and F. Min. 2018. Adsorption of alkylamine cations on montmorillonite (001) surface: A density functional theory study. *Applied Clay Science* 152: 249–258.

Peng, C., G. Wang, L. Qin, S. Luo, F. Min, and X. Zhu. 2020. Molecular dynamics simulation of NH_4-montmorillonite interlayer hydration: Structure, energetics, and dynamics. *Applied Clay Science* 195: 105657.

Penland, R.B., S. Mizushima, C. Curran, and J.V. Quagliano. 1957. Infrared absorption spectra of inorganic coordination complexes. X. Studies of some metal-urea complexes. *Journal of the American Chemical Society* 79: 1575–1578.

Pereira, T.R., D.A. Laird, M.L. Thompson et al. 2008. Role of smectite quasicrystal dynamics in adsorption of dinitrophenol. *Soil Science Society of America Journal* 72: 347–354.

Pereira, E.I., F.B. Minussi, C.C.T. da Cruz, A.C.C. Benardi, and C. Ribeiro. 2012. Urea-montmorillonite-extruded nanocomposites: A novel slow-release material. *Journal of Agricultural and Food Chemistry* 60: 5267–5272.

Pérez-Santano, A., R. Trujillano, C. Belver, A. Gil, and M.A. Vicente. 2005. Effect if the intercalation conditions of a montmorillonite with octadecylamine. *Journal of Colloid and Interface Science* 284: 239–244.

Pfirrmann, G., G. Lagaly, and A. Weiss. 1973. Phase transitions in complexes of nontronite with *n*-alkanols. *Clays and Clay Minerals* 21: 239–247.

Pham, T.R. and G.W. Brindley. 1970. Methylene blue absorption by clay minerals. Determination of surface areas and cation exchange capacities (Clay-organic studies XVIII). *Clays and Clay Minerals* 18: 203–212.

Philen, O.D., Jr., S.B. Weed, and J.B. Weber. 1970. Estimation of surface charge density of mica and vermiculite by competitive adsorption of diquat^{2+} *vs.* paraquat^{2+}. *Soil Science Society of America Proceedings* 34: 527–531.

Philen, O.D., Jr., S.B. Weed, and J.B. Weber. 1971. Surface charge characterization of layer silicates by competitive adsorption of two organic divalent cations. *Clays and Clay Minerals* 19: 295–302.

Pimchan, P., N. Khaorapapong, M. Sohmiya, and M. Ogawa. 2014. In situ complexation of 8-hydroxyquinoline and 4,4'-bipyridine with zinc(II) in the interlayer space of montmorillonite. *Applied Clay Science* 95: 310–316.

Pinnavaia, T.J., P.L. Hall, S.S. Cady, and M.M. Mortland. 1974. Aromatic radical cation formation on the intracrystal surfaces of transition metal layer lattice silicates. *Journal of Physical Chemistry* 78: 994–999.

Pinnavaia, T.J. and M.M. Mortland. 1971. Interlamellar metal complexes of layer silicates. 1. Copper(II)-arene complexes on montmorillonite. *Journal of Physical Chemistry* 75: 3957–3962.

Pleysier, J.L. and A. Cremers. 1975. Stability of silver-thiourea complexes in montmorillonite clay. *Journal of the Chemical Society, Faraday Transactions* 71: 256–264.

Pleysier, J.L. and A.S.R. Juo. 1980. A single-extraction method using silver-thiourea for measuring exchangeable cations and effective CEC in soils with variable charge. *Soil Science* 129: 205–211.

Praus, P., M. Veteška, and M. Pospíšil. 2011. Adsorption of phenol and aniline on natural and organically modified montmorillonite: Experiment and molecular modelling. *Molecular Simulation* 37: 964–974.

Prost, R., T. Koutit, A. Benchara, and E. Huard. 1998. State and location of water adsorbed on clay minerals: Consequences of the hydration and swelling-shrinkage phenomena. *Clays and Clay Minerals* 46: 117–131.

Pusino, A., G. Micera, A. Premoli, and C. Gessa. 1989. D-glucosamine sorption on Cu(II)-montmorillonite as the protonated and neutral species. *Clays and Clay Minerals* 37: 377–380.

Qin, C., W. Zhang, B. Wang, X. Chen, K. Xia, and Y. Gao. 2018. DNA facilitates the sorption of polycyclic aromatic hydrocarbons on montmorillonites. *Environmental Science & Technology* 52: 2694–2703.

Qu, X., P. Liu, and D. Zhu. 2008. Enhanced sorption of polycyclic aromatic hydrocarbons to tetra-alkyl ammonium modified smectites via cation-π interactions. *Environmental Science & Technology* 42: 1109–1116.

Qu, X., Y. Zhang, H. Li, S. Zheng, and D. Zhu. 2011. Probing the specific sorption sites on montmorillonite using nitroaromatic compounds and hexafluorobenzene. *Environmental Science & Technology* 45: 2209–2216.

Quirk, J.P. 1955. Significance of surface areas calculated from water vapor sorption isotherms by use of the B.E.T. equation. *Soil Science* 80: 423–430.

Quirk, J.P. 1994. Interparticle forces: A basis for the interpretation of soil physical behavior. *Advances in Agronomy* 53: 121–183.

Quirk, J.P. and R.S. Murray. 1999. Appraisal of the ethylene glycol monoethyl ether method for measuring hydratable surface area of clays and soils. *Soil Science Society of America Journal* 63: 839–849.

Radul, N.M. and F.D. Ovcharenko. 1971. Method for determining the orientation of adsorbed alcohol molecules in the interlayer space of montmorillonite. *Ukrainskiy Khimicheskiy Zhurnal* 37: 775–777.

Raman, K.V. and M.M. Mortland. 1969. Proton transfer reactions at clay mineral surfaces. *Soil Science Society of America Journal* 33: 313–317.

Rana, K., S.A. Boyd, B.J. Teppen, H. Li, C. Liu, and C.T. Johnston. 2009. Probing the microscopic hydrophobicity of smectite surfaces: A vibrational spectroscopic study of dibenzo-p-dioxin sorption to smectite. *Physical Chemistry, Chemical Physics* 11: 2976–2985.

Raupach, M. and L.J. Janik. 1976. The orientation of ornithine and 6-aminohexanoic acid adsorbed on vermiculite from polarized I.R. ATR spectra. *Clays and Clay Minerals* 24: 127–133.

Rawajfih, Z. and N. Nsour. 2006. Characteristics of phenol and chlorinated phenols sorption onto surfactant-modified bentonite. *Journal of Colloid and Interface Science* 298: 39–49.

Reynolds, R.C., Jr. 1965. An X-ray study of an ethylene glycol-montmorillonite complex. *American Mineralogist* 50: 990–1001.

Reynolds, R.C., Jr. 1969. Orientation of ethylene glycol monoethyl ether molecules on montmorillonite. *American Mineralogist* 54: 562–567.

Richards, S. and A. Bouazza. 2007. Phenol adsorption in organo-modified basaltic clay and bentonite. *Applied Clay Science* 37: 133–142.

Ristori, G.G., M. De Nobili, and P. Fusi. 1991. Non-biological degradation of oxamide adsorbed on homoionic montmorillonite. *Applied Clay Science* 6: 171–179.

Ristori, G.G., P. Fusi, and M. Franci. 1987. Complexes of montmorillonite with thioamides. *Applied Clay Science* 2: 145–154.

Ristori, G.G., E. Sparvoli, P. Fusi, J.P. Quirk, and C. Martelloni. 1985. Measurement of total and external surface area of homoionic smectites by *p*-nitrophenol adsorption. *Mineralogica et Petrographica Acta* 29A: 137–143.

Ristori, G.G., E. Sparvoli, L. Landi, and C. Martelloni. 1989. Measurement of specific surface areas of soils by *p*-nitrophenol adsorption. *Applied Clay Science* 4: 521–532.

Roberts, A.L., G.B. Street, and D. White. 1964. The mechanism of phenol adsorption by organo-clay derivatives. *Journal of Applied Chemistry* 14: 261–265.

Roberts, M.G., H. Li, B.J. Teppen, and S.A. Boyd. 2006. Sorption of nitroaromatics by ammonium- and organic ammonium-exchanged smectite: Shifts from adsorption/complexation to a partition-dominated process. *Clays and Clay Minerals* 54: 426–434.

Rowland, R.A. and E.J. Weiss. 1963. Bentonite-methylamine complexes. *Clays and Clay Minerals* 10: 460–468.

Ruan, X., L. Zhu, and B. Chen. 2008. Adsorptive characteristics of the siloxane surfaces of reduced-charge bentonites saturated with tetramethylammonium cation. *Environmental Science & Technology* 42: 7911–7917.

Ruiz-Conde, A., Ruiz-Amil, A., J.L. Perez-Rodriguez, P.J. Sanchez-Soto, and F.A. de la Cruz. 1997. Interaction of vermiculite with aliphatic amides (formamide, acetamide and propionamide): Formation and study of interstratified phases in the transformation of Mg- to NH_4-vermiculite. *Clays and Clay Minerals* 45: 311–326.

Ruiz-Hitzky, E. and B. Casal. 1978. Crown ether intercalations with phyllosilicates. *Nature* 276: 596–597.

Rupert, J.P. 1973. Electron spin resonance spectra of interlamellar Cu(II)-arene complexes on montmorillonite. *Journal of Physical Chemistry* 77: 784–790.

Russell, J.D., M.I. Cruz, and J.L. White. 1968. The adsorption of 3-aminotriazole by montmorillonite. *Journal of Agricultural and Food Chemistry* 16: 21–24.

Saltzman, S. and S. Yariv. 1975. Infrared study of the sorption of phenol and p-nitrophenol by montmorillonite. *Soil Science Society of America Journal* 39: 474–479.

Santos, A., J. Gonzalez-Garmendia, and A. Rodriguez. 1969. Complejos de vermiculita con aminas. *Anales de Química* 65: 433–442.

Santos, A., A. Rodriguez, J. Gonzales-Garmendia, and J. Barrios. 1970. Espectros infrarrojos de muestras homoiónicas de montmorillonita y vermiculita y sus complejos interlaminares con aminas. *Réunion Hispano-Belga de Minerales de la Arcilla, Madrid* 87–97.

Sarkar, B., Y. Xi, M. Megharaj et al. 2012. Bioreactive organoclay: A new technology for environmental remediation. *Critical Reviews in Environmental Science and Technology* 42: 435–488.

Scholtzová, E., D. Tunega, J. Madejová, H. Pálková, and P. Komadel. 2013. Theoretical and experimental study of montmorillonite intercalated with tetramethylammonium cation. *Vibrational Spectroscopy* 66: 123–131.

Schoonheydt, R.A. and C.T. Johnston. 2013. Surface and interface chemistry of clay minerals. In *Handbook of Clay Science*, 2nd edition. Developments in Clay Science, Vol. 5A, (Eds.) F. Bergaya and G. Lagaly, pp. 139–172. Amsterdam, The Netherlands: Elsevier.

Schoonheydt, R.A., F. Velghe, R. Baerts, and J.B. Uytterhoeven. 1979. Complexes of diethylenetriamine (dien) and tetraethylenepentamine (tetren) with Cu(II) and Ni(II) on hectorite. *Clays and Clay Minerals* 27: 269–278.

Schrader, M.E. and S. Yariv. 1990. Wettability of clay minerals. *Journal of Colloid and Interface Science* 136: 85–94.

Searle, P.L. 1986. The measurement of soil exchange properties using the single extraction, silver thiourea method: An evaluation using a range of New Zealand soils. *Australian Journal of Soil Research* 24: 193–200.

Seki, Y., Y. Ide, T. Okada, and M. Ogawa. 2015. Concentration of 2-phenylphenol by organoclays from aqueous sucrose solution. *Applied Clay Science* 109–110: 65–67.

Senturk, H.B., D. Ozdes, A. Gundogdu, C. Duran, and M. Soylak. 2009. Removal of phenol from aqueous solutions by adsorption onto organomodified Tirebolu bentonite: Equilibrium, kinetic and thermodynamic study. *Journal of Hazardous Materials* 172: 353–362.

Serratosa, J.M. 1966. Infrared analysis of the orientation of pyridine molecules in clay complexes. *Clays and Clay Minerals* 14: 385–391.

Serratosa, J.M. 1968a. Infrared study of the orientation of chlorobenzene sorbed on pyridinium-montmorillonite. *Clays and Clay Minerals* 16: 93–97.

Serratosa, J.M. 1968b. Infrared study of benzonitrile (C_6H_5-CN)-montmorillonite complexes. *American Mineralogist* 53: 1244–1251.

Servais, A., J.J. Fripiat, and A. Léonard. 1962. Étude de l'adsorption des amines par les montmorillonites. I. Les processus chimiques. *Bulletin de la Société Chimique de France* 617–625.

Shah, S., M.R. Jan, M. Zeeshan, and M. Imran. 2017. Kinetic, equilibrium and thermodynamic studies for sorption of 2,4-dichlorophenol onto surfactant modified Fullers' earth. *Applied Clay Science* 143: 227–233.

Shah, K.J., S.-Y. Pan, A.D. Shukla, D.O. Shah, and P.-C. Chiang. 2018. Mechanism of organic pollutants sorption from aqueous solution by cationic tunable organoclays. *Journal of Colloid and Interface Science* 529: 90–99.

Shakir, K., H.F. Ghoneimy, A.F. Elkafrawy, S.G. Beheir, and M. Refaat. 2008. Removal of catechol from aqueous solutions by adsorption onto organophilic-bentonite. *Journal of Hazardous Materials* 150: 765–773.

Sharmasarkar, S., W.F. Jaynes, and G.F. Vance. 2000. BTEX sorption by montmorillonite organo-clays: TMPA, ADAM, HDTMA. *Water, Air, and Soil Pollution* 119: 257–273.

Shen, Y.-H. 2002. Removal of phenol from water by adsorption-flocculation using organobentonite. *Water Research* 36: 1107–1114.

Shen, Y.-H. 2004. Phenol sorption by organoclays having different charge characteristics. *Colloids and Surfaces A: Physicochemical and Engineering Aspects* 232: 143–149.

Shen, T. and M. Gao. 2019. Gemini surfactant modified organo-clays for removal of organic pollutants from water. *Chemical Engineering Journal* 375: 121910.

Shen, W., L. Li., H. Zhou, Q. Zhou, M. Chen, and J. Zhu. 2018. Effects of charge density on the hydration of siloxane surface of montmorillonite: A molecular dynamics simulation study. *Applied Clay Science* 159: 10–15.

Sheng, G. and S.A. Boyd. 2000. Polarity effect on dichlorobenzene sorption by hexadecyltrimethylammonium-exchanged clays. *Clays and Clay Minerals* 48: 43–50.

Sheng, G., C.T. Johnston, B.J. Teppen, and S.A. Boyd. 2002. Adsorption of dinitrophenol herbicides from water by montmorillonites. *Clays and Clay Minerals* 50: 25–34.

Sheng, G., S. Xu, and S.A. Boyd. 1997. Surface heterogeneity of trimetylphenylammonium-smectite as revealed by adsorption of aromatic hydrocarbons from water. *Clays and Clay Minerals* 45: 659–669.

Shiga, Y. 1961. Studies on the complexes of montmorines with urea and its derivatives. I. The influence of exchangeable cations on the interlayer adsorption of urea by montmorines. *Soil Science and Plant Nutrition (Tokyo)* 7: 119–124.

Shimoyama, A. and W.D. Johns. 1971. Catalytic conversion of fatty acids to petroleum-like paraffins and their maturation. *Nature Physical Science* 232: 140–144.

Shoval, S. and S. Yariv. 1979. The interaction between roundup (glyphosate) and montmorillonite. Part II. Ion exchange and sorption of *iso*-propylammonium by montmorillonite. *Clays and Clay Minerals* 27: 29–38.

Siantar, D.P., B.A. Feinberg, and J.J. Fripiat. 1994. Interactions between organic and inorganic pollutants in the clay interlayer. *Clays and Clay Minerals* 42: 187–196.

Sieskind, O. and R. Wey. 1958a. Influence du pH sur l'adsorption d'amines aliphatiques normales par la montmorillonite-H. *Comptes Rendus de l'Academie des Sciences, Paris* 247: 74–76.

Sieskind, O. and R. Wey. 1958b. Sur l'adsorption d'amines aliphatiques par la montmorillonite. *Bulletin Groupe Français des Argiles* 10: 9–14.

Silva, I.A., F.K.A. Sousa, R.R. Menezes, G.A. Neves, L.N.L. Santana, and H.C. Ferreira. 2014. Modification of bentonites with nonionic surfactants for use in organic-based drilling fluids. *Applied Clay Science* 95: 371–377.

Slabaugh, W.H. and R.W. Vasofsky. 1975. Adsorption of xylene by organoclays. *Clays and Clay Minerals* 23: 458–461.

Slade, P.G., J.P. Quirk, and K. Norrish. 1991. Crystalline swelling of smectite samples in concentrated NaCl solutions in relation to layer charge. *Clays and Clay Minerals* 39: 234–238.

Slade, P.G., M. Raupach, and W.W. Emerson. 1978. The ordering of cetylpyridinium bromide on vermiculite. *Clays and Clay Minerals* 26: 125–134.

Slade, P.G., P.K. Schultz, and E.R.T. Tiekink. 1989. Structure of a 1,4-diazabicyclo[2,2,2]octane-vermiculite intercalate. *Clays and Clay Minerals* 37: 81–88.

Slade, P.G. and P.A. Stone. 1983. Structure of a vermiculite-aniline intercalate. *Clays and Clay Minerals* 31: 200–206.

Slaný, M., L. Jankovič, and J. Madejová. 2019. Structural characterization of organo-montmorillonites prepared from a series of primary alkylamines salts: Mid-IR and near-IR study. *Applied Clay Science* 176: 11–20.

Smith, C.R. and K. Yates. 1972. Kinetic evidence for predominant oxygen protonation of amides. *Canadian Journal of Chemistry* 50: 771–773.

Smith, J.A., P.R. Jaffé, and C.T. Chiou. 1990. Effect of ten quaternary ammonium cations on tetrachloromethane sorption to clay from water. *Environmental Science & Technology* 24: 1167–1172.

Šnirková, S., E. Jóna, L. Lajdová et al. 2009. Ni-exchanged montmorillonite with methyl-, dimethyl- and trimethylamine and their thermal properties. *Journal of Thermal Analysis and Calorimetry* 96: 63–66.

Sohn, J.R., J.S. Han, and J.S Lim. 2005. Infrared spectroscopic study of alkyl aldehydes adsorbed on cations supported by layer silicate. *Applied Clay Science* 252: 858–865.

Sohn, J.R. and J.T. Kim. 2000. Infrared study of alky ketones adsorbed on the interlamellar surface of montmorillonite. *Langmuir* 16: 5430–5434.

Solomon, D.H., B.C. Loft, and J.D. Swift. 1968. Reactions catalysed by minerals. IV. The mechanism of the benzidine blue reaction on silicate minerals. *Clay Minerals* 7: 389–397.

Soma, M. and Y. Soma. 1988a. Intercalation of tetrathiafulvalene (TTF) into montmorillonite by reaction with the interlayer cations. *Chemistry Letters* 405–408.

Soma, Y. and M. Soma. 1988b. Adsorption of benzidines and anilines on Cu- and Fe-montmorillonites studied by resonance Raman spectroscopy. *Clay Minerals* 23: 1–12.

Soma, Y., M. Soma, Y. Furukawa, and I. Harada. 1987. Reactions of thiophene and methylthiophenes in the interlayer of transition-metal ion-exchanged montmorillonite studied by resonance Raman spectroscopy. *Clays and Clay Minerals* 35: 53–59.

Soma, Y., M. Soma, and I. Harada. 1983. Resonance Raman spectra of benzene adsorbed on Cu^{2+}-montmorillonite: Formation of poly(*p*-phenylene) cations in the interlayer of the clay mineral. *Chemical Physics Letters* 99: 153–156.

Soma, Y., M. Soma, and I. Harada. 1984. Resonance Raman spectroscopic study of cation radicals of aromatic molecules formed in the interlayer of transition-metal exchanged montmorillonites. *Journal of Inclusion Phenomena* 2: 135–142.

Soma, Y., M. Soma, and I. Harada. 1986. The oxidative polymerization of aromatic molecules in the interlayer of montmorillonites studied by resonance Raman spectroscopy. *Journal of Contaminant Hydrology* 1: 95–106.

Sor, K. and W.D. Kemper. 1959. Estimation of hydratable surface area of soils and clays from the amount of absorption and retention of ethylene glycol. *Soil Science Society of America Proceedings* 23: 105–110.

Sposito, G. 1984. *The Surface Chemistry of Soils*. New York: Oxford University Press.

Sposito, G. and R. Prost. 1982. Structure of water adsorbed on smectites. *Chemical Reviews* 82: 554–573.

Spyrou, K., G. Potsi, E.K. Diamanti et al. 2014. Towards novel multifunctional pillared nanostructures: Effective intercalation of adamantylamine in graphene oxide and smectite clays. *Advanced Functional Materials* 24: 5841–5850.

Środoń, J. and D.K. McCarty. 2008. Surface area and layer charge of smectite from CEC and EGME/H_2O-retention measurements. *Clays and Clay Minerals* 56: 155–174.

Steinfink, H., E.J. Weiss, D.J. Haase, and R.A. Rowland 1963. An X-ray diffraction study of a hexamethylenediamine-vermiculite complex. *Proceedings International Clay Conference, Stockholm* 1: 343–348.

Stevens, J.J. and S.J. Anderson. 1996a. Orientation of trimethylphenylammonium (TMPA) on Wyoming montmorillonite: Implications for sorption of aromatic compounds. *Clays and Clay Minerals* 44: 132–141.

Stevens, J.J. and S.J. Anderson. 1996b. An FTIR study of water sorption on TMA- and TMPA-montmorillonites. *Clays and Clay Minerals* 44: 142–150.

Stevens, J.J., S.J. Anderson, and S.A. Boyd. 1996. FTIR study of competitive water-arene sorption on tetramethylammonium- and tetrmethylphenylammonium-montmorillonites. *Clays and Clay Minerals* 44: 88–95.

Stockmeyer, M. and K. Kruse. 1991. Adsorption of Zn and Ni ions and phenol and diethylketones by bentonites of different organophilicities. *Clay Minerals* 26: 431–434.

Stoessel, F., J.L. Guth, and R. Wey. 1977. Polymerisation de benzene en polyparaphenylene dans une montmorillonite cuivrique. *Clay Minerals* 12: 255–259.

Stul, M.S. and J.B. Uytterhoeven. 1975. Interlamellar sorption of ethanol on montmorillonite clays with different layer charges. *Journal of the Chemical Society, Faraday Transactions* 71: 1396–1401.

Stul, M.S., J.B. Uytterhoeven, J. De Bock, and P.L. Huyskens. 1979. The adsorption of n-aliphatic alcohols from dilute aqueous solutions on RNH_3-monmorillonites. II. Interlamellar association of the adsorbate. *Clays and Clay Minerals* 27: 377–386.

Stutzmann, T. and B. Siffert. 1977. Contribution to the adsorption mechanism of acetamide and polyacrylamide on to clays. *Clays and Clay Minerals* 25: 392–406.

Su, J., H.F. Lin, Q.-P. Wang, Z.-M. Xie, and Z.-L. Chen. 2011. Adsorption of phenol from aqueous solutions by organomontmorillonite. *Desalination* 269: 163–169.

Sutherland, H.H. and D.M.C. MacEwan. 1961. Organic complexes of vermiculite. *Clay Minerals Bulletin* 4: 229–233.

Suzuki, Y., Y. Tenma, Y. Nishioka, and J. Kawamata. 2012. Efficient nonlinear optical properties of dyes confined in interlayer nanospaces of clay minerals. *Chemistry – An Asian Journal* 7: 1170–1179.

Svensson, P.D. and S. Hansen. 2010. Intercalation of smectite with liquid ethylene glycol – Resolved in time and space by synchrotron X-ray diffraction. *Applied Clay Science* 48: 358–367.

Swoboda, A.R. and G.W. Kunze. 1968. Reactivity of montmorillonite surfaces with weak organic bases. *Soil Science Society of America Proceedings* 32: 806–811.

Szczerba, M., Z. Kłapyta, and A. Kalinichev. 2014. Ethylene glycol intercalation in smectites. Molecular dynamics simulation studies. *Applied Clay Science* 91: 87–97.

Tahoun, S. and M.M. Mortland. 1966a. Complexes of montmorillonite with primary, secondary, and tertiary amides: I. Protonation of amides on the surface of montmorillonite. *Soil Science* 102: 248–254.

Tahoun, S.A. and M.M. Mortland. 1966b. Complexes of montmorillonite with primary, secondary, and tertiary amides: II. Coordination of amides on the surface of montmorillonite. *Soil Science* 102: 314–321.

Takagi, S., T. Shimada, M. Eguchi et al. 2002. High-density adsorption of cationic porphyrins on clay layer surfaces without aggregation: The size-matching effect. *Langmuir* 18: 2265–2272.

Taramasso, M. and F. Veniale. 1969. Gas chromatographic performance of long chain alkylammonium complexes with 'beidellite-type' clay minerals. *Chromatographia* 2: 239–241.

Tarasevich, Y.I. and V.E. Doroshenko. 1983. Interaction of triphenylcarbinol with montmorillonite. *Kolloidnyi Zhurnal* 45: 1167–1170.

Tarasevich, Y.L., V.R. Telichkun, and F.D. Ovcharenko. 1968. Infrared spectra of acetonitrile adsorbed on montmorillonite. *Doklady Akademii Nauk., S.S.S.R.* 182: 141–143.

Tennakoon, D.T.B., J.M. Thomas, and M.J. Tricker. 1974b. Surface and intercalate chemistry of layered silicates. Part II. An iron-57 Mössbauer study of the role of lattice-substituted iron in the benzidine blue reaction of montmorillonite. *Journal of the Chemical Society Dalton Transactions* 2211–2215.

Tennakoon, D.T.B., J.M. Thomas, M.J. Tricker, and J.O. Williams. 1974a. Surface and intercalate chemistry of layered silicates. Part I. General introduction and uptake of benzidine and related organic molecules by montmorillonite. *Journal of the Chemical Society Dalton Transactions* 2207–2211.

Tensmeyer, L.G., R.W. Hoffmann, and G.W. Brindley. 1960. Infrared studies of some complexes between ketones and calcium montmorillonite. Clay-organic studies. Part III. *Journal of Physical Chemistry* 64: 1655–1662.

Tettenhorst, R.T., C.W. Beck, and G. Brunton. 1962. Montmorillonite-polyalcohol complexes. *Clays and Clay Minerals* 9: 500–519.

Theng, B.K.G. 1971. Mechanisms of formation of colored clay-organic complexes: A review. *Clays and Clay Minerals* 19: 383–390.

Theng, B.K.G. 1982. Clay-polymer interactions: Summary and perspectives. *Clays and Clay Minerals* 30: 1–10.

Theng, B.K.G. 1995. On measuring the specific surface area of clays and soils by adsorption of *para*-nitrophenol: Use and limitations. In *Clays: Controlling the Environment. Proceedings of the 10th International Clay Conference, 1993, Adelaide*, (Eds.) G.J. Churchman, R.W. Fitzpatrick, and R.A. Eggleton, pp. 304–310. Melbourne, Australia: CSIRO Publishing.

Theng, B.K.G. 2012. *Formation and Properties of Clay-Polymer Complexes*, 2nd edition. Amsterdam, The Netherlands: Elsevier.

Theng, B.K.G. 2018. *Clay Mineral Catalysis of Organic Reactions*. Boca Raton, FL: CRC Press.

Theng, B.K.G., J. Aislabie, and R. Fraser. 2001. Bioavailability of phenanthrene intercalated into an alkylammonium-montmorillonite clay. *Soil Biology & Biochemistry* 33: 845–848.

Theng, B.K.G., D.J. Greenland, and J.P. Quirk. 1967. Adsorption of alkylammonium cations by montmorillonite. *Clay Minerals* 7: 1–17.

Theng, B.K.G., R.H. Newman, and J.S. Whitton. 1998. Characterization of an alkylammonium-montmorillonite-phenanthrene intercalation complex by carbon-13 nuclear magnetic resonance spectroscopy. *Clay Minerals* 33: 221–229.

Theng, B.K.G., G.G. Ristori, C.A. Santi, and H.J. Percival. 1999. An improved method for determining the specific surface areas of topsoils with varied organic matter content, texture and clay mineral composition. *European Journal of Soil Science* 50: 309–316.

Tiller, K.G. and L.H. Smith. 1990. Limitations of EGME retention to estimate the surface area of soils. *Australian Journal of Soil Research* 28: 1–26.

Tipton, T. and L.E. Gerdom. 1992. Polymerization and transalkylation reactions of toluene on Cu(II)-montmorillonite. *Clays and Clay Minerals* 40: 429–435.

Touillaux, R., P. Salvador, C. Vandermeersche, and J.J. Fripiat. 1968. Study of water layers adsorbed on Na and Ca montmorillonite by the pulsed nuclear magnetic resonance technique. *Israel Journal of Chemistry* 6: 337–349.

Tricker, M.J., D.T.B. Tennakoon, J.M. Thomas, and S.H. Graham. 1975. Novel reactions of hydrocarbon complexes of metal-substituted sheet silicates: Thermal dimerisation of *trans*-stilbene. *Nature* 253: 110–111.

Trigo, C., R. Celis, M.C. Hermosin, and J. Cornejo. 2009. Organoclay-based formulations to reduce the environmental impact of the herbicide diuron in olive groves. *Soil Science Society of America Journal* 73: 1652–1657.

Ugochukwu, U. 2017. Measurement of surface area of modified clays by ethylene glycol monoethyl ether method. *Asian Journal of Chemistry* 29: 1891–1896.

Van Bladel, R. and A. Moreale. 1974. Adsorption of fenuron and monuron (substituted ureas) by two montmorillonite clays. *Soil Science Society of America Journal* 38: 244–249.

Van Damme, H., M. Crespin, F. Obrecht, M.I. Cruz, and J.J. Fripiat. 1978. Acid-base and complexation behavior of porphyrins on the intracrystal surface of swelling clays: Meso-tetraphenylporphyrin and meso-tetra (4-pyridyl)porphyrin on montmorillonites. *Journal of Colloid and Interface Science* 66: 43–54.

Van Damme, H., F. Obrecht, and M. Letellier. 1984. Intercalation of tetrathiafulvalene in smectite clays: Evidence for charge transfer interactions. *Nouveau Journal de Chimie* 8: 681–683.

Van de Poel, D., P. Cloos, J. Helsen, and E. Jannini. 1973. Comportement particulier du benzene adsorbé sur la montmorillonite cuivrique. *Bulletin du Groupe français des Argiles* 25: 115–126.

van Olphen, H. 1969. Determination of surface areas of clays – evaluation of methods. In *Surface Area Determination, Proceedings International Symposium*, (Eds.) D.H. Everett and R.H. Ottewill, pp. 255–271. London: Butterworth.

Vansant, E.F. and G. Peeters. 1978. The exchange of alkylammonium ions on Na-laponite. *Clays and Clay Minerals* 26: 279–284.

Vansant, E.F. and J.B. Uytterhoeven. 1972. Thermodynamics of the exchange of *n*- alkylammonium ions on Na-montmorillonite. *Clays and Clay Minerals* 20: 47–54.

Vansant, E.F. and J.B. Uytterhoeven. 1973. The adsorption of aromatic, heterocyclic and cyclic ammonium cations by montmorillonite. *Clay Minerals* 10: 61–69.

Vasudevan, D., T.A. Arey, D.R. Dickstein et al. 2013. Nonlinearity of cationic aromatic amine sorption to aluminosilicates and soils: Role of intermolecular cation-π interactions. *Environmental Science & Technology* 47: 14119–14127.

Villabona-Estupiñán, S., J. de A. Rodrigues, and R.S.V. Nascimento. 2017. Understanding the clay-PEG (and hydrophobic derivatives) interactions and their effect on clay hydration and dispersion: A comparative study. *Applied Clay Science* 143: 89–100.

Viraraghavan, T. and F.D.M. Alfaro. 1998. Adsorption of phenol from wastewater by peat, fly ash and bentonite. *Journal of Hazardous Materials* 57: 59–70.

Walker, G.F. 1957. On the differentiation of vermiculites and smectites in clays. *Clay Minerals Bulletin* 3: 154–163.

Walker, G.F. 1958. Reactions of expanding-lattice clay minerals with glycerol and ethylene glycol. *Clay Minerals Bulletin* 3: 302–313.

Walker, G.F. and W.G. Garrett. 1961. Complexes of vermiculite with amino acids. *Nature* 191: 1389.

Walter, D., D. Saehr, and R. Wey. 1990. Les complexes montmorillonite-Cu(II)-benzène: Une contribution. *Clay Minerals* 25: 343–354.

Wang, G., L. Ran, J. Xu et al. 2021. Technical development of characterization methods provides insights into clay mineral-water interactions: A comprehensive review. *Applied Clay Science* 206: 106088.

Wang, G., S. Zhang, Y. Hua et al. 2017. Phenol and/or Zn^{2+} adsorption by single- or dual-cation organomontmorillonites. *Applied Clay Science* 140: 1–9.

Wang, L., S. Liu, T. Wang, and D. Sun. 2011. Effect of poly(oxypropylene)diamine adsorption on hydration and dispersion of montmorillonite particles in aqueous solution. *Colloids and Surfaces A: Physicochemical and Engineering Aspects* 381: 41–47.

Wang, X.-C. and C. Lee. 1993. Adsorption and desorption of aliphatic amines, amino acids and acetate by clay minerals and marine sediments. *Marine Chemistry* 44: 1–23.

Wanyika, H. 2014. Controlled release of agrochemicals intercalated into montmorillonite interlayer space. *The Scientific World Journal*, Article ID 656287: 9.

Warren, D.S., A.L. Clark, and R. Perry. 1986. A review of clay-aromatic interactions with a view to their use in hazardous waste disposal. *Science of the Total Environment* 54: 157–172.

Weaver, C.E. 1958. The effects and geologic significance of potassium 'fixation' by expandable clay minerals derived from muscovite, biotite, chlorite and volcanic material. *American Mineralogist* 43: 839–861.

Weed, S.B. and J.B. Weber. 1968. The effect of adsorbent charge on the competitive adsorption of divalent organic cations by layer-silicate minerals. *American Mineralogist* 53: 478–490.

Weismiller, R.A. 1970. Effect of N-ethylacetamide (NEA) upon some physicochemical properties of montmorillonite and vermiculite. *Dissertation Abstracts* 30: 4874-B.

Weiss, A. 1969. Organic derivatives of clay minerals, zeolites, and related minerals. In *Organic Geochemistry*, (Eds.) G. Eglinton and M.T.J. Murphy, pp. 737–781. Berlin, Germany: Springer Verlag.

Weiss, A., A. Mehler, and U. Hofmann. 1956. Zur Kenntnis von organophilem Vermikulit. *Zeitschrift für Naturforschung* 11B: 431–434.

Weiss, A. and G. Roloff. 1963. Die Rolle organischer Derivate von glimmerartigen Schichtsilikaten bei der Bildung von Erdöl. *Proceedings International Clay Conference, Stockholm* 2: 373–378.

Weiss, Z., M. Valášková, M. Křistová, P. Čapková, and M. Pospíšil. 2003. Intercalation and grafting of vermiculite with octadecylamine using low-temperature melting. *Clays and Clay Minerals* 51: 555–565.

Weissmahr, K.W., S.B. Haderlein, and R.P. Schwarzenbach. 1997. *In situ* spectroscopic investigations of adsorption mechanisms of nitroaromatic compounds at clay minerals. *Environmental Science & Technology* 31: 240–247.

White, J.L. 1976. Protonation and hydrolysis of s-triazines by Ca-montmorillonite as influenced by substitutions at the 2-, 4- and 6-positions. In *Proceedings of the International Clay Conference 1975*, (Ed.) S.W. Bailey, pp. 391–398. Wilmette, IL: Applied Publishing.

Whitton, J.S. and G.J. Churchman. 1987. Standard methods for mineral analysis of soil survey samples for characterisation and classification in NZ Soil Bureau. *NZ Soil Bureau Scientific Report* 79: 27.

Wibulswas, R., D.A. White, and R. Rautiu. 1999. Adsorption of phenolic compounds from water by surfactant-modified pillared clays. *Transactions of the Institution of Chemical Engineers. Part B* 77: 88–92.

Willemsen, J.A.R., S.C.B. Myneni, and I.C. Bourg. 2019. Molecular dynamics simulations of the adsorption of phthalate esters on smectite clay surfaces. *Journal of Physical Chemistry C* 123: 13624–136336.

Witthuhn, B., T. Pernyeszi, P. Klauth, H. Vereecken, and E. Klumpp. 2005. Sorption of 2,4-dichlorophenl on organoclays constructed for soil bioremediation. *Colloids and Surfaces A: Physicochemical and Engineering Aspects* 265: 81–87.

Wolfe, T.A., T. Demirel, and E.R. Baumann. 1985. Interaction of aliphatic amines with montmorillonite to enhance adsorption of organic pollutants. *Clays and Clay Minerals* 33: 301–311.

Xie, G., P. Luo, M. Deng et al. 2017. Investigation of the inhibition mechanism of the number of primary amine groups of alkylamines on the swelling of bentonite. *Applied Clay Science* 136: 43–50.

Xie, G., Y. Xiao, M. Deng, Y. Luo, and P. Luo. 2020. Low molecular weight branched polyamine as a clay swelling inhibitor and its inhibition mechanisms: Experiment and density functional theory simulation. *Energy Fuels* 34: 2169–2177.

Xu, H., Y. Wan, H. Li, S. Zheng, and D. Zhu. 2011. Sorption of aromatic ionizable organic compounds to montmorillonites modified by hexadecyltrimethyl ammonium and polydiallyldimethyl ammonium. *Journal of Environmental Quality* 40: 1895–1902.

Xu, L., M. Zhang, and L. Zhu. 2014. Adsorption-desorption behavior of naphthalene onto CDMBA modified bentonite: Contribution of the π-π interaction. *Applied Clay Science* 100: 29–34.

Xu, L. and L. Zhu. 2007. Structures of hexamethonium exchanged bentonite and the sorption characteristics for phenol. *Colloids and Surfaces A: Physicochemical and Engineering Aspects* 307: 1–6.

Xu, W., C.T. Johnston, P. Parker, and S.F. Agnew. 2000. Infrared study of water sorption on Na-, Li-, Ca-, and Mg-exchanged (SWy-1 and SAz-1) montmorillonite. *Clays and Clay Minerals* 48: 120–131.

Xu, Y., M.A. Khan, F. Wang, M. Xia, and W. Lei. 2018. Novel multi amine-containing Gemini surfactant modified montmorillonite as adsorbents for removal of phenols. *Applied Clay Science* 162: 204–213.

Xue, G., M. Gao, Z. Gu, Z. Luo, and Z. Hu. 2013. The removal of *p*-nitrophenol from aqueous solutions by adsorption using Gemini surfactants modified montmorillonites. *Chemical Engineering Journal* 218: 223–231.

Yamamoto, K., T. Shiono, R. Yoshimura, Y. Matsui, and M. Yoneda. 2018. Influence of hydrophilicity on adsorption of caffeine onto montmorillonite. *Adsorption Science & Technology* 36: 967–981.

Yamanaka, S., F. Kanamaru, and M. Koizumi. 1974. Role of interlayer cations in the formation of acetonitrile-montmorillonite complexes. *Journal of Physical Chemistry* 78: 42–44.

Yamanaka, S., F. Kanamaru, and M. Koizumi. 1975. Studies on the orientation of acrylonitrile adsorbed on interlamellar surfaces of montmorillonite. *Journal of Physical Chemistry* 79: 1285–1288.

Yang, Q., M. Gao, Z. Luo, and S. Yang. 2016. Enhanced removal of bisphenol A from aqueous solution by organo-montmorillonites modified with novel Gemini pyridinium surfactants containing long alkyl chain. *Chemical Engineering Journal* 285: 27–38.

Yang, Q., M. Gao, and W. Zang. 2017. Comparative study of 2,4,6-trichlorophenol adsorption by montmorillonite functionalized with surfactants differing in the number of head group and alkyl chain. *Colloids and Surfaces A: Physicochemical and Engineering Aspects* 520: 805–816.

Yang, S., M. Gao, Z. Luo, and Q. Yang. 2015. The characterization of organo-montmorillonite modified with a novel aromatic-containing Gemini surfactant and its comparative adsorption for 2-naphthol and phenol. *Chemical Engineering Journal* 268: 125–134.

Yao, H., J. Zhu, A.B. Morgan, and C.A. Wilkie. 2002. Crown ether-modified clays and their polystyrene nanocomposites. *Polymer Engineering and Science* 42: 1808–1814.

Yariv, S. 1992. The effect of tetrahedral substitution of Si by Al on the surface acidity of the oxygen plane of clay minerals. *International Reviews in Physical Chemistry* 11: 345–375.

Yariv, S. and H. Cross (Eds.). 2002. *Organo-Clay Complexes and Interactions.* New York: Marcel Dekker.

Yariv, S. and L. Heller. 1970. Sorption of cyclohexylamine by montmorillonite. *Israel Journal of Chemistry* 8: 935–945.

Yariv, S., L. Heller, Z. Sofer, and W. Bodenheimer. 1968. Sorption of aniline by montmorillonite. *Israel Journal of Chemistry* 6: 741–756.

Yariv, S., J.D. Russell, and V.C. Farmer. 1966. Infrared study of the adsorption of benzoic acid and nitrobenzene in montmorillonite. *Israel Journal of Chemistry* 4: 201–213.

Yıldız, N., R. Gönülşen, H. Koyuncu, and A. Çalımlı. 2005. Adsorption of benzoic acid and hydroquinone by organically modified bentonites. *Colloids and Surfaces A: Physicochemical and Engineering Aspects* 260: 87–94.

Yılmaz, N. and S. Yapar. 2004. Adsorption properties of tetradecyl- and hexadecyl- trimethylammonium bentonites. *Applied Clay Science* 27: 223–228.

Yotsuji, K., Y. Tachi, H. Sakuma, and K. Kawamura. 2021. Effect of interlayer cations on montmorillonite swelling: Comparison between molecular dynamic simulations and experiments. *Applied Clay Science* 204: 106034.

Yu, J.-Y., M.-Y. Shin, J.-H. Noh, and J.-J. Seo. 2004. Adsorption of phenol and chlorophenols on Ca-montmorillonite in aqueous solutions. *Geosciences Journal* 8: 185–189.

Yu, M., M. Gao, T. Shen, and J. Wang. 2018. Organo-vermiculites modified by low-dosage Gemini surfactants with different spacers for adsorption toward *p*-nitrophenol. *Colloids and Surfaces A* 553: 601–611.

Yuan, G.D., B.K.G. Theng, G.J. Churchman, and W.P. Gates. 2013. Clays and clay minerals for pollution control. In *Handbook of Clay Science*, 2nd edition. Developments in Clay Science, Vol. 5B, (Eds.) F. Bergaya and G. Lagaly, pp. 587–644. Amsterdam, The Netherlands: Elsevier.

Zaghouane-Boudiaf, H., M. Boutahala, S. Sahnoun, C. Tiar, and F. Gomri. 2014. Adsorption characteristics, isotherm, kinetics, and diffusion of modified natural bentonite for removing the 2,4,5-trichlorophenol. *Applied Clay Science* 90: 81–87.

Zhang, W., Y. Ding, S.A. Boyd, B.J. Teppen, and H. Li. 2010. Sorption and desorption of carbamazepine from water by smectite clays. *Chemosphere* 81: 954–960.

Zhang, L., L. Luo, and S. Zhang. 2011. Adsorption of phenanthrene and 1,3-dinitrobenzene on cation-modified clay minerals. *Colloids and Surfaces A: Physicochemical and Engineering Aspects* 377: 278–283.

Zhang, S., Z. Qiu, and W. Huang. 2015. Water adsorption properties of Xiazijie montmorillonite intercalated with polyamine. *Clay Minerals* 50: 537–548.

Zhang, S., J.J. Sheng, and Z. Qiu. 2016. Water adsorption on kaolinite and illite after polyamine adsorption. *Journal of Petroleum Science and Engineering* 142: 13–20.

Zhao, N., F. Ju, Q. Song, H. Pan, and H. Ling. 2022. A simple empirical model for phenanthrene adsorption on soil clay minerals. *Journal of Hazardous Materials* 429: 127849.

Zhao, S., H. Jia, G. Nulaji, H. Gao, F. Wang, and C. Wang. 2017. Photolysis of polycyclic aromatic hydrocarbons (PAHs) on Fe^{3+}-montmorillonite surface under visible light: Degradation kinetics, mechanism, and toxicity assessments. *Chemosphere* 184: 1346–1354.

Zheng, S., Z. Sun, Y. Park, G.A. Ayoko, and R.L. Frost. 2013. Removal of bisphenol A from wastewater by Ca-montmorillonite modified with selected surfactants. *Chemical Engineering Journal* 234: 416–422.

Zhou, C.H., C.J. Li, W.P. Gates, T.T. Zhu, and W.H. Yu. 2019. Co-intercalation of organic cations/amide molecules into montmorillonite with tunable hydrophobicity and swellability. *Applied Clay Science* 179: 105157.

Zhou, Q., R. Zhu, S.C. Parker, J. Zhu, H. He, and M. Molinari. 2015. Modelling the effects of surfactant loading level on the sorption of organic contaminants on organoclays. *RSC Advances* 5: 47022–47030.

Zhu, D., B.E. Herbert, M.A. Schlautman, E.R. Carraway, and J. Hur. 2004. Cation-π bonding: A new perspective on the sorption of polycyclic aromatic hydrocarbons to mineral surfaces. *Journal of Environmental Quality* 33: 1322–1330.

Zhu, L., B. Chen, and X. Shen. 2000. Sorption of phenol, *p*-nitrophenol, and aniline to dual-cation organobentonites from water. *Environmental Science & Technology* 34: 468–475.

Zhu, L., X. Ren, and S. Yu. 1998. Use of cetyltrimethylammonium bromide-bentonite to remove organic contaminants of varying polar character from water. *Environmental Science & Technology* 32: 3374–3378.

Zhu, L. and R. Zhu. 2008. Surface structure of $CTMA^+$ modified bentonite and their sorptive characteristics towards organic compounds. *Colloids and Surfaces A: Physicochemical and Engineering Aspects* 320: 19–24.

Zhu, R., Q. Chen, Q. Zhou, Y. Xi, J. Zhu, and H. He. 2016. Adsorbents based on montmorillonite for contaminant removal from water: A review. *Applied Clay Science* 123: 239–258.

Zhu, R., W. Hu, Z. You, F. Ge, and K. Tian. 2012. Molecular dynamics simulation of TCDD adsorption on organo-montmorillonite. *Journal of Colloid and Interface Science* 377: 328–333.

Zhu, R., J. Zhao, F. Ge et al. 2014. Restricting layer collapse enhances the adsorption capacity of reduced-charge organoclays. *Applied Clay Science* 88–89: 73–77.

Zhu, R., L. Zhu, J. Zhu, and L. Xu. 2008. Structure of surfactant-clay complexes and their sorptive characteristics toward HOCs. *Separation and Purification Technology* 63: 156–162.

Zhu, R., Q. Zhou, J. Zhu, Y. Xi, and H. He. 2015. Organo-clays as sorbents of hydrophobic organic contaminants: Sorptive characteristics and approaches to enhancing sorption capacity. *Clays and Clay Minerals* 63: 199–221.

3 Interactions of Clay Minerals with Organic Compounds of Biological and Environmental Importance

3.1 INTERACTIONS WITH ORGANIC PESTICIDES

3.1.1 INTRODUCTION

Here we use the term 'pesticides' to denote a variety of organic compounds – commonly of synthetic origin – capable of controlling such pests as weeds (herbicides), insects (insecticides), and fungi (fungicides). The most important pesticides, in terms of global production and usage, are herbicides. Depending on their predominant charge characteristics, herbicides may be divided into three broad classes: uncharged (non-ionic), positively charged (cationic), and negatively charged (anionic). Since many organic pesticides have a complex chemical structure, they are frequently referred to by their common or registered trade names. The structures of some important pesticides, other than *s*-triazines, are shown in Scheme 3.1.

Cationic pesticides, such as the bipyridilium halides (Weed and Weber 1969) and chlordimeform (Hermosin and Perez Rodriguez 1981), are generally applied as their aqueous salt solution. Being completely ionized, they are adsorbed by clays (and soils) through a cation-exchange process; that is, by replacing the inorganic counterions initially present at the silicate surface. Anionic pesticides, such as 2,4-D (Weber et al. 1965) and triasulfuron (Pusino et al. 2000) tend to be negatively adsorbed by, or repelled from, negatively charged clay surfaces. At low solution pH, however, the acidic functional groups in the molecule, such as aminocyclopyrachlor (Cabrera et al. 2012), may accept protons to form the corresponding uncharged species which can then be taken up through various mechanisms described in Chapter 2.

In the pH range commonly encountered in soils, non-ionic pesticides are taken up and retained by clays predominantly by cation-dipole interactions, either directly or through a water bridge, depending on the hydration energy of the counterion (Sheng et al. 2002). Under acidic conditions, however, such uncharged pesticides as aminotriazole (Morillo et al. 1997a), atrazine (Celis et al. 1997; Herwig et al. 2001; Davies and Jabeen 2003), and fluridone (Yaron-Marcovich et al. 2004) can be protonated and hence behave as cations, while under alkaline conditions such compounds as *s*-triazines (Weber 1966) and mecoprop (Lambert 2018) would display anionic properties. In addition, the basicity (pK_a) of the compounds can markedly influence adsorption behaviour (Pereira et al. 2007). To complicate matters further, some pesticides and antibiotics may have two or more pK_a values. Thus, ciprofloxacin with two pK_a values behaves as a cation below pH 6, as a zwitterion between pH 6 and 9, and as an anion above pH 9 (Wang et al. 2011), while glyphosate (Sprankle et al. 1975) and tetracycline (Figueroa et al. 2004) have four pK_a values. Pesticide uptake is also influenced by solution ionic strength because inorganic cations compete directly with cationic and basic pesticides for adsorption sites.

The interactions of montmorillonite with some carbamate- and urea-based herbicides have already been mentioned in Chapter 2. Here we outline the formation and properties of clay-pesticide

DOI: 10.1201/9781003080244-3

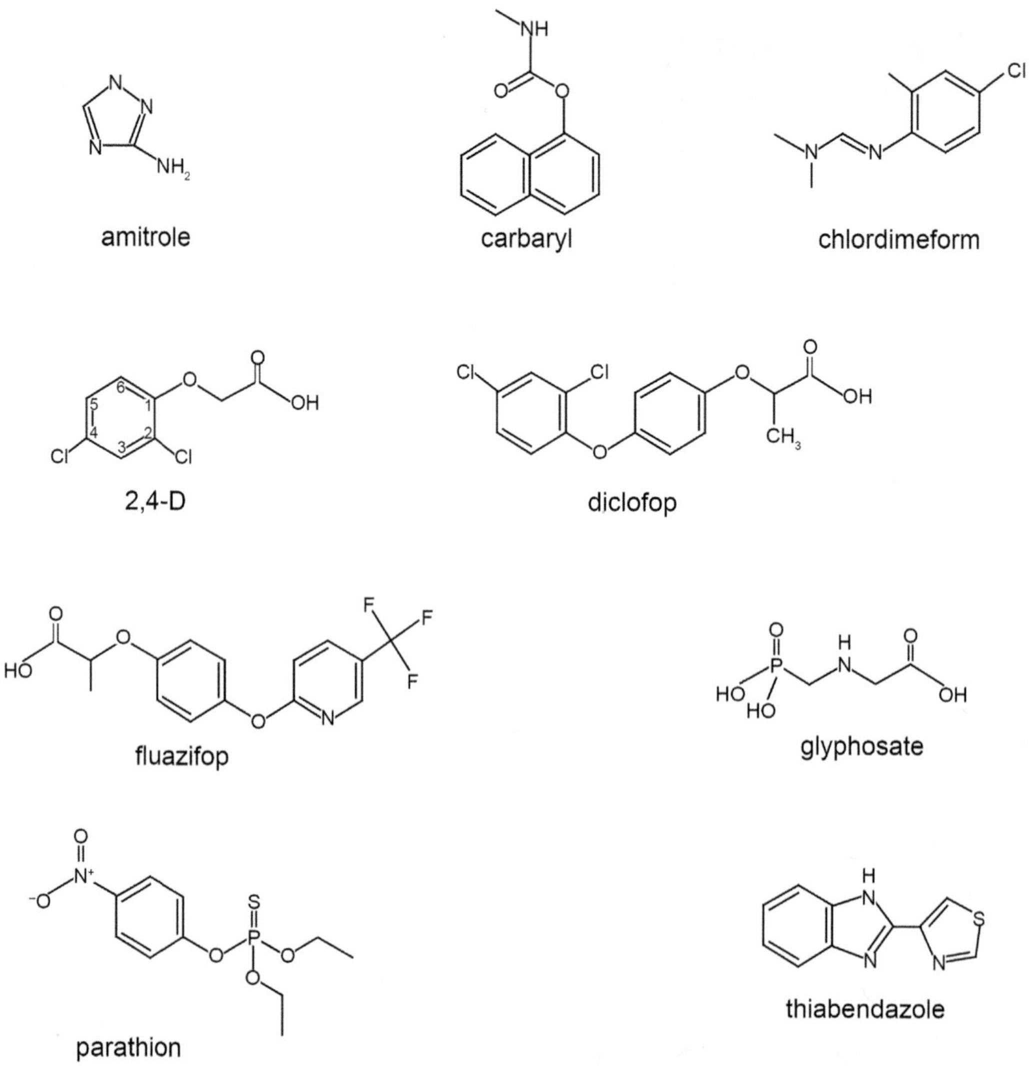

SCHEME 3.1 Molecular structures of some common pesticides.

complexes. For more details we refer to the reviews by Bailey and White (1970), Weber (1970a), Gerstl and Yaron (1981), Lagaly (2001), Cornejo et al. (2008), and Shattar et al. (2016). The reviews by Theng et al. (1999) and Site (2001) contain useful information on the soil-pesticide interaction, while those by Lambert (2018) and Awad et al. (2019) focus on the interactions of organic pollutants, including pesticides, with natural and modified clay minerals.

3.1.2 Uncharged (Non-Ionic) Pesticides

In the non-ionic class of pesticides, *s*ymmetrical or *s*-triazines have attracted much attention due to their widespread use as weed control agents in soil (LeBaron et al. 2008; Mudhoo and Garg 2011). These commercial herbicides comprise the chloro-, methylthio-, and methoxy-*s*-triazines, the molecules of which share a common heterocyclic ring structure differing only in the type of substituent at the 2-, 4-, and 6-positions (Scheme 3.2). The interactions of *s*-triazines with clay minerals are

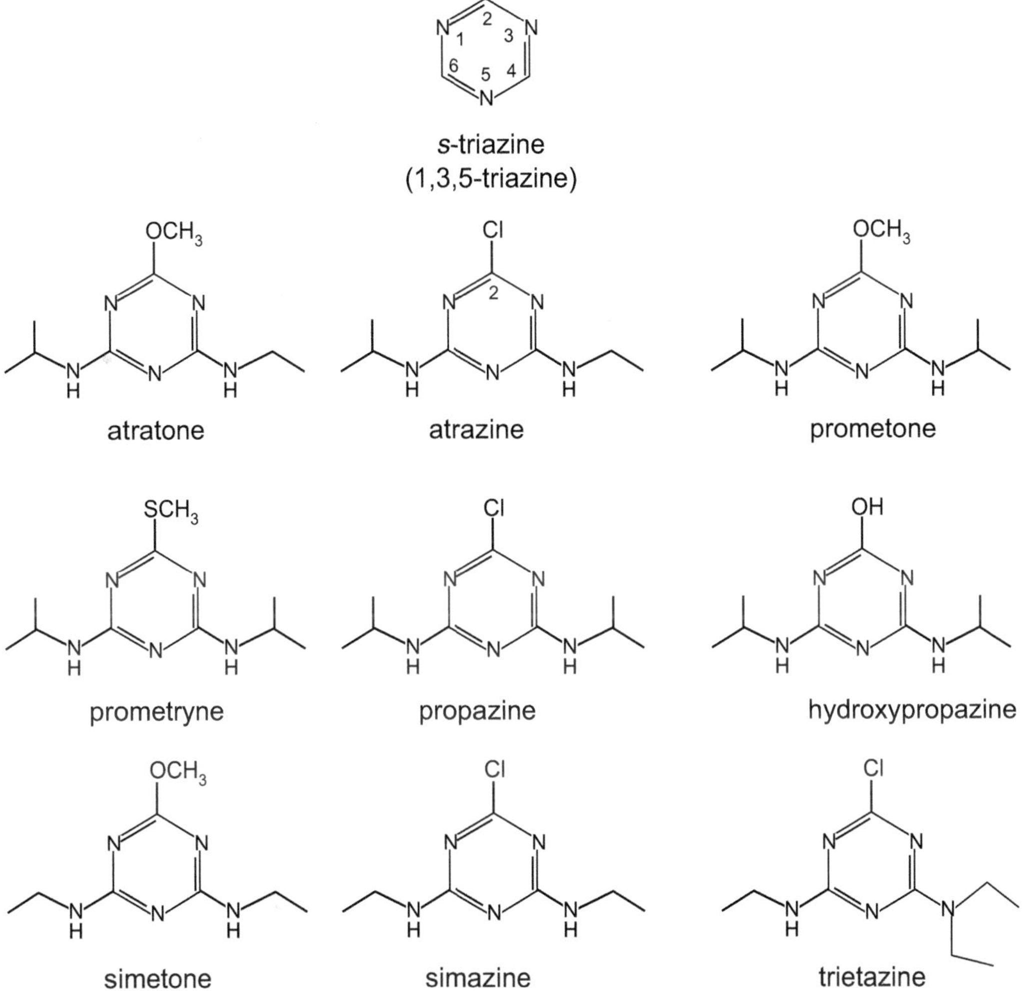

SCHEME 3.2 Molecular structures of some *s*-triazine herbicides.

determined more by the nature and location of the substituent groups than by the heterocyclic ring (Weber 1966, 1970b).

Frissel and Bolt (1962) made an early attempt at understanding the mechanisms underlying the clay-pesticide interaction by measuring the adsorption from water of various herbicides by montmorillonite, illite, and kaolinite as a function of pH and ionic strength. They noted that such chloro-*s*-triazines, as simazine and trietazine, were adsorbed as the corresponding uncharged (non-ionic) species at pH ≥ 7 but as the cations under acidic conditions. They further suggested that van der Waals forces were responsible for uptake between pH 8 and 10, while in the region of pH 2.9 to 8 both coulombic and van der Waals interactions were operative. The extent of adsorption decreased in the order montmorillonite > illite > kaolinite. Electrolyte addition affected adsorption by modifying the hydration properties of both the clay mineral and the herbicide. High electrolyte concentrations (> 0.1 M) can enhance pesticide adsorption through a 'salting-out' effect (Turner and Rawling 2001) by inducing the formation of better-ordered quasi-crystals and reducing interlayer swelling (Li et al. 2006) as well as promoting face-to-face aggregation of the silicate layers (Nennemann et al. 2001a). The influence of smectite interlayer hydration on the adsorption of uncharged pesticides, such as atrazine, has been described by Chappell et al. (2005).

That *s*-triazine adsorption to clay minerals increases with falling suspension pH has since been substantiated by other workers as does the finding that montmorillonite can take up more than either illite or kaolinite (Weber 1966; Yamane and Green 1972; Weber and Weed 1968; Hance 1969; Terce and Calvet 1978; Fruhstorfer et al. 1993; Polati et al. 2006). Indeed, in some instances, the adsorption of *s*-triazines by kaolinite is so small as to be undetectable (Frissel and Bolt 1962; Talbert and Fletchall 1965; Weber 1966).

The enhanced uptake by montmorillonite, as compared with illite and kaolinite, may be ascribed to the large (hydratable) surface area of montmorillonite, and its propensity for intercalating uncharged and cationic pesticides (Hermosin and Perez Rodriguez 1981; Haderlein et al. 1996; Cox et al. 2000; Chen et al. 2009). Thus, prior acid treatment/activation of smectites can enhance pesticide uptake (Hearon et al. 2019) by increasing the specific surface area and pore volume of the samples (Theng 2018). Similarly, the creation of a large interlayer micropore volume by intercalating montmorillonite with oligomeric iron or aluminium polyoxo cations lies behind the large capacity of 'pillared interlayered clays' (PILC) for removing various herbicides (Konstantinou et al. 2000; Undabeytia et al. 2000; Nennemann et al. 2001b; Abate and Masini 2005) and fungicides (Jalil et al. 2013) from aqueous solution. Only limited interlayer penetration, however, would be expected with non-ionic pesticides unless the silicate layers have already been separated as shown by Na^+-montmorillonite in water or dilute salt solutions (Undabeytia et al. 1999). On the other hand, if the pesticide molecule occurs in the protonated form (at low suspension pH), or when acid-treated smectite is used as the sorbent, intercalation can occur by a cation exchange process (Lombardi et al. 2006; Paul et al. 2010).

Early on, Weber et al. (1965) reported that prometone was intercalated by both Na^+- and H^+-montmorillonite. As the amount adsorbed increased so did the basal spacing from ~1.30 to ~1.85 nm, indicating that the intercalated molecules rearranged from a position in which the *s*-triazine ring was parallel (flat orientation) to one in which the ring was tilted at an increasingly higher angle to the silicate layer. A parallel interlayer orientation has also been proposed by Sheng et al. (2002) for 4,6-dinitro-*o*-cresol (DNOC), while Li et al. (2003) reported that the basal spacing of K^+-montmorillonite increased with DNOC intercalation. Similarly, thiabendazole and benzimidazole adopt a planar (monolayer) configuration on intercalation into montmorillonite (Lombardi et al. 2006).

Earlier, Weber (1966) examined the influence of molecular structure and pH on the interaction of 13 *s*-triazines with Na^+-montmorillonite. For a given herbicide, adsorption increased with a fall in suspension pH, reaching a maximum when the pH approached the pK_a of the pesticide and then decreased. Interestingly, the maximum amount adsorbed depended on the nature of the substituent in the 2-position of the *s*-triazine ring (Scheme 3.2), decreasing in the order $-SCH_3 > -OCH_3 > -OH > -Cl$ as represented by prometryne > prometone > hydroxypropazine > propazine. Similar substituent effects have been observed by Lambert (1967) for 2,6-dinitro-4-alkylsulphoryl-N,N-diallylanilines and by Briggs (1969) for substituted phenylureas and alkyl-N-phenyl carbamates for the soil-pesticide interaction.

Another important parameter controlling the adsorption of *s*-triazines is their solubility in water. In this regard, Bailey et al. (1968) have reported that the adsorption from water of six *s*-triazine compounds by Na^+-montmorillonite conforms to the Freundlich equation, given by Equation 3.1,

$$x/m = K_F C^{1/n} \qquad \text{(Equation 3.1)}$$

or its linear form (Equation 3.2)

$$\log x/m = \log K_F + 1/n \log \mathrm{C} \qquad \text{(Equation 3.2)}$$

where x is the amount of solute adsorbed; m is the unit weight of clay; C is the equilibrium solute concentration; K_F and n are constants. K_F may be regarded as the adsorption capacity which, in this instance, decreased in the order of simetone > atratone > prometone > trietazine > propazine > atrazine. Except for atrazine, this is also the order of their respective solubility in water. Similarly, Bowman and Sans (1977) found an inverse relationship between adsorption capacity (of Na^+- and Ca^{2+}-montmorillonite) and insecticide water solubility for parathion, methylparathion, fenitrothion, and aminoparathion.

Laird et al. (1992) reported that log (K_F) values for the sorption of atrazine to smectites correlated with the surface charge density and surface area of the minerals. Subsequently, Aggarwal et al. (2006) found that the low charge-density saponite had a larger sorption capacity for triazine herbicides than the high charge-density beidellite. As already elaborated in Chapter 2, this finding may be ascribed to saponite having a relatively large siloxane surface available for accommodating triazines. Likewise, Sheng et al. (2002) reported that low-charge montmorillonite (SWy-2) could take up much more DNOC from water than its high-charge (SAz-1) counterpart.

So far, we have confined our remarks to interactions in aqueous media where *s*-triazines may be protonated (Laird et al. 1992; Herwig et al. 2001). Protonation may also occur when the pesticides are added (to air-dry montmorillonite) as a solution in chloroform or ethanol. Here the active protons derive from the dissociation of residual water molecules coordinated to the exchangeable cations. The protonation and hydrolysis of some *s*-triazines, adsorbed to montmorillonite from organic solvents, have also been observed by Cruz et al. (1968) using infrared spectroscopy. Thus, the spectra of hydroxypropazine and prometone (adsorbed to NH_4^+- and H^+-montmorillonite) are essentially identical with those of the corresponding *s*-triazine cations. This finding strongly indicates that the uncharged herbicide molecules are protonated on adsorption to the clay mineral surface. The spectra also provide evidence for (acid) hydrolysis at the 2-position of the *s*-triazine ring (Scheme 3.2) in that both propazine and prometone are transformed to hydroxypropazine. Likewise, atrazine is converted into its protonated hydroxy derivative which then binds to montmorillonite (Russell et al. 1968a), while dimepiperate gives rise to α-methylstyrene and piperidine (Pusino et al. 1993).

The infrared spectra of dry montmorillonite, exchanged with different cations, and containing varying amounts of amitrole (3-amino-1,2,4-triazole) also indicate that much of the retained pesticide is in the protonated form (Russell et al. 1968b). The extent of protonation depends on the nature of the exchangeable cation, being highest with Al^{3+}-montmorillonite and decreasing in roughly the same order as the polarizing power of the cation. The importance of protonation in the adsorption of amitrole by montmorillonite extends to aqueous systems (Nearpass 1970). Little adsorption by Ca^{2+}-montmorillonite is observed at pH > 5 when amitrole is largely present as the uncharged species. On the other hand, appreciable adsorption occurs at acid pH, or with H^+-montmorillonite. Ikhsan et al. (2005b) have also found that at low solute concentrations (< 0.1 mM) amitrole is predominantly adsorbed as a cation. Besides being protonated, amitrole adsorbed to Ni^{2+}- and Cu^{2+}-montmorillonite can be coordinated (chelated) to the counterion.

Strong bonding, chelation, and protonation also characterize the interactions of atrazine (Sawhney and Singh 1997; Herwig et al. 2001; Wu et al. 2021), carbaryl (Fusi et al. 1986) fluazifop-butyl and fluazifop (Fusi et al. 1988), amitrole (Morillo et al. 1991), imazamethabenz-methyl (Pusino et al. 1995), and rimsulfuron (Calamai et al. 1997) with Al^{3+}-, Cu^{2+}- and Fe^{3+}-smectites.

Bailey et al. (1968) also measured the adsorption of substituted ureas (3-phenylurea, fenuron, monuron, and diuron) by Na^+- and H^+-montmorillonite in aqueous suspensions. As might be expected, the H^+-exchanged clay was much more reactive than its Na^+-saturated counterpart. In common with *s*-triazines, the 'affinity' of these urea-based pesticides for montmorillonite, as indicated by the magnitude of K_F in Equation 3.1, is related to their solubility in water. Here again, solution pH has a profound influence on adsorption. For example, the herbicides IPC (isopropyl-N-phenylcarbamate), dicryl [N-(3,4-dichlorophenyl) methacrylamide], and solan (pentanochlor) fail to interact with Na^+-montmorillonite (pH ~6.80) but are adsorbed by the H^+-exchanged clay (pH ~3.35).

These observations would suggest that within a given chemical family (or an analogous series) of pesticides, adsorption is controlled by the water solubility of the compounds, while between families the basicity (pK_a) of pesticides is an important factor influencing their adsorption. When the exchange positions of smectites are occupied by polyvalent and transition metal cations, coordination to the counterion, either directly or indirectly through a water bridge, appears to be the dominant mechanism underlying the adsorption of uncharged pesticides.

The interaction between ethyl N,N-di-*n*-propylthiolcarbamate (EPTC) and montmorillonite exchanged with different (inorganic) cations has been investigated by Mortland and Meggitt (1966). Infrared spectroscopy indicates that the C=O and C–N stretching frequencies of the adsorbed herbicide decrease and increase, respectively, as compared with those in the free liquid. Since the extent of these shifts depends on the nature of the exchangeable cation, EPTC is apparently coordinated to the cation through the oxygen of the carbonyl, rather than the nitrogen group, since the latter is sterically hindered by attached bulky propyl groups (cf. Scheme 2.4). Likewise, N,N-dimethylurea adsorbed to montmorillonite by coordination of the carbonyl group with the exchangeable cation (Davies and Jabeen 2002). Earlier, Bosetto et al. (1993) noted that the strength of coordination was directly correlated with the polarizing power of the counterion.

Coordination through the C=O group has also been proposed by Fusi et al. (1986, 1989) for carbaryl and isoxaben and by Cox et al. (1994) for methomyl. In the case of azinphos-methyl and methidathion, the various functional groups (C=O, P=S, N=N) in the insecticide molecules may coordinate to the interlayer counterions at the same time (Martín and Camazano 1984; Romero-Taboada and Cancela 2004). Coordination of the carbonyl group of malathion (Bowman et al. 1970), the P=O group of phenamiphos (Rodríguez et al. 1988), and the thioester carbonyl group of dimepiperate (Pusino et al. 1993) to the exchangeable cation in montmorillonite, either directly or through a water bridge has also been proposed on the basis of infrared spectroscopic evidence. In the case of malathion, the downward shift in the C=O stretching frequency of the adsorbed insecticide was greater with dehydrated montmorillonite than when the sample contained some interlayer water. Overall, the presence of water is conducive to pesticide intercalation.

3.1.3 Negatively Charged (Anionic) Pesticides

Among the anionic pesticides that Bailey et al. (1968) investigated were picloram, phenylalkanoic and benzoic acids together with their derivatives, all of which possess a carboxyl group. None of these herbicides were adsorbed by Na^+-montmorillonite, while the acid-treated clay mineral showed both positive and negative adsorption. More recently, Marco-Brown et al. (2014) reported that picloram adsorption by montmorillonite increased in the order pH 7 < pH 5 < pH 3. Infrared spectroscopy indicated the formation of inner-sphere complexes involving the pyridinic nitrogen and carboxyl group of the herbicide.

Earlier, Frissel and Bolt (1962) reported that 2,4-D and 2,4,5-T were negatively adsorbed (repelled) by Na^+-montmorillonite between pH 10 and 4 although (positive) uptake occurred at pH < 4. The transition from negative to positive adsorption probably coincided with the formation of the uncharged species, accompanied by the rapid disappearance of the anionic form below pH 4. Some uptake occurred with illite at pH > 4, possibly involving positively charged sites at (particle) edges. Adsorption also increased at high electrolyte concentrations owing to the salting-out effect. Very little, or even negative, adsorption of 2,4-D by Na^+-montmorillonite and kaolinite at near neutral and alkaline pH was reported by Weber et al. (1965), Schwartz (1967), and Haque and Sexton (1968). Other anionic herbicides, such as MCPA (2-methyl-4 chlorophenoxyacetic acid), DNC (3,5-dinitro-*o*-cresol), DNBP (4,6-dinitro-*o*-sec-butylphenol), picric and picolinic acids, behaved in a similar manner (Frissel 1961).

The interactions of the herbicides fluazifop, diclofop, diclofop-methyl, and acifluorfen with Al^{3+}-, Fe^{3+}-, and Cu^{2+}-exchanged montmorillonites have been investigated by Pusino and co-workers (Micera et al. 1988; Pusino et al. 1989a, 1991). Infrared and electron spin resonance spectroscopy indicates that the pyridine-nitrogen atom of fluazifop can accept a proton, while the carboxylate group of the molecule is directly coordinated to the exchangeable cation. The latter mode of bonding also applies to diclofop whereas diclofop-methyl interacts with the counterion through a water bridge. The formation of inner-sphere complexes between acifluorfen and the various counterions can lead to the loss of herbicidal activity.

Among the anionic herbicides, glyphosate (*N*-phosphomethyl glycine) occupies a special place because of its widespread use and efficacy in controlling weeds (Franz et al. 1997; Duke and Powles 2008; Okada et al. 2016). Having four pK_a values, glyphosate can display different pH-dependent charge characteristics. As a result, its interaction with clay minerals is highly influenced by solution/suspension pH (Khoury et al. 2010).

Shoval and Yariv (1979) reported that glyphosate could intercalate into Al^{3+}- and Fe^{3+}-montmorillonite as the uncharged or even positively charged species in which the carboxyl and phosphonate groups were coordinated to the counterions directly or through a water bridge. The formation of coordination/chelation complexes with interlayer Cu^{2+} ions was also suggested by Glass (1987) although the affinity of the Cu-glyphosate complex for montmorillonite might be less than that of the free herbicide (Morillo et al. 1997b). Earlier, McConnell and Hossner (1985, 1989) found that glyphosate adsorption decreased as solution pH increased although intercalation could still occur with Al^{3+}-montmorillonite. Damonte et al. (2007) have proposed that glyphosate adsorbs to montmorillonite through ligand exchange (cf. Equations 1.1 and 1.2) involving the phosphonate group, giving rise to monodentate and bidentate complexes, while the positively charged amino group may interact directly with the mineral surface (Khoury et al. 2010; Pereira et al. 2020). Recent computational modelling (Wang et al. 2023) similarly indicates that at pH 7 both glyphosate and aminomethylphosphonic acid can adsorb to Ca^{2+}-exchanged and acid-treated montmorillonite through electrostatic and H-bonding interactions. Nevertheless, we would suggest that ligand exchange involving the clay mineral edge surface is equally, if not more, important.

3.1.4 Positively Charged (Cationic) Pesticides

Among the compounds in this class, diquat (DQ) and paraquat (PQ) are perhaps the best known and most widely investigated (cf. Figure 2.1). The two pyridilium rings in both compounds are coplanar, a configuration that appears to be necessary for herbicidal activity. The influence of surface charge density of clay minerals on the simultaneous and preferential adsorption of DQ and PQ from solution has been described in Chapter 2.

Weber et al. (1965) reported that both herbicides adsorbed to Na^{+}-montmorillonite and kaolinite by replacing an equivalent amount of the sodium counterions initially present. Although cation exchange is clearly the dominant mechanism underlying the clay-herbicide interaction, Haque et al. (1970) have suggested that charge transfer between the organic cation and the (anionic) silicate framework may also be involved in addition to the non-specific van der Waals interactions (Knight and Tomlinson 1967).

For both montmorillonite and kaolinite, the adsorption isotherms are of the H- (high-affinity) class (Giles et al. 1960), characterized by a region in which all the added herbicide is adsorbed, followed by a short shoulder in which the extent of adsorption is related to the herbicide concentration in solution, finally reaching a plateau approaching the cation exchange capacity of the clay adsorbent (Weed and Weber 1969; Knight and Denny 1970; Hayes et al. 1978; Sidhoum et al. 2013).

X-ray diffraction of DQ and PQ complexes with montmorillonite indicates that the compounds are intercalated with the plane of the rings lying parallel to the silicate layers with some keying into the ditrigonal cavities in the siloxane surface (Weber et al. 1965; Knight and Denny 1970). In vermiculite, on the other hand, the pyridinium rings of paraquat are slightly tilted and partly keyed (Raupach et al. 1979; Rytwo et al. 1996). Since no intercalation occurs with kaolinite, adsorption here is confined to external basal and edge surfaces. As a result, paraquat is more strongly retained by montmorillonite than kaolinite. Repeated extraction using 1 M $BaCl_2$ solutions, for example, removed ~80% of adsorbed PQ from kaolinite but only ~5% from montmorillonite (Weber and Weed 1968) although benzalkonium can apparently replace montmorillonite-adsorbed PQ (Ilari et al. 2021). The general features of the uptake and retention of bipyridilium cations by clays and soils, outlined earlier, have been substantiated by Knight and Tomlinson (1967) and Sidhoum et al. (2013). Since little desorption would occur under field conditions, the phytotoxicity of soil-applied diquat and paraquat is likely to be short-lived.

Similarly, the adsorption of chlordimeform to clay minerals (montmorillonite, kaolinite, illite, vermiculite) is a cation exchange process (Hermosin and Perez Rodriguez 1981). X-ray diffraction indicates that this positively charged pesticide intercalates into montmorillonite as a flat lying molecule. No interlayer penetration is detected with the other three mineral species. A cation exchange mechanism is also operative in the intercalation of thiabendazole by montmorillonite with little, if any, desorption being detectable. X-ray diffraction and molecular modelling indicate that this cationic fungicide is intercalated as a monolayer of flatly oriented molecules (Lombardi et al. 2003; Gamba et al. 2017). Similarly, nickel and cobalt complexes of thiabendazole can intercalate into Na^{+}-montmorillonite through a cation exchange mechanism (Ennajih et al. 2012). The nature of the exchangeable cation in montmorillonite also plays an important role in controlling the loading and release of intercalated thiabendazole (Cavalcanti et al. 2019).

The interactions of pesticides with organically modified smectites ('organoclays') have been the topic of many investigations as they provide a means of enhancing the uptake, and controlling the release, of pesticides. Commonly used organic modifiers are short- and long-chain alkylammonium ions such as decyl- and octadecyl-ammonium (Vimond-Laboudigue et al. 1996; Carrizosa et al. 2001, 2004; Saha et al. 2013) and quaternary ammonium ions such as dodecyl-, hexadecyl- and octadecyl-trimethylammonium (Zhao et al. 1996; Celis et al. 2002; Sánchez-Martín et al. 2006; Liu et al. 2012; Huang et al. 2013; Park et al. 2014; Gamba et al. 2015; Gámiz et al. 2015; Grundgeiger et al. 2015; Mo et al. 2015; Shirzad-Siboni et al. 2015; Wu et al. 2020). Other cationic organic modifiers include thioflavin T (Undabeytia et al. 2000), L-carnitine, choline, L-cysteine dimethyl ester, hexadimethrine, thiamine, spermine, and tyramine (Cruz-Guzmán et al. 2004, 2005; Celis et al. 2007; Gámiz et al. 2010, 2012, 2015; Wang et al. 2019a).

Organically modified smectites are invariably superior to their pristine counterparts in their ability to take up pesticides (and pollutants) from solution. This finding may be explained in terms of the hydrophilic-to-hydrophobic transformation of the mineral surface together with the increased availability of the interlayer space to pesticide molecules (Undabeytia et al. 2000; Carrizosa et al. 2001; Rytwo et al. 2005; Lazorenko et al. 2020; Salcedo et al. 2021). The use of organically modified clays for the reduced leaching and slow-release formulation of herbicides, and the removal of pollutants and pharmaceuticals from water, has been summarized by El-Nahhal et al. (2001), Nir et al. (2006), Park et al. (2011), Yuan et al. (2013), Zhu et al. (2016), and Undabeytia et al. (2021).

3.2 INTERACTIONS WITH ANTIBIOTICS, PHARMACEUTICALS, AND TOXINS

Clays and clay minerals have served as both excipients and active carriers of antibiotics, as adsorbents of pharmaceuticals, as carriers of drugs, and as components in cosmetics (Carretero et al. 2006; Aguzzi et al. 2007; Viseras et al. 2010, 2019, 2021; Jayrajsinh et al. 2017; Massaro et al. 2018; Thiebault 2020; Gomes et al. 2021), while the clay-antibiotic interaction has the potential of controlling and reducing the release of antimicrobial agents into the environment (Thiele-Bruhn 2003; Sarmah et al. 2006; Sukul and Spiteller 2007; Kim et al. 2022).

Like pesticides, antibiotics may be cationic, uncharged (neutral) or anionic, while some compounds are amphoteric containing both basic and acidic functional groups. The interactions of a particular class of antibiotics with clays (and soils) might therefore be expected to be analogous to those shown by the corresponding pesticides. Although relatively little work has been carried out on the clay-antibiotic interaction, the available data indicate that this is so.

Early on, Siminoff and Gottlieb (1951) noted that cationic streptomycin was readily taken up by bentonite, illite, and soil whereas little, if any, adsorption occurred with neutral and anionic antibiotics (Gottlieb and Siminoff 1952). Similarly, Pinck et al. (1961) and Pinck (1962) reported that montmorillonite could take up appreciable amounts of cationic antibiotics including streptomycin, dihydrostreptomycin, neomycin, and kanamycin. By contrast, adsorption of anionic and uncharged types such as penicillin, chloramphenicol, and cycloheximide was either low or undetectable. In the case of amphoteric antibiotics (bacitracin, aureomycin, terramycin) their adsorption was, as expected, influenced by solution pH. In terms of clay mineral species, adsorption decreased in the order montmorillonite > vermiculite > illite > kaolinite. Furthermore, basic and amphoteric antibiotics can intercalate into montmorillonite to form single- and double-layer complexes with a basal spacing of ~1.40 and ~1.72 nm, respectively.

The adsorption of tetracyclines by expanding layer silicates, especially smectites, has been the focus of many investigations. In general, clay-bound antibiotics (tetracycline, ciprofloxacin, chlorhexidine acetate) must be released if they are to function as antibacterial agents (Meng et al. 2009; Parolo et al. 2010; Ghadiri et al. 2013; Saha et al. 2014; Rivera et al. 2016; Lobato-Aguilar et al. 2018; Chahardahmasoumi et al. 2019; Hamilton et al. 2019). By the same token, the antimicrobial activity of tetracycline is appreciably diminished in the presence of montmorillonite (Lv et al. 2019).

Tetracycline occurs and behaves as a cation below pH 3.3, as a zwitterion between pH 3.3 and 7.68, and as an anion above pH 7.68. The clay-tetracycline interaction is therefore influenced by solution pH (and ionic strength) with adsorption decreasing as pH and salinity increase (Sithole and Guy 1987; Figueroa et al. 2004; Parolo et al. 2008; Essington et al. 2010; Wang et al. 2010a; Zhao et al. 2012). Thus, the mechanism underlying the adsorption of tetracycline (and oxytetracycline) at acid pH is one of cation exchange which may be supplemented by hydrophobic interactions and cation complexation under slightly acid-to-neutral conditions (Kulshrestha et al. 2004; Li et al. 2010; Aristilde et al. 2010, 2016; Zhao et al. 2012). Hydrophobic interactions, however, would control the adsorption of oxytetracycline to vermiculite that has been modified with phosphatidylcholine (Liu et al. 2017a). The structures of some antibiotics, pharmaceuticals, and toxins are shown in Scheme 3.3.

X-ray diffraction analysis of the interlayer montmorillonite-tetracycline complex (formed under acid conditions) shows a basal spacing of ~2.2. nm (Porubcan et al. 1978; Li et al. 2010; Parolo et al. 2013; Wu et al. 2019a) while a slightly lower value (~2 nm) was reported by Kulshrestha et al. (2004) for a complex with low-charge (SWy-2) montmorillonite. Li et al. (2010) have proposed that tetracycline adopts a tilted orientation in which the dimethylammonium group of the antibiotic is electrostatically attached to the negatively charged interlayer surface. This orientation, however, may change from a twisted to an extended conformation for complexes formed under alkaline conditions (Chang et al. 2009; Li et al. 2010).

Besides tetracyclines, the fluoroquinolone (FQ) antibiotics have attracted wide attention in terms of their interaction with clay minerals. Within the FQ group, ciprofloxacin (Scheme 3.3) occupies a central place. Since the (calculated) adsorption energies of tetracycline and ciprofloxacin with respect to the montmorillonite layer are similar in magnitude, these antibiotics would compete for interlayer sites without showing a preferential relationship (Wu et al. 2019b). On the other hand, the zwitterionic form of ciprofloxacin is apparently preferred to that of oxytetracycline. This might be because the cationic and anionic groups of the former compound are separated by a larger distance, maximizing electrostatic attraction to the clay surface while minimizing repulsive interactions (Carrasquillo et al. 2008).

Since ciprofloxacin has two pK_a values at ~6 and ~8.8, its interaction with smectites is dependent on solution pH with uptake being high at acidic pH and declining under alkaline conditions. Ciprofloxacin adsorption to montmorillonite at pH < 8.8 may be ascribed to a cation exchange mechanism while uptake at pH > 8.8 is probably due to complex formation, probably involving the ketone group, with the interlayer counterion supplemented by H-bonding to the siloxane surface.

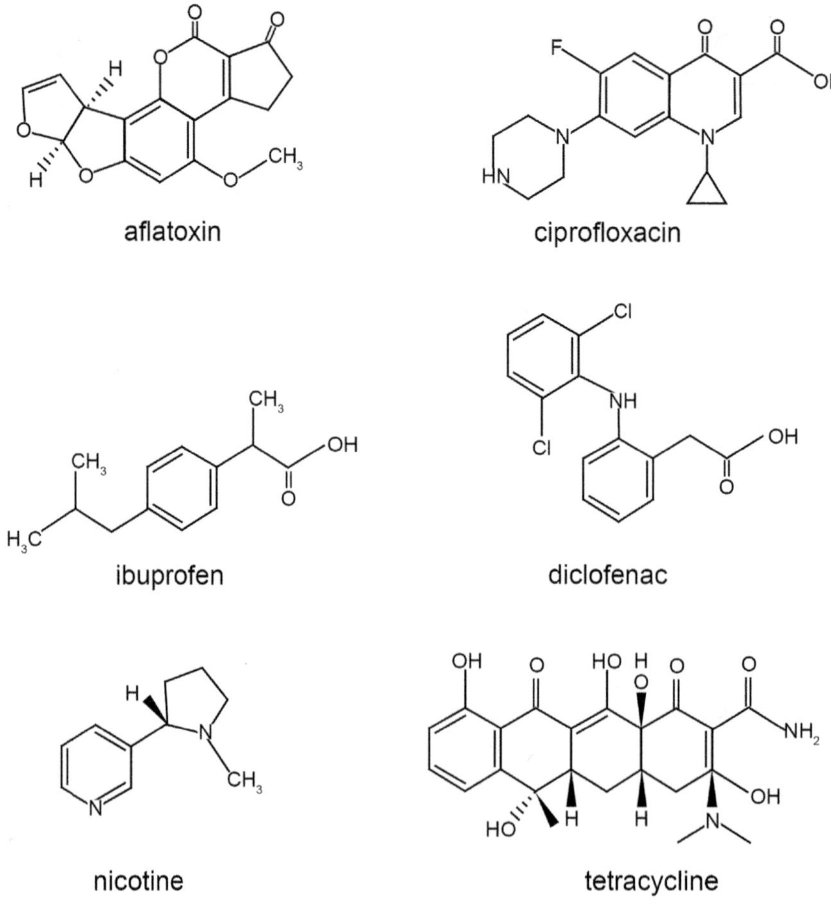

aflatoxin ciprofloxacin

ibuprofen diclofenac

nicotine tetracycline

SCHEME 3.3 Molecular structures of some antibiotics and pharmaceuticals.

Like tetracycline, intercalated ciprofloxacin apparently adopts a tilted conformation with the degree of tilting being sensitive to pH (Wang et al. 2010b; Wu et al. 2010; Jalil et al. 2015; Valdés et al. 2017; dos Santos et al. 2018).

More recently, Fashina and Deng (2023) have proposed that pyocyanin ($pK_a = 4.9$) intercalates into smectite as a protonated molecule at $pH < pK_a$ and as a neutral zwitterionic species at $pH > pK_a$. Accordingly, at $pH < 4.9$ this cytotoxic pigment is taken up by cation exchange with the interlayer counterions. Under slightly acidic to neutral pH conditions adsorption occurs mainly through cation-dipole interactions, while with Cu^{2+}-smectite pyocyanin can form a coordination complex.

Complex formation of norfloxacin with added and surface-adsorbed Cu(II) has also been proposed by Pei et al. (2011). Similarly, enrofloxacin can form inner-sphere surface complexes with K^+-montmorillonite (Yan et al. 2012) while levofloxacin interacts with iron-pillared montmorillonite through coordination of its ketone and carboxylate functional groups with an iron atom (Liu et al. 2015). In the case of nalidixic acid, intercalation into montmorillonite apparently occurs through coordination of the keto oxygen or C=N group in the pyridine ring with the exchangeable cation (Wu et al. 2013).

We might mention that the interlayer hydration status of smectites can exert a strong influence on the intercalation of antibiotics. All things being equal, the less hydrated saponite is a better sorbent (of lincomycin) than montmorillonite and sorption decreases in the order Ca^{2+}- $<$ K^+- $<$ Cs^+-smectite (Wang et al. 2009). In addition, intercalation of chlorhexidine acetate into montmorillonite

(He et al. 2006) and of ciprofloxacin into rectorite leads to surface and interlayer dehydration (Jiang et al. 2013).

The principles and mechanisms underlying the clay-antibiotic interaction are also applicable to the formation of complexes between clay minerals and pharmaceuticals/drugs. For example, the adsorption (and intercalation) by montmorillonite of codeine, diazepam, and oxazepam from solution at pH 2.5 is a cation-exchange process, and hence is comparable in magnitude. At pH 7.5, however, only codeine is protonated, and much more of this compound is adsorbed than the other two psychoactive drugs (Thiebault and Boussafir 2019).

Cation exchange is also the dominant mechanism underlying the interaction of montmorillonite with aminopyridines (Ikhsan et al. 2005a), amitriptyline (Chang et al. 2014), atenolol (Chang et al. 2019), buformin HCl (Fejér et al. 2002), chlorhexidine diacetate (He et al. 2006; Holešová et al. 2013; Lobato-Aguilar et al. 2018), chlorpheniramine (Li et al. 2011a; Lv et al. 2012), doxepin (Thiebault et al. 2015, 2016), diphenhydramine (Li et al. 2011b), hydralazine HCl (Sánchez-Martín et al. 1988), isoniazid (Carazo et al. 2018), metroprolol (Thiebault et al. 2016), nortriptyline (Sadri et al. 2018), promethazine chloride (Fejér et al. 2002; Gereli et al. 2006), propranolol (Orta et al. 2019), propylfenothiazine chloride (Fejér et al. 2002), sotalol HCl (Sánchez-Camazano et al. 1987), tamoxifen (Silva et al. 2019, 2020a), thiamine HCl (Joshi et al. 2009a), timolol maleate (Joshi et al. 2009b), and tramadol HCl (Chen et al. 2010; Thiebault et al. 2015, 2016). Adsorption and intercalation of these pharmaceuticals are accompanied by the replacement of an equivalent amount of inorganic counterions initially present on the clay surface. The process commonly follows pseudo second-order kinetics and gives rise to L-class (Langmuir) isotherms (Giles et al. 1960).

As might be expected, the interactions of acidic pharmaceuticals with clay minerals are highly dependent on solution pH (Styszko et al. 2015). Behera et al. (2012) observed that adsorption of ibuprofen (to kaolinite and montmorillonite) under acid conditions increased with ionic strength, attaining a maximum at pH 3. More recently, Vallova et al. (2022) observed that the uptake of some drugs from aqueous solution by KSF montmorillonite (a commercial, acid-treated clay) decreased in the order diclofenac ($pK_a = 4.15$) > ibuprofen ($pK_a = 4.91$) > paracetamol ($pK_a = 9.38$). Using titanium oxide-pillared montmorillonite as adsorbent, Chauhan et al. (2020) reported the following order of drug removal: imipramine > diclofenac > paracetamol > amoxicillin. Similarly, montmorillonite modified with ionic liquids as well as hexadecyltrimethylammonium (HDTMA) and alkylpyridinium ions was effective in removing diclofenac from aqueous solution (Ghemit et al. 2019; Wu et al. 2019c; França et al. 2020). Earlier, Jin et al. (2014) reported that HDTMA-modified bentonite was six times more effective in removing amoxicillin from aqueous solution than the raw mineral. More recently, Kryuchkova et al. (2021) found uptake by stearyltrimethylammonium-exchanged montmorillonite to fall in the order carbamazepine > ibuprofen > paracetamol. Malvar et al. (2020) have suggested that ibuprofen and its metabolites are taken up by octadecylammonium-montmorillonite through electrostatic interaction and solute partitioning.

The use of organically modified montmorillonite as both an antimicrobial agent and an adsorbent/carrier of antibiotics and pharmaceuticals has attracted a great deal of attention. In this regard, interlayer complexes of montmorillonite with cetylpyridinium (CP) ions can act as antimicrobial agents. The antibiotic activity of these complexes could be ascribed to intercalated CP since the parent (unmodified) montmorillonite was inactive, and very little CP was desorbed from the complex (Herrera et al. 2000; Malachová et al. 2009; Özdemir et al. 2010, 2013). Thus, CP^+-montmorillonite by itself (Ke et al. 2014) or in combination with Cu^{2+}-ZnO (Ma et al. 2010b), and sodium lauroyl sarcosinate (Yapar et al. 2017) can function as antibiotics. Cetylpyridinium-modified clay minerals, including montmorillonite, are also efficient in taking up fungicides (Andrades et al. 2004). More recently, Nippes et al. (2022) used a commercial organo-bentonite to remove ivermectin from aqueous solution at natural pH.

Bujdáková et al. (2018) reported that montmorillonite intercalated with dodecyltrimethylammonium ions was more active (against *Escherichia coli* and *Staphylococcus aureus*) than the complex with either tetrabutylphosphonium or tetrabutylammonium ions. Ohashi et al. (2006,

2009) also found that an interlayer complex of montmorillonite with silver-imidazole chelate and 6-benzylaminopurine showed antimicrobial and antifungal activity. The same is true of montmorillonite modified with quaternary imidazolium salts (Kleyi et al. 2016), while complexes of bentonite with non-ionic surfactants (Brij 30 and Igepal CO-890) can detoxify aflatoxin B_1 in chickens (Tzou et al. 2019). On the other hand, Iannuccelli et al. (2018) have suggested that gentamicin must be released from its complex with bentonite before it can act as an antibiotic. The release of doxorubicin and ibuprofen from their respective complexes with bentonite has been reported by Hosseini et al. (2018) and Alkrad et al. (2022).

As we would expect, organically modified montmorillonites ('organoclays') are much better at taking up pharmaceuticals than their pristine (raw) counterparts. The same is true for organo-kaolinite (Sun et al. 2017) and organo-vermiculite (Liu et al. 2017b). Interestingly, the adsorption of various drugs to organoclays commonly gives rise to curvilinear isotherms, at least at low equilibrium solute concentrations. This finding has been explained in terms of solute partitioning into the interlayer quaternary ammonium cations supplemented by electrostatic and van der Waals interactions (De Oliveira and Guégan 2016; De Oliveira et al. 2017; Mártinez-Costa et al. 2018; Ghemit et al. 2019; Phuekphong et al. 2020; Kryuchkova et al. 2021). Similar results have been reported for vermiculite intercalated with phosphatidylcholine (Liu et al. 2017a) as well as montmorillonite modified with polyoxyethylene (20) oleyl ether (De Oliveira et al. 2018), octadecylammonium (Martín et al. 2019), and octyl phenyl polyoxyethylene ether (Yan et al. 2020), and beidellite modified with tetraalkylphosphonium ions (Jankovič et al. 2021).

In closing we would mention the formation of complexes between clay minerals and some fungal and plant toxins. In this regard, the interaction of aflatoxin B_1, produced by *Aspergillus* spp, with montmorillonite has received much attention as complex formation has long been known to alleviate its toxicity (Masimango et al. 1979). This finding has since been verified using raw and modified smectites from different countries and geological origins (Dixon et al. 2008; Carraro et al. 2014; Aisawa et al. 2018; D'Ascanio et al. 2019; Gan et al. 2019; Wang et al. 2020). The role of clay minerals (montmorillonite, palygorskite, sepiolite) in the adsorption and degradation of aflatoxin B_1 (and zearalenone) has recently been reviewed by Zhang et al. (2022).

X-ray diffraction, infrared spectroscopy, and molecular modelling indicate that aflatoxin adsorbs to montmorillonite by cation-dipole interactions (between the interlayer counterion and carbonyl group), either directly or through a water bridge (cf. Chapter 2). Furthermore, the capacity of smectites for adsorbing aflatoxin decreases as layer charge density increases. In this regard, octahedrally substituted smectites are superior to tetrahedrally substituted species (Deng et al. 2010; Barrientos-Velázquez et al. 2016; Szczerba et al. 2022). Subsequent computational studies have confirmed that ion-dipole interactions control the adsorption of aflatoxin to smectite (Deng and Szczerba 2011; Kang et al. 2016). We would also point out that surfactant-modified montmorillonite and related organoclays are much better at taking up aflatoxin (and other mycotoxins) than the parent clay mineral (Wang et al. 2018a, 2018b; Saeed et al. 2020a; Zhang et al. 2020a).

Barr and co-workers (Evcim and Barr 1955; Barr and Arnista 1957a, 1957b) made the early attempt at measuring the capacity of various clay minerals for adsorbing various alkaloids and a diphteria toxin. Palygorskite is superior to kaolinite and halloysite because of its open structure, enclosing channels into which the alkaloid compounds can be accommodated. The isotherms for the adsorption of strychnine, atropine, and quinine may be described by the Langmuir equation which, in the linear form, is given by Equation 3.3,

$$\frac{C}{x/m} = \frac{1}{K_L V_m} + \frac{C}{V_m} \qquad \text{(Equation 3.3)}$$

where C is the equilibrium concentration; x is the amount adsorbed per unit weight m of clay; K_L is the Langmuir constant; V_m is the monolayer capacity of the adsorbent. Conformity of the data to Equation 3.3 is indicated by the straight line observed when $C/(x/m)$ is plotted against C.

Hendricks (1941) earlier reported that brucine and codeine were intercalated by H^+-montmorillonite through proton transfer and physical (van der Waals) interactions. Because of their large size, however, the 'cover-up' effect (cf. Chapter 4) would come into play since not all of the alkaloids could be protonated. On this basis, strychnine, atropine, and quinine would be affected because their molecular weight (in the order of 300) is comparable with that of brucine and codeine.

Nicotine (MW = 162), on the other hand, can neutralize stoichiometric amounts of hydrogen ions held by various clay minerals (Khan and Singhal 1967; Singhal and Singh 1976) because the molecule occupies less than the equivalent area per exchange site. By the same token, nicotine can enter the interlayer space of saponite (Suárez Barrios et al. 1996). The high adsorption of nicotine by acid-activated montmorillonite, reported by Ilić et al. (2019) and Ibrahim et al. (2020), is similarly indicative of surface protonation and electrostatic interaction. Here again, more nicotine can be removed from aqueous solution by dodecylammonium-exchanged bentonite than by the raw or unmodified mineral (Akçay and Yudakoç 2008).

Schönsee et al. (2021) have measured the adsorption of 52 phytotoxic compounds (alkaloids, terpenoids, steroids) by kaolinite and montmorillonite. For both minerals the maximum amount adsorbed is much smaller than their respective cation exchange capacity. Montmorillonite shows greater and stronger uptake (of the cationic and uncharged forms) than kaolinite although adsorption is apparently confined to external particle surfaces, involving cation exchange, cation-dipole, and cation-π interactions.

3.3 INTERACTIONS WITH AMINO ACIDS AND PEPTIDES

Because amino acids (and peptides) play a vital role in biochemical processes and organic matter transformation in soil, great interest attaches to the interactions of these compounds with clay minerals. We might also add that complexes of montmorillonite with amino acids (glycine, arginine, histidine) are capable of sequestering carbon dioxide (Pires et al. 2018).

The presence of acid (carboxyl) and basic (amino) functional groups in the same structure makes it possible for amino acids to exist as cations, zwitterions (or dipolar ions), or anions depending on the prevailing pH conditions. For monoamino monocarboxylic acids, the relationship between the different ionic forms may be described by the following equilibria,

$$\underset{\substack{\text{Cation}\\ pH < pI}}{R{-}CH(NH_3^+){-}COOH} \underset{}{\overset{K_1}{\rightleftharpoons}} \underset{\substack{\text{Zwitterion}\\ pH = pI}}{R{-}CH(NH_3^+){-}COO^-} \overset{K_2}{\rightleftharpoons} \underset{\substack{\text{Anion}\\ pH > pI}}{R{-}CH(NH_2){-}COO^-}$$

where R is hydrogen or alkyl; pI is the 'isoelectric' pH which for the most part ranges from 5 to 6. For monoamino dicarboxylic acids (e.g., aspartic acid) the first two dissociation constants (pK_1 and pK_2) refer to the first and second carboxyl group, respectively, while pK_3 denotes the constant for the NH_3^+ group. Similarly, for diamino monocarboxylic acids (e.g., arginine), pK_1 is the constant for the carboxyl group while pK_2 and pK_3 are identifiable with the first and second amino groups, respectively.

Hedges and Hare (1987) have reported that both kaolinite and montmorillonite have a high capacity for taking up basic (cationic) amino acids. In the case of neutral (zwitterionic) amino acids, however, montmorillonite is a much better adsorbent than kaolinite. Silva et al. (2020b) found similarly with respect to the hectorite- and kaolinite-alanine interaction, while Benetoli et al. (2007) noted that bentonite and kaolinite could take up more basic than neutral amino acids.

Since montmorillonite is relatively unreactive towards anionic amino acids, our focus here is on complex formation with cationic and zwitterionic amino acids in which cation exchange and proton transfer have a controlling influence. Stable coordination complexes may also form when strongly chelating cations such as copper, cobalt, and zinc occupy exchange positions at the clay mineral surface (Dashman and Stotzky 1985). In addition, the basicity and molecular size of amino acids would influence complex formation since these factors determine both the extent of van der Waals contact and the ease with which amino acids would enter the interlayer space.

In investigating the adsorption of some amino acids (and proteins) by Na^+-montmorillonite at acid pH, Talibudeen (1954) found that compound basicity was more important than molecular size. For example, arginine ($pK_3 = 12.48$ guanidium) could displace more sodium counterions than glycine ($pK_2 = 9.78$). Because of steric hindrance, however, the protein salmine with 40 arginine residues per molecule, was less efficient than arginine in displacing sodium ions. We would also expect that the 'cover-up' effect (Hendricks 1941) would apply here since the bulky salmine can cover more than the area per exchange site.

Subsequently Sieskind (1960a) determined the isotherms for the adsorption (at pH 2) of some amino acids by montmorillonite. The curves for type I amino acids show a plateau value close to the cation exchange capacity (CEC) of the clay mineral. On the other hand, the isotherms for type II do not level off to a discernible plateau while the amount adsorbed at the limit of solute concentration is appreciably less than the CEC. In type I amino acids the carboxyl and amino groups at the extremities of the chain are separated by at least two carbon atoms as in β-amino butyric acid, γ-amino butyric acid, and ε-amino caproic acid, all of which have a pK_2 value greater than 10.5. The compounds of type II have a pK_2 near 9.8 and are represented by such α-amino acids as glycine, α-alanine, α-amino butyric acid, and norvaline.

Figure 3.1 shows the relationship between the amount of amino acid retained (against water washing), corresponding to the initial part of the isotherms, and solution/suspension pH (Sieskind

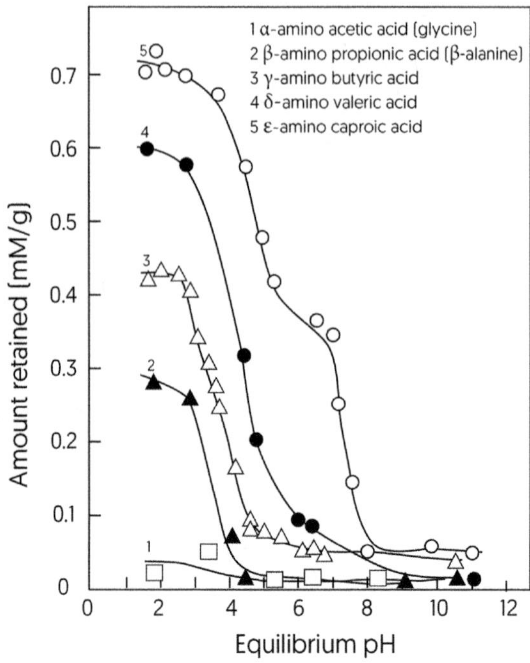

FIGURE 3.1 Relationship between the amount of amino acids retained (against washing with water) and the equilibrium (suspension) pH. (From Sieskind, O. and Wey, R., *C.R. Acad. Sci. Paris*, 248, 1652–1655, 1959.)

and Wey 1959). Adsorption is highest at pH ~2 and steeply declines at pH > 3, showing an abrupt change in slope near pK_1. A similar but less marked change also occurs in the alkaline pH range, near pK_2, as McLaren et al. (1958) have reported for α-alanine, and Nkoh et al. (2022) for cysteine and glutamic acid. Subsequently, Siffert and Kessaissia (1978) found that the retention of amino acid by montmorillonite was influenced by nature of the exchangeable counterions.

Ovcharenko et al. (1969) also found that adsorption of lysine and norleucine by various cation-exchanged montmorillonites decreased with a rise in solution pH with the highest uptake being recorded for the sodium-exchanged clay. At pH 2 and for a given initial solute concentration, adsorption decreased in the order $Na^+ > K^+ > Mn^{2+} > Zn^{2+} > Cr^{3+} > Ca^{2+} > Fe^{3+}$. No explanation was given for this observation although extensive interlayer expansion of Na^+-montmorillonite would facilitate amino acid intercalation. The successful intercalation of the protonated forms of glycine, L-alanine, L-tryptophane, L-histidine, L-methionine, and L-lysine into Na^+-montmorillonite at pH 4 has been reported by Kollár et al. (2003). In the case of transition metal-exchanged clays, complex formation would be favoured as Di Leo (2000) has proposed for the interlayer adsorption of cationic glycine to Cd^{2+}-montmorillonite. Similarly, Bodenheimer and Heller (1967) observed that uptake by montmorillonite (Wyoming) of amphoteric and basic amino acids such as glycine, α-alanine, methionine, and lysine from dilute solutions (< 2 mM g^{-1} added) at pH > 5 was enhanced in the presence of copper.

Parbhakar et al. (2007) have suggested that lysine adsorbs to Na^+-montmorillonite first as a cation form and then as a zwitterion. Spectroscopic analyses and molecular simulations strongly indicate that lysine is primarily adsorbed through electrostatic interactions between the terminal $-NH_3^+$ group of the molecule and the negatively charged interlayer surface of montmorillonite (Dong et al. 2018; Zhu et al. 2019). Accordingly, the affinity of (cationic) lysine to layer silicates, in general, is much stronger than that of zwitterionic and anionic amino acids (Yeasmin et al. 2014). Using acid-treated montmorillonite and attenuated total reflectance infrared (ATR-IR) spectroscopy, Kitadai et al. (2009) found that lysine was adsorbed in the cationic form over the pH range of 4.9 to 9.7. They further suggested that intercalated lysine adopted an upright conformation with its side-chain amino group pointing to the silicate surface. This suggestion, however, needs to be confirmed by X-ray diffraction analysis.

Although basicity is clearly important, the influence of molecular weight on adsorption is by no means inconsequential as the data of Sieskind and Wey (1959) would indicate. Figure 3.2 shows that the amount of amino acid retained (at pH 2) is linearly related to the number of carbon atoms (n_c) separating the terminal functional groups, rather than to molecular weight (MW) *per se*. For example, the points for *α*-amino acetic acid (glycine) ($n_c = 1$; MW 75), *α*-amino butyric acid ($n_c = 1$; MW 103), and *α*-amino caproic acid ($n_c = 1$; MW 131) are coincident, as are those for *β*-alanine ($n_c = 2$; MW 89) and β-amino butyric acid. This observation suggests that proximity of the COOH and NH_3^+ groups in the amino acid cation is more important than the length of the alkyl substituent in the α-position.

In measuring the adsorption of glycine and alanine by Ca^{2+}- and Mg^{2+}-montmorillonite over the pH range of 4 to 9, Kalra et al. (2000) noted a weak maximum near pH 7, while for the L-cysteine-bentonite system Faghihian and Nejati-Yazdinejad (2009) reported maximum uptake at pH 4. Benincasa et al. (2000) found that the adsorption of zwitterionic glycine to montmorillonite was markedly influenced by the nature of the interlayer counterions, increasing in the order $Ca^{2+} < Na^+ < Cu^{2+}$. Subsequently, Ramos and Huertas (2013) used X-ray diffraction and ATR-IR spectroscopy to characterize the adsorption of glycine to K^+-montmorillonite over the pH range of 2 to 12. They suggested that at low solute concentration the carboxylate group of zwitterionic glycine adsorbed to the edge surface of the clay mineral through electrostatic interactions. When this surface was saturated, glycine entered the interlayer space by replacing the K^+ counterions.

A more quantitative analysis of the clay-amino acid interaction was used by Greenland et al. (1962, 1965a, 1965b) who determined the isotherms for the adsorption of glycine, glycine peptides, and other naturally occurring amino acids by montmorillonite (and illite) saturated with calcium,

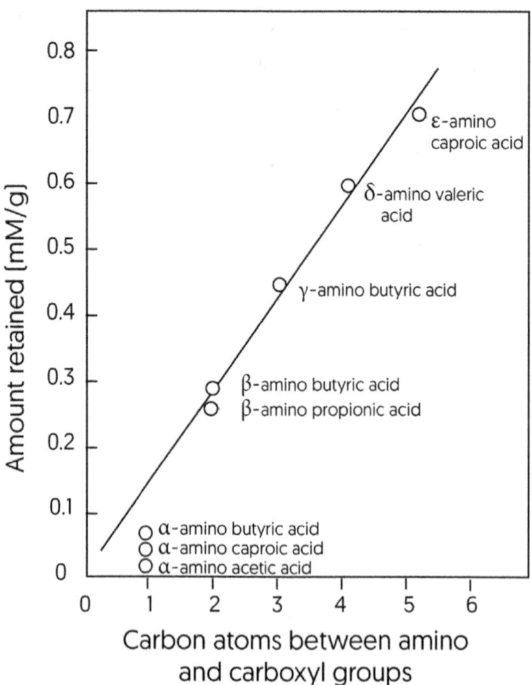

FIGURE 3.2 Relationship between the amount retained (at pH 2) and the number of carbon atoms separating the terminal functional groups in the amino acid molecules. (From Sieskind, O. and Wey, R., *C.R. Acad. Sci. Paris*, 248, 1652–1655, 1959.)

hydrogen, and sodium ions. The results and conclusions of this investigation have been summarized by Theng (1974).

In the case of Ca^{2+}-montmorillonite, the isotherms belong to the C-class (Giles et al. 1960) for which the amount adsorbed increased linearly with equilibrium solute concentration (Figures 3.3). Apparently, the number of energetically equivalent sites at the clay surface did not decrease as adsorption progressed. The formation of fresh adsorption sites was ascribed to the ability of the solute (rather than the solvent) to expand the interlayer space of montmorillonite. In line with this suggestion, the basal spacing increased from 1.9 nm for the parent clay to 2.2–2.4 nm for the complexes with glycine and its peptides. The resultant increase in interlayer volume was more than adequate to accommodate glycine, glycyl glycine, and diglycyl glycine, and no net loss of interlayer water was necessary. For triglycyl glycine (tetraglycine), however, some interlayer water was lost when more than half the total amount of solute was adsorbed. Furthermore, the amino acids and peptides were largely taken up as the corresponding zwitterions since very little, if any, desorption of calcium ions occurred. These observations are strongly indicative of solute partitioning between the bulk and surface-adsorbed ('Stern-layer') water.

Similarly, Kalra et al. (2003) found that adsorption of glycine and its peptides (pH 7; 23°C) to Ca^{2+}-montmorillonite increased in the order glycyl glycine < diglycyl glycine < triglycyl glycine. Density functional theory (DFT) and two-dimensional correlation analysis further suggested that glycine, serine, and arginine adsorbed to Ca^{2+}-montmorillonite through cation exchange, van der Waals, and hydrophilic interactions (Li et al. 2019a). Atomic force microscopy and molecular simulations by Moro et al. (2019, 2020) indicated that L- and di-L-alanine preferentially interact with the brucite-like sheet of clinochlore.

A partitioning mechanism was also proposed by Theng (1972) for the adsorption of *n*-alkylammonium ions by allophane (cf. Figure 7.3). Likewise, Dashman and Stotzky (1982)

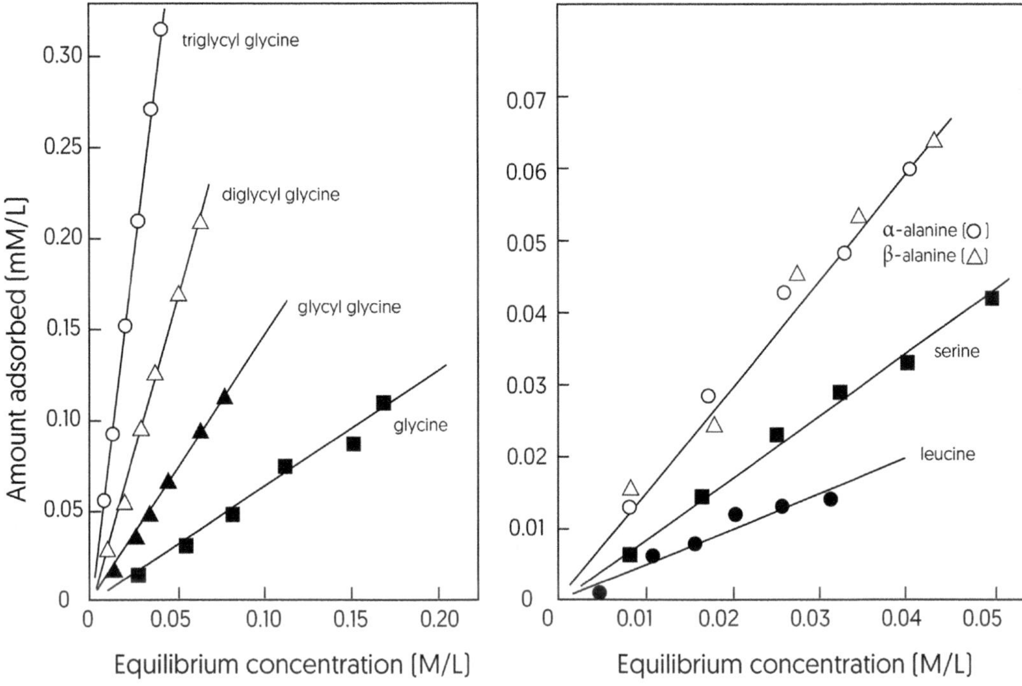

FIGURE 3.3 Isotherms for the adsorption of various amino acids by Ca^{2+}-montmorillonite at 25°C. (From Greenland, D.J. et al., *Trans. Faraday Soc.*, 61, 2024–2035, 1965b.)

recorded C-class isotherms for the adsorption of arginine, cysteine, and proline to H^+-, Zn^{2+}-, and Al^{3+}-montmorillonite. Closely linear isotherms also obtained for the adsorption of microcystin-LR, a monocyclic heptapeptide, to montmorillonite, illite, and kaolinite (Liu et al. 2019).

The isotherms for the adsorption of zwitterionic glycine and its peptides by Na^+-montmorillonite, however, were of the L-class, presumably because solute partitioning between the bulk and Stern-layer water varied with the amount adsorbed. Similarly, adsorption of neutral and acidic amino acid and peptide zwitterions by H^+-montmorillonite gave rise to Langmuir-type isotherms. The same applied to the uptake of basic amino acid and peptide cations by Na^+- and Ca^{2+}-montmorillonite (and illite). This finding is consistent with uptake being mediated by proton transfer (for the H^+-exchanged mineral) and by cation exchange (in the case of Na^+- and Ca^{2+}-exchanged clays). Likewise, Cloos et al. (1966) have suggested that two types of adsorbed species were formed when neutral solutions of glycine, glycyl glycine, and β-alanine were added to H^+-montmorillonite. The first type, formed by protonation of the zwitterion, was strongly held against repeated washing with water, while the second type consisted of associated zwitterions and could be removed by washing.

X-ray diffraction analysis indicated that the amino acids of Figure 3.3 together with arginine, carnosine, histidine, and lysine were intercalated as a single layer, giving basal spacings of 1.25–1.36 nm (Greenland et al. 1965a, 1965b) in agreement with the results of other workers (Talibudeen 1954; Sieskind 1960b; Chassin 1969). Slightly larger spacings were reported by Mallakpour and Dinari (2011, 2013) for complexes of a commercial Na^+-montmorillonite (Cloisite) with alanine, iso-leucine, leucine, methionine, phenylalanine, and valine as well as the protonated forms of aspartic acid, glutamic acid, serine, and tyrosine.

Greenland et al. (1965a, 1965b) noted that the observed interlayer separation (Δ-values) was smaller than the minimum van der Waals thickness of the intercalated amino acid. This apparent shortening of contact distances could be accounted for by 'keying' of the intercalated molecules into the ditrigonal depressions of the siloxane surface (cf. Chapter 2). Using a cadmium-rich montmorillonite, Di Leo (2000) have suggested that more than one monolayer of glycine molecules could be intercalated from a concentrated (1 M) solution of the amino acid. Dong et al. (2018) also found that the interlayer spacing of montmorillonite-lysine complexes varied with the amount of amino acid intercalated. Similarly, Block et al. (2015) observed that the basal spacing of montmorillonite complexes with small peptides increased as intercalation progressed.

Some time ago, Bodenheimer and Heller (1967) reported that uptake of some amphoteric and basic amino acids (glycine, α-alanine, lysine) from dilute solutions (< 2 mM/g added) at pH > 5 by montmorillonite (from Wyoming) was enhanced in the presence of copper. This finding was ascribed to the formation of copper-glycine/alanine complexes in aqueous solution and in the interlayer space of montmorillonite. The strong adsorption of lysine to Cu^{2+}-montmorillonite was attributed to the formation of a copper-lysine coordination complex involving the amino group of the lysine zwitterion. Subsequently, Fu et al. (1996) reported on the intercalation from aqueous solution of (preformed) planar complexes of $Cu(lysine)_2^{2+}$ and $Cu(histidine)_2^{2+}$ into saponite through a cation exchange mechanism.

That zwitterionic and basic amino acids can form coordination complexes with copper (and other transition metal ions) in montmorillonite has been clearly demonstrated by Jang and Condrate (1972a, 1972b) using infrared spectroscopy. The interlayer copper-valine complex, formed under acid conditions, is apparently of the monodentate type (Scheme 3.4a), while the complex formed at pH 8.2 has a bidentate structure (Scheme 3.4b). The formation of positively charged complexes between Cu^{2+}-montmorillonite and valine has also been proposed by Nagy and Kónya (2004). Even glycine may form strong, stable complexes with Cu^{2+}-smectite (Benincasa et al. 2000).

For complexes of lysine with montmorillonite containing divalent transition metal ions (M^{2+}), Jang and Condrate (1972b) have proposed the following structure for the intercalated species (Scheme 3.5) based on infrared spectroscopy measurements and assuming that the α-amino, rather than the ε-amino, group of lysine is involved. Pandey et al. (2015) have suggested similarly for complexes of L-leucine with Cu^{2+}-montmorillonite, while cysteine apparently interacts with Cu^{2+}- and Hg^{2+}-montmorillonite through its thiol group (Brigatti et al. 1999; Malferrari et al. 2010).

Like montmorillonite, vermiculite can form interlayer complexes with amino acids. As early as 1952, Barshad noted that vermiculites could intercalate several amino acids from their respective aqueous solutions. The extent of interlayer expansion (intracrystalline swelling) is related to the nature of the exchangeable cation, the magnitude of the dipole moment, and dielectric constant of the interlayer solution. Vermiculite crystals can also show extensive swelling after immersion in strong solutions of glycine, β-alanine, γ-amino butyric acid, and ε-amino caproic acid in their respective zwitterionic forms (Walker and Garrett 1961). Subsequently, Rausell-Colom and Salvador (1971a, 1971b) measured the basal spacing of two vermiculites, immersed in solutions of γ-amino butyric and ε-amino caproic acids. Beyond a critical value of solute concentration, extensive swelling occurred giving interlayer separations > 20 nm as previously reported by Walker and Garrett (1961).

$$\left[\begin{array}{l} H_2O \quad NH_2{-}CH{-}CH(CH_3)_2 \\ \quad Cu \quad\quad | \\ H_2O \quad OH_2 \quad C(O)O \end{array}\right]^+ \quad \left[\begin{array}{l} H_2O \quad NH_2{-}CH{-}CH(CH_3)_2 \\ \quad Cu \quad\quad | \\ H_2O \quad O{-}C{=}O \end{array}\right]^+$$

(a) (b)

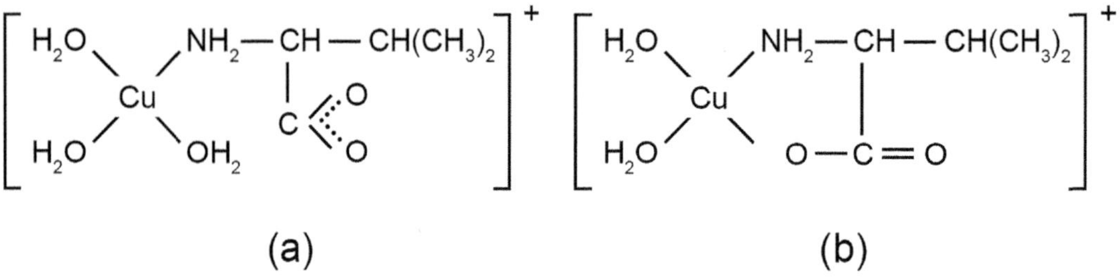

SCHEME 3.4 Proposed structures of interlayer copper-valine complexes in montmorillonite: (a) monodentate and (b) bidentate.

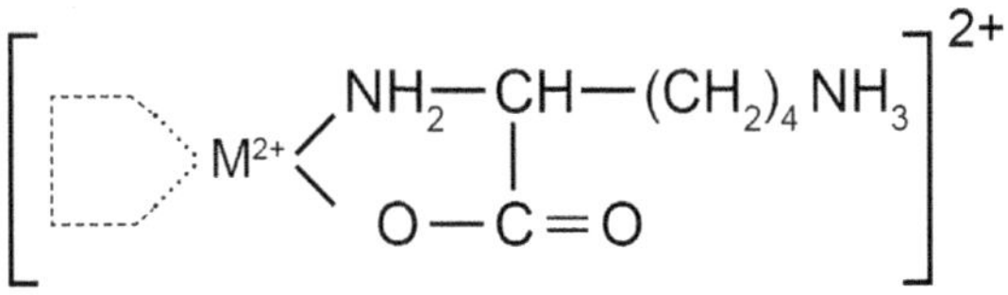

SCHEME 3.5 Proposed structures of the interlayer complex formed between divalent transition metal counterions of montmorillonite and lysine.

This type of swelling, however, does not take place uniformly in that some of the layers (within a crystal) remained unexpanded even in saturated solutions of the amino acids. Basal spacing values, recorded below the point of extensive swelling (< 2.1 nm), would indicate intercalation of a single and a double layer of amino acids adopting different orientation, from one in which the molecular chain is nearly parallel to one in which this chain is inclined at a high angle to the silicate layer. Likewise, the interlayer complex of vermiculite with ε-amino caproic acid (6-amino hexanoic acid), having a basal spacing of 1.732 nm, is consistent with intercalation of a double layer of amino acid cations in which the carbon zig-zag plane is tilted at an angle of ~50° to the silicate layer (Kanamaru and Vand 1970). A similar configuration has been proposed by Parbhakar et al. (2007) for a montmorillonite-lysine complex. We might also mention that lysine can form a double-layer complex on intercalation into vermiculite as does 6-aminohexanoic acid, while ornithine tends to intercalate as a single layer (Raupach et al. 1975; Slade et al. 1976).

Amino acid adsorption to kaolinite is confined to external particle surfaces unless the interlayer space of the mineral has previously been expanded by intercalation of suitable guest molecules (Lagaly et al. 2006). Density functional theory (DFT) analysis by Kasprzhitskii et al. (2022) indicates that H-bonding between the carboxyl group of amino acids (glycine, alanine, valine, leucine, and isoleucine) and the hydroxyl surface of kaolinite (cf. Figure 1.4) is the dominant mechanism, while H-bonding between the nitrogen atom of amino acids and surface hydroxyl plays a subsidiary role in the interaction. On the other hand, Benincasa et al. (2002) have suggested that Cd-cysteine complexes interact with the siloxane surface of kaolinite and may even penetrate the interlayers of disordered kaolinite.

Little doubt, however, attaches to the intercalation of amino acids into kaolinite that has been expanded by treatment with certain compounds. Thus, Sato (1999) were able to intercalate zwitterionic glycine, β-alanine, γ-aminobutyric acid, and δ-aminovaleric acid into expanded ('hydrated') kaolinite. The intercalated amino acids formed a monolayer complex with their alkyl chain tilted toward the interlayer surface. Infrared spectroscopy would indicate that the molecules were H-bonded to both the aluminol and siloxane surface of kaolinite. On the other hand, serine is apparently intercalated into hydrated kaolinite as a flattened monolayer with a basal spacing of ~1.13 nm (Wang et al. 2017a). Chapter 6 has more details on the formation and properties of complexes between kaolinite and organic compounds.

We would also point out that clay minerals can serve as supports and catalysts for the synthesis of polypeptides. The accumulated literature on this topic has been summarized by Theng (2018).

3.4 INTERACTIONS WITH PURINES, PYRIMIDINES, NUCLEOSIDES, AND NUCLEOTIDES

In his classic paper on the clay-organic interaction, Hendricks (1941) proposed that purines and nucleosides were intercalated by H^+-montmorillonite through a proton transfer reaction, giving basal spacings that were related to molecular configuration. For example, adenine and guanine complexes have spacings of 1.24 nm, indicating intercalation of a single layer of purine cations that are co-planarly arranged. Guanine was apparently intercalated as the *enol*, rather than the *keto*, form. On the other hand, 3-methylcytosine was intercalated as the *keto* tautomer. Since a coplanar arrangement was not realizable, the basal spacing (1.29 nm) of the 3-methylcytosine complex exceeded that of the complex with guanine. Complexes with adenosine and guanosine (in which a ribofuranose group substitutes at the 9-position in the purine ring), gave still larger basal spacings (1.35 nm). These values, however, were only 0.10 nm larger than would be required for a strictly coplanar atomic arrangement. The intercalated species apparently adopted the flattest possible configuration, allowing maximum van der Waals contact to be made between the nucleosides and the silicate surface.

Subsequently, Brindley and co-workers (Lailach et al. 1968a, 1968b; Lailach and Brindley 1969; Thompson and Brindley 1969) made a systematic investigation into the interactions of purines, pyrimidines, and nucleosides with montmorillonite (and illite) as a function of pH, exchangeable cation type, and molecular constitution. They used Li^+-, Na^+-, Mg^{2+}-, and Ca^{2+}-exchanged montmorillonites and initial solute concentrations of ~1 mM such that no more than ~25% of the exchange positions could be occupied by the organic cations (under acidic conditions). Figures 3.4 and 3.5 illustrate the results for Ca^{2+}-montmorillonite.

The pH-uptake curves show that besides being dependent on pH conditions, complex formation is influenced by the basicity and constitution of the organic molecules. When the solution pH is lower than the pK_a of the organic compound (R), the amount of protonated compound (RH^+) increases, leading to the formation of RH^+-montmorillonite. Figure 3.5b shows that below pH ~5.5 the amount of calcium ions released far exceeds the quantity of adenine adsorbed. This difference may be ascribed to proton uptake by adenine. Little adsorption occurs at pH > 7, and the original cation-saturated montmorillonites, by and large, preserve their identity.

The effect of molecular constitution is reflected in the observation that, all things being equal, the steep decline in (percentage) adsorption occurs at a higher equilibrium pH as basicity increases. This point is illustrated by the sequence 6-chloropurine ($pK_a = 0.8$) < purine ($pK_a = 2.39$) < adenine ($pK_a = 4.22$) and the sequence inosine ($pK_a = 1.2$) < guanosine ($pK_a = 1.6$) < adenosine ($pK_a = 3.45$). Aromaticity is also an important molecular parameter influencing uptake. Thus, the sequence of hypoxanthine ($pK_a = 1.98$) < 5-amino-6-methyluracil (AMU; $pK_a = 3.28$) < adenine ($pK_a = 4.22$)

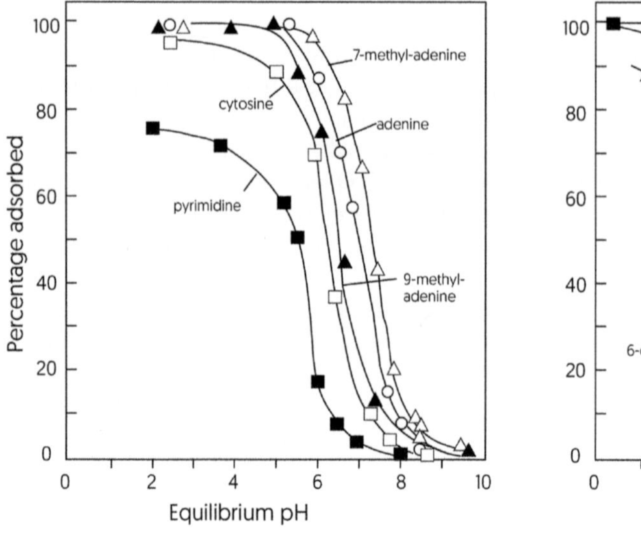

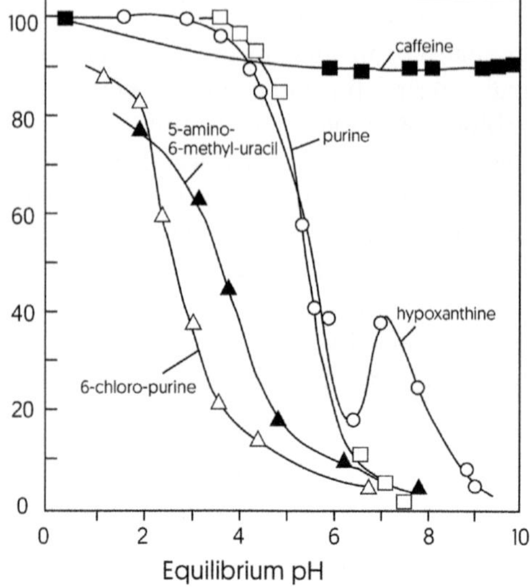

FIGURE 3.4 Influence of equilibrium (suspension) pH on the percentage adsorption of some purines and pyrimidines by Ca^{2+}-montmorillonite. (From Lailach et al. 1968a, Figures 1d and 2d. Reproduced with kind permission of The Clay Minerals Society, publisher of *Clays and Clay Minerals*.)

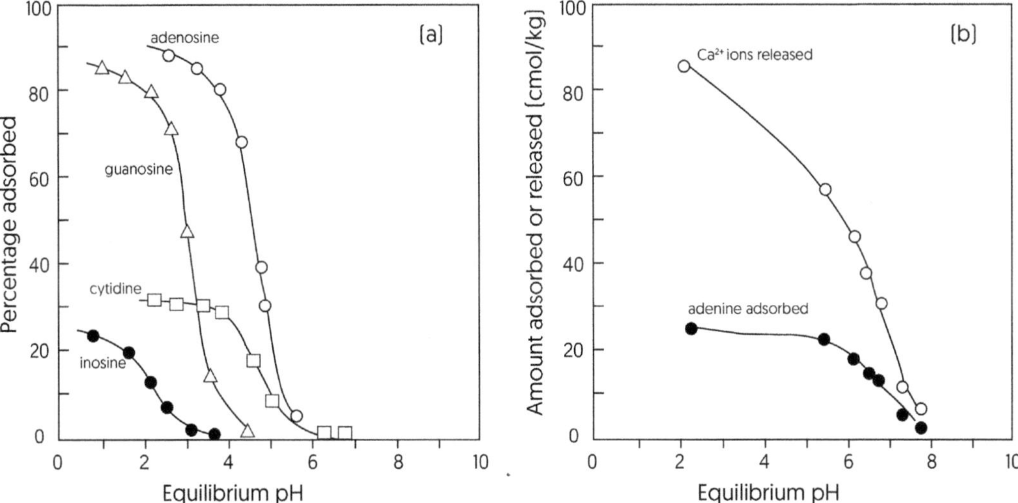

FIGURE 3.5 (a) influence of suspension pH on the (percentage) adsorption of some nucleosides by Ca^{2+}-montmorillonite; (b) influence of equilibrium (suspension) pH on the amount of adenine adsorbed by, and Ca^{2+} ions released from, Ca^{2+}-montmorillonite. (From Lailach et al. 1968a, Figures 3d and 4b. Reproduced with kind permission of The Clay Minerals Society, publisher of *Clays and Clay Minerals.*)

is anticipated, whereas the observed order is AMU < hypoxanthine < adenine. This behaviour is attributed to AMU being non-aromatic in both the neutral and cationic forms, while both hypoxanthine and adenine are aromatic. Interestingly, hypoxanthine shows a rise in adsorption in the range of pH 7 to 9, probably due to the formation of a complex with the exchangeable cation.

The influence of molecular size shows itself in the markedly strong adsorption of caffeine above pH 7, despite its weak basic character ($pK_a = 0.61$), while the effect of molecular structure is indicated by the larger uptake of (non-planar) nucleosides in relation to purines and pyrimidines of lower molecular weight. The basal spacing of ~1.33 nm, shown by the adenosine complex (after drying over P_2O_5), is consistent with the tendency for intercalated organic species to adopt a flat conformation.

The pH isotherms for the adsorption of purines and pyrimidines by Co^{2+}- and Ni^{2+}-exchanged montmorillonite at pH < 6, and by Cu^{2+}-montmorillonite at pH < 3, are similar to those shown by the Ca^{2+}-saturated mineral, indicative of a cation exchange process (Lailach et al. 1968b). Under weakly acidic and weakly basic conditions, coordination complexes are formed with the transition metal cations, the strength of which decreases in the order Cu >> Ni > Co >> Ca. Chelation is indicated by the presence of well-defined maxima in the isotherms in the range of pH 7 to 9 (Co and Ni) and at pH > 4 (Cu) and by pronounced colour changes in the case of Co and Cu. There is evidence to indicate that the N-3 and N-9 positions in the purine structure (Scheme 3.6) are involved in complex formation with transition metal cations.

In a subsequent study, Lailach and Brindley (1969) described the co-adsorption of certain purines and pyrimidines from aqueous solutions by Na^+- and Ca^{2+}-montmorillonite at pH 1 to 6. For example, thymine by itself does not adsorb to montmorillonite but does so in the presence of adenine or hypoxanthine, both of which are intercalated. Similarly, the presence of adenine can enhance thymine and uracil adsorption (Lagaly 1984; Samii and Lagaly 1987). Lailach and Brindley (1969) have suggested that the underlying mechanism is one of H-bonding between the protonated base (adenine or hypoxanthine) and the uncharged base (thymine). Intermolecular hydrogen bonding of this type is reminiscent of the formation of hemisalt complexes described in Chapter 2. The basal spacing of co-adsorption complexes at ~1.25 nm indicates intercalation of a single layer of cation-base assemblages in a flat configuration.

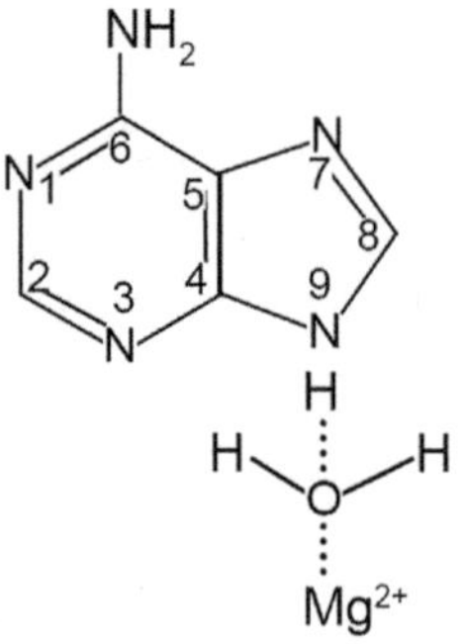

SCHEME 3.6 Interaction between nucleic acid bases (e.g., adenine) and Mg^{2+} counterions in the interlayer space of montmorillonite through water-bridging.

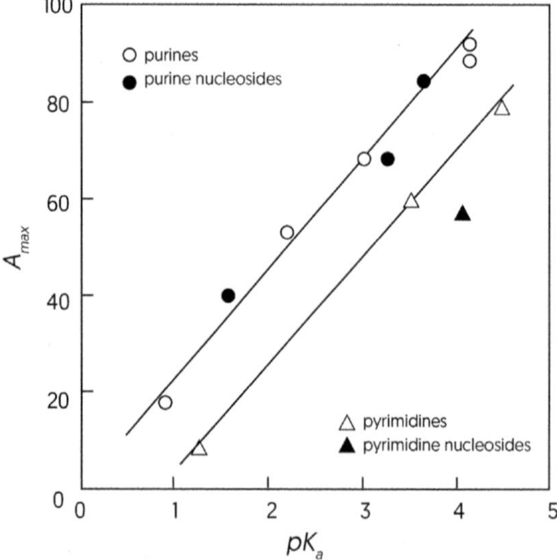

FIGURE 3.6 Relationship between A_{max}, denoting the maxima in the adsorption *vs* pH isotherms of Na^+-illite, and the pK_a values for some purines, pyrimidines, and their respective nucleosides. (From Thompson and Brindley 1969, Figure 1. Reproduced with permission of the Mineralogical Society of America publisher of *American Mineralogist*.)

The adsorption of purines, pyrimidines, and nucleosides from aqueous solutions by Na^+-, Mg^{2+}-, and Cu^{2+}-illite occurs by mechanisms like those controlling uptake by montmorillonite (Thompson and Brindley 1969). Since illite is a non-expanding mineral, however, adsorption is confined to external particle surfaces. Under acidic conditions, cation exchange and proton transfer reactions predominate, while the pH isotherms show well-defined maxima. With respect to Na^+-illite, the height of the maxima (A_{max}) is linearly related to the pK_a of the respective heterocyclic base or nucleoside (Figure 3.6). Under comparable experimental conditions, the value of A_{max} increases with molecular weight in line with the enhanced contribution of van der Waals interactions to the adsorption energy.

In the case of Cu^{2+}-illite, the maximum amount adsorbed correlates with the electron-releasing ability of the group attached at the 6-position of the purine ring (Scheme 3.6). Having only a single ring structure, pyrimidines do not appear to form a complex with copper ions. Rather, their adsorption to Cu^{2+}-illite resembles that shown by the Mg^{2+}- and Na^{+}-exchanged minerals, indicative of a cation-exchange mechanism.

Benetoli et al. (2008) have measured the adsorption of nucleic acid bases to different clay minerals at pH 2 and 7.2 under simulated sea-water conditions. Montmorillonite can take up much more adenine, cytosine, thymine, and uracil than kaolinite since adsorption to the latter mineral is restricted to external particle surfaces. Using density functional theory, Robinson et al. (2007) have suggested that thymine and uracil can form hydrogen bonds with the octahedral and tetrahedral surface sites of dickite. Benetoli et al. (2008) have further noted that uptake by montmorillonite is greater at pH 2 than at pH 7.2 since nucleic acid bases are positively charged under acid conditions. Likewise, Hashizume and Fujii (2016) reported that montmorillonite could take up to five times more adenine at pH 2.5 than at pH 8.5. Carneiro et al. (2011) found adsorption to decrease in the order adenine ≈ cytosine > thymine > uracil as Lagaly (1984) had described earlier.

Hashizume et al. (2010) have made a detailed analysis of the adsorption of adenine, cytosine, and uracil by Mg^{2+}-montmorillonite at pH ~ 8. Figure 3.7 shows that for all three bases, the isotherms are essentially linear (C-class) indicating that fresh (interlayer) sites are formed as intercalation progresses. Since hardly any Mg^{2+} ions are desorbed, the (uncharged) organic bases are presumably coordinated to interlayer counterions through a water bridge (cf. Figure 2.16) as indicated in Scheme 3.6 for adenine. A similar mechanism has been suggested by Sciascia et al. (2011) as regards adenine and cytosine and by Akyuz and Akyuz (2018) with respect to 5-chlorouracil. Interestingly, Gururani et al. (2012) obtained L-class isotherms for the adsorption of adenine to divalent cation-exchanged montmorillonite. Villafañe-Barajas et al. (2018) reported similarly for Na^{+}-montmorillonite and Yamamoto et al. (2018) for the montmorillonite-caffeine interaction.

Figure 3.7 shows that adsorption (at comparable equilibrium solute concentration) decreases in the order adenine > cytosine > uracil, while the basicity of the compounds follows the opposite order. This finding is consistent with the proposed mode of interaction since the less basic the solute (adenine) the greater its ability to coordinate to Mg^{2+}. At comparable solute concentrations, adsorption is greatest for adenine which has the largest molecular weight and the lowest solubility in water. X-ray diffraction further suggests that the intercalated bases are oriented with their heterocyclic rings parallel to the silicate surface as Perezgasga et al. (2005) have suggested earlier. Density functional theory would indicate that thymine and uracil adsorb to kaolinite through cation (Na^{+})-oxygen interaction (Michalkova et al. 2011).

The clay-nucleotide interaction has attracted considerable attention because of its possible role in the formation of biopolymers on the early Earth (Theng 2018). Early on, Graf and Lagaly (1980) reported that the capacity of clay minerals (beidellite, illite, kaolinite, montmorillonite) for adsorbing adenosine-5-phosphates increased in the order AMP << ADP < ATP. They suggested

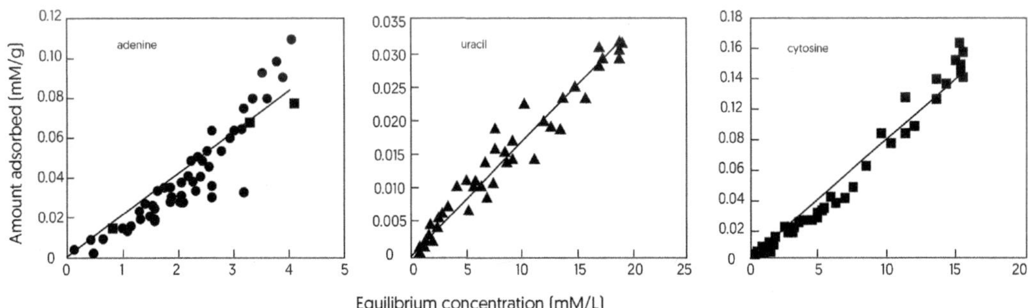

FIGURE 3.7 Isotherms for the adsorption of adenine, cytosine, and uracil to Mg^{2+}-montmorillonite at pH ~8. (From Hashizume et al. 2010, Figure 1. Reproduced with kind permission of The Mineralogical Society of the UK and Ireland publisher of *Clay Minerals*.)

that adsorption was mediated by ligand exchange between hydroxyl groups, exposed at clay particle edges, and nucleotide phosphate groups (cf. Equation 1.1). The very low uptake of AMP was ascribed to steric hindrance in that the bulky adenosine part of the nucleotide prevented close contact to be established between clay edge surface and the phosphate group of AMP. The pivotal role that the phosphate group plays in the nucleotide-clay interaction has also been emphasized by Hashizume and Theng (2007) and Pedreira-Segade et al. (2018a).

Nucleotides (and RNA) could also adsorb to divalent cation-exchanged montmorillonites by means of cation-bridging (Liebmann et al. 1982; Rishpon et al. 1982; Ferris et al. 1989; Pedreira-Segade et al. 2018b; Oliveira et al. 2021; Wang et al. 2021). Indeed, this mechanism lies behind the interaction between anionic organic polymers and smectite surfaces (Theng 2012). In this regard, transition metal cations are especially effective in acting as a bridge between the anionic group of the solute and the silicate surface (Hao et al. 2019). Under acid conditions, nucleotides are protonated and can adsorb through electrostatic and van der Waals interactions (Feuillie et al. 2013; Mignon et al. 2020).

3.5 INTERACTIONS WITH FATTY ACIDS AND FATS

There is much interest in the clay-fatty acid interaction because clay minerals can catalyze the conversion of carboxylic acids into petroleum hydrocarbons (Jurg and Eisma 1964; Shimoyama and Johns 1971; Johns 1979; Theng 2018), although fatty acids would not be strongly attracted to negatively charged clay surfaces. Except for formic acid, however, alkane carboxylic acids show limited dissociation in aqueous media and occur as uncharged compounds. Furthermore, clay-fatty acid complexes are usually obtained in the absence of water. Thus, the preparation of clay mineral complexes with fatty acids (Brindley and Moll 1965), like those with *n*-alcohols (Brindley and Ray 1964), remains something of an art.

Yariv and Shoval (1982), for example, obtained complexes of montmorillonite with acetic, lauric, and stearic acids by contacting the clay with CCl_4 solutions of the organic acids. An alternative procedure is to mix dry montmorillonite with the fatty acid, notably stearic acid, in a ball mill or blender (Gonzalez et al. 2014; Del Angel et al. 2015; Hernandez et al. 2016; Bu et al. 2017). A widely used and facile method is to introduce the fatty acid into montmorillonite that has previously been expanded by intercalation of long-chain quaternary ammonium ions (Freitas et al. 2007; Ma et al. 2007; Das et al. 2011; Rooj et al. 2012) or the sodium salt of 4-dodecylbenzene sulfonic acid (Sarier et al. 2010).

The formation of two types of complexes between montmorillonite and (saturated) fatty acids has been described and analyzed by Brindley and Moll (1965). The basal spacings of 'short-spacing' (type I) complexes are less than 1.7 nm, while the corresponding values of the 'long-spacing' (type II) complexes exceed 2 nm. Type I complexes were prepared by reacting the heat-treated (250°C, 45 mins) clay samples directly with the fatty acids (acetic, butyric, caproic, caprylic, pelargonic). Type II complexes were obtained by expanding the montmorillonite layers with *n*-hexanol or *n*-octanol before introducing the appropriate fatty acids (capric, undecylic, lauric, tridecylic, myristic, palmitic, stearic). These complexes were formed at temperatures above the melting point of the free acids.

The fatty acids in type II complexes are apparently intercalated as single layers of molecules inclined at a high angle, and with their carboxyl groups attached, to the silicate surface. Brindley and Moll (1965) have further suggested that the terminal methyl groups of the fatty acids are arranged pairwise in a head-to-tail fashion with some lateral displacement due to the bulky carboxyl end groups (Figure 3.8). Besides giving rise to good intermolecular packing, the high angle of tilt (~70°) allows van der Waals and O–H···O–Si interactions to be established. The probability that the carboxyl group of stearic acid can penetrate the ditrigonal cavity in the siloxane surface has been proposed by Lu et al. (2010). We might also mention that stearic acid intercalated into a halloysite-dimethylsulphoxide complex can partially adopt a tilted conformation (Zhang et al. 2020b).

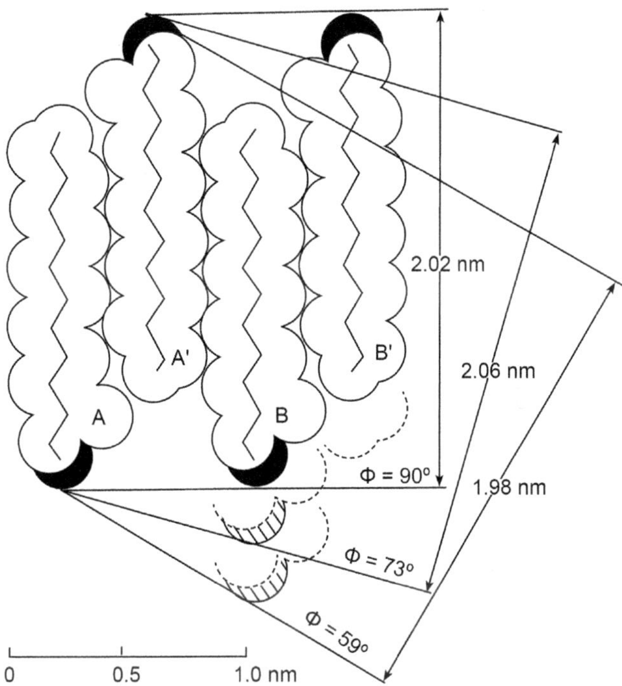

FIGURE 3.8 Diagram showing the pairwise close-packing of fatty acid molecules with head-to-tail arrangement in the interlayer space of montmorillonite. Broken lines show the various displacements of the B,B' pair with respect to the A,A' pair. Spacings of terminal planes are indicated as well as the corresponding angles of inclination, φ, of the chain axes with respect to the terminal planes. Hydrogen atoms in terminal hydroxyls are shaded. (From Brindley and Moll 1965, Figure 1. Reproduced with permission of the Mineralogical Society of America publisher of *American Mineralogist*.)

As regards type I complexes, no regular trend in basal spacing is discernible with an increase in fatty acid chain length. In the absence of adsorption data, it is also difficult to remark on the mechanism underlying the intercalation of short-chain fatty acids. By analogy with benzoic acid (Yariv et al. 1966), however, it seems likely that the compounds are adsorbed through cation- or water-bridging, involving the carboxyl or carboxylate group of the molecule and the exchangeable cation (Yariv and Shoval 1982). Cation-bridging or 'inner-sphere coordination' (cf. Figure 5.1) has also been proposed by Kang and Xing (2007) for the interaction of some dicarboxylic acids with freeze-dried Na^+-montmorillonite at pH 4. There is also evidence to indicate that some short-chain fatty acids (acetic, formic, chloroacetic, oxalic) interact with NH_4^+-montmorillonite by a cation-exchange mechanism (Gautier et al. 2009). Similarly, Nishikiori et al. (2010) have suggested that linoleic acid adsorbs to allophane by electrostatic interactions.

Brindley and Moll (1965) have also examined the stability of the long-spacing complexes against heat treatment and solvent extraction. Curiously, the samples are stable above, but not below, the melting point of the respective free fatty acid. Washing with ethanol causes the basal spacing of the complexes to collapse although some high spacings (1.3–1.4 nm) persist even after heating the samples to 250°C for 70 hours. Čapková et al. (2006) found that octadecylammonium stearate, intercalated into Na^+-montmorillonite, gradually decomposed under ambient conditions. Interestingly, fulvic acid could intercalate into complexes of montmorillonite with palmityl palmitate and stearic acid (Dubbin et al. 2014).

For complexes of Ca^{2+}-montmorillonite with stearic and behenic acids, Sieskind and Ourisson (1971) have suggested that the part which is resistant to solvent (benzene) extraction is associated with clay particle edges. The amount retained, however, is higher than can be accounted for by this mechanism alone. It seems probable that a portion of the retained stearic acid is linked to the planar clay surface through cation-bridging. In synthetic hectorite with Ni^{2+} substituting for Al^{3+} in the octahedral sheet, Sieskind and Siffert (1972) have proposed that the stearate anion adsorbs to clay particle edges through ligand exchange (cf. Figure 5.2B). This mechanism has also been suggested by Sarier et al. (2010) for the interaction of Na^+-montmorillonite with stearic and lauric acids.

In this context, we would also mention the interaction of K-Cu chlorophyllin with montmorillonite exchanged with Na, Cs, Mg, Co, Cu, and Al cations (Kaufherr et al. 1971). The organic solute consists of chlorophyllin in the tri-carboxylated and keto forms and an aliphatic carboxylate salt derived from the phytyl group. During uptake of this mixture, potassium replaces the exchangeable cations (except cesium), while the anionic group of the molecule interacts with the clay particle edge surface. Chemical and infrared data indicate that the three components of the mixture are taken up selectively and to different extents by the various cationic forms of the montmorillonite. Cs^+-montmorillonite adsorbs chlorophyllin only, while the other clay samples take up varying amounts of the carboxylic acid salts.

The importance of clay minerals as a concentrating substrate/medium for fatty acids present in natural bodies of water has been investigated by Meyers and Quinn (1971). Using a saline solution (comparable to sea water), the isotherm for the adsorption of fatty acids, such as heptadecanoic acid, by montmorillonite was linear up to a concentration of 120 μg/L. The amount adsorbed from sea water, however, is consistently less than from artificial saline solution. This discrepancy is ascribed to interference and competition from other organic substances dissolved in sea water and to the presence of ionic species other than sodium and chloride in the natural environment of the sea.

We would also mention the interactions of clay minerals with triglycerides (fats) which are esters of glycerol and fatty acids. For example, tristearin is formed by condensing glycerol with three stearic acid molecules (Scheme 3.7). The intercalation of symmetric triglycerides into smectites (beidellite, hectorite, montmorillonites) has been investigated by Weiss and Roloff (1966) using X-ray diffraction techniques. Interlayer complex formation here, like that involving long-chain fatty acids, is rather problematic. Nor are the results always reproducible because the intercalated fats may partially undergo hydrolysis, giving rise to a mixture of mono-, di-, and triglycerides, glycerol, and fatty acids. Interestingly, vermiculite fails to intercalate triglycerides.

On the other hand, no difficulty attaches to intercalating fats into smectite (and vermiculite) that have previously been exchanged with *n*-alkylammonium ions. In this instance, regular complexes are formed showing rational orders of basal reflections. Indeed, the basal spacing of complexes is related to the number of carbon atoms in the fatty acid moiety of the triglyceride, the length of the alkylammonium cation, and the surface charge density of the smectite sample (Figure 3.9).

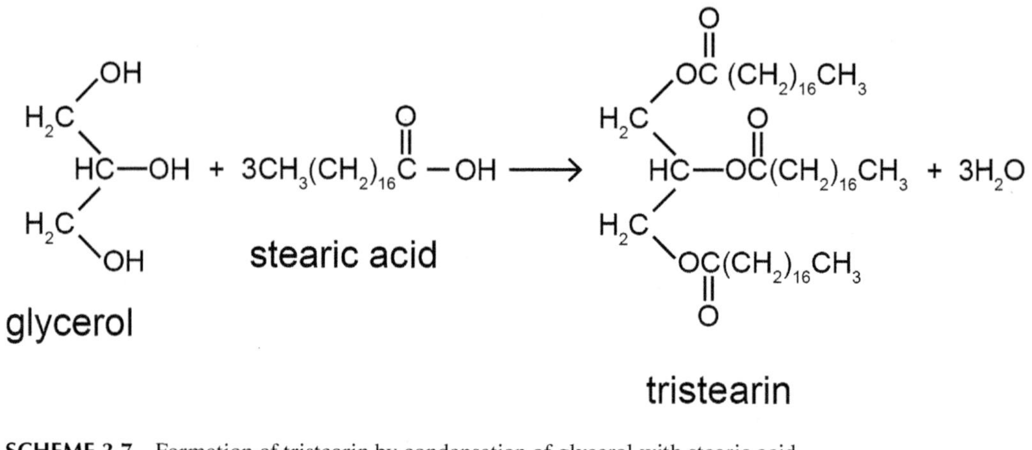

$$\underset{\text{glycerol}}{H_2C(OH)-HC(OH)-H_2C(OH)} + \underset{\text{stearic acid}}{3CH_3(CH_2)_{16}C(=O)-OH} \longrightarrow \underset{\text{tristearin}}{H_2C(OC(=O)(CH_2)_{16}CH_3)-HC(OC(=O)(CH_2)_{16}CH_3)-H_2C(OC(=O)(CH_2)_{16}CH_3)} + 3H_2O$$

SCHEME 3.7 Formation of tristearin by condensation of glycerol with stearic acid.

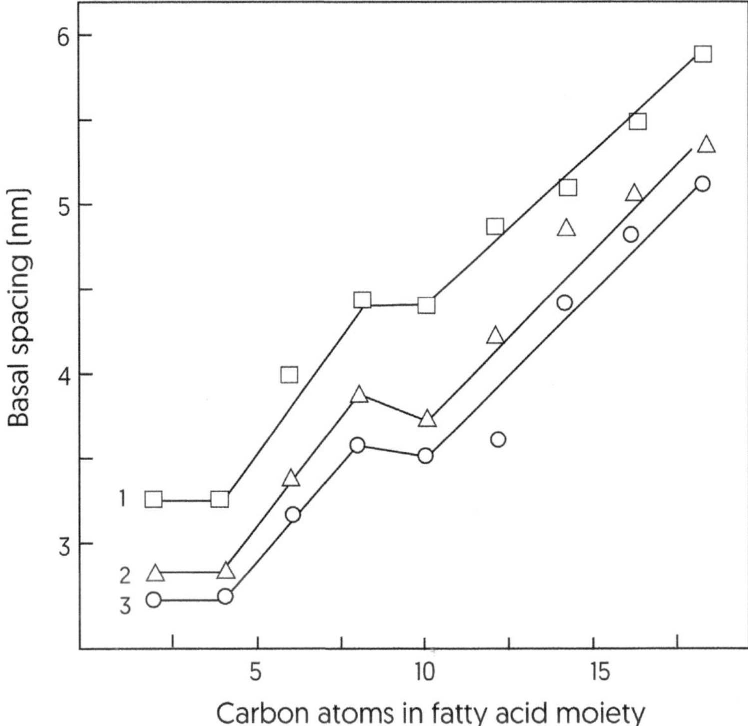

FIGURE 3.9 Relationship between the basal spacings of *n*-alkylammonium-exchanged complexes of beidellite with symmetric triglycerides and number of carbon atoms in the fatty acid moiety: (1) *n*-octadecylammonium; (2) *n*-dodecylammonium; (3) *n*-decylammonium. (From Weiss, A. and Roloff, G., *Proc. Int. Clay Conf., Jerusalem*, 1, 263–275, 1966.)

The break in the curves for C_{10} (tricaprin), whose melting point is close to room temperature, would indicate that interlayer expansion is temperature-dependent. Below the melting point of the fat, there is a small but regular increase in basal spacing with temperature, while near the melting point the spacing abruptly decreases (Figure 3.10). This transition is quite pronounced for the high members of the series and resembles that shown by interlayer complexes with *n*-alcohols (cf. Figure 2.9). In this instance, however, the phase change is reversible and unlikely to be caused by a net loss of material. Rather, the transition is probably due to a change in triglyceride configuration near or at the melting point of the molecule as Brindley and Ray (1964) have proposed for complexes with *n*-alcohols.

Because of its large capacity for adsorbing fatty acids and triglycerides, montmorillonite can potentially impede fat absorption, hydrolysis, and digestion in mammalian guts. As a result, lipid excretion through bowel movement is enhanced, while obesity can be attenuated if not prevented (Xu et al. 2016; Dening et al. 2019). In this regard, sepiolite appears to be superior to montmorillonite (Gutman et al. 2020). Furthermore, when mice on a high-fat diet (HFD) are supplied with montmorillonite, the composition of their gut microbiota changes in favour of reducing body weight gain (Xu et al. 2017; Lee et al. 2018; Joyce et al. 2020).

3.6 INTERACTIONS WITH SACCHARIDES

Bradley (1945) noted that sucrose could intercalate into montmorillonite, yielding a complex with a basal spacing of ~ 1.83 nm. Subsequently, Greenland (1956a) investigated the interactions of mono

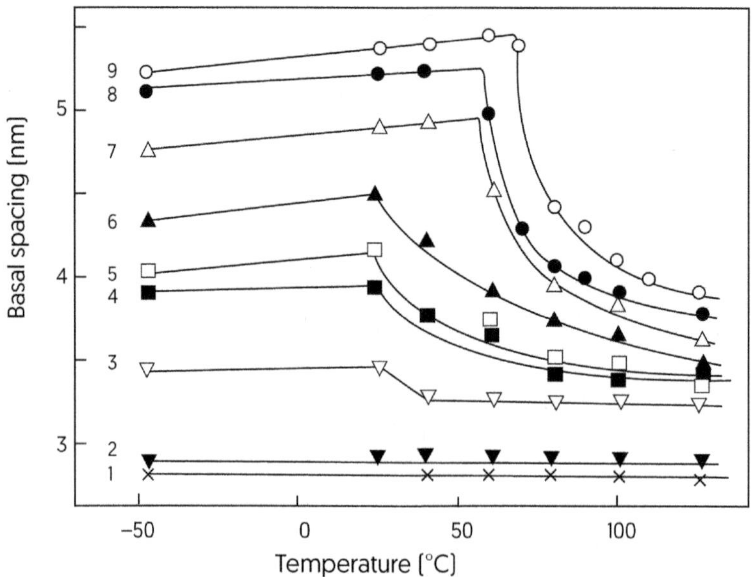

FIGURE 3.10 Temperature-dependent variation in basal spacing of beidellite complexes with *n*-dodecylammonium ions following intercalation of some symmetric triglycerides: (1) triacetin; (2) tributyrin; (3) tricaproin; (4) tricaprylin; (5) tricaprin; (6) trilaurin; (7) trimyristin; (8) tripalmitin; and (9) tristearin. (From Weiss, A. and Roloff, G., *Proc. Int. Clay Conf., Jerusalem*, 1, 263–275, 1966.)

and oligo-saccharides and their derivatives with H^+-, Na^+-, K^+-, and Ca^{2+}-exchanged montmorillonites. The X-ray diffraction data indicated the formation of more than one type of complex, each type being segregated rather than either randomly or regularly interstratified within a given clay particle. The relative amount of single- and double-layer complexes could be estimated from the relative intensity of the basal reflections, assuming that equal quantities of each type give rise to two sets of reflections of equal intensity and that the intensity of the higher orders of reflection relative to that of the first order varies in the same way as in a single-type system.

The dependence of adsorption on sugar concentration is consistent with a physical adsorption process. The extent of adsorption, however, does not appear to go beyond the intercalation of a double layer of sugar molecules, even when the sugar concentration is increased to 20%. The influence of the interlayer counterion is also evident. With glucose, for example, adsorption increases in the order $K^+ < Ca^{2+} < Mg^{2+} < H^+ < Na^+$ under comparable conditions of sugar and clay concentrations. Omitting hydrogen and sodium, this order corresponds to that of the polarizing power of the cations, indicative of ion-dipole interactions as the underlying mechanism (cf. Chapter 2). The position of hydrogen may be ascribed to the partial occupation of interlayer exchange sites by aluminium, while that of sodium is related to the propensity of Na^+-montmorillonite particles to expand, and hence facilitate intercalation.

Since the extent of adsorption depends on both the sugar concentration and the nature of the interlayer cation, these factors must be kept constant in comparing the affinity of various saccharides for the silicate surface. On the basis that saccharides capable of forming a double-layer complex from dilute solutions are strongly adsorbed, while sugars requiring high initial concentrations for complex formation are weakly held, Greenland (1956a) has concluded that (a) methylated, carboxyl- and amino-substituted glucoses are more strongly adsorbed than unsubstituted sugars, and (b) disaccharides have a greater affinity for montmorillonite than monosaccharides.

Table 3.1 compares Δ-values with minimum molecular thicknesses of the intercalated saccharides derived from the dimension of the pyranose and furanose rings found in crystals of some of

TABLE 3.1
Interlayer Complexes of Montmorillonite with Various Saccharides and Related Compounds Together with Interlayer Distance (Δ-Value) and Minimum van der Waals Thickness of the Molecules.

Saccharide/compound (in decreasing order of affinity)	Δ-value (nm) Single layer	Double layer	Minimum molecular thickness (nm)
2,4,6-Trimethyl glucose	0.51	0.92	0.53
4,6-Dimethyl glucose	0.56	1.09	0.53
Arabinose	n.a.	0.90	n.d.
Ascorbic acid	n.a.	0.83	n.d.
Fucose	n.a.	0.94	n.d.
N-acetylglucosamine	n.a.	0.81	n.d.
Cellobiose	0.50	0.88	0.51
Lactose	0.48	0.87	0.53
Maltose	0.54	0.93	0.54
Melibiose	0.53	0.88	0.53
Galacturonolactone	0.58	0.85	n.d.
Raffinose	0.58	0.91	n.d.
Rhamnose	0.64	1.28	0.59
Sucrose	0.56	0.91	0.56
α-Methyl glucoside	0.58	n.a.	0.57
Sorbose	0.53	n.a.	0.52
Erythritol	0.52	0.91	n.d.
Galactose	0.58; 0.50	n.a.	0.53
Heptulose	0.62	n.a.	n.d.
i-Inositol	0.49	n.a.	n.d.
Xylose	0.47; 0.51	0.96	0.48; 0.52
β-Glucose	0.48	0.93	0.47
Mannose	0.67		1.04
Mannitol	0.50	n.a.	n.d.
Gluconolactone	0.57	n.a.	n.d.
Glucuronolactone	0.57	n.a.	n.d.
Glucosamine	0.48	n.a.	0.47; 0.53

Source: Greenland, D.J., *J. Soil Sci.*, 7, 329–334, 1956.
The complexes were prepared using a 2.5% suspension of Ca^{2+}-montmorillonite in a 1% solution of the saccharide.
n.a. = not applicable; n.d. = not determined

the sugars. Pauling's (1960) van der Waals radii (with hydrogen equals 0.12 nm) were used to estimate the size of the appropriate groups around the rings. There is clearly good agreement between Δ-value and minimum molecular thickness for most of the compounds examined. This finding further supports the view that the smallest observed spacing for any complex is ascribable to the presence of a single layer of sugar molecules in the interlayer region, with the molecules lying as flat as possible between two opposing silicate layers. Only with rhamnose and 4,6-dimethylglucose does the second (double-layer) Δ-value correspond to twice or nearly twice the first (single-layer) value. With the other sugars the second Δ-value is slightly less than twice the first. This indicates that where a double layer of sugar molecules is intercalated the layers tend to interpenetrate. It is also of interest to note that the Δ-values for single-layer complexes with β-glucose, 2,4,6trimethyl glucose, and cellobiose are not much different from the corresponding minimum molecular thicknesses. Since, in other respects, their dimensions are widely different, this observation provides strong evidence for the flat configuration of the intercalated molecules.

Chemical analysis further indicates that methylated sugars are more strongly adsorbed than their unsubstituted counterparts or sugars containing amino or carboxyl groups (Greenland 1956b). The relatively strong adsorption of methylated sugars may be related to the absence of O–H···O–Si hydrogen bonding and the tendency of such compounds to polymerize (Bell and Northcote 1950). Furthermore, glucose is strongly hydrated in aqueous solutions (Taylor and Rowlinson 1955) and hence cannot displace water molecules from the montmorillonite surface as easily as its methylated counterpart. Molecular dynamics simulations by Aristilde et al. (2017) suggest that glucose can 'capture' the interlayer water in Na^+-montmorillonite and hence deplete the cation-associated water, while ^{23}Na NMR spectroscopy indicates that the mobility of sodium counterions decreases as glucose intercalation progresses.

The pH isotherms for the adsorption of gluconic acid, glucuronic acid, and glucosamine by Ca^{2+}-montmorillonite show a defined maximum in the range of pH 4 to 5 (Greenland 1956b). The isotherm for glucosamine closely resembles that of purines and pyrimidines (Figure 3.4) and can be similarly interpreted. The high adsorption (of glucosammonium) at pH 3–6 may be ascribed to cation exchange and proton transfer at planar surfaces (Parfitt 1972). Using Cu^{2+}-montmorillonite, Pusino et al. (1989b) have proposed that at pH > 7 glucosamine is coordinated to copper through its amino and adjacent (deprotonated) hydroxyl groups. Sugar acids, on the other hand, tend to adsorb to clay particle edges, and their increased uptake at pH > 7 may be ascribed to the formation of insoluble salts or ligand exchange involving edge aluminium ions.

Mitra et al. (1957) noted that less glucose was adsorbed by Ca^{2+}-bentonite than by its natural (Na^+-saturated) counterpart. As already remarked on, the natural material would swell in water, allowing the sugar molecules to enter the interlayer space. Adsorption to the calcium-exchanged material conforms to the Freundlich isotherm and may be ascribed to cation-dipole interactions. In comparison with montmorillonite, kaolinite has a very low affinity for sugars. Indeed, Davis and Worrall (1971) reported that glucose was negatively adsorbed (repelled) by kaolinite although Jepson and Williams (1972) could measure a small (positive) uptake from aqueous solution.

In closing this section, we might mention that clay minerals can promote the formation of humic-like substances (melanoidins) through the Maillard reaction between glucose and various amino acids (Hedges 1978). In this regard, the nature of the exchangeable cation is as important as the clay surface in influencing the process (Arfaioli et al. 1999; Bosetto et al. 2002; Gonzalez and Laird 2004). Subsequently, Vicente Vilas et al. (2010) have proposed that the melanoidin, arising from the reaction of glucose with L-tyrosine and L-glutamic acid, is partially intercalated into, and partially associated with, the external particle surface of the montmorillonite substrate.

3.7 INTERACTIONS WITH DYES

The clay-dye interaction has attracted a great deal of attention in terms of its chemistry, underlying mechanisms, and potential in removing dye molecules from industrial effluents and wastewater as the following reviews would attest (Ogawa and Kuroda 1995; Lagaly et al. 2006; Liu and Zhang 2007; Bujdák 2015; Ismadji et al. 2015; Pandey and Ramontja 2016; Adeyemo et al. 2017; Ngulube et al. 2017; Anastopoulos et al. 2018; Fizir et al. 2018; Teepakakorn et al. 2019; Gahlot et al. 2020; López-Rodríguez et al. 2021; Kausor et al. 2022; Cunha et al. 2023). Among the large variety of basic (cationic) dyes, methylene blue (Scheme 3.8) has long held the limelight because of its wide usage, toxicity, and persistence (Yariv 2002; Bujdák 2006; Kausar et al. 2018; Khan et al. 2022).

The clay-methylene blue (MB) interaction has mostly been investigated using smectite as the clay component although other layer silicates (e.g., kaolinite, sepiolite, vermiculite) have also featured. Besides being concerned with the underlying mechanism, much research into complex formation

Methylene blue

Rhodamine 6G

Acridine orange

Crystal violet

azo functional group

$C_8AzoC_{10}N^+$

Methyl red

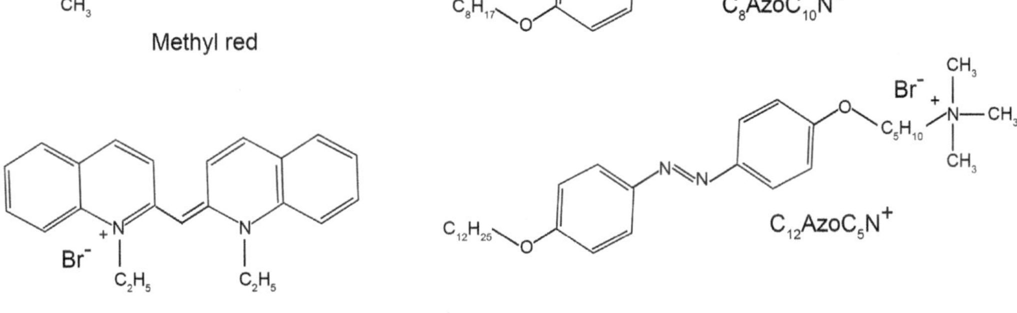

I,I'–diethyl–2–2'–cyanine

SCHEME 3.8 Molecular structures of some cationic dyes.

has been directed at estimating the cation exchange capacity and specific surface area of clay minerals (and soils) (Kipling and Wilson 1960; Ramachandran et al. 1962; Nevins and Weintritt 1967; Brindley and Thompson 1970; Hang and Brindley 1970; Rytwo et al. 1991; Kahr and Madsen 1995; Wang et al. 1996; Avena et al. 2001; Yukselen and Kaya 2008; Yener et al. 2012; Hegyesi et al. 2017). The data, however, have not always been self-consistent.

Although the primary mechanism is one of cation exchange (at planar surfaces), H-bonding of the tertiary amine group of MB to aluminol groups at the edge surface of smectite particles, either directly or through a water bridge, is also involved (Fang et al. 2018). Metachromasy is another characteristic of the clay-MB interaction. This feature refers to the shift of the principal (monomer) absorption band (at 670 nm) to shorter wavelengths as uptake progresses due to the formation of dimers, trimers, and higher aggregates (Bergmann and O'Konski 1963; Cenens and Schoonheydt 1988; Schoonheydt and Heughebaert 1992). On the other hand, Yariv and co-workers (Yariv and Lurie 1971; Cohen and Yariv 1984; Garfinkel-Shweky and Yariv 1999; Yariv 2002) have explained the metachromatic effect in terms of the interaction between the π electrons of the dye and the siloxane surface of smectite. Thus, the absence of metachromasy during the adsorption of rhodamine 6G and B to montmorillonite might be because the bulky $C_6H_5COOC_2H_5$ substituent lies nearly perpendicular to the xanthene skeleton (Scheme 3.8), hindering close surface-dye interaction (Grauer et al. 1984, 1987). In this context, we might also mention that the mobility of *n*-alkyl acridine orange cations, adsorbed to montmorillonite, is apparently influenced by steric hindrance as well as hydrophobic interactions between the alkyl chains (Yamagishi and Soma 1981).

Also relevant is the work by Ras et al. (2004) on the orientation of octadecyl rhodamine B (RhB18) in Langmuir-Blodgett monolayers. They noted that in films with Na^+-saponite, both the phenyl and xanthene groups of RhB18 are oriented parallel to the clay mineral surface lying parallel to the substrate. This conformation would be conducive to molecular aggregation as Tapia Estévez et al. (1993) have suggested during the adsorption of rhodamine 6G by Wyoming montmorillonite. Likewise, Bujdák (2006) has argued for solute aggregation as the dominant mechanism underlying metachromasy.

The smectite-MB interaction is influenced by many factors, including exchangeable cation species (Bodenheimer and Heller 1968; Chen et al. 1999; Czímerová et al. 2004b; Ma et al. 2004; Bilgiç 2005), solution ionic strength and pH (Narine and Guy 1981; Hajjaji and Alami 2009; Feddal et al. 2014), solute aggregation (Bergmann and O'Konski 1963; Cenens and Schoonheydt 1988; Aznar et al. 1992; Schoonheydt and Heughebaert 1992; Gessner et al. 1994; Jacobs and Schoonheydt 1999; Bujdák et al. 2002; Czímerová et al. 2004a), clay layer charge density (Bujdák et al. 2001; Neumann et al. 2002; Kovář et al. 2009; Pentrák et al. 2012), and surface molecular configuration/orientation (Hang and Brindley 1970; Yu et al. 2000; Bujdák et al. 2003). There is general agreement, however, that MB adsorption to both smectite and kaolinite follows pseudo second-order kinetics and conforms to the Langmuir isotherm (Bilgiç 2005; Gürses et al. 2006; Almeida et al. 2009; Ma et al. 2010a; Sarma et al. 2011; Tehrani-Bagha et al. 2011; Rida et al. 2013; Chang et al. 2016; Mouni et al. 2018). We might also add that MB can compete well with crystal violet (Rytwo et al. 1993) and *n*-alkyltrimethylammonium surfactants (Breen and Loughlin 1994) for adsorption sites on montmorillonite. On the other hand, thioflavin T can displace MB from the montmorillonite surface (Margulies et al. (1988).

Like methylene blue, such cationic dyes as acridine orange, azo and cyanine dyes, crystal violet, malachite green, methyl green, and rhodamine (Scheme 3.8) interact with clay minerals through a cation-exchange mechanism, forming interlayer complexes where the dye cations may adopt various orientations. Moreover, adsorption is influenced by solution pH, ionic strength, and temperature as well as by solute concentration and aggregation, and may be described by the Langmuir or Freundlich isotherm (Cohen and Yariv 1984; Ogawa and Ishikawa 1998; Iwasaki et al. 2000; Miyamoto et al. 2000; Rytwo et al. 2000; Ogawa et al. 2003; Jiang et al. 2008; Lv et al. 2011; Günay et al. 2013; Berez et al. 2014; Sanoğlan et al. 2014; Fang et al. 2018; Marco-Brown et al. 2018; El Haouti et al. 2019; Paredes-Quevedo et al. 2021; Ullah et al. 2021). Interestingly, intercalated

$C_8AzoC_{10}N^+$ and $C_{12}AzoC_5N^+$ azo dyes (Scheme 3.8) in montmorillonite show reversible *cis-trans* isomerization on irradiation with ultraviolet and visible light (Ogawa and Ishikawa 1998). The same is true for azobenzene intercalated into a dialkyldimethylammonium-exchanged fluor-tetrasilicic mica (Ogawa et al. 1999).

The adsorption of rhodamine dyes to clay minerals has already been remarked on. Gemeay (2002) also noted that the adsorption capacity of Na^+-montmorillonite with respect to rhodamine 6G (Rh6G) decreased with an increase in solute concentration (and temperature) because of solute aggregation. Morimoto et al. (2012), however, have suggested that rhodamine B (RhB) does not show appreciable aggregation during its adsorption to synthetic Zn-substituted smectites. Molecular modelling would indicate the formation of three types of RhB dimers in the interlayer space of Wyoming montmorillonite (Pospíšil et al. 2003). Molecular aggregation has also been observed during the adsorption of rhodamine 3B and 6G by smectite (López Arbeloa et al. 1997, 2002a, 2002b)

The adsorption of rhodamine and other basic dyes to various layer silicates has been the topic of many investigations. As in smectite, uptake to kaolinite (and vermiculite) is influenced by dye concentration, pH, and temperature, while the process commonly follows pseudo second-order kinetics and the Langmuir isotherm (Hu et al. 2006; Selvam et al. 2008; Nandi et al. 2009a, 2009b; Khan et al. 2012; Duman et al. 2015; Ma et al. 2016; Hebert et al. 2021). Montmorillonite can intercalate rhodamine B as either a monolayer or a bilayer depending on solute concentration, but adsorption of (zwitterionic) RhB to kaolinite and palygorskite is confined to external particle surfaces (Rao et al. 2020). As a result, the capacity of montmorillonite for taking up RhB far exceeds that of kaolinite (Bhattacharyya et al. 2014).

In all instances, the underlying mechanism is one of cation exchange although hydrophobic (and van der Waals) interactions would contribute as Harris et al. (2006) have proposed for the adsorption of 9-aminoacridine, 3,6-diaminoacridine, azure A, and safranin O to kaolinite. Castellini et al. (2008) have suggested similarly for the kaolinite-malachite green interaction in which the particle edges appear to play an important role. Thus, dye adsorption is inhibited when the edge surface is 'masked' by treating the kaolinite sample with sodium hexametaphosphate. Other examples of the kaolinite-dye interaction are given in Chapter 6.

Teepakakorn et al. (2018) also made the interesting observation that the stability of Rh6G was markedly enhanced by complexation to a natural montmorillonite through quenching of the dye photoexcited state by mineral impurities in the clay sample. Likewise, Wu et al. (2019d) observed that the light stability and luminous intensity of Rh6G greatly increased on intercalation into montmorillonite, as did Kohno et al. (2009) for an interlayer complex of montmorillonite with anthocyanin.

The interactions of clay minerals with acid (anionic) dyes have received relatively little attention since the molecules would be repelled from the negatively charged mineral surface. Nevertheless, adsorption can and does happen when the charge on the dye molecules has been neutralized or reversed. Thus, Yermiyahu et al. (2007) measured appreciable uptake of Congo red from aqueous solution at pH 1.2 by Na^+-montmorillonite through a cation exchange mechanism. Similarly, Li et al. (2009) reported that kaolinite had a high capacity for adsorbing acid brown 75 at pH 1.0. Subsequently, Elsherbiny (2013) found that the adsorption of acid fuchsin to Na^+-montmorillonite conformed to the Freundlich isotherm (Equation 3.1). Karaoğlu et al. (2010) and Sharma et al. (2016), respectively, have reported that the rate of reactive blue 221 adsorption to kaolinite and that of methyl blue to montmorillonite follows pseudo second-order kinetics. More recently, Pajak (2021) has suggested that reactive red 198 and acid red 18 interact with smectite through H-bonding between Si–OH and Al–OH groups (exposed at particle edge surface) and –NH, $-NH_2$ and –OH groups of the anionic dyes. Furthermore, prior acid treatment of the clay sample leads to a marked enhancement of adsorption capacity.

On the other hand, anionic (and cationic) dyes are readily adsorbed by smectites that have been modified by intercalation of (large) organic cations and surfactants, or by grafting with organosilanes (Table 3.2). The corresponding isotherms are commonly of the Langmuir type, rather than

TABLE 3.2

Adsorption and Removal of Various Dyes by Clay Minerals Previously Modified by Intercalation of Organic Compounds, Ionic Liquids, and Surfactants or by Grafting with Organosilanes

Clay mineral	Organic modifier	Dye species	Reference (in chronological order)
Montmorillonite	Cetylpyridinium	Crystal violet, Orange II, Phenol red	Lee et al. (2001)
Bentonite	Dodecyltrimethylammonium	Acid blue 193	Özcan et al. (2004)
Montmorillonite	Cetyltrimethylammonium	Rhodamine 6G	Sasai et al. (2004)
Montmorillonite	Hexadecyltrimethylammonium	Amido naphthol red G	Wang et al. (2004)
Bentonite	Benzyltrimethylammonium	Acid blue 193	Özcan et al. (2005)
Sepiolite	Dodecylethyldimethylammonium	Acid blue 193	Özcan et al. (2006)
Hectorite	Cetyldimethylbenzylammonium	C.I. Reactive orange 122	Baskaralingam et al. (2007)
Bentonite	Dodecyltrimethylammonium	Reactive blue 19	Özcan et al. (2007)
Bentonite	Cetyltrimethylammonium	Direct red 2	Zohra et al. (2007)
Bentonite	Hexadecyltrimethylammonium	Acid orange 10	Jović-Jovičić et al. (2008)
Montmorillonite	Cetylpyridinium	Eriochrome black T, Orange II, Methyl orange, Thioflavin T, Methylene blue, Crystal violet	Shin (2008)
Palygorskite (Attapulgite)	Hexadecyltrimethylammonium	Congo red	Chen and Zhao (2009)
Montmorillonite	Chitosan	Basic blue 9, Basic blue 66, Basic yellow 1	Monvisade and Siriphannon (2009)
Bentonite	Cetyltrimethylammonium	Acid green 25	Koswojo et al. (2010)
Montmorillonite, Bentonite	Diphosphonium, bis-imidazolium	Telon red, Telon blue, Telon orange	Makhoukhi et al. (2010, 2015)
Bentonite	Cetylpyridinium	Reactive red 120	Tabak et al. (2010)
Montmorillonite	Octadecyltrimethylammonium	Acid red G	Tong et al. (2010)
Bentonite	Cetyltrimethylammonium	Acid blue 129	Yesi et al. (2010)
Montmorillonite	Cetyltrimethylammonium, sodium stearate, and mixture	Methyl orange	Chen et al. (2011)
Bentonite	Gemini surfactant, cetyl-trimethylammonium	Methyl orange	Kan et al. (2011)
Montmorillonite	Octyl- and dodecyl-trimethylammonium	Annatto dye	Kohno et al. (2011)
Bentonite	Polyepichlorohydrin-dimethylamine	Acid scarlet GR, Acid dark blue 2G	Li et al. (2011c)
Montmorillonite	Gemini surfactants	Methyl orange	Liu et al. (2011)

Bentonite	Hexadecyltrimethylammonium	Orange II, Orange G	Ma et al. (2011)
Hectorite	Cetylpyridinium	Neutral red	Yue et al. (2011)
Bentonite	Hexadecyltrimethylammonium	Reactive blue 19	Ikhtiyarova et al. (2012)
Kaolinite	Rarasaponin	Malachite green	Suwandi et al. (2012)
Bentonite	Benzylhexadecyldimethylammonium, dimethyldioctadecylammonium, benzylstearyldimethylammonium	Phenol red, Direct blue 53	Nguyen et al. (2013)
Bentonite	Gemini surfactant	Methyl orange	Wang et al. (2013)
Montmorillonite	Dodecyl sulfobetaine	Methylene blue	Fan et al. (2014)
Montmorillonite	Gemini surfactant	Methyl orange, Congo red	Gu et al. (2014)
Al_{13}-pillared montmorillonite	Trimethylchlorosilane	Orange II	Qin et al. (2014)
Bentonite	Hexadecyltrimethylammonium	Methylene blue, Crystal violet, Rhodamine B	Anirudhan and Ramachandran (2015)
Bentonite	Chitosan	Methylene blue	Bulut and Karaer (2015)
Montmorillonite	Octadecylamine, aminopropyltriethoxysilane	Basic green 4, Basic yellow 28	Hassani et al. (2015)
Montmorillonite (K10)	1-Methyl, 3-hexadecyl imidazolium	Amaranth dye	Lawal and Moodley (2015)
Vermiculite	Ethylenediamine	Remazol Brilliant blue RN	de Queiroga et al. (2016)
Bentonite	Chitosan	Amaranth red, methylene blue	Dotto et al. (2016)
Montmorillonite	Hexadecyltrimethylammonium	Methyl red	Omidi-Khaniabadi et al. (2016)
Montmorillonite	Cetyltrimethylammonium	Remazol Black B	Huang et al. (2017)
Palygorskite/Sepiolite	3-Aminopropyltriethoxysilane	Methylene blue, metanil yellow	Moreira et al. (2017)
Bentonite	Cetyltrimethylammonium	Congo red	Pandey and De (2017)
Montmorillonite	Hexadecylpyridinium, hexadecyl-trimethylammonium	Methylene blue, malachite green	Ullah et al. (2017)
Montmorillonite	Octylphenol polyoxyethylene ether	Methylene blue	Wang et al. (2017b)
Vermiculite	Gemini surfactant	Methyl orange	Zang et al. (2017)
Montmorillonite	Hexadecyltrimethylammonium, chitosan	Congo red, Crystal violet	Zhu et al. (2017)
Montmorillonite	1-Butyl-3-methylimidazolium	Fluorescein	Belbel et al. (2018)
Bentonite	Tetradecyl-, hexadecyl-, and octadecyl-trimethylammonium	Remazol Blue RN	Brito et al. (2018)
Montmorillonite	Tetra-*n*-butylammonium	Toluidine blue, Malachite green, Congo red	Chicinaş et al. (2018)
Montmorillonite	Hexadecylpyridinium	Methyl red	Luo et al. (2018)

(Continued)

Table 3.2 ***(Continued)***

Clay mineral	Organic modifier	Dye species	Reference (in chronological order)
Montmorillonite	Gemini surfactants	Methyl orange	Ren et al. (2018)
Cloisite (bentonite)	Cetyltrimethylammonium	Eosin	Kooli et al. (2019)
	Chitosan	Reactive red 136	Li et al. (2019b)
Montmorillonite	Cetylpyridinium, alkyl dimethyl- benzylammonium	Reactive red 120	Mahmoodi et al. (2019)
Bentonite	Sodium dodecyl sulfate	Methyl violet	Mohammed and Al-Heetimi (2019)
Montmorillonite	Gemini surfactants	Methyl orange	Peng et al. (2019)
Bentonite	3-Aminopropyltrimethoxysilane	Reactive violet 5R	Queiroga et al. (2019)
Rectorite	Cetyltrimethylammonium	Acid orange	Shen et al. (2019)
Vermiculite	Gemini surfactants	Rhodamine B	Wang et al. (2019b)
Bentonite	Hexadecyltrimethylammonium	Methylene blue, Bezathren red, Telon blue	Baouch et al. (2020)
Bentonite	Organochlorosilanes	Sudan dyes	Saeed et al. (2020b)
Bentonite	Ethylene sulfide and ethylenediamine	Remazol blue RN	Santos et al. (2020)
Bentonite	*n*-Alkylammonium	Rhodamine 6G	Intachai et al. (2021)
Montmorillonite	Cetyltrimethylammonium	Orange G	Ouachtak et al. (2021)
Bentonite	Ionic liquid	Congo red	Ozola-Davidane et al. (2021)
Palygorskite (Attapulgite)	3-Aminopropyltriethoxysilane	Congo red	Yang et al. (2021)
Montmorillonite	Docosyl-trimethylammonium, hyamine, and sodium dodecyl sulfate	Crystal violet	Sarabadan et al. (2021, 2022)

linear, while adsorption increases as solution pH decreases. As already pointed out with respect to the organoclay-drug interaction, these observations indicate that other mechanisms, such as electrostatic, H-bonding, and π-π interactions are involved besides solute partitioning between solution and surface (cf. Figure 2.19). The water solubility and molecular weight of the dyes can also influence their removal by organoclays.

REFERENCES

Abate, G. and J.C. Masini. 2005. Adsorption of atrazine, hydroxyatrazine, deethylatrazine, and deisopropylatrazine onto Fe(III)polyhydroxy cations intercalated vermiculite and montmorillonite. *Journal of Agricultural and Food Chemistry* 53: 1612–1619.

Adeyemo, A.A., I.O. Adeoye, and O.S. Bello. 2017. Adsorption of dyes using different types of clay: A review. *Applied Water Science* 7: 543–568.

Aggarwal, V., H. Li, and B.J. Teppen. 2006. Triazine adsorption by saponite and beidellite clay minerals. *Environmental Toxicology and Chemistry* 25: 392–399.

Aguzzi, C., P. Cerezo, C. Viseras, and C. Caramella. 2007. Use of clays as drug delivery systems: Possibilities and limitations. *Applied Clay Science* 36: 22–36.

Aisawa, S., N. Takahishi, S. Kikuchi, H. Hirahara, and E. Narita. 2018. Adsorption of aflatoxin B_1 on natural clays in aqueous solutions. *Clay Science* 22: 53–59.

Akçay, G. and K. Yudakoç. 2008. Removal of nicotine and its pharmaceutical derivatives from aqueous solution by raw bentonite and dodecylammonium-bentonite. *Journal of Scientific & Industrial Research* 67: 451–454.

Akyuz, S. and T. Akyuz. 2018. Investigation of adsorption of 5-chlorouracil onto montmorillonite: An IR and Raman spectroscopic study. *Applied Clay Science* 164: 54–57.

Alkrad, J.A., S. Al-Sammarraie, E.Z. Dahmash, N.A. Qinna, and A.Y. Naser. 2022. Formulation and release kinetics of ibuprofen-bentonite tablets. *Clay Minerals* 57: 172–182.

Almeida, C.A.P., N.A. Debacher, A.J. Downs, L. Cottet, and C.A.D. Mello. 2009. Removal of methylene blue from colored effluents by adsorption on montmorillonite clay. *Journal of Colloid and Interface Science* 332: 46–53.

Anastopoulos, I., A. Mittal, M. Usman et al. 2018. A review on halloysite-based adsorbents to remove pollutants in water and wastewater. *Journal of Molecular Liquids* 269: 855868.

Andrades, M.S., M.S. Rodríguez-Cruz, M.J. Sánchez-Martín, and M. Sánchez-Camazano. 2004. Effect of the modification of natural clay minerals with hexadecylpyridinium cation on the adsorption-desorption of fungicides. *International Journal of Environmental Analytical Chemistry* 84: 133–141.

Anirudhan, T.S. and M. Ramachandran. 2015. Adsorptive removal of basic dyes from aqueous solutions by surfactant modified bentonite clay (organoclay): Kinetic and competitive adsorption isotherm. *Process Safety and Environmental Protection* 95: 215–225.

Arfaioli, P., O.L. Pantani, M. Bosetto, and G.G. Ristori. 1999. Influence of clay minerals and exchangeable cations on the formation of humic-like substances (melanoidins) from D-glucose and L-tyrosine. *Clay Minerals* 34: 487–497.

Aristilde, L., S.M. Galdi, S.E. Kelch, and T.G. Aoki. 2017. Sugar-influenced water diffusion, interaction, and retention in clay interlayer nanopores probed by theoretical simulations and experimental spectroscopies. *Advances in Water Resources* 106: 24–38.

Aristilde, L., B. Lanson, J. Miéhé-Brendlé, C. Marichal, and L. Charlet. 2016. Enhanced interlayer trapping of a tetracycline antibiotic within montmorillonite layers in the presence of Ca and Mg. *Journal of Colloid and Interface Science* 464: 153–159.

Aristilde, L., C. Marichal, J. Miéhé-Brendlé, B. Lanson, and L. Charlet. 2010. Interactions of oxytetracycline with a smectite clay: A spectroscopic study with molecular simulation. *Environmental Science & Technology* 44: 7839–7845.

Avena, M.J., L.E. Valenti, V. Pfaffen, and C.P. De Pauli. 2001. Methylene blue dimerization does not interfere in surface-area measurements of kaolinite and soils. *Clays and Clay Minerals* 49: 168–173.

Awad, A.M., S.M.R. Shaikh, R. Jalab et al. 2019. Adsorption of organic pollutants by natural and modified clays: A comprehensive review. *Separation and Purification Technology* 228: 115719.

Aznar, A.J., B. Casal, E. Ruiz-Hitzky et al. 1992. Adsorption of methylene blue on sepiolite gels: Spectroscopic and rheological studies. *Clay Minerals* 27: 101–108.

Bailey, G.W. and J.L. White. 1970. Factors influencing the adsorption, desorption, and movement of pesticides in soil. *Residue Reviews* 32: 29–92.

Bailey, G.W., J.L. White, and T. Rothberg. 1968. Adsorption of organic herbicides by montmorillonite: Role of pH and chemical character of adsorbate. *Soil Science Society of America Proceedings* 32: 222–234.

Baouch, Z., K.I. Benabadji, and B. Bouras. 2020. Adsorption of different dyes from aqueous solutions using organo-clay composites. *Physical Chemistry Research* 8: 767–787.

Barr, M. and E.S. Arnista. 1957a. Adsorption studies on clays. I. The adsorption of two alkaloids by activated attapulgite, halloysite, and kaolin. *Journal of the American Pharmaceutical Association* 46: 486–489.

Barr, M. and E.S. Arnista. 1957b. Adsorption studies on clays. III. The adsorption of diphtheria toxin by activated attapulgites, halloysite, and kaolin. *Journal of the American Pharmaceutical Association* 46: 493–497.

Barrientos-Velázquez, A.L., A.M. Cardona, L. Liu, T. Phillips, and Y. Deng. 2016. Influence of layer charge origin and layer charge density of smectites on their aflatoxin adsorption. *Applied Clay Science* 131–133: 281–289.

Barshad, I. 1952. Factors affecting the interlayer expansion of vermiculite and montmorillonite with organic substances. *Soil Science Society of America Proceedings* 16: 176–182.

Baskaralingam, P., M. Pulikesi, V. Ramamurthi, and S. Sivanesan. 2007. Modified hectorites and adsorption studies of a reactive dye. *Applied Clay Science* 37: 207–214.

Behera, S.K., S.Y. Oh, and H.S. Park. 2012. Sorptive removal of ibuprofen from water using selected soil minerals and activated carbon. *International Journal of Environmental Science and Technology* 9: 85–94.

Belbel, A., M. Kharroubi, J.-M. Janot et al. 2018. Preparation and characterization of homoionic montmorillonite modified with ionic liquid: Application in dye adsorption. *Colloids and Surfaces A* 558: 219–227.

Bell, D.J. and D.H. Northcote. 1950. Structure of a cell-wall polysaccharide of baker's yeast. *Journal of the Chemical Society* 1944–1947.

Benetoli, L.O.B., H. de Santana, C.T.B.V. Zaia, and D.A.M. Zaia. 2008. Adsorption of nucleic acid bases on clays: An investigation using Langmuir and Freundlich isotherms and FT-IR spectroscopy. *Monatshefte für Chemie* 139: 753–761.

Benetoli, L.O.B., C.M.D. de Souza, K.L. da Silva et al. 2007. Amino acid interaction with and adsorption on clays: FT-IR and Mössbauer spectroscopy and X-ray diffractometry investigations. *Origins of Life and Evolution of Biospheres* 37: 479–493.

Benincasa, E., M.F. Brigatti, C. Lugli, L. Medici, and L. Poppi. 2000. Interaction between glycine and Na-, Ca- and Cu-rich smectites. *Clay Minerals* 35: 635–641.

Benincasa, E., M.F. Brigatti, D. Malferrari, L. Medici, and L. Poppi. 2002. Sorption of Cd-cysteine complexes by kaolinite. *Applied Clay Science* 21: 191–201.

Berez, A., F. Ayari, N. Abidi, G. Schäfer, and M. Trabelsi-Ayadi. 2014. Adsorption-desorption processes of azo dye on natural bentonite: Batch experiments and modelling. *Clay Minerals* 49: 747–763.

Bergmann, K. and C.T. O'Konski. 1963. A spectroscopic study of methylene blue monomer, dimer, and complexes with montmorillonite. *Journal of Physical Chemistry* 67: 2169–2177.

Bhattacharyya, K.G., S. SenGupta, and G.K. Sarma. 2014. Interactions of the dye, rhodamine B with kaolinite and montmorillonite in water. *Applied Clay Science* 99: 7–17.

Bilgiç, C. 2005. Investigation of the factors affecting organic cation adsorption on some silicate minerals. *Journal of Colloid and Interface Science* 281: 33–38.

Block, K.A., A. Trusiak, A. Katz et al. 2015. Exfoliation and intercalation of montmorillonite by small peptides. *Applied Clay Science* 107: 173–181.

Bodenheimer, W. and L. Heller. 1967. Sorption of α-amino acids by copper montmorillonite. *Clay Minerals* 7: 167–176.

Bodenheimer, W. and L. Heller. 1968. Sorption of methylene blue by montmorillonite saturated with different cations. *Israel Journal of Chemistry* 6: 307–314.

Bosetto, M., P. Arfaioli, and P. Fusi. 1993. Interactions of alachlor with homoionic montmorillonites. *Soil Science* 155: 105–113.

Bosetto, M., P. Arfaioli, and O.L. Pantani. 2002. Study of the Maillard reaction products formed by glycine and D-glucose on different mineral substrates. *Clay Minerals* 37: 195–204.

Bowman, B.T., R.S. Adams, Jr., and S.W. Fenton. 1970. Effect of water upon malathion adsorption onto five montmorillonite systems. *Journal of Agricultural and Food Chemistry* 18: 723–727.

Bowman, B.T. and W.W. Sans. 1977. Adsorption of parathion, fenitrothion, methyl parathion, aminoparathion and paraoxon by Na^+, Ca^{2+}, and Fe^{3+} montmorillonite suspensions. *Soil Science Society of America Proceedings* 41: 514–519.

Bradley, W.F. 1945. Molecular associations between montmorillonite and some polyfunctional organic liquids. *Journal of the American Chemical Society* 67: 975–981.

Breen, C. and H. Loughlin. 1994. The competitive adsorption of methylene blue on to Na-montmorillonite from binary solution with *n*-alkyltrimethylammonium surfactants. *Clay Minerals* 29: 775–783.

Brigatti, M.F., C. Lugli, S. Montorsi, and L. Poppi. 1999. Effects of exchange cations and layer-charge location on cysteine retention by smectites. *Clays and Clay Minerals* 47: 664–671.

Briggs, G.G. 1969. Molecular structure of herbicides and their sorption by soils. *Nature* 223: 1288.

Brindley, G.W. and W.F. Moll, Jr. 1965. Complexes of natural and synthetic Ca-montmorillonites with fatty acids (Clay-organic studies IX). *American Mineralogist* 50: 1355–1370.

Brindley, G.W. and S. Ray. 1964. Complexes of Ca-montmorillonite with primary monohydric alcohols (Clay-organic studies VIII). *American Mineralogist* 49: 106–115.

Brindley, G.W. and T.D. Thompson. 1970. Methylene blue absorption by montmorillonites. Determinations of surface areas and exchange capacities with different initial cation saturations (Clay-organic studies XIX). *Israel Journal of Chemistry* 8: 409–415.

Brito, D.F., E.C. da Silva Filho, M.G. Fonseca, and M. Jaber. 2018. Organophilic bentonites obtained by microwave heating as adsorbents for anionic dyes. *Journal of Environmental Chemical Engineering* 6: 7080–7090.

Bu, H., P. Yuan, H. Liu, D. Liu, and X. Zhou. 2017. Thermal decomposition of long-chain fatty acids and its derivative in the presence of montmorillonite. *Journal of Thermal Analysis and Calorimetry* 128: 1661–1669.

Bujdák, J. 2006. Effect of the layer charge of clay minerals on optical properties of organic dyes. A review. *Applied Clay Science* 34: 58–73.

Bujdák, J. 2015. Hybrid systems based on organic dyes and clay minerals: Fundamentals and potential applications. *Clay Minerals* 50: 549–571.

Bujdák, J., M. Janek, J. Madejová, and P. Komadel. 2001. Methylene blue interactions with reduced-charge smectites. *Clays and Clay Minerals* 49: 244–254.

Bujdák, J., N. Iyi, and T. Fujita. 2002. The aggregation of methylene blue in montmorillonite dispersions. *Clay Minerals* 37: 121–133.

Bujdák, J., N. Iyi, Y. Kaneko, and R. Sasai. 2003. Molecular orientation of methylene blue cations adsorbed on clay surfaces. *Clay Minerals* 38: 561–572.

Bujdáková, H., V. Bujdáková, H. Májeková-Koščová et al. 2018. Antimicrobial activity of organoclays based on quaternary alkylammonium and alkylphosphonium surfactants and montmorillonite. *Applied Clay Science* 158: 21–28.

Bulut, Y. and H. Karaer. 2015. Adsorption of methylene blue from aqueous solution by crosslinked chitosan/bentonite composite. *Journal of Dispersion Science and Technology* 36: 61–67.

Cabrera, A., C. Trigo, L. Cox et al. 2012. Sorption of the herbicide aminocyclopyrachlor by cation-modified clay minerals. *European Journal of Soil Science* 63: 694–700.

Calamai, L., O. Pantani, A. Pusino, C. Gessa, and P. Fusi. 1997. Interaction of rimsulfuron with smectites. *Clays and Clay Minerals* 45: 23–27.

Čapková, P., M. Pospíšil, M. Valášková et al. 2006. Structure of montmorillonite cointercalated with stearic acid and octadecylamine: Modeling, diffraction, IR spectroscopy. *Journal of Colloid and Interface Science* 300: 264–269.

Carazo, E., A. Borrego-Sánchez, R. Sánchez-Espejo et al. 2018. Kinetic and thermodynamic assessment on isoniazid/montmorillonite adsorption. *Applied Clay Science* 165: 82–90.

Carneiro, C.E.A., G. Berndt, I.G. de Souza, Jr. et al. 2011. Adsorption of adenine, cytosine, thymine, and uracil on sulfide-modified montmorillonite: FT-IR, Mössbauer and EPR spectroscopy and X-ray diffractometry studies. *Origins of Life and Evolution of Biospheres* 41: 453–468.

Carraro, A., A. De Giacomo, M.L. Giannossi et al. 2014. Clay minerals as adsorbents of aflatoxin M_1 from contaminated milk and effects on milk quality. *Applied Clay Science* 88–89: 92–99.

Carrasquillo, A.J., G.L. Bruland, A.A. Mackay, and D. Vasudevan. 2008. Sorption of ciprofloxacin and oxytetracycline zwitterions to soils and soil minerals: Influence of compound structure. *Environmental Science & Technology* 42: 7634–7642.

Carretero, M.I., C.S.F. Gomez, and F. Tateo. 2006. Clays and human health. In *Handbook of Clay Science*. Developments in Clay Science, Vol. 1, (Eds.) F. Bergaya, B.K.G. Theng, and G. Lagaly, pp. 717–741. Amsterdam, The Netherlands: Elsevier.

Carrizosa, M.J., M.C. Hermosin, W.C. Koskinen, and J. Cornejo. 2004. Interactions of two sulfonylurea herbicides with organoclays. *Clays and Clay Minerals* 52: 643–649.

Carrizosa, M.J., W.C. Koskinen, M.C. Hermosin, and J. Cornejo. 2001. Dicamba adsorption-desorption on organoclays. *Applied Clay Science* 18: 223–231.

Castellini, E., R. Andreoli, G. Malavasi, and A. Pedone. 2008. Deflocculant effects on the surface properties of kaolinite investigated through malachite green adsorption. *Colloids and Surfaces A: Physicochemical and Engineering Aspects* 329: 31–37.

Cavalcanti, G.R.S., M.G. Fonseca, E.C. da Silva Filho, and M. Jaber. 2019. Thiabendazole/bentonite hybrids as controlled release systems. *Colloids and Surfaces B: Biointerfaces* 176: 249–255.

Celis, R., J. Cornejo, M.C. Hermosin, and W.C. Koskinen. 1997. Sorption-desorption of atrazine and simazine by model soil colloidal components. *Soil Science Society of America Journal* 61: 436–443.

Celis, R., M.C. Hermosin, M.J. Carrizosa, and J. Cornejo. 2002. Inorganic and organic clays as carriers for controlled release of the herbicide hexazinone. *Journal of Agricultural and Food Chemistry* 50: 2324–2330.

Celis, R., C. Trigo, G. Facenda, M.D.C. Hermosin, and J. Cornejo. 2007. Selective modification of clay minerals for the adsorption of herbicides widely used in olive groves. *Journal of Agricultural and Food Chemistry* 55: 6650–6658.

Cenens, J. and R.A. Schoonheydt. 1988. Visible spectroscopy of methylene blue on hectorite, Laponite B, and Barasym in aqueous suspension. *Clays and Clay Minerals* 36: 214–224.

Chahardahmasoumi, S., M.N. Sarvi, and S.A.H. Jalali. 2019. Modified montmorillonite nanosheets as a nanocarrier with smart pH-responsive control on the antimicrobial activity of tetracycline upon release. *Applied Clay Science* 178: 105135.

Chang, J., J. Ma, Q. Ma et al. 2016. Adsorption of methylene blue onto Fe_3O_4/activated montmorillonite nanocomposite. *Applied Clay Science* 119: 132–140.

Chang, P.-H., W.-T. Jiang, Z, Li et al. 2014. Mechanism of amitriptyline adsorption on Ca-montmorillonite. *Journal of Hazardous Materials* 277: 44–52.

Chang, P.-H., W.-T. Jiang, B. Sarkar, W. Wang, and Z. Li. 2019. The triple mechanisms of atenolol adsorption on Ca-montmorillonite: Implication on pharmaceutical wastewater treatment. *Materials* 12: 2858. https://doi.org/10.3390/ma12182858.

Chang, P.-H., Z. Li, W.-T. Jiang, and J.-S. Jean. 2009. Adsorption and intercalation of tetracycline by swelling clay minerals. *Applied Clay Science* 46: 27–36.

Chappell, M.A., D.A. Laird, M.L. Thompson et al. 2005. Influence of smectite hydration and swelling on atrazine sorption behavior. *Environmental Science & Technology* 39: 3150–3156.

Chassin, P. 1969. Adsorption du glycocolle par la montmorillonite. *Bulletin du Groupe français des Argiles* 21: 71–88.

Chauhan, M., V.K. Saini, and S. Suthar. 2020. Ti-pillared montmorillonite clay for adsorptive removal of amoxicillin, imipramine, diclofenac-sodium, and paracetamol from water. *Journal of Hazardous Materials* 399: 122832.

Chen, D., J. Chen, X. Luan, H. Ji, and Z. Xia. 2011. Characterization of anion-cationic surfactants modified montmorillonite and its application for the removal of methyl orange. *Chemical Engineering Journal* 171: 1150–1158.

Chen, G., J. Pan, B. Han, and H. Yan. 1999. Adsorption of methylene blue on montmorillonite. *Journal of Dispersion Science and Technology* 20: 1179–1187.

Chen, H., X. He, X. Rong et al. 2009. Adsorption and biodegradation of carbaryl on montmorillonite, kaolinite and goethite. *Applied Clay Science* 46: 102–108.

Chen, H. and J. Zhao. 2009. Adsorption study for removal of Congo red anionic dye using organo-attapulgite. *Adsorption* 15: 381–389.

Chen, Y., A. Zhou, B. Liu, and J. Liang. 2010. Tramadol hydrochloride/montmorillonite composite: Preparation and controlled drug release. *Applied Clay Science* 49: 108–112.

Chicinaş, R.P., H. Bedelean, R. Stefa, and A. Măicăneanu. 2018. Ability of a montmorillonite clay to interact with cationic and anionic dyes in aqueous solutions. *Journal of Molecular Structure* 1154: 187–195.

Cloos, P., B. Calicis, J.J. Fripiat, and K. Makay. 1966. Adsorption of amino acids and peptides by montmorillonite. 1. Chemical and X-ray diffraction studies. *Proceedings of the International Clay Conference, Jerusalem* 1: 223–232.

Cohen, R. and S. Yariv. 1984. Metachromasy in clay minerals: Sorption of acridine orange by montmorillonite. *Journal of the Chemical Society, Faraday Transactions 1* 80: 1705–1715.

Cornejo, J., R. Celis, I. Pavlovic, and M.A. Ulibarri. 2008. Interactions of pesticides with clays and layered double hydroxides: A review. *Clay Minerals* 43: 155–175.

Cox, L., R. Celis, M.C. Hermosin, and J. Cornejo. 2000. Natural soil colloids to retard simazine and 2,4-D leaching in soil. *Journal of Agricultural and Food Chemistry* 48: 93–99.

Cox, L., M.C. Hermosin, and J. Cornejo. 1994. Interactions of methomyl with montmorillonites. *Clay Minerals* 29: 767–774.

Cruz, M., J.L. White, and J.D. Russell. 1968. Montmorillonite-*s*-triazine interactions. *Israel Journal of Chemistry* 6: 315–323.

Cruz-Guzmán, M., R. Celis, M.C. Hermosín, and J. Cornejo. 2004. Adsorption of the herbicide simazine by montmorillonite modified with natural organic cations. *Environmental Science & Technology* 38: 180–186.

Cruz-Guzmán, M., R. Celis, M.C. Hermosín, W.C. Koskinen, and J. Cornejo. 2005. Adsorption of pesticides from water by functionalized organobentonites. *Journal of Agricultural and Food Chemistry* 53: 7502–7511.

Cunha, R., P. Trigueiro, M.D.M. Orta Cuevas et al. 2023. The stability of anthocyanins and their derivatives through clay minerals: Revising the current literature. *Minerals* 13: 268. https://doi.org/10.3390/min13020268.

Czímerová, A., J. Bujdák, and A. Gáplovský. 2004a. The aggregation of thionine and methylene blue dye in smectite dispersion. *Colloids and Surfaces A: Physicochemical and Engineering Aspects* 243: 89–96.

Czímerová, A., L. Jankovič, and J. Bujdák. 2004b. Effect of the exchangeable cations on the spectral properties of methylene blue in clay dispersions. *Journal of Colloid and Interface Science* 274: 126–132.

Damonte, M., R.M. Torres Sánchez, and M. dos Santos Afonso. 2007. Some aspects of the glyphosate adsorption on montmorillonite and its calcined form. *Applied Clay Science* 36: 86–94.

Das, A., K.W. Stöckelhuber, R. Jurk, D. Jehnichen, and G. Heinrich. 2011. A general approach to rubber-montmorillonite nanocomposites: Intercalation of stearic acid. *Applied Clay Science* 51: 117–125.

D'Ascanio, V., D. Greco, E. Menicagli et al. 2019. The role of geological origin of smectites and of their physico-chemical properties on aflatoxin adsorption. *Applied Clay Science* 181: 105209.

Dashman, T. and G. Stotzky. 1982. Adsorption and binding of amino acids on homoionic montmorillonite and kaolinite. *Soil Biology & Biochemistry* 14: 447–456.

Dashman, T. and G. Stotzky. 1985. Physical properties of homoionic montmorillonite and kaolinite complexes with amino acids and peptides. *Soil Biology & Biochemistry* 17:189–195.

Davies, J.E.D. and N. Jabeen. 2002. The adsorption of herbicides and pesticides on clay minerals and soils. Part 1. Isoproturon. *Journal of Inclusion Phenomena and Macrocyclic Chemistry* 43: 329–336.

Davies, J.E.D. and N. Jabeen. 2003. The adsorption of herbicides and pesticides on clay minerals and soils. Part 2. Atrazine. *Journal of Inclusion Phenomena and Macrocyclic Chemistry* 46: 57–64.

Davis, G.A. and W.E. Worrall. 1971. Adsorption of water by clays. *Transactions of the British Ceramic Society* 70: 71–75.

Del Angel, C., A.B. Morales, F. Navarro-Pardo et al. 2015. Mechanical and rheological properties of polypropylene/bentonite composites with stearic acid as an interface modifier. *Journal of Applied Polymer Science* 132: 42264.

Deng, Y., A.L. Barrientos Velázquez, F. Billes, and J.B. Dixon. 2010. Bonding mechanisms between aflatoxin B_1 and smectite. *Applied Clay Science* 50: 92–98.

Deng, Y. and M. Szczerba. 2011. Computational evaluation of bonding between aflatoxin B_1 and smectite. *Applied Clay Science* 54: 26–33.

Dening, T.J., P. Joyce, M. Kovalainen, H. Gustafsson, and C.A. Prestidge. 2019. Spray dried smectite clay particles as a novel treatment against obesity. *Pharmaceutical Research* 36: 21. https://doi.org/10.1007/11095-018-2552-9.

De Oliveira, T., E. Fernandez, L. Fougère et al. 2018. Competitive association of antibiotics with a clay mineral and organoclay derivatives as a control of their lifetimes in the environment. *ACS Omega* 3: 15332–15342.

De Oliveira, T. and R. Guégan. 2016. Coupled organoclay/micelle action for the adsorption of diclofenac. *Environmental Science & Technology* 50: 10209–10215.

De Oliveira, T., R. Guégan, T. Thiebault et al. 2017. Adsorption of diclofenac onto organoclays: Effects of surfactant and environmental (pH and temperature) conditions. *Journal of Hazardous Materials* 323: 558–566.

de Queiroga, L.N.F., P.K. Soares, M.G. Fonseca, and F.J.V.E. de Oliveira. 2016. Experimental design investigation for vermiculite modification: Intercalation reaction and application for dye removal. *Applied Clay Science* 126: 113–121.

Di Leo, P. 2000. A nuclear magnetic resonance (NMR) and Fourier-transform infrared (FTIR) study of glycine speciation on a Cd-rich montmorillonite. *Clays and Clay Minerals* 48: 495–502.

Dixon, J.B., I. Kannewischer, M.G. Tenorio Arvide, and A.L. Barrientos Velázquez. 2008. Aflatoxin sequestration in animal feeds by quality-labeled smectite clays: An introductory plan. *Applied Clay Science* 40: 201–208.

Dong, F., Y. Guo, M. Liu, L. Zhou, Q. Zhou, and H. Li. 2018. Spectroscopic evidence and molecular simulation investigation of the bonding interaction between lysine and montmorillonite: Implications for the distribution of soil organic nitrogen. *Applied Clay Science* 159: 3–9.

dos Santos, E.C., W.P. Gates, L. Michels et al. 2018. The pH influence on the intercalation of the bioactive agent ciprofloxacin in fluorohectorite. *Applied Clay Science* 166: 288–298.

Dotto, G.L., F.K. Rodrigues, E.H. Tanabe et al. 2016. Development of chitosan/bentonite hybrid composite to remove hazardous anionic and cationic dyes from colored effluents. *Journal of Environmental Chemical Engineering* 4: 3230–3239.

Dubbin, W.E., J.P. Vetterlein, and J.L. Jonsson. 2014. Fatty acids promote fulvic acid intercalation by montmorillonite. *Applied Clay Science* 97–98: 53–61.

Duke, S.O. and S.B. Powles 2008. Glyphosate: A once-in-century herbicide. *Pest Management Science* 64: 319–325.

Duman, O., S. Tunç, and T.G. Polat. 2015. Determination of adsorptive properties of expanded vermiculite for the removal of C.I. basic red 9 from aqueous solution: Kinetic, isotherm and thermodynamic studies. *Applied Clay Science* 109–110: 22–32.

El Haouti, R., H. Ouachtak, A. El Guerdaoui et al. 2019. Cationic dyes adsorption by Na-montmorillonite nano clay: Experimental study combined with theoretical investigation using DFT-based descriptors and molecular dynamics simulations. *Journal of Molecular Liquids* 290: 111139.

El-Nahhal, Y., T. Undabeytia, T. Polubesova, Y.G. Mishael, S. Nir, and B. Rubin. 2001. Organo-clay formulations of pesticides: Reduced leaching and photodegradation. *Applied Clay Science* 18: 309–326.

Elsherbiny, A.S. 2013. Adsorption kinetics and mechanism of acid dye onto montmorillonite from aqueous solutions: Stopped-flow measurements. *Applied Clay Science* 83–84: 56–62.

Ennajih, H., H. Gueddar, A.E. Kadib, R. Bouhfid, M. Bousmina, and E.M. Essassi. 2012. Intercalation of nickel and cobalt thiabendazole complexes into montmorillonite. *Applied Clay Science* 65–66: 139–142.

Essington, M.E., J. Lee, and Y. Seo. 2010. Adsorption of antibiotics by montmorillonite and kaolinite. *Soil Science Society of America Journal* 74: 1577–1588.

Evcim, N. and M. Barr. 1955. Adsorption of some alkaloids by different clays. *Journal of the American Pharmaceutical Association* 44: 570–573.

Faghihian, H. and M. Nejati-Yazdinejad. 2009. Equilibrium study of adsorption of L-cysteine by natural bentonite. *Clay Minerals* 44: 125–133.

Fan, H., L. Zhou, X. Jiang, Q. Huang, and W. Lang. 2014. Adsorption of Cu^{2+} and methylene blue on dodecyl sulfobetaine surfactant-modified montmorillonite. *Applied Clay Science* 95: 150–158.

Fang, Y., A. Zhou, W. Yang et al. 2018. Complex formation via hydrogen bonding between rhodamine B and montmorillonite in aqueous solution. *Scientific Reports* 8: 229 https://doi.org/10.1038/s41598-017-18057-8.

Fashina, B. and Y. Deng. 2023. The effect of layer charge origin and density, type of interlayer cations, and occupancy of the octahedral sheet of smectites on the adsorption of pyocyanin. *Applied Clay Science* 237: 106884.

Feddal, I., A. Ramdani, S. Taleb, E.M. Gaigneaux, N. Batis, and N. Ghaffour. 2014. Adsorption capacity of methylene blue, an organic pollutant, by montmorillonite clay. *Desalination and Water Treatment* 52: 2654–2661.

Fejér, I., M. Kata, I. Erõs, and I. Dékány. 2002. Interaction of monovalent cationic drugs with montmorillonite. *Colloid and Polymer Science* 280: 372–379.

Ferris, J.P., G. Ertem, and V.K. Agarwal. 1989. The adsorption of nucleotides and polynucleotides on montmorillonite clay. *Origins of Life and Evolution of the Biosphere* 19: 153–164.

Feuillie, C., I. Daniel, L.J. Michot, and U. Pedreira-Segade. 2013. Adsorption of nucleotides onto Fe-Mg-Al rich swelling clays. *Geochimica et Cosmochimica Acta* 120: 97–108.

Figueroa, R.A., A. Leonard, and A.A. MacKay. 2004. Modeling tetracycline antibiotic sorption to clays. *Environmental Science & Technology* 38: 476–483.

Fizir, M., P. Dramou, N.S. Dahiru, W. Ruya, T. Huang, and H. He. 2018. Halloysite nanotubes in analytical sciences and in drug delivery: A review. *Microchimica Acta* 185: 389. https://doi.org/10.1007/s00604-018-2908-1.

França, D.B., P. Trigueiro, E.C. Silva Filho, M.G. Fonseca, and M. Jaber. 2020. Monitoring diclofenac adsorption by organophilic alkylpyridinium bentonites. *Chemosphere* 242: 125109.

Franz, J.E., M.K. Mao, and J.A. Sikorski. 1997. *Glyphosate: A Unique Global Herbicide.* ACS Monograph 189, p. 653. Washington, DC: American Chemical Society.

Freitas, A.F., M.F. Mendes, and G.L.V. Coelho. 2007. Thermodynamic study of fatty acids adsorption on different adsorbents. *Journal of Chemical Thermodynamics* 39: 1027–1037.

Frissel, M.J. 1961. *The Adsorption of Some Organic Compounds, Especially Herbicides, on Clay Minerals.* Ph.D. Thesis, Verslagen Landbouwkundige Onderzoekingen NR 67.3, Wageningen, The Netherlands.

Frissel, M.J. and G.H. Bolt. 1962. Interaction between certain ionizable organic compounds (herbicides) and clay minerals. *Soil Science* 94: 284–291.

Fruhstorfer, P., R.J. Schneider, L. Weil, and R. Niessner. 1993. Factors influencing the adsorption of atrazine on montmorillonitic and kaolinitic clays. *Science of the Total Environment* 138: 317–328.

Fu, L., B.M. Weckhuysen, A.A. Verberckmoes, and R.A. Schoonheydt. 1996. Clay intercalated Cu(II) amino acid complexes: Synthesis, spectroscopy and catalysis. *Clay Minerals* 31: 491–500.

Fusi, P., M. Franci, and M. Bosetto. 1988. Interaction of fluazifop-butyl and fluazifop with smectites. *Applied Clay Science* 3: 63–73.

Fusi, P., M. Franci, and G.G. Ristori. 1989. Interactions of isoxaben with montmorillonite. *Applied Clay Science* 4: 235–245.

Fusi, P., G.G. Ristori, and M. Franci. 1986. Interactions of carbaryl with homoionic montmorillonite. *Applied Clay Science* 1: 375–383.

Gahlot, R., K. Taki, and M. Kumar. 2020. Efficacy of nanoclays as the potential adsorbent for dyes and metal removal from the wastewater: A review. *Environmental Nanotechnology, Monitoring & Management* 14: 100339.

Gamba, M., F.M. Flores, J. Madejová, and R.M. Torres Sánchez. 2015. Comparison of imazalil removal onto montmorillonite and nanomontmorillonite and adsorption surface sites involved: An approach for agricultural wastewater treatment. *Industrial & Engineering Chemistry Research* 54: 1529–1538.

Gamba, M., P. Kovář, M. Pospíšil, and R.M. Torres Sánchez. 2017. Insight into thiabendazole interaction with montmorillonite and organically modified montmorillonites. *Applied Clay Science* 137: 59–68.

Gámiz, B., R. Celis, M.C. Hermosin, and J. Cornejo. 2010. Organoclays as soil amendments to increase the efficacy and reduce the environmental impact of the herbicide fluometuron in agricultural soils. *Journal of Agricultural and Food Chemistry* 58: 7893–7901.

Gámiz, B., R. Celis, M.C. Hermosin, J. Cornejo, and C.T. Johnston. 2012. Preparation and characterization of spermine-exchanged montmorillonite and interaction with the herbicide fluometuron. *Applied Clay Science* 58: 8–15.

Gámiz, B., M.C. Hermosin, J. Cornejo, and R. Celis. 2015. Hexadimethrine-montmorillonite nanocomposite: Characterization and application as a pesticide adsorbent. *Applied Clay Science* 332: 606–613.

Gan, F., X. Hang, Q. Huang, and Y. Deng. 2019. Assessing and modifying China bentonites for aflatoxin adsorption. *Applied Clay Science* 168: 348–354.

Garfinkel-Shweky, D. and S. Yariv. 1999. Metachromasy in clay dye systems: The adsorption of acridine orange by Na-beidellite. *Clay Minerals* 34: 459–467.

Gautier, M., F. Muller, J.-M. Bény, L. Le Forestier, P. Albéric, and P. Baillif. 2009. Interactions of ammonium smectite with low molecular weight carboxylic acids. *Clay Minerals* 44: 207–219.

Gemeay, A.H. 2002. Adsorption characteristics and the kinetics of the cation exchange of rhodamine-6G with Na^+-montmorillonite. *Journal of Colloid and Interface Science* 251: 235–241.

Gereli, G., Y. Seki, I.M. Kuşoğlu, and K. Yurdakoç. 2006. Equilibrium and kinetics for the sorption of promethazine hydrochloride onto K10 montmorillonite. *Journal of Colloid and Interface Science* 299: 155–162.

Gerstl, Z. and B. Yaron. 1981. Attapulgite-pesticide interactions. *Residue Reviews* 78: 69–99.

Gessner, F., C.C. Schmitt, and M.G. Neumann. 1994. Time-dependent spectrophotometric study of the interaction of basic dyes with clays. 1. Methylene blue and neutral red on montmorillonite and hectorite. *Langmuir* 10: 3749–3753.

Ghadiri, M., H. Hau, W. Chrzanowski, H. Agus, and R. Rohanizadeh. 2013. Laponite clay as a carrier for *in situ* delivery of tetracycline. *RSC Advances* 3: 20193–20201.

Ghemit, R., A. Makhloufi, N. Djebri, A. Flilissa, L. Zeroual, and M. Boutahala. 2019. Adsorptive removal of diclofenac and ibuprofen from aqueous solution by organobentonite: Study in single and binary systems. *Groundwater for Sustainable Development* 8: 520–529.

Giles, C.H., T.H. MacEwan, S.N. Nakhwa, and D. Smith. 1960. Studies in adsorption. XI. A system of classification of solution adsorption isotherms and its use in diagnosis of adsorption mechanisms and in measurement of specific surface areas of solids. *Journal of the Chemical Society* 3973–3993.

Glass, R.L. 1987. Adsorption of glyphosate by soils and clay minerals. *Journal of Agricultural and Food Chemistry* 35: 497–500.

Gomes, C., M. Rautureau, J. Poustis, and J. Gomes. 2021. Benefits and risks of clays and clay minerals to human health from ancestral to current times: A synoptic overview. *Clays and Clay Minerals* 69: 612–632.

Gonzalez, J.M. and D.A. Laird. 2004. Role of smectites and Al-substituted goethites in the catalytic condensation of arginine and glucose. *Clays and Clay Minerals* 52: 443–450.

Gonzalez, L., P. Lafleur, T. Lozano et al. 2014. Mechanical and thermal properties of polypropylene/montmorillonite nanocomposites using stearic acid as both an interface and a clay surface modifier. *Polymer Composites* 35: 1–9.

Gottlieb, D. and P. Siminoff. 1952. The production and role of antibiotics in the soil. II. Chloromycetin. *Phytopathology* 42: 91–97.

Graf, G. and G. Lagaly. 1980. Interaction of clay minerals with adenosine-5-phosphates. *Clays and Clay Minerals* 28: 12–18.

Grauer, Z., D. Avnir, and S. Yariv. 1984. Adsorption characteristics of rhodamine 6G on montmorillonite and Laponite, elucidated from electronic absorption and emission spectra. *Canadian Journal of Chemistry* 62: 1889–1894.

Grauer, Z., A.B. Malter, S. Yariv, and D. Avnir. 1987. Sorption of rhodamine B by montmorillonite and Laponite. *Colloids and Surfaces* 25: 41–65.

Greenland, D.J. 1956a. The adsorption of sugars by montmorillonite. I. X-ray studies. *Journal of Soil Science* 7: 319–328.

Greenland, D.J. 1956b. The adsorption of sugars by montmorillonite. II. Chemical studies. *Journal of Soil Science* 7: 329–334.

Greenland, D.J., R.H. Laby, and J.P. Quirk. 1962. Adsorption of glycine and its di-, tri-, and tetra-peptides by montmorillonite. *Transactions of the Faraday Society* 58: 829–841.

Greenland, D.J., R.H. Laby, and J.P. Quirk. 1965a. Adsorption of amino acids and peptides by montmorillonite and illite. Part 1. Cation exchange and proton transfer. *Transactions of the Faraday Society* 61: 2013–2023.

Greenland, D.J., R.H. Laby, and J.P. Quirk. 1965b. Adsorption of amino acids and peptides by montmorillonite and illite. Part 2. Physical adsorption. *Transactions of the Faraday Society* 61: 2024–2035.

Grundgeiger, E., Y.H. Lim., R.L. Frost, G.A. Ayoko, and Y. Xi. 2015. Application of organo-beidellites for the adsorption of atrazine. *Applied Clay Science* 105–106: 252–258.

Gu, Z., M. Gao, Z. Luo, G. Xue, L. Lu, and Y. Liu. 2014. Gemini surfactant modified montmorillonite as highly efficient adsorbent for anionic dyes. *Separation Science and Technology* 49: 2878–2889.

Günay, A., B. Ersoy, S. Dikmen, and A. Evcin. 2013. Investigation of equilibrium, kinetic, thermodynamic and mechanism of basic blue 16 adsorption by montmorillonitic clay. *Adsorption* 19: 757–768.

Gürses, A., Ç. Doğar, M. Yalçın, M. Açıkyız, R. Bayrak, and S. Karaca. 2006. The adsorption kinetics of the cationic dye, methylene blue, onto clay. *Journal of Hazardous Materials* B131: 217–228.

Gururani, K., C.K. Pant, and H.D. Pathak. 2012. Surface interaction of adenine on montmorillonite clay in presence and absence of divalent cations in relevance to chemical evolution. *International Journal of Scientific and Technological Research* 1: 106–109.

Gutman, R., M. Rauch, A. Neuman et al. 2020. Sepiolite clay attenuates the development of hypercholesterolemia and obesity in mice fed a high-fat high-cholesterol diet. *Journal of Medicinal Food* 23: 289–296.

Haderlein, S.B., K.W. Weissmahr, and R.P. Scharzenbach. 1996. Specific adsorption of nitroaromatic explosives and pesticides to clay minerals. *Environmental Science & Technology* 30: 612–622.

Hajjaji, M. and A. Alami. 2009. Influence of operating conditions on methylene blue uptake by a smectite rich clay fraction. *Applied Clay Science* 44: 127–129.

Hamilton, A.R., M. Roberts, G.A. Hutcheon, and E.E. Gaskell. 2019. Formulation and antibacterial properties of clay-mineral-tetracycline and -doxycycline composites. *Applied Clay Science* 179: 105148.

Hance, R.J. 1969. Influence of pH, exchangeable cation and the presence of organic matter on the adsorption of some herbicides by montmorillonite. *Canadian Journal of Soil Science* 49: 357–364.

Hang, P.T. and G.W. Brindley. 1970. Methylene blue absorption by clay minerals. Determination of surface areas and cation exchange capacities (Clay-organic studies XVIII). *Clays and Clay Minerals* 18: 203–212.

Hao, J., M. Mokhtari, U. Pedreira-Segade, L.J. Michot, and I. Daniel. 2019. Transition metals enhance the adsorption of nucleotides onto clays: Implications for the origin of life. *ACS Earth and Space Chemistry* 3: 100–119.

Haque, R., S. Lilley, and W.R. Coshow. 1970. Mechanism of adsorption of diquat and paraquat on montmorillonite surface. *Journal of Colloid and Interface Science* 33: 185–188.

Haque, R. and R. Sexton. 1968. Kinetic and equilibrium study of the adsorption of 2,4-dichlorophenoxy acetic acid on some surfaces. *Journal of Colloid and Interface Science* 27: 818–827.

Harris, R.G., B.B. Johnson, and J.D. Wells. 2006. Studies on the adsorption of dyes to kaolinite. *Clays and Clay Minerals* 54: 435–448.

Hashizume, H. and K. Fujii. 2016. Relationship between adsorption and desorption of adenine by montmorillonite. *Clay Science* 20: 21–25.

Hashizume, H. and B.K.G. Theng. 2007. Adenine, adenosine, ribose and 5'-AMP adsorption to allophane. *Clays and Clay Minerals* 55: 599–605.

Hashizume, H., S. van der Gaast, and B.K.G. Theng. 2010. Adsorption of adenine, cytosine, uracil, ribose, and phosphate by Mg-exchanged montmorillonite. *Clay Minerals* 45: 469–475.

Hassani, A., A. Khataee, S. Karaca, M. Karaca, and M. Kiranşan. 2015. Adsorption of two cationic dyes from water with modified nanoclay: A comparative study by using central composite design. *Journal of Environmental Chemical Engineering* 3: 2738–2749.

Hayes, M.H.B., M.E. Pick, and B.A. Toms. 1978. The influence of organocation structure on the adsorption of mono- and bipyridinium cations by expanding lattice clay minerals: I. Adsorption by Na^+-montmorillonite. *Journal of Colloid and Interface Science* 65: 254–265.

He, H., D. Yang, P. Yuan, W. Shen, and R.L. Frost. 2006. A novel organoclay with antibacterial activity prepared from montmorillonite and Chlorhexidini Acetas. *Journal of Colloid and Interface Science* 297: 235–243.
Hearon, S.E., M. Wang, and T.D. Phillips. 2019. Strong adsorption of dieldrin by parent and processed montmorillonite clays. *Environmental Toxicology and Chemistry* 39: 517–525.
Hebert, J., L. Wang, X. Wang et al. 2021. Mechanisms of safranin O interaction with 1:1 layered clay minerals. *Separation Science and Technology* 56: 1985–1995.
Hedges, J.I. 1978. The formation and clay mineral reactions of melanoidins. *Geochimica et Cosmochimica Acta* 42: 69–76.
Hedges, J.I. and P.E. Hare. 1987. Amino acid adsorption by clay minerals in distilled water. *Geochimica et Cosmochimica Acta* 51: 255–259.
Hegyesi, N., R.T. Vad, and B. Pukánszky. 2017. Determination of the specific surface area of layered silicates my methylene blue adsorption: The role of structure, pH and layer charge. *Applied Clay Science* 146: 50–55.
Hendricks, S.B. 1941. Base exchange of the clay mineral montmorillonite for organic cations and its dependence upon adsorption due to van der Waals forces. *Journal of Physical Chemistry* 45: 65–81.
Hermosin, M.C. and J.L. Perez Rodriguez. 1981. Interaction of chlordimeform with clay minerals. *Clays and Clay Minerals* 29: 143–152.
Hernandez, Y., T. Lozano, A.B. Morales et al. 2016. Improvement of toughness properties of polypropylene filled with nanobentonite using stearic acid as interface modifier. *Journal of Composite Materials* 0: 1–8. https://doi.org/10.1177/0021998316644852.
Herrera, P., R.C. Burghardt, and T.D. Phillips. 2000. Adsorption of *salmonella enteritidis* by cetylpyridinium-exchanged montmorillonite clays. *Veterinary Microbiology* 74: 259–272.
Herwig, U., E. Klumpp, H.-D. Narres, and M.J. Schwuger. 2001. Physicochemical interactions between atrazine and clay minerals. *Applied Clay Science* 18: 211–222.
Holešová, S., M. Samlíková, E. Pazdziora, and M. Valášková. 2013. Antibacterial activity of organomontmorillonites and organovermiculites prepared using chlorhexidine diacetate. *Applied Clay Science* 83–84: 17–23.
Hosseini, F., F. Hosseini, S.M. Jafari, and A. Taheri. 2018. Bentonite nanoclay-based drug-delivery systems for treating melanoma. *Clay Minerals* 53: 53–63.
Hu, Q.H., S.Z. Qiao, F. Haghseresht, M.A. Wilson, and G.Q. Lu. 2006. Adsorption study for removal of basic red dye using bentonite. *Industrial & Engineering Chemistry Research* 45: 733–738.
Huang, A., Z. Huang, Y. Dong et al. 2013. Controlled release of phoxim from organobentonite based formulation. *Applied Clay Science* 80–81: 63–68.
Huang, P., A. Kazlauciunas, R. Menzel, and L. Lin. 2017. Determining the mechanism and efficiency of industrial dye adsorption through facile structural control of organo-montmorillonite adsorbents. *ACS Applied Materials & Interfaces* 9: 26383–26391.
Iannuccelli, V., E. Maretti, A. Bellini et al. 2018. Organo-modified bentonite for gentamicin topical application: Interlayer structure and *in vivo* permeation. *Applied Clay Science* 158: 158–168.
Ibrahim, Z., Y. Mohsen, J. Toufaily et al. 2020. Modeling of adsorption isotherms and competitive adsorption breakthroughs of nicotine/pyridine removal from aqueous solution by activated montmorillonite. *Journal of Water and Environmental Nanotechnology* 5: 114–128.
Ikhsan, J., M.J. Angove, J.D. Wells, and B.B. Johnson. 2005a. Surface complexation modeling of the sorption of 2-, 3-, and 4-aminopyridine by montmorillonite. *Journal of Colloid and Interface Science* 284: 383–392.
Ikhsan, J., J.D. Wells, B.B. Johnson, and M.J. Angove. 2005b. Sorption of 3-amino-1,2,4-triazole and Zn(II) onto montmorillonite. *Clays and Clay Minerals* 53: 137–146.
Ikhtiyarova, G.A., A.S. Özcan, Ö. Gök, and A. Özcan. 2012. Characterization of natural- and organo-bentonite by XRD, SEM, FT-IR and thermal analysis techniques and its adsorption behaviour in aqueous solutions. *Clay Minerals* 47: 31–44.
Ilari, R., M. Etcheverry, C.V. Waiman, and G.P. Zanini. 2021. A simple cation exchange model to assess the competitive adsorption between the herbicide paraquat and the biocide benzalkonium chloride. *Colloids and Surfaces A: Physicochemical and Engineering Aspects* 611: 125797.
Ilić, I., N. Jović-Jovičić, P. Banković et al. 2019. Adsorption of nicotine from aqueous solutions on montmorillonite and acid-modified montmorillonite. *Science of Sintering* 51: 93–100.
Intachai, S., C. Suppaso, and N. Khaorapapong. 2021. A novel process for intercalating alkylammonium ions in a Thai bentonite and its effect on adsorption performance. *Clays and Clay Minerals* 69: 477–488.
Ismadji, S., F.E. Soetaredjo, and A. Ayucitra. 2015. *Clay Minerals for Environmental Remediation*. Springer Briefs in Green Chemistry for Sustainability. Cham: Springer. https://doi.org/10.1007/978-3-319-16712-1-2.
Iwasaki, M., M. Kita, K. Ito, A. Kohno, and K. Fukunishi. 2000. Intercalation characteristics of 1,1'-diethyl-2,2'-cyanine and other cationic dyes in synthetic saponite: Orientation in the interlayer. *Clays and Clay Minerals* 48: 392–399.

Jacobs, K.Y. and R.A. Schoonheydt. 1999. Spectroscopy of methylene blue-smectite suspensions. *Journal of Colloid and Interface Science* 220: 103–111.

Jalil, M.E.R., M. Baschini, and K. Sapag. 2015. Influence of pH and antibiotic solubility on the removal of ciprofloxacin from aqueous media using montmorillonite. *Applied Clay Science* 114: 69–76.

Jalil, M.E.R., R.S. Vieira, D. Azevedo, M. Baschini, and K. Sapag. 2013. Improvement in the adsorption of thiabendazole by using aluminum pillared clays. *Applied Clay Science* 71: 55–63.

Jang, S.D. and R.A. Condrate, Sr. 1972a. The infrared spectra of valine adsorbed on Cu-montmorillonite. *American Mineralogist* 57: 494–498.

Jang, S.D. and R.A. Condrate, Sr. 1972b. The I.R. spectra of lysine adsorbed on several cation-substituted montmorillonites. *Clays and Clay Minerals* 20: 79–82.

Jankovič, L., P. Škorňa, D.M. Rodriguez, E. Scholtzová, and D. Tunega. 2021. Preparation, characterization and adsorption properties of tetraalkylphosphonium organobeidellites. *Applied Clay Science* 204: 105989.

Jayrajsinh, S., G. Shankar, Y.K. Agrawal, and L. Bakre. 2017. Montmorillonite nanoclay as a multifaceted drug-delivery carrier: A review. *Journal of Drug Delivery Science and Technology* 39: 200–209.

Jepson, W.B. and J.R. Williams. 1972. The adsorption of water by clays. *Clay Minerals* 9: 275–279.

Jiang, Y.-X., H.-J. Xu, D.-W. Liang, and Z.-F. Tong. 2008. Adsorption of basic violet 14 from aqueous solution on bentonite. *Comptes Rendus Chimie* 11: 125–129.

Jiang, W.-T., C.-J. Wang, and Z. Li. 2013. Intercalation of ciprofloxacin accompanied by dehydration in rectorite. *Applied Clay Science* 74: 74–80.

Jin, X., S. Zha, S. Li, and Z. Chen. 2014. Simultaneous removal of mixed contaminants by organoclays – amoxicillin and Cu(II) from aqueous solution. *Applied Clay Science* 102: 196–201.

Johns, W.D. 1979. Clay mineral catalysis and petroleum generation. *Annual Review of Earth and Planetary Sciences* 7: 1983–1988.

Joshi, G.V., H.A. Patel, B.D. Kevadiya, and H.C. Bajaj. 2009a. Montmorillonite intercalated with vitamin B_1 as drug carrier. *Applied Clay Science* 45: 248–253.

Joshi, G.V., B.D. Kevadiya, H.A. Patel, H.C. Bajaj, and R.V. Jasra. 2009b. Montmorillonite as a drug delivery system: Intercalation and in vitro release of timolol maleate. *International Journal of Pharmaceutics* 374: 53–57.

Jović-Jovićić, N., A. Milutinović-Nicolić, I. Gržetić, and D. Jovanović. 2008. Organobentonite as efficient textile dye sorbent. *Chemical Engineering and Technology* 31: 567–574.

Joyce, P., T.J. Dening, T.R. Meola et al. 2020. Contrasting anti-obesity effects of smectite clays and mesoporous silica in Sprague-Dawley rats. *ACS Applied Bio Materials* 3: 7779–7788.

Jurg, J.W. and E. Eisma. 1964. Petroleum hydrocarbons: Generation from fatty acid. *Science* 144: 1451–1452.

Kahr, G. and F.T. Madsen. 1995. Determination of the cation exchange capacity and the surface area of bentonite, illite and kaolinite by methylene blue adsorption. *Applied Clay Science* 9: 327–336.

Kalra, S., C.K. Pant, H.D. Pathak, and M.S. Mehata. 2000. Adsorption of glycine and alanine on montmorillonite with or without coordinated divalent cations. *Indian Journal of Biochemistry & Biophysics* 37: 341–346.

Kalra, S., C.K. Pant, H.D. Pathak, and M.S. Mehata. 2003. Studies on the adsorption of peptides of glycine/alanine on montmorillonite clay with or without co-ordinated divalent cations. *Colloids and Surfaces A: Physicochemical and Engineering Aspects* 212: 43–50.

Kan, T., X. Jiang, L. Zhou et al. 2011. Removal of methyl orange from aqueous solutions using a bentonite modified with a new Gemini surfactant. *Applied Clay Science* 54: 184–187.

Kanamaru, F. and V. Vand. 1970. The crystal structure of a clay-organic complex of 6-amino hexanoic acid and vermiculite. *American Mineralogist* 55: 1550–1561.

Kang, F., Y. Ge, X. Hu et al. 2016. Understanding the sorption mechanisms of aflatoxin B_1 to kaolinite, illite, and smectite clays via a comparative computational study. *Journal of Hazardous Materials* 320: 80–87.

Kang, S. and B. Xing. 2007. Adsorption of dicarboxylic acids by clay minerals as examined by *in situ* ATR-FTIR and *ex situ* DRIFT. *Langmuir* 23: 7024–7031.

Karaoğlu, M.H., M. Doğan, and M. Alkan. 2010. Kinetic analysis of reactive blue 221 adsorption on kaolinite. *Desalination* 256: 154–165.

Kasprzhitskii, A., G. Lazorenko, D.S. Kharytonau, M.A. Osipenko, A.A. Kasach, and I.I. Kurilo. 2022. Adsorption mechanism of aliphatic amino acids on kaolinite surfaces. *Applied Clay Science* 226: 106566.

Kaufherr, N., S. Yariv, and L. Heller. 1971. The effect of exchangeable cations on the sorption of chlorophyllin by montmorillonite. *Clays Clay Minerals* 19: 193–200.

Kausar, A., M. Iqbal, A. Javed et al. 2018. Dyes adsorption using clay and modified clay: A review. *Journal of Molecular Liquids* 256: 395–407.

Kausor, M.A., S.S. Gupta, K.G. Bhattacharyya, and D. Chakrabortty. 2022. Montmorillonite and modified montmorillonite as adsorbents for removal of water soluble organic dyes: A review on current status of the art. *Inorganic Chemistry Communications* 143: 109686.

Ke, Y.L., L.F. Jiao, Z.H. Song et al. 2014. Effects of cetylpyridinium-montmorillonite, as alternative to antibiotic, on the growth performance, intestinal microflora and mucosal architecture of weaned pigs. *Animal Feed Science and Technology* 198: 257–262.

Khan, I., K. Saeed, I. Zekker et al. 2022. Review on methylene blue: Its properties, uses, toxicity and photodegradation. *Water* 14: 242. https://doi.org/10.3390/w14020242.

Khan, S. and J.P. Singhal. 1967. Titrations of hydrogen clays with nicotine. *Soil Science* 104: 427–432.

Khan, T.A., S. Dahiya, and I. Ali. 2012. Use of kaolinite as adsorbent: Equilibrium, dynamics and thermodynamic studies on the adsorption of rhodamine B from aqueous solution. *Applied Clay Science* 69: 58–66.

Khoury, G.A., T.C. Gehris, L. Tribe, R.M. Torres Sánchez, and M. dos Santos Afonso. 2010. Glyphosate adsorption on montmorillonite: An experimental and theoretical study of surface complexes. *Applied Clay Science* 50: 167–175.

Kim, M., K.-M. Roh, J. Kim, M. Seo, I.-M. Kang, and J. Kim. 2022. Sustained release drug delivery potential of antibiotic-montmorillonite composites prepared using montmorillonite mined in Gampo, Republic of Korea. *Clays and Clay Minerals* 70: 62–71.

Kipling, J.J. and R.B. Wilson. 1960. Adsorption of methylene blue in the determination of surface areas. *Journal of Applied Chemistry* 10: 109–113.

Kitadai, N., T. Yokoyama, and S. Nakashima. 2009. *In situ* ATR-IR investigation of L-lysine adsorption on montmorillonite. *Journal of Colloid and Interface Science* 338: 395–401.

Kleyi, P.E., S.S. Ray, A.L.K. Abia, E. Ubomba-Jaswa, J. Wesley-Smith, and A. Maity. 2016. Preparation and evaluation of quaternary imidazolium-modified montmorillonite for disinfection of drinking water. *Applied Clay Science* 127–128: 95–104.

Knight, B.A.G. and P.J. Denny. 1970. The interaction of paraquat with soil: Adsorption by an expanding lattice clay mineral. *Weed Research* 10: 40–48.

Knight, B.A.G. and T.E. Tomlinson. 1967. The interaction of paraquat (1;1'-dimethyl 4:4'-dipyridilium dichloride) with mineral soils. *Journal of Soil Science* 18: 233–243.

Kohno, Y., M. Inagawa, S. Ikoma et al. 2011. Stabilization of a hydrophobic natural dye by intercalation into organo-montmorillonite. *Applied Clay Science* 54: 202–205.

Kohno, Y., R. Kinoshita, S. Ikoma et al. 2009. Stabilization of natural anthocyanin by intercalation into montmorillonite. *Applied Clay Science* 42: 519–523.

Kollár, T., I. Pálinkó, Z. Kónya, and I. Kiricsi. 2003. Intercalating amino acid guests into montmorillonite host. *Journal of Molecular Structure* 651–653: 335–340.

Konstantinou, I.K., T.A. Albanis, D.E. Petrakis, and P.J. Pomonis. 2000. Removal of herbicides from aqueous solutions by adsorption on Al-pillared clays, Fe-Al pillared clays and mesoporous alumina aluminum phosphates. *Water Research* 34: 3123–3136.

Kooli, F., S. Rakass, Y. Liu et al. 2019. Eosin removal by cetyltrimethylammonium-cloisites: Influence of the surfactant solution type and regeneration properties. *Molecules* 24: 3015. https://doi.org/10.3390/molecules24163015.

Koswojo, R., R.P. Utomo, Y.-H. Ju et al. 2010. Acid Green 25 removal from wastewater by organo-bentonite from Pacitan. *Applied Clay Science* 48: 81–86.

Kovář, P., M. Pospíšil, P. Malý et al. 2009. Molecular modeling of surface modification of Wyoming and Cheto montmorillonite by methylene blue. *Journal of Molecular Modeling* 15: 1391–1396.

Kryuchkova, M., S. Batasheva, F. Akhatova et al. 2021. Pharmaceuticals removal by adsorption with montmorillonite nanoclay. *International Journal of Molecular Sciences* 22: 9670. https://doi.org/10.3390/ijms22189670.

Kulshrestha, P., R.F. Giese, and D.S. Aga. 2004. Investigating the molecular interactions of oxytetracycline in clay and organic matter: Insights on factors affecting its mobility in soil. *Environmental Science & Technology* 38: 4097–4105.

Lagaly, G. 1984. Clay-organic interactions. *Philosophical Transactions of the Royal Society of London A* 311: 315–332.

Lagaly, G. 2001. Pesticide-clay interactions and formulations. *Applied Clay Science* 18: 205–209.

Lagaly, G., M. Ogawa, and I. Dékány. 2006. Clay mineral organic interactions. In *Handbook of Clay Science*. Developments in Clay Science, Vol. 1, (Eds.) F. Bergaya, B.K.G. Theng, and G. Lagaly, pp. 309–377. Amsterdam, The Netherlands: Elsevier.

Lailach, G.E. and G.W. Brindley. 1969. Specific co-absorption of purines and pyrimidines by montmorillonite (Clay-organic studies XV). *Clays Clay Minerals* 17: 95–100.

Lailach, G.E., T.D. Thompson, and G.W. Brindley. 1968a. Absorption of pyrimidines, purines, and nucleosides by Li-, Na-, Mg-, and Ca-montmorillonite (Clay-organic studies XII). *Clays Clay Minerals* 16: 285–293.

Lailach, G.E., T.D. Thompson, G.W. Brindley. 1968b. Absorption of pyrimidines, purines, and nucleosides by Co-, Ni-, Cu-, and Fe(III)-montmorillonite (Clay-organic studies XVI). *Clays Clay Minerals* 16: 295–301.

Laird, D.A., E. Barriuso, R.H. Dowdy, and W.C. Koskinen. 1992. Adsorption of atrazine on smectites. *Soil Science Society of America Journal* 56: 62–67.

Lambert, J.-F. 2018. Organic pollutant adsorption on clay minerals. In *Surface and Interface Chemistry of Clay Minerals*. Developments in Clay Science, Vol. 9, (Eds.) R. Schoonheydt, C.T. Johnston, and F. Bergaya, pp. 195–253. Amsterdam, The Netherlands: Elsevier.

Lambert, S.M. 1967. Functional relationship between sorption in soil and chemical structure. *Journal of Agricultural and Food Chemistry* 15: 572–576.

Lawal, I.A. and B. Moodley. 2015. Synthesis, characterization and application of imidazolium based ionic liquid modified montmorillonite sorbents for the removal of amaranth dye. *RSC Advances* 5: 61913–61924.

Lazorenko, G., A. Kasprzhitskii, and V. Yavna. 2020. Comparative study of the hydrophobicity of organo-montmorillonite modified with cationic, amphoteric and nonionic surfactants. *Minerals* 10: 732. https://doi.org/10.3390/min10090732.

LeBaron, H.M., J.E. McFarland, and O.C. Burnside. 2008. In *The Triazine Herbicides: 50 Years of Revolutionizing Agriculture*, (Eds.) H.M. LeBaron, J.E. McFarland, and O.C. Burnside, pp. 1–12. Amsterdam, The Netherlands: Elsevier.

Lee, E.-S., E.-J. Song, S.-Y. Lee et al. 2018. Effects of bentonite Bgp35-p on the gut microbiota of mice fed a high-fat diet. *Journal of the Science of Food and Agriculture* 98: 4369–4373.

Lee, S.H., D.I. Song, and Y.W. Jeon. 2001. An investigation of the adsorption of organic dyes onto organo-montmorillonite. *Environmental Technology* 22: 247–254.

Li, H., G. Sheng, B.J. Teppen, C.T. Johnston, and S.A. Boyd. 2003. Sorption and desorption of pesticides by clay minerals and humic acid-clay complexes. *Soil Science Society of America Journal* 67: 122–131.

Li, H., B.J. Teppen, D.A. Laird, C.T. Johnston, and S.A. Boyd. 2006. Effect of increasing potassium chloride and calcium chloride ionic strength on pesticide sorption by potassium- and calcium-smectite. *Soil Science Society of America Journal* 70: 1889–1895.

Li, H.-L., L. Bian, F.-Q. Dong et al. 2019a. DFT and 2D-CA methods unravelling the mechanism of interfacial interaction between amino acids and Ca-montmorillonite. *Applied Clay Science* 183: 105356.

Li, J., J. Cai, L. Zhong, H. Cheng, H. Wang, and Q. Ma. 2019b. Adsorption of reactive red 136 onto chitosan/montmorillonite intercalated composite from aqueous solution. *Applied Clay Science* 167: 9–22.

Li, J.-T., Y.-L. Song, and H. Chen. 2009. Removal of C.I. acid brown 75 from aqueous solution by kaolinite. *Toxicological & Environmental Chemistry* 91: 1069–1072.

Li, Q., Y. Su, Q.-Y. Yue, and B.-Y. Gao. 2011c. Adsorption of acid dyes onto bentonite modified with polycations: Kinetic study and process design to minimize the contact time. *Applied Clay Science* 53: 760–765.

Li, Z., P.-H. Chang, J.-S. Jean, W.-T. Jiang, and H. Hong. 2011a. Mechanism of chlorpheniramine adsorption on Ca-montmorillonite. *Colloids and Surfaces A: Physicochemical and Engineering Aspects* 385: 213–218.

Li, Z., P.-H. Chang, W.-T. Jiang, J.-S. Jean, H. Hong, and L. Liao. 2011b. Removal of diphenhydramine from water by swelling clay minerals. *Journal of Colloid and Interface Science* 360: 227–232.

Li, Z., V.M. Kolb, W.-T. Jiang, and H. Hong. 2010. FTIR and XRD investigations of tetracycline intercalation in smectites. *Clays and Clay Minerals* 58: 462–474.

Liebmann, P., G. Loew, S. Burt, J. Lawless, and R.D. MacElroy. 1982. Interaction of metal ions and nucleotides: Possible mechanisms for the adsorption of nucleotides on homoionic bentonite clays. *Inorganic Chemistry* 21: 1586–1594.

Liu, B., X. Wang, B. Yang, and R. Sun. 2011. Rapid modification of montmorillonite with novel Gemini surfactants and its adsorption for methyl orange. *Materials Chemistry and Physics* 139: 1220–1226.

Liu, N., M.-X. Wang, M.-M. Liu et al. 2012. Sorption of tetracycline on organo-montmorillonites. *Journal of Hazardous Materials* 225–226: 28–35.

Liu, P. and L. Zhang. 2007. Adsorption of dyes from aqueous solutions or suspensions with clay nano-adsorbents. *Separation and Purification Technology* 58: 32–39.

Liu, S., P. Wu, L. Yu et al. 2017a. Preparation and characterization of organo-vermiculite based on phosphatidylcholine and adsorption of two typical antibiotics. *Applied Clay Science* 137: 160–167.

Liu, S., P. Wu, M. Chen et al. 2017b. Amphoteric modified vermiculites as adsorbents for enhancing removal of organic pollutants: Bisphenol A and tetrabromobisphenol A. *Environmental Pollution* 228: 277–286.

Liu, Y., C. Dong, H. Wei, W. Yuan, and K. Li. 2015. Adsorption of levofloxacin onto an iron-pillared montmorillonite (clay mineral): Kinetics, equilibrium and mechanism. *Applied Clay Science* 118: 301–307.

Liu, Y.-L., H.W. Walker, and J.J. Lenhart. 2019. Adsorption of mycrocystin-LR onto kaolinite, illite and montmorillonite. *Chemosphere* 220: 696–705.
Lobato-Aguilar, H., J.A. Uribe-Calderón, W. Herrera-Kao et al. 2018. Synthesis, characterization and chlorhexidine release from either montmorillonite or palygorskite modified organoclays for antibacterial applications. *Journal of Drug Delivery Science and Technology* 46: 452–460.
Lombardi, B., M. Baschini, and R.M. Torres Sánchez. 2003. Optimization of parameters and loadsorption mechanism of thiabendazole fungicide by a montmorillonite of North Patagonia, Argentina. *Applied Clay Science* 24: 43–50.
Lombardi, B.M., R.M. Torres Sánchez, P. Eloy, and M. Genet. 2006. Interaction of thiabendazole and benzimidazole with montmorillonite. *Applied Clay Science* 33: 59–65.
López Arbeloa, F., R. Chaudhuri, T. Arbeloa López, and I. López Arbeloa. 2002a. Aggregation of rhodamine 3B adsorbed in Wyoming montmorillonite aqueous suspensions. *Journal of Colloid and Interface Science* 246: 281–287.
López Arbeloa, F., V. Martínez Martínez, J. Bañuelos Prieto, and I. López Arbeloa. 2002b. Adsorption of rhodamine 3B dye on saponite colloidal particles in aqueous suspensions. *Langmuir* 18: 2658–2664.
López Arbeloa, F., M.J. Tapia Estevez, T. López Arbeloa, and I. López Arbeloa. 1997. Spectroscopic study of the adsorption of rhodamine 6G on clay minerals in aqueous suspensions. *Clay Minerals* 32: 97–106.
López-Rodríguez, D., B. Micó-Vicent, J. Jordán-Núñez, and M. Bonet-Aracil. 2021. Uses of nanoclays and adsorbents for dye recovery: A textile industry review. *Applied Science* 11: 11422. https://doi.org/10.3390/app112311422.
Lu, L., J. Cai, and R.L. Frost. 2010. Desorption of stearic acid upon surfactant adsorbed montmorillonite: A thermogravimetric study. *Journal of Thermal Analysis and Calorimetry* 100: 141–144.
Luo, W., K. Sasaki, and T. Hirajima. 2018. Influence of the pre-dispersion of montmorillonite on organic modification and the adsorption of perchlorate and methyl red anions. *Applied Clay Science* 154: 1–9.
Lv, G., Z. Li, L. Elliott, M.J. Schmidt, M.P. MacWilliams, and B. Zhang. 2019. Impact of tetracycline-clay interactions on bacterial growth. *Journal of Hazardous Materials* 370: 91–97.
Lv, G., Z. Li, W.-T. Jiang, P.-H. Chang, J.-S. Jean, and K.-H. Lin. 2011. Mechanism of acridine orange removal from water by low-charge swelling clays. *Chemical Engineering Journal* 174: 603–611.
Lv, G., L. Liu, Z. Li, L. Liao, and M. Liu. 2012. Probing the interactions between chlorpheniramine and 2:1 phyllosilicates. *Journal of Colloid and Interface Science* 374: 218–225.
Ma, J., B. Cui, J. Dai, and D. Li. 2011. Mechanism of adsorption of anionic dye from aqueous solutions onto organobentonite. *Journal of Hazardous Materials* 186: 1758–1765.
Ma, J., Y.-Z. Jia, Y. Jing, J.-H. Sun, Y. Yao, and X.-H. Wang. 2010a. Equilibrium models and kinetic for the adsorption of methylene blue on Co-hectorites. *Journal of Hazardous Materials* 175: 965–969.
Ma, L., Y. Xi, H. He, G.A. Ayoko, R. Zhu, and J. Zhu. 2016. Efficiency of Fe-montmorillonite on the removal of rhodamine B and hexavalent chromium from aqueous solution. *Applied Clay Science* 120: 9–15.
Ma, Y.-L., Q.-F. Li, L.-Q. Zhang, and Y.-P. Wu. 2007. Role of stearic acid in preparing EPDM/clay nanocomposites by melt compounding. *Polymer Journal* 39: 48–54.
Ma, Y.-L., Z.-R. Xu, T. Guo, and P. You. 2004. Adsorption of methylene blue on Cu(II)-exchanged montmorillonite. *Journal of Colloid and Interface Science* 280: 283–288.
Ma, Y.-L., B. Yang, T. Guo, and L. Xie. 2010b. Antibacterial mechanism of Cu^{2+}–ZnO/cetylpyridinium-montmorillonite *in vitro*. *Applied Clay Science* 50: 348–353.
Mahmoodi, N.M., A. Taghizadeh, M. Taghizadeh, and M. Azimi. 2019. Surface modified montmorillonite with cationic surfactants: Preparation, characterization, and dye adsorption from aqueous solution. *Journal of Environmental Chemical Engineering* 7: 103243.
Makhoukhi, B., M. Djab, and M.A. Didi. 2015. Adsorption of Telcon dyes onto bis-imidazolium modified bentonite in aqueous solutions. *Journal of Environmental Chemical Engineering* 3: 1384–1392.
Makhoukhi, B., M.A. Didi, H. Moulessehoul, A. Azzouz, and D. Villemin. 2010. Diphosphonium ion-exchanged montmorillonite for Telon dye removal from aqueous media. *Applied Clay Science* 50: 354–361.
Malachová, K., P. Praus, Z. Pavlíčková, and M. Turicová. 2009. Activity of antibacterial compounds immobilized on montmorillonite. *Applied Clay Science* 43: 364–368.
Malferrari, D., M.F. Brigatti, A. Marcelli, W. Chu, and Z. Wu. 2010. Effect of temperature on Hg-cysteine complexes in vermiculite and montmorillonite. *Applied Clay Science* 50: 12–18.
Mallakpour, S. and M. Dinari. 2011. Preparation and characterization of new organoclays using natural amino acids and Cloisite Na^{+}. *Applied Clay Science* 51: 353–359.
Mallakpour, S. and M. Dinari. 2013. Preparation, characterization, and thermal properties of organoclay hybrids based on trifunctional natural amino acids. *Journal of Thermal Analysis and Calorimetry* 111: 611–618.
Malvar, J.L., J. Martín, M.D.M. Orta et al. 2020. Simultaneous and individual adsorption of ibuprofen metabolites by a modified montmorillonite. *Applied Clay Science* 189: 105529.

Marco-Brown, J.L., M.M. Areco, R.M. Torres Sánchez, and M.D.S. Afonso. 2014. Adsorption of picloram herbicide on montmorillonite: Kinetic and equilibrium studies. *Colloids and Surfaces A: Physicochemical and Engineering Aspects* 449: 121–128.

Marco-Brown, J.L., L. Guz, M.S. Olivelli et al. 2018. New insights on crystal violet dye adsorption on montmorillonite: Kinetics and surface complexes studies. *Chemical Engineering Journal* 333: 495–504.

Margulies, L., H. Rozen, and S. Nir. 1988. Model for competitive adsorption of organic cations on clays. *Clays and Clay Minerals* 36: 270–276.

Martín, M.J.S. and M.S. Camazano. 1984. Aspects of the adsorption of azinphos-methyl by smectites. *Journal of Agricultural and Food Chemistry* 32: 720–725.

Martín, J., M.D.M. Orta, S. Medina-Carrasco, J.L. Santos, I. Aparicio, and E. Alonso. 2019. Evaluation of a modified mica and montmorillonite for the adsorption of ibuprofen from aqueous media. *Applied Clay Science* 171: 29–37.

Martinez-Costa, J.I., R. Leyva-Ramos, and E. Padilla-Ortega. 2018. Sorption of diclofenac from aqueous solution on an organobentonite and adsorption of cadmium on organobentonite saturated with diclofenac. *Clays and Clay Minerals* 66: 515–528.

Masimango, N., J. Remacle, and J. Ramaut. 1979. Elimination, par des argiles gonflantes, de l'aflatoxine B1des milieux contaminés. *Annales de la Nutrition et de l'Alimentation* 33: 137–147.

Massaro, M., C.G. Colletti, G. Lazzara, and S. Riela. 2018. The use of some clay minerals as natural resources for drug carrier applications. *Journal of Functional Biomaterials* 9: 58. https://doi.org/10.3390/jfb9040058.

McConnell, J.S. and L.R. Hossner. 1985. pH-dependent adsorption isotherms of glyphosate. *Journal of Agricultural and Food Chemistry* 33: 1075–1078.

McConnell, J.S. and L.R. Hossner. 1989. X-ray diffraction and infrared spectroscopic studies of adsorbed glyphosate. *Journal of Agricultural and Food Chemistry* 37: 555–560.

McLaren, A.D., G.H. Peterson, and I. Barshad. 1958. The adsorption reactions of enzymes and proteins on clay minerals. IV. Kaolinite and montmorillonite. *Soil Science Society of America Proceedings* 22: 239–244.

Meng, N., N.-L. Zhou, S.-Q. Zhang, and J. Shen. 2009. Controlled release and antibacterial activity chlorhexidine acetate (CA) intercalated in montmorillonite. *International Journal of Pharmaceutics* 382: 45–49.

Meyers, P.A. and J.G. Quinn. 1971. Fatty acid-clay mineral association in artificial and natural sea water solutions. *Geochimica et Cosmochimica Acta* 35: 628–632.

Micera, G., A. Pusino, C. Gessa, and S. Petretto. 1988. Interaction of fluazifop with Al-, Fe^{3+}, and Cu^{2+}-saturated montmorillonite. *Clays and Clay Minerals* 36: 354–358.

Michalkova, A., T.L. Robinson, and J. Leszczynski. 2011. Adsorption of thymine and uracil on 1:1 clay mineral surfaces; comprehensive *ab initio* study on influence of sodium cation and water. *Physical Chemistry Chemical Physics* 13: 7862–7881.

Mignon, P., G. Corbin, S.L. Crom, V. Marry, J. Hao, and I. Daniel. 2020. Adsorption of nucleotides on clay surfaces: Effects of mineral composition, pH and solution salts. *Applied Clay Science* 190: 105544.

Mitra, S.P., S.G. Misra, and N. Panda. 1957. Adsorption of glucose by calcium bentonite. *Proceedings of the National Academy of Science, India* 26A: 72–74.

Miyamoto, N., R. Kawai, K. Kuroda, and M. Ogawa. 2000. Adsorption and aggregation of a cationic cyanine dye on layered clay minerals. *Applied Clay Science* 16: 161–170.

Mo, J., L. Dao, L. Chen et al. 2015. Structural effects of organobentonites on controlled release of pretilachlor. *Applied Clay Science* 115: 150–156.

Mohammed, S.S. and D.T.A. Al-Heetimi. 2019. Adsorption of methyl violet dye from aqueous solution by Iraqi bentonite and surfactant-modified Iraqi bentonite. *Ibn Al Haitham Journal for Pure and Applied Science* 32: 28–42.

Monvisade, P. and P. Siriphannon. 2009. Chitosan intercalated montmorillonite: Preparation, characterization and cationic dye adsorption. *Applied Clay Science* 42: 427–431.

Moreira, M.A., K.J. Ciuffi, V. Rives et al. 2017. Effect of chemical modification of palygorskite and sepiolite by 3-aminopropyltriethoxysilane on adsorption of cationic and anionic dyes. *Applied Clay Science* 135: 394–404.

Morillo, E., J.L. Perez-Rodriguez, and C. Maqueda. 1991. Mechanisms of interaction between montmorillonite and 3-aminotriazole. *Clay Minerals* 26: 269–279.

Morillo, E., J.L. Perez-Rodriguez, P. Rodriguez-Rubio, and C. Maqueda. 1997a. Interaction of aminotriazole with montmorillonite and Mg-vermiculite at pH 4. *Clay Minerals* 32: 307–313.

Morillo, E., T. Undabeytia, and C. Maqueda. 1997b. Adsorption of glyphosate on the clay mineral montmorillonite: Effect of Cu(II) in solution and adsorbed on the mineral. *Environmental Science & Technology* 31: 3588–3592.

Morimoto, K., K. Tamura, C.S. Pascua, T. Hatta, J. Ye, and H. Yamada. 2012. Photodegradation of rhodamine B in aqueous suspension of synthetic Zn-substituted phyllosilicates. *Clay Science* 16: 41–47.

Moro, D., G. Ulian, and G. Valdrè. 2019. Amino acids-clay interaction at the nano-atomic scale: The L-alanine-chlorite system. *Applied Clay Science* 172: 28–39.

Moro, D., G. Ulian, and G. Valdrè. 2020. Nanoscale oligopeptide adsorption behaviour on chlorite as revealed by scanning probe microscopy and density functional simulations. *Applied Clay Science* 197: 105777.

Mortland, M.M. and W.R. Meggitt. 1966. Interaction of ethyl-N,N-di-*n*-propyl-thiolcarbamate (EPTC) with montmorillonite. *Journal of Agricultural and Food Chemistry* 14: 126–129.

Mouni, L., L. Belkhiri, J.-C. Bollinger et al. 2018. Removal of methylene blue from aqueous solutions by adsorption on kaolin: Kinetic and equilibrium studies. *Applied Clay Science* 153: 38–45.

Mudhoo, A. and V.K. Garg. 2011. Sorption, transport and transformation of atrazine in soils, minerals and composts: A review. *Pedosphere* 21: 11–25.

Nagy, N.M. and J. Kónya. 2004. The adsorption of valine on cation-exchanged montmorillonites. *Applied Clay Science* 25: 57–69.

Nandi, B.K., A. Goswami, and M.K. Purkait. 2009a. Removal of cationic dyes from aqueous solutions by kaolin: Kinetic and equilibrium studies. *Applied Clay Science* 42: 583–590.

Nandi, B.K., A. Goswami, and M.K. Purkait. 2009b. Adsorption characteristics of brilliant green dye on kaolin. *Journal of Hazardous Materials* 161: 387–395.

Narine, D.R. and R.D. Guy. 1981. Interactions of some large organic cations with bentonite in dilute aqueous systems. *Clays and Clay Minerals* 29: 205–212.

Nearpass, D.C. 1970. Exchange adsorption of 3-amino-1,2,4-triazole by montmorillonite. *Soil Science* 109: 77–84.

Nennemann, A., S. Kulbach, and G. Lagaly. 2001a. Entrapping pesticides by coagulating smectites. *Applied Clay Science* 18: 285–298.

Nennemann, A., Y. Mishael, S. Nir et al. 2001b. Clay-based formulations of metolachlor with reduced leaching. *Applied Clay Science* 18: 265–275.

Neumann, M.G., F. Gessner, C.C. Schmitt, and R. Sartori. 2002. Influence of the layer charge and clay particle size on the interactions between the cationic dye methylene blue and clays in an aqueous suspension. *Journal of Colloid and Interface Science* 255: 254–259.

Nevins, M.J. and D.J. Weintritt. 1967. Determination of cation exchange capacity by methylene blue adsorption. *American Chemical Society Bulletin* 46: 587–592.

Ngulube, T., J.R. Gumbo, V. Masindi, and A. Maity. 2017. An update on synthetic dyes adsorption onto clay-based minerals: A state-of-art review. *Journal of Environmental Management* 191: 35–57.

Nguyen, V.N., T.D.C. Nguyen, T.P. Dao, H.T. Tran, D.B. Nguyen, and D.H. Ahn. 2013. Synthesis of organoclays and their application for the adsorption of phenolic compounds from aqueous solution. *Journal of Industrial and Engineering Chemistry* 19: 640–644.

Nippes, R.P., P.D. Macruz, T.F. Coslop, D. Molinari, and M.H.N.O. Scalianti. 2022. Removal of ivermectin from aqueous media using commercial, bentonite-based organophilic clay as an adsorbent. *Clay Minerals* 57: 21–30.

Nir, S., Y. El-Nahhal, T. Undabeytia et al. 2006. Clays and pesticides. In *Handbook of Clay Science*. Developments in Clay Science, Vol. 1, (Eds.) F. Bergaya, B.K.G. Theng, and G. Lagaly, pp. 677–691. Amsterdam, The Netherlands: Elsevier.

Nishikiori, H., K. Kobayashi, S. Kubota, N. Tanaka, and T. Fujii. 2010. Removal of detergents and fats from wastewater using allophane. *Applied Clay Science* 47: 325–329.

Nkoh, J.N., Z. Hong, H. Lu, J. Li, and R. Xu. 2022. Adsorption of amino acids by montmorillonite and gibbsite: Adsorption isotherms and spectroscopic analysis. *Applied Clay Science* 219: 106437.

Ogawa, M., M. Hama, and K. Kuroda. 1999. Photochromism of azo benzene in the hydrophobic interlayer spaces of dialkyldimethylammoiun-fluor-tetrasilicic mica films. *Clay Minerals* 34: 213–220.

Ogawa, M., T. Ishii, N. Miyamoto, and K. Kuroda. 2003. Intercalation of a cationic azobenzene into montmorillonite. *Applied Clay Science* 22: 179–185.

Ogawa, M. and A. Ishikawa. 1998. Controlled microstructures of amphiphilic cationic azobenzene-montmorillonite intercalation compounds. *Journal of Materials Chemistry* 8: 463–467.

Ogawa, M. and K. Kuroda. 1995. Photofunctions of intercalation compounds. *Chemical Reviews* 95: 399–438.

Ohashi, F., H. Abe, S. Ueda, T. Taguri, T. Urakawa, and L. Duclaux. 2006. Structural model calculation for antimicrobial agent derived from montmorillonite intercalated by Ag^{+}-chelate of imidazole. *Clay Science* 12(Supplement 2): 21–24.

Ohashi, F., S. Ueda, T. Taguri, S. Kawachi, and H. Abe. 2009. Antimicrobial activity and thermostability of silver 6-benzylaminopurine montmorillonite. *Applied Clay Science* 46: 296–299.

Okada, E., J.L. Costa, and F. Bedmar. 2016. Adsorption and mobility of glyphosate in different soils under no-till and conventional tillage. *Geoderma* 263: 78–85.
Oliveira, L.H.D., P. Trigueiro, B. Rigaud et al. 2021. When RNA meets montmorillonite: Influence of the pH and divalent cations. *Applied Clay Science* 214: 106234.
Omidi-Khaniabadi, Y., B. Kamarehei, H. Nourmoradi, M. Jourvand, H. Basiri, and S. Heidari. 2016. Hexadecyl trimethyl ammonium bromide-modified montmorillonite as a low-cost sorbent for the removal of methyl red from liquid medium. *International Journal of Engineering Transactions A: Basics* 29: 60–67.
Orta, M.D.M., J. Martín, S. Medina-Carrasco, J.L. Santos, I. Aparicio, and E. Alonso. 2019. Adsorption of propranolol onto montmorillonite: Kinetic, isotherm and pH studies. *Applied Clay Science* 173: 107–114.
Ouachtak, H., A.E. Guerdaoui, R. Haounati et al. 2021. Highly efficient and fast batch adsorption of orange G dye from polluted water using superb organo-montmorillonite: Experimental study and molecular dynamics investigation. *Journal of Molecular Liquids* 335: 116560.
Ovcharenko, F.D., N.V. Vdovenko, V.P. Tchichkun, and Y.L. Tarasevich. 1969. Adsorption of amino acids by montmorillonite. *Ukrainskii Khimicheskii Zhurnal* 35: 123–128.
Özcan, A.S., B. Erdem, and A. Özcan. 2004. Adsorption of acid blue 193 from aqueous solutions onto Na-bentonite and DTMA-bentonite. *Journal of Colloid and Interface Science* 280: 44–54.
Özcan, A.S., B. Erdem, and A. Özcan. 2005. Adsorption of Acid Blue 193 from aqueous solutions onto BTMA-bentonite. *Colloids and Surfaces A: Physicochemical and Engineering Aspects* 266: 73–81.
Özcan, A.S., Ç. Ömeroğlu, Y. Erdoğan, and A.S. Özcan. 2007. Modification of bentonite with a cationic surfactant: An adsorption study of textile dye reactive blue 19. *Journal of Hazardous Materials* 140: 173–179.
Özcan, A.S., E.M. Öncü, and A.S. Özcan. 2006. Adsorption of acid blue 193 from aqueous solutions onto DEDMA-sepiolite. *Journal of Hazardous Materials* B129: 244–252.
Özdemir, G., M.H. Limoncu, and S. Yapar. 2010. The antibacterial effect of heavy metal and cetylpyridinium-exchanged montmorillonites. *Applied Clay Science* 48: 319–323.
Özdemir, G., S. Yapar, and M.H. Limoncu. 2013. Preparation of cetylpyridinium montmorillonite for antibacterial applications. *Applied Clay Science* 72: 201–205.
Ozola-Davidane, R., J. Burlakovs, T. Tamm, S. Zeltkalne, and A.E. Krauklis. 2021. Bentonite-ionic liquid composites for Congo red removal from aqueous solutions. *Journal of Molecular Liquids* 337: 116373.
Pajak, M. 2021. Adsorption capacity of smectite clay and its thermal and chemical modification for two anionic dyes: Comparative study. *Water, Air, & Soil Pollution* 232: 83. https://doi.org/10.1007/s11270-021-05032-3.
Pandey, P. and N. De. 2017. Surfactant-induced changes in physicochemical characters of bentonite clay. *International Research Journal of Pure & Applied Chemistry* 15: 1–11.
Pandey, P., C.K. Pant, K. Gururani et al. 2015. Affinity of smectite and divalent metal ions (Mg^{2+}, Ca^{2+}, Cu^{2+}) with L-leucine: An experimental and theoretical approach relevant to astrobiology. *Origins of Life and Evolution of Biospheres* 45: 411–426.
Pandey, S. and J. Ramontja. 2016. Natural bentonite clay and its composites for dye removal: Current state and future potential. *American Journal of Chemistry and Applications* 3: 8–19.
Parbhakar, A., J. Cuadros, M.A. Sephton, W. Dubbin, B.J. Coles, and D. Weiss. 2007. Adsorption of L-lysine on montmorillonite. *Colloids and Surfaces A: Physicochemical and Engineering Aspects* 307: 142–149.
Paredes-Quevedo, L.C., C. González-Caicedo, J.A. Torres-Luna, and J.G. Carriazo. 2021. Removal of a textile azo-dye (Basic red 46) in water by efficient adsorption on a natural clay. *Water, Air, & Soil Pollution* 232: 4. https://doi.org/10.1007/s11270-020-04968-2.
Parfitt, R.L. 1972. Adsorption of charged sugars by montmorillonite. *Soil Science* 113: 417–421.
Park, Y., G.A. Ayoko, and R.L. Frost. 2011. Application of organoclays for the adsorption of recalcitrant organic molecules from aqueous media. *Journal of Colloid and Interface Science* 354: 292–305.
Park, Y., Z. Sun, G.A. Ayoko, and R.L. Frost. 2014. Removal of herbicides from aqueous solutions by modified forms of montmorillonite. *Journal of Colloid and Interface Science* 415: 127–132.
Parolo, M.E., M.J. Avena, G. Pettinari, I. Zajonkovsky, J.M. Vallés, and M.T. Baschini. 2010. Antimicrobial properties of tetracycline and minocycline-montmorillonites. *Applied Clay Science* 49: 194–199.
Parolo, M.E., M.J. Avena, M.C. Savini, M.T. Baschini, and V. Nicotra. 2013. Adsorption and circular dichroism of tetracycline on sodium and calcium-montmorillonites. *Colloids and Surfaces A: Physicochemical and Engineering Aspects* 417: 57–64.
Parolo, M.E., M.C. Savini, J.M. Vallés, M.T. Baschini, and M.J. Avena. 2008. Tetracycline adsorption on montmorillonite: pH and ionic strength effects. *Applied Clay Science* 40: 179–186.
Paul, B., D. Yang, X. Yang, X. Ke, R. Frost, and H. Zhu. 2010. Adsorption of the herbicide simazine on moderately acid-activated beidellite. *Applied Clay Science* 49: 80–83.
Pauling, L. 1960. *The Nature of the Chemical Bond*, 3rd edition. Ithaca, NY: Cornell University Press.

Pedreira-Segade, U., J. Hao, A. Razafitianamaharavo et al. 2018a. How do nucleotides adsorb onto clays? *Life* 8: 59. https://doi.org/10.3390/life8040059.

Pedreira-Segade, U., L.J. Michot, and I. Daniel. 2018b. Effects of salinity on the adsorption of nucleotides onto phyllosilicates. *Physical Chemistry Chemical Physics* 20: 1938–1952.

Pei, Z.-G., X.-Q. Shan, S.-Z. Zhang et al. 2011. Insight to ternary complexes of co-adsorption of norfloxacin and Cu(II) onto montmorillonite at different pH using EXAFS. *Journal of Hazardous Materials* 186: 842–848.

Peng, S., T. Mao, C. Zheng et al. 2019. Polyhydroxyl gemini surfactant-modified montmorillonite for efficient removal of methyl orange. *Colloids and Surfaces A* 578: 123602.

Pentrák, M., A. Czímerová, J. Madejová, and P. Komadel. 2012. Changes in layer charge of clay minerals upon acid treatment as obtained from their interactions with methylene blue. *Applied Clay Science* 55: 100–107.

Pereira, R.C., A.C.S. da Costa, F.F. Ivashita, A. Paesano, Jr., and D.A.M. Zaia. 2020. Interaction between glyphosate and montmorillonite in the presence of artificial seawater. *Heliyon* 6: e03532.

Pereira, T.R., D.A. Laird, C.T. Johnston, B.J. Teppen, H. Li, and S.A. Boyd. 2007. Mechanism of dinitrophenol herbicide sorption by smectites in aqueous suspensions at varying pH. *Soil Science Society of America Journal* 71: 1476–1481.

Perezgasga, L., A. Serrato-Díaz, A. Negrón-Mendoza, L. de Pablo Galán, and F.G. Mosqueira. 2005. Sites of adsorption of adenine, uracil, and their corresponding derivatives on sodium montmorillonite. *Origins of Life and Evolution of Biospheres* 35: 91–110.

Phuekphong, A.F., K.J. Imwiset, and M. Ogawa. 2020. Organically modified bentonite as a efficient and reusable adsorbent for triclosan removal from water. *Langmuir* 36: 9025–9034.

Pinck, L.A. 1962. Adsorption of proteins, enzymes and antibiotics by montmorillonite. *Clays and Clay Minerals* 9: 520–529.

Pinck, L.A., W.F. Holton, and F.E. Allison. 1961. Antibiotics in soils I. Physico-chemical studies of antibiotic-clay complexes. *Soil Science* 91: 22–28.

Pires, J., J. Juźków, and M.L. Pinto. 2018. Amino acid modified montmorillonite clays as sustainable materials for carbon dioxide adsorption and separation. *Colloids and Surfaces A* 544: 105–110.

Polati, S., S. Angioi, V. Gianotti, F. Gosetti, and M.C. Gennaro. 2006. Sorption of pesticides on kaolinite and montmorillonite as a function of hydrophilicity. *Journal of Environmental Science and Health, Part B* 41: 333–344.

Porubcan, L.S., C.J. Serna, J.L. White, and S.L. Hem. 1978. Mechanism of adsorption of clindamycin and tetracycline by montmorillonite. *Journal of Pharmaceutical Sciences* 67: 1081–1087.

Pospíšil, M., P. Čapková, H. Weissmannová et al. 2003. Structure analysis of montmorillonite intercalated with rhodamine B: Modeling and experiment. *Journal of Molecular Modeling* 9: 39–46.

Pusino, A., I. Braschi, and C. Gessa. 2000. Adsorption and degradation of triasulfuron on homoionic montmorillonites. *Clays and Clay Minerals* 48: 19–25.

Pusino, A., A. Gelsomino, and C. Gessa. 1995. Adsorption mechanisms of imazamethabenz-methyl on homoionic montmorillonite. *Clays and Clay Minerals* 43: 346–352.

Pusino, A., W. Liu, and C. Gessa. 1993. Dimepiperate adsorption and hydrolysis on Al^{3+}-, Fe^{3+}-, Ca^{2+}-, and Na^{+}-montmorillonite. *Clays and Clay Minerals* 41: 335–340.

Pusino, A., G. Micera, and C. Gessa. 1991. Interaction of the herbicide acifluorfen with montmorillonite: Formation of insoluble Fe^{3+}, $Al^{3+,}$ $Cu^{2+,}$ and Ca^{2+} complexes. *Clays and Clay Minerals* 39: 50–53.

Pusino, A., G. Micera, C. Gessa, and S. Petretto. 1989a. Interaction of diclofop and diclofop-methyl with Al^{3+}-, Fe^{3+}-, and Cu^{2+}-saturated montmorillonite. *Clays and Clay Minerals* 37: 558–562.

Pusino, A., G. Micera, A. Premoli, and C. Gessa. 1989b. D-glucosamine sorption on Cu(II)-montmorillonite as the protonated and neutral species. *Clays and Clay Minerals* 37: 377–380.

Qin, Z., P. Yuan, S. Yang, D. Liu, H. He, and J. Zhu. 2014. Silylation of Al_{13}-intercalated montmorillonite with trimethoxychlorosilane and their adsorption for orange II. *Applied Clay Science* 99: 229–236.

Queiroga, L.N.F., M.B.B. Pereira, L.S. Silva et al. 2019. Microwave bentonite silylation for dye removal: Influence of the solvent. *Applied Clay Science* 168: 478–487.

Ramachandran, V.S., K.P. Kacker, and N.K. Patwardhan. 1962. Adsorption of dyes by clay minerals. *American Mineralogist* 47: 165–169.

Ramos, M.E. and F.J. Huertas. 2013. Adsorption of glycine on montmorillonite in aqueous solutions. *Applied Clay Science* 80–81: 10–17.

Rao, W., P. Piliouras, X. Wang et al. 2020. Zwitterionic dye rhodamine B (RhB) uptake on different types of clay. *Applied Clay Science* 197: 105790.

Ras, R.H., J. Németh, C.T. Johnston, I. Dékány, and R.A. Schoonheydt. 2004. Orientation and conformation of octadecyl rhodamine B in hybrid Langmuir-Blodgett monolayers containing clay minerals. *Physical Chemistry Chemical Physics* 6: 5347–5352.

Raupach, M., W.W. Emerson, and P.G. Slade. 1979. The arrangement of paraquat bound by vermiculite and montmorillonite. *Journal of Colloid and Interface Science* 69: 398–408.

Raupach, M., P.G. Slade, L. Janik, and E.W. Radoslovich. 1975. A polarized infrared and X-ray study of lysine-vermiculite. *Clays and Clay Minerals* 23: 181–186.

Rausell-Colom, J.A. and P.S. Salvador. 1971a. Complexes vermiculite-amino acides. *Clay Minerals* 9: 139–149.

Rausell-Colom, J.A. and P.S. Salvador. 1971b. Gélification de vermiculite dans des solutions d'acide γ-aminobutyrique. *Clay Minerals* 9: 193–208.

Ren, H.-P., S.-P. Tian, M. Zhu et al. 2018. Modification of montmorillonite by Gemini surfactants with different chain lengths and its adsorption behavior for methyl orange. *Applied Clay Science* 151: 29–36.

Rida, K., S. Bouraoui, and S. Hadnine. 2013. Adsorption of methylene blue from aqueous solution by kaolin. *Applied Clay Science* 83–84: 99–105.

Rishpon, J., P.J. O'Hara, N. Lahav, and J.G. Lawless. 1982. Interaction between ATP, metal ions, glycine, and several minerals. *Journal of Molecular Evolution* 18: 179–184.

Rivera, A., L. Valdés, J. Jiménez et al. 2016. Smectite as ciprofloxacin delivery system: Intercalation and temperature-controlled release properties. *Applied Clay Science* 124–125: 150–156.

Robinson, T.L., A. Michalkova, L. Gorb, and J. Leszczynski. 2007. Hydrogen bonding of thymine and uracil with surface of dickite: An *ab initio* study. *Journal of Molecular Structure* 844–845: 48–58.

Rodríguez, J.M., A.J. López, and S. Bruque. 1988. Interaction of phenamiphos with montmorillonite. *Clays and Clay Minerals* 36: 284–288.

Romero-Taboada, E. and G.D. Cancela. 2004. Methidathion complexes with homoionic and surfactant-modified montmorillonites. *Journal of Environmental Science and Health, Part B* 39: 551–564.

Rooj, S., A. Das, K.W. Stöckelhuber et al. 2012. Pre-intercalation of long chain fatty acid in the interlayer space of layered silicates and preparation of montmorillonite/natural rubber nanocomposites. *Applied Clay Science* 67–68: 50–56.

Russell, J.D., M. Cruz, and J.L. White. 1968b. The adsorption of 3-aminotriazole by montmorillonite. *Journal of Agricultural and Food Chemistry* 16: 21–24.

Russell, J.D., M. Cruz, J.L. White et al. 1968a. Mode of chemical degradation of *s*-triazines by montmorillonite. *Science* 160: 1340–1342.

Rytwo, G., Y. Gonen, S. Afuta, and S. Dultz. 2005. Interactions of pendimethalin with organo-montmorillonite complexes. *Applied Clay Science* 28: 67–77.

Rytwo, G., S. Nir, M. Crespin, and L. Margulies. 2000. Adsorption and interactions of methyl green with montmorillonite and sepiolite. *Journal of Colloid and Interface Science* 222: 12–19.

Rytwo, G., S. Nir, and L. Margulies. 1993. Competitive adsorption of methylene blue and crystal violet to montmorillonite. *Clay Minerals* 28: 139–143.

Rytwo, G., S. Nir, and L. Margulies. 1996. Adsorption and interactions of diquat and paraquat with montmorillonite. *Soil Science Society of America Journal* 60: 601–610.

Rytwo, G., C. Serban, S. Nir, and L. Margulies. 1991. Use of methylene blue and crystal violet for determination of exchangeable cations in montmorillonite. *Clays and Clay Minerals* 39: 551–555.

Sadri, S., B.B. Johnson, M. Ruyter-Hooley, and M.J. Angove. 2018. The adsorption of nortriptyline on montmorillonite, kaolinite and gibbsite. *Applied Clay Science* 165: 64–70.

Saeed, M., K. Farooq, M. Nafees, M. Arshad, M.S. Akhter, and A. Waseem. 2020a. Green and eco-friendly removal of mycotoxins with organo-bentonites; isothermal, kinetic, and thermodynamic studies. *CLEAN – Soil, Air, Water* 48: 1900427. https://doi.org/10.1002/clen.201900427.

Saeed, M., M. Munir, M. Nafees, S.S.A. Shah, H. Ullah, and A. Waseem. 2020b. Synthesis, characterization and applications of silylation based grafted bentonites for the removal of Sudan dyes: Isothermal, kinetic and thermodynamic studies. *Microporous and Mesoporous Materials* 291: 109697.

Saha, A., A. Shabeer Tp, V.T. Gajbhiye, S. Gupta, and R. Kumar. 2013. Removal of mixed pesticides from aqueous solutions using organoclays: Evaluation of equilibrium and kinetic model. *Bulletin of Environmental Contamination and Toxicology* 91: 111–116.

Saha, K., B.S. Butola, and M. Joshi. 2014. Synthesis and characterization of chlorhexidine acetate drug-montmorillonite intercalates for antibacterial applications. *Applied Clay Science* 101: 477–483.

Salcedo, M.F., A.Y. Mansilla, S.L. Colman, M.J. Iglesias, V.A. Alvarez, and C.A. Casalongué. 2021. Efficacy of an organically modified bentonite to adsorb 2,4-dichlorophenoxy acetic acid(2,4-D) and prevent its phytotoxicity. *Journal of Environmental Management* 297: 113427.

Samii, A.M. and G. Lagaly. 1987. Adsorption of nuclein bases on smectites. In *Proceedings of the International Clay Conference, Denver, 1985*, (Eds.) L.G. Schultz, H. van Olphen, and F.A. Mumpton, pp. 363–369. Bloomington, IN: The Clay Minerals Society.

Sánchez-Camazano, M., M.J. Sánchez-Martín, M.T. Vicente, and A. Dominguez-Gil. 1987. Adsorption-desorption of sotalol hydrochloride by Na-montmorillonite. *Clay Minerals* 22: 121–128.

Sánchez-Martín, M.J., M.S. Rodríguez-Cruz, M.S. Andrades, and M. Sanchez-Camazano. 2006. Efficiency of different clay minerals modified with a cationic surfactant in the adsorption of pesticides: Influence of clay type and pesticide hydrophobicity. *Applied Clay Science* 31: 216–228.

Sánchez-Martín, M.J., M. Sánchez-Camazano, M.L. Sayalero, and A. Dominguez-Gil. 1988. Physicochemical study of the interaction of montmorillonite with hydralazine hydrochloride, a cardiovascular drug. *Applied Clay Science* 3: 53–61.

Sanoğlan, S., S. Gürbüz, T. Ipeksaç, M.G. Seden, and M. Erol. 2014. Pararosaniline and crystal violet tagged montmorillonite for latent fingerprint investigation. *Applied Clay Science* 87: 235–244.

Santos, S.S.G., D.B. França, L.R.C. Castellano et al. 2020. Novel modified bentonites applied to the removal of an anionic azo-dye from aqueous solution. *Colloid and Surfaces A* 585: 134152.

Sarabadan, M., H. Bashiri, and S.M. Mousavi. 2021. Modelling, kinetic and equilibrium studies of crystal violet adsorption on modified montmorillonite by sodium dodecyl sulfate and hyamine surfactants. *Clay Minerals* 56: 16–27.

Sarabadan, M., H. Bashiri, and S.M. Mousavi. 2022. Efficient removal of crystal violet from solution by montmorillonite modified with docosyl-trimethylammonium chloride and sodium dodecyl sulfate: Modelling, kinetics and equilibrium studies. *Clay Minerals* 57: 7–20.

Sarier, N., E. Onder, and S. Ersoy. 2010. The modification of Na-montmorillonite by salts of fatty acids: An easy intercalation process. *Colloids and Surfaces A: Physicochemical and Engineering Aspects* 371: 40–49.

Sarma, G., S. Sengupta, and K.G. Bhattacharyya. 2011. Methylene blue adsorption on natural and modified clays. *Separation Science and Technology* 46: 1602–1614.

Sarmah, A.K., M.T. Meyer, and A.B.A. Boxall. 2006. A global perspective on the use, sales, exposure pathways, occurrence, fate and effects of veterinary antibiotics (VA) in the environment. *Chemosphere* 65: 725–759.

Sasai, R., N. Iyi, T. Fujita et al. 2004. Luminescence properties of Rhodamine 6G intercalated in surfactant/clay hybrid thin solid films. *Langmuir* 20: 4715–4719.

Sato, M. 1999. Preparation of kaolinite-amino acid intercalates derived from hydrated kaolinite. *Clays and Clay Minerals* 47: 793–802.

Sawhney, B.L. and S.S. Singh. 1997. Sorption of atrazine by Al- and Ca-saturated smectite. *Clays and Clay Minerals* 45: 333–338.

Schönsee, C.D., F.E. Wettstein, and T.D. Bucheli. 2021. Phytotoxin sorption to clay minerals. *Environmental Sciences Europe* 33: 36. https://doi.org/10.1186/s12302-021-00469-z.

Schoonheydt, R.A. and L. Heughebaert. 1992. Clay adsorbed dyes: Methylene blue on laponite. *Clay Minerals* 27: 91–100.

Schwartz, H.G., Jr. 1967. Adsorption of selected pesticides on activated carbon and mineral surfaces. *Environmental Science & Technology* 1: 332–337.

Sciascia, L., M.L.T. Liveru, and M. Merli. 2011. Kinetic and equilibrium studies for the adsorption of acid nucleic bases onto K10 montmorillonite. *Applied Clay Science* 53: 657–668.

Selvam, P.P., S. Preethi, B. Basakaralingam, N. Thinakaram, A. Sivasamy, and S. Sivanesan. 2008. Removal of rhodamine B from aqueous solution by adsorption onto sodium montmorillonite. *Journal of Hazardous Materials* 155: 39–44.

Sharma, P., D.J. Borah, P. Das, and M.R. Das. 2016. Cationic and anionic dye removal from aqueous solution using montmorillonite clay: Evaluation of adsorption parameters and mechanism. *Desalination and Water Treatment* 57: 8372–8388.

Shattar, S.F.A., N.A. Zakaria, and K.Y. Foo. 2016. Feasibility of montmorillonite-assisted adsorption process for the effective treatment of organo-pesticides. *Desalination and Water Treatment* 57: 13645–13677.

Shen, Y., X. Yu, and Y. Wang. 2019. Facile synthesis of modified rectorite (M-REC) for effective removal of anionic dye from water. *Journal of Molecular Liquids* 278: 12–18.

Sheng, G., C.T. Johnston, B.J. Teppen, and S.A. Boyd. 2002. Adsorption of dinitrophenol herbicides from water by montmorillonites. *Clays and Clay Minerals* 50: 25–34.

Shimoyama, A. and W.D. Johns. 1971. Catalytic conversion of fatty acids into petroleum-like paraffins and their maturation. *Nature Physical Science* 232: 140–144.

Shin, W.S. 2008. Competitive sorption of anionic and cationic dyes onto cetylpyridinium-modified montmorillonite. *Journal of Environmental Science and Health, Part A* 43: 1459–1470.

Shirzad-Siboni, M., A. Khataee, A. Hassani, and S. Karaca. 2015. Preparation, characterization and application of a CTAB-modified nanoclay for the adsorption of an herbicide from aqueous solutions: Kinetic and equilibrium studies. *Comptes Rendus Chimie* 18: 204–214.

Shoval, S. and S. Yariv. 1979. The interaction between roundup (glyphosate) and montmorillonite. Part I. Infrared study of the sorption of glyphosate by montmorillonite. *Clays and Clay Minerals* 27: 19–28.

Sidhoum, D.A., M.M. Socías-Viciana, M.D. Ureña-Amate, A. Derdour, E. González-Pradas, and N. Debbagh-Boutarbouch. 2013. Removal of paraquat from water by an Algerian bentonite. *Applied Clay Science* 83–84: 441–448.

Sieskind, O. 1960a. Étude des complexes d'adsorption formé entre la montmorillonite-H et certains acides aminés: Isothermes d'adsorption à pH 2 et à 20 °C. *Comptes Rendus de l'Académie des Sciences* 250: 2228–2230.

Sieskind, O. 1960b. Sur les complexes d'adsorption formés en milieu acide entre la montmorillonite-H et certains acides aminés: Leur structure. *Comptes Rendus de l'Académie des Sciences* 250: 2392–2393.

Sieskind, O. and G. Ourisson. 1971. Interactions argile-matière organique: Formation de complexes entre la montmorillonite et les acides stéarique et béhénique. *Comptes Rendus de l'Académie des Sciences* 272: 1885–1888.

Sieskind, O. and B. Siffert. 1972. Formation d'un carboxylate de surface entre l'acide stéarique et une hectorite nickelifére: Localisation de l'acide gras sur le réseau silicate. *Comptes Rendus de l'Académie des Sciences* 274: 973–976.

Sieskind, O. and R. Wey. 1959. Sur l'adsorption d'acides aminés par la montmorillonite-H. Influence de la position relative des deux fonctions -NH_2 et -COOH. *Comptes Rendus de l'Académie des Sciences* 248: 1652–1655.

Siffert, B. and S. Kessaissia. 1978. Contribution au mécanisme d'adsorption des α-amino-acides par la montmorillonite. *Clay Minerals* 13: 255–270.

Silva, D.T.C., I.E.S. Arruda, L.M. França et al. 2019. Tamoxifen/montmorillonite system – effect of the experimental conditions. *Applied Clay Science* 180: 105142.

Silva, F.D.C., L.C.B. Lima, E.C. Silva-Filho, M.G. Fonseca, J.-F. Lambert, and M. Jaber. 2020a. A comparative study of alanine adsorption and condensation to peptides in two clay minerals. *Applied Clay Science* 192: 105617.

Silva, D.T.C., M.G. Fonseca, A. Borrego-Sánchez et al. 2020b. Adsorption of tamoxifen on montmorillonite surface. *Microporous and Mesoporous Materials* 297: 110012.

Siminoff, P. and D. Gottlieb. 1951. The production and role of antibiotics in the soil. I. The fate of streptomycin. *Phytopathology* 41: 420–430.

Singhal, J.P. and R.P. Singh. 1976. Studies on the adsorption of nicotine on kaolinites. *Soil Science and Plant Nutrition* 22: 35–41.

Site, A.D. 2001. Factors affecting sorption of organic compounds in natural sorbent/water systems and sorption coefficients for selected pollutants: A review. *Journal of Physical and Chemical Reference Data* 30: 187–439.

Sithole, B.B. and R.D. Guy. 1987. Models for tetracycline in aquatic environments. I. Interaction with bentonite clay systems. *Water, Air, and Soil Pollution* 32: 303–314.

Slade, P.G., M.I. Telleria, and E.W. Radoslovich. 1976. The structures of ornithine-vermiculite and 6-aminohexanoic acid-vermiculite. *Clays and Clay Minerals* 24: 134–141.

Sprankle, P., W.F. Meggitt, and D. Penner. 1975. Adsorption, mobility, and microbial degradation of glyphosate in the soil. *Weed Science* 23: 229–234.

Styszko, K., K. Nosek, M. Motak, and K. Bester. 2015. Preliminary selection of clay minerals for the removal of pharmaceuticals, bisphenol A and triclosan in acidic and neutral aqueous solutions. *Comptes Rendus Chimie* 18: 1134–1142.

Suárez Barrios, M., M.A. Vicente-Rodríguez, and J.M. Martín Pozas. 1996. Intercalation compounds between nicotine and a high surface area saponite. *Journal of Inclusion Phenomena and Molecular Recognition in Chemistry* 24: 263–272.

Sukul, P. and M. Spiteller. 2007. Fluoroquinolone antibiotics in the environment. *Reviews of Environmental Contamination and Toxicology* 191: 131–162.

Sun, K., Y. Shi, X. Wang, J. Rasmussen, Z. Li, and J. Zhu. 2017. Organokaolin for the uptake of pharmaceuticals diclofenac and chloramphenicol from water. *Chemical Engineering Journal* 330: 1128–1136.

Suwandi, A.C., N. Indraswati, and S. Ismadji. 2012. Adsorption of N-methylated diaminotriphenilmethane dye (malachite green) on natural rarasaponin modified kaolin. *Desalination and Water Treatment* 41: 342–355.

Szczerba, M., Y. Deng, and M. Kowalik-Hyla. 2022. Molecular modeling to predict the optimal mineralogy of smectites as binders of aflatoxin. *Clays and Clay Minerals* 70: 824–836.

Tabak, A., N. Baltas, B. Afsin, M. Emirik, B. Caglar, and E. Eren. 2010. Adsorption of reactive red 120 from aqueous solutions by cetylpyridinium-bentonite. *Journal of Chemical Technology & Biotechnology* 85: 1199–1207.

Talbert, R.E. and O.H. Fletchall. 1965. The adsorption of some *s*-triazines in soils. *Weeds* 13: 46–52.

Talibudeen, O. 1954. Complex formation between montmorillonoid clays and amino acids and proteins. *Transactions of the Faraday Society* 51: 582–590.
Tapia Estévez, M.J., F. López Arbeloa, T. López Arbeloa, and I. López Arbeloa. 1993. Absorption and fluorescence properties of rhodamine 6G adsorbed on aqueous suspensions of Wyoming montmorillonite. *Langmuir* 9: 3629–3634.
Taylor, J.B. and J.S. Rowlinson. 1955. Thermodynamic properties of aqueous solutions of glucose. *Transactions of the Faraday Society* 51: 1183–1192.
Teepakakorn, A.P., S. Bureekaew, and M. Ogawa. 2018. Adsorption-induced dye stability of cationic dyes on clay nanosheets. *Langmuir* 34: 14069–14075.
Teepakakorn, A.P., T. Yamaguchi, and M. Ogawa. 2019. The improved stability of molecular guests by the confinement into nanospaces. *Chemistry Letters* 48: 398–409.
Tehrani-Bagha, A.R., H. Nikkar, N.M. Mahmoodi, M. Markazi, and F.M. Menger. 2011. The sorption of cationic dyes onto kaolin: Kinetic, isotherm and thermodynamic studies. *Desalination* 266: 274–280.
Terce, M. and R. Calvet. 1978. Adsorption of several herbicides by montmorillonite, kaolinite and illite clays. *Chemosphere* 4: 365–370.
Theng, B.K.G. 1972. Adsorption of ammonium and some primary *n*-alkylammonium cations by soil allophane. *Nature* 238: 150–151.
Theng, B.K.G. 1974. Complexes of clay minerals with amino acids and peptides. *Chemie der Erde* 33: 125–144.
Theng, B.K.G. 2012. *Formation and Properties of Clay-Polymer Complexes*, 2nd edition. Amsterdam, The Netherlands: Elsevier.
Theng, B.K.G. 2018. *Clay Mineral Catalysis of Organic Reactions*. Boca Raton, FL: CRC Press.
Theng, B.K.G., R.S. Kookana, and A. Rahman. 1999. Environmental concerns of pesticides in soil and ground water and management strategies in Oceania. In *Soils and Groundwater Pollution and Remediation: Asia, Africa, and Oceania*, (Eds.) P.M. Huang and I.K. Iskandar, pp. 42–79. Boca Raton, FL: Lewis Publishers.
Thiebault, T. 2020. Raw and modified clays and clay minerals for the removal of pharmaceutical products from aqueous solutions: State of the art and future perspectives. *Critical Reviews in Environmental Science and Technology* 50: 1451–1514.
Thiebault, T. and M. Boussafir. 2019. Adsorption mechanisms of psychoactive drugs onto montmorillonite. *Colloid and Interface Science Communications* 30: 100183.
Thiebault, T., M. Boussafir, L. Le Forestier, C. Le Milbeau, L. Monnin, and R. Guégan. 2016. Competitive adsorption of a pool of pharmaceuticals onto a raw clay mineral. *RSC Advances* 6: 65257–65265.
Thiebault, T., R. Guégan, and M. Boussafir. 2015. Adsorption mechanisms of emerging micro-pollutants with a clay mineral: Case of tramadol and doxepin pharmaceutical products. *Journal of Colloid and Interface Science* 453: 1–8.
Thiele-Bruhn, S. 2003. Pharmaceutical antibiotic compounds in soils – a review. *Journal of Plant Nutrition and Soil Science* 166: 145–167.
Thompson, T.D. and G.W. Brindley. 1969. Absorption of pyrimidines, purines and nucleosides by Na-, Mg-, and Cu(II)-illite. (Clay-organic studies XVI). *American Mineralogist* 54: 858–868.
Tong, D.S., C.H.C. Zhou, Y. Lu, H. Yu, G.F. Zhang, and W.H. Yu. 2010. Adsorption of Acid Red G dye on octadecyl trimethylammonium montmorillonite. *Applied Clay Science* 50: 427–431.
Turner, A. and M.C. Rawling. 2001. The influence of salting out on the sorption of neutral organic compounds in estuaries. *Water Research* 35: 4379–4389.
Tzou, Y.-M., Y.-T. Chan, S.-E. Chen et al. 2019. Use 3-D tomography to reveal structural modification of bentonite-enriched clay by nonionic surfactants: Application of organo-clay composites to detoxify aflatoxin B1 in chickens. *Journal of Hazardous Materials* 375: 312–319.
Ullah, H., M. Nafees, F. Iqbal, M.S. Awan, A. Shah, and A. Waseem. 2017. Adsorption kinetics of malachite green and methylene blue from aqueous solutions using surfactant-modified organoclays. *Acta Chimica Slovenica* 64: 449–460.
Ullah, S., A.U. Rahman, F. Ullah et al. 2021. Adsorption of malachite green dye onto mesoporous natural inorganic clays: Their equilibrium isotherm and kinetic studies. *Water* 13: 965. https://doi.org/103390/w13070965.
Undabeytia, T., S. Nir, T. Polubesova, G. Rytwo, E. Morillo, and C. Maqueda. 1999. Adsorption-desorption of chlordimeform on montmorillonite: Effect of clay aggregation and competitive adsorption with cadmium. *Environmental Science & Technology* 33: 864–869.
Undabeytia, T., S. Nir, and B. Rubin. 2000. Organo-clay formulations of the hydrophobic herbicide norflurazon yield reduced leaching. *Journal of Agricultural and Food Chemistry* 48: 4767–4773.

Undabeytia, T., U. Shuali, S. Nir, and B. Rubin. 2021. Applications of chemically modified clay minerals and clays to water purification and slow release formulations of herbicides. *Minerals* 11: 9. https://dx.doi.org/10.3390/min11010009.

Valdés, L., I. Pérez, L.C. de Ménorval, E. Altshuler, J.O. Fossum, and A. Rivera. 2017. A simple way for targeted delivery of an antibiotic: *In vivo* evaluation of a nanoclay-based composite. *PLoS One* 12(11): e0187879. https://doi.org/10.1371/journal.pone.0187879.

Vallova, S., E. Plevova, K. Smutna et al. 2022. Removal of analgesics from aqueous solutions onto montmorillonite KSF. *Journal of Thermal Analysis and Calorimetry* 147: 1973–1981.

Vicente Vilas, V., B. Mathiasch, J. Huth et al. 2010. Synthesis and characterization of the hybrid clay-based material montmorillonite-melanoidin: A potential soil model. *Soil Science Society of America Journal* 74: 2239–2245.

Villafañe-Barajas, S.A., J.P.T. Baú, M. Colín-García et al. 2018. Salinity effects on the adsorption of nucleic acid compounds on Na-montmorillonite: A prebiotic chemistry experiment. *Origins of Life and Evolution of Biospheres* 48: 181–200.

Vimond-Laboudigue, A., M.-H. Baron, J.-C. Merlin, and R. Prost. 1996. Processus d'adsorption du dinoseb sur l'hectorite- et la vermiculite-decylammonium. *Clay Minerals* 31: 95–111.

Viseras, C., E. Carazo, A. Borrego-Sánchez et al. 2019. Clay minerals in skin drug delivery. *Clays and Clay Minerals* 67: 59–71.

Viseras, C., P. Cerezo, R. Sánchez, I. Salcedo, and C. Aguzzi. 2010. Current challenges in clay minerals for drug delivery. *Applied Clay Science* 48: 291–295.

Viseras, C., R. Sánchez-Espero, R. Palumbo et al. 2021. Clays in cosmetics and personal-care products. *Clays and Clay Minerals* 69: 561–575.

Walker, G.F. and W.G. Garrett. 1961. Complexes of vermiculite with amino acids. *Nature* 191: 1389.

Wang, C., Y. Ding, B.J. Teppen, S.A. Boyd, C. Song, and H. Li. 2009. Role of interlayer hydration in lincomycin sorption by smectite clays. *Environmental Science & Technology* 43: 6171–6176.

Wang, C.-C., L.-C. Juang, C.-K. Lee, T.-C. Hsu, J.-F. Lee, and H.-P. Chao. 2004. Effects of exchanged surfactant cations on the pore structure and adsorption characteristics of montmorillonite. *Journal of Colloid and Interface Science* 280: 27–35.

Wang, C.-J., Z. Li, and W.-T. Jiang. 2011. Adsorption of ciprofloxacin on 2:1 dioctahedral clay minerals. *Applied Clay Science* 53: 723–728.

Wang, C.-J., Z. Li, and W.-T. Jiang, J.-S. Jean, and C.-C. Liu. 2010b. Cation exchange interaction between antibiotic ciprofloxacin and montmorillonite. *Journal of Hazardous Materials* 183: 309–314.

Wang, G., C. Lian, Y. Xi, Z. Sun, and S. Zheng. 2018b. Evaluation of nonionic surfactant modified montmorillonite as mycotoxins adsorbent for aflatoxin B_1 and zearalenone. *Journal of Colloid and Interface Science* 518: 48–56.

Wang, G., Y. Miao, Z. Sun, and S. Zheng. 2018a. Simultaneous adsorption of aflatoxin B_1 and zearalenone by mono- and di-alkyl cationic surfactants modified montmorillonites. *Journal of Colloid and Interface Science* 511: 67–76.

Wang, G., S. Wang, Z. Sun, S. Zheng, and Y. Xi. 2017b. Structures of nonionic surfactant modified montmorillonites and their enhanced adsorption capacities toward a cationic organic dye. *Applied Clay Science* 148: 1–10.

Wang, J., J. Hu, and S. Zhang. 2010a. Studies on the sorption of tetracycline onto clays and marine sediment from seawater. *Journal of Colloid and Interface Science* 349: 578–582.

Wang, M., S.E. Hearon, N.M. Johnson, and T.D. Phillips. 2019a. Development of broad-acting clays for the tight adsorption of benz[α]pyrene and aldicarb. *Applied Clay Science* 168: 196–202.

Wang, M.K., S.L. Wang, and W.M. Wang. 1996. Rapid estimation of cation-exchange capacities of soils and clays with methylene blue exchange. *Soil Science Society of America Journal* 60: 138–141.

Wang, Y., X. Jiang, L. Zhou et al. 2013. A comparison of new Gemini surfactant modified clay with its monomer modified one: Characterization and application in methyl orange removal. *Journal of Chemical & Engineering Data* 58: 1760–1771.

Wang, Z., W. Zheng, Z. Zhang et al. 2017a. Formation of a kaolinite-serine intercalation compound via exchange of the pre-intercalated transition molecules in kaolinite with serine. *Applied Clay Science* 135: 378–385.

Wang, J., M. Gao, T. Shen, M. Yu, Y. Xiang, and J. Liu. 2019b. Insights into the efficient adsorption of Rhodamine B on tunable organo-vermiculites. *Journal of Hazardous Materials* 366: 501–511.

Wang, M., S.E. Hearon, and T.D. Phillips. 2020. A high-capacity bentonite clay for the sorption of aflatoxins. *Food Additives & Contaminants: Part A* 37: 332–341.

Wang, M., K.J. Rivenbark, and T.D. Phillips. 2023. Adsorption and detoxification of glyphosate and aminomethylphosphonic acid by montmorillonite clays. *Environmental Science and Pollution Research* 30: 11417–11430.

Wang, Z., Z. Huang, J. Han, G. Xie, and J. Liu. 2021. Polyvalent metal ion promotes adsorption of DNA oligonucleotides by montmorillonite. *Langmuir* 37: 1037–1044.

Weber, J.B. 1966. Molecular structure and pH effects on the adsorption of 13 *s*-triazine compounds on montmorillonite clay. *American Mineralogist* 51: 1657–1670.

Weber, J.B. 1970a. Mechanisms of adsorption of s-triazines by clay colloids and factors affecting plant availability. *Residue Reviews* 32: 93–130.

Weber, J.B. 1970b. Adsorption of s-triazines by montmorillonite as a function of pH and molecular structure. *Soil Science Society of America Proceedings* 34: 401–404.

Weber, J.B., P.W. Perry, and R.P. Upchurch. 1965. The influence of temperature and time on the adsorption of paraquat, diquat, 2,4-D, and prometone by clays, charcoal, and an anion-exchange resin. *Soil Science Society of America Proceedings* 29: 678–687.

Weber, J.B. and S.B. Weed. 1968. Adsorption and desorption of diquat, paraquat, and prometone by montmorillonitic and kaolinitic clay minerals. *Soil Science Society of America Proceedings* 32: 485–487.

Weed, S.B. and J.B. Weber. 1969. The effect of cation exchange capacity on the retention of diquat^{2+} and paraquat^{2+} by three-layer type clay minerals. I. Adsorption and release. *Soil Science Society of America Proceedings* 33: 379–382.

Weiss, A. and G. Roloff. 1966. Über die Einlagerung symmetrischer Triglyceride in quellungsfähige Schichtsilicate. *Proceedings of the International Clay Conference, Jerusalem* 1: 263–275.

Wu, C., X. Lou, A. Huang, M. Zhang, and L. Ma. 2020. Thermodynamics and kinetics of pretilachlor adsorption: Implication to controlled release from organobentonites. *Applied Clay Science* 190: 105566.

Wu, L., X. Bao, H. Zhong, Y. Pan, G. Lv, and L. Liao. 2019d. Preparation of a novel clay/dye composite and its application in contaminant detection. *Clays and Clay Minerals* 67: 244–251.

Wu, L., Q. Wang, N. Tang, and L. Gao. 2019c. Preparation of ionic liquids/montmorillonite composites and its application for diclofenac sodium removal. *Journal of Contaminant Hydrology* 220: 1–5.

Wu, M., S. Zhao, R. Jing et al. 2019b. Competitive adsorption of antibiotic tetracycline and ciprofloxacin on montmorillonite. *Applied Clay Science* 180: 105175.

Wu, M., S. Zhao, M. Tang et al. 2019a. Adsorption of sulfamethoxazole and tetracycline on montmorillonite in single and binary systems. *Colloids and Surfaces A* 575: 264–270.

Wu, Q., Z. Li, and H. Hong. 2013. Adsorption of the quinolone antibiotic nalidixic acid onto montmorillonite and kaolinite. *Applied Clay Science* 74: 66–73.

Wu, Q., Z. Li, H. Hong, K. Yin, and L. Tie. 2010. Adsorption and intercalation of ciprofloxacin on montmorillonite. *Applied Clay Science* 50: 204–211.

Wu, J., W. Zhang, C. Li, and E. Hu. 2021. Effects of Fe(III) and Cu(II) on the sorption of s-triazine herbicides on clay minerals. *Journal of Hazardous Materials* 418: 126232.

Xu, P., S. Dai, J. Wang et al. 2016. Preventive obesity agent montmorillonite adsorbs dietary lipids and enhances lipid excretion from the digestive tract. *Scientific Reports* 6: 19659. https://doi.org/10.1038/srep19659.

Xu, P., F. Hong, J. Wang et al. 2017. Microbiome remodeling *via* the montmorillonite adsorption-excretion axis prevents obesity-related metabolic disorders. *EBioMedicine* 16: 251–261.

Yamagishi, A. and M. Soma. 1981. Aliphatic tail effects on adsorption of acridine orange cation on a colloidal surface of montmorillonite. *Journal of Physical Chemistry* 85: 3090–3092.

Yamamoto, K., T. Shiono, R. Yoshimura, Y. Matsui, and M. Yoneda. 2018. Influence of hydrophilicity on adsorption of caffeine onto montmorillonite. *Adsorption Science & Technology* 36: 967–981.

Yamane, V.K. and R.E. Green. 1972. Adsorption of ametrine and atrazine on an Oxisol, montmorillonite, and charcoal in relation to pH and solubility effects. *Soil Science Society of America Proceedings* 36: 58–64.

Yan, H., P. Zhang, X. Chen et al. 2020. Preparation and characterization of octyl phenyl polyoxyethylene ether modified organo-montmorillonite for ibuprofen-controlled release. *Applied Clay Science* 189: 105519.

Yan, W., S. Hu, and C. Jing. 2012. Enrofloxacin sorption on smectite clays: Effects of pH, cations, and humic acid. *Journal of Colloid and Interface Science* 372: 141–147.

Yang, S., F. Zhao, Q. Sang et al. 2021. Investigation of 3-aminopropyltriethoxysilane modifying attapulgite for Congo red removal: Mechanisms and site energy distribution. *Powder Technology* 383: 74–83.

Yapar, S., M. Ateş, and G. Özdemir. 2017. Preparation and characterization of sodium lauroyl sarcosinate adsorbed on cetylpyridinium-montmorillonite as a possible antibacterial agent. *Applied Clay Science* 150: 16–22.

Yariv, S. 2002. Staining of clay minerals and visible absorption spectroscopy of dye-clay complexes. In *Organo-Clay Complexes and Interactions*, (Eds.) S. Yariv and H. Cross, pp. 463–566. New York: Marcel Dekker.

Yariv, S. and D. Lurie. 1971. Metachromasy in clay minerals. Part 1. Sorption of methylene-blue by montmorillonite. *Israel Journal of Chemistry* 9: 537–552.

Yariv, S., J.D. Russell, and V.C. Farmer. 1966. Infrared study of the adsorption of benzoic acid and nitrobenzene in montmorillonite. *Israel Journal of Chemistry* 4: 201–213.

Yariv, S. and S. Shoval. 1982. The effects of thermal treatment on associations between fatty acids and montmorillonite. *Israel Journal of Chemistry* 22: 259–265.

Yaron-Marcovich, D., S. Nir, and Y. Chen. 2004. Fluridone adsorption-desorption on organo-clays. *Applied Clay Science* 24: 167–175.

Yeasmin, S., B. Singh, R.S. Kookana, M. Farrell, D.L. Sparks, and C.T. Johnston. 2014. Influence of mineral characteristics on the retention of low molecular weight organic compounds: A batch sorption-desorption and ATR-FTIR study. *Journal of Colloid and Interface Science* 432: 246–257.

Yener, N., C. Biçer, M. Önal, and Y. Sarikaya. 2012. Simultaneous determination of cation exchange capacity and surface area of acid activated bentonite powders by methylene blue sorption. *Applied Surface Science* 258: 2534–2539.

Yermiyahu, Z., I. Lapides, and S. Yariv. 2007. Thermo-visible-absorption spectroscopy study of the protonated Congo-red montmorillonite complex. *Applied Clay Science* 37: 1–11.

Yesi, F.P.S., Y.-H. Ju, F.E. Soetaredjo, and S. Ismadji. 2010. Adsorption of acid blue 129 from aqueous solutions onto raw and surfactant-modified bentonite: Application of temperature-dependent forms of adsorption isotherms. *Adsorption Science & Technology* 28: 847–868.

Yu, C.-H., S.Q. Newton, M.A. Norman, D.M. Miller, L. Schäfer, and B.J. Teppen. 2000. Molecular dynamics simulations of the adsorption of methylene blue at clay mineral surfaces. *Clays and Clay Minerals* 48: 665–681.

Yuan, G.D., B.K.G. Theng, G.J. Churchman and W.P. Gates. 2013. Clays and clay minerals for pollution control. In *Handbook of Clay Science*, 2nd edition. Developments in Clay Science, Vol. 5B, (Eds.) F. Bergaya and G. Lagaly, pp. 587–644. Amsterdam, The Netherlands: Elsevier.

Yue, D., Y. Jing, J. Ma, C. Xia, X. Yin, and Y. Jia. 2011. Removal of neutral red from aqueous solution by using modified hectorite. *Desalination* 267: 9–15.

Yukselen, Y. and A. Kaya. 2008. Suitability of the methylene blue test for surface area, cation exchange capacity and swell potential determination of clayey soils. *Engineering Geology* 102: 38–45.

Zang, W., M. Gao, T. Shen, F. Ding, and J. Wang. 2017. Facile modification of homoionic vermiculites by a Gemini surfactant: Comparative adsorption exemplified by methyl orange. *Colloids and Surfaces A* 533: 99–108.

Zhang, N., X. Han, Y. Zhao et al. 2022. Removal of aflatoxin B_1 and zearalenone by clay mineral materials: In the animal industry and environment. *Applied Clay Science* 228: 106614.

Zhang, W., L. Zhang, X. Jiang, X. Liu, Y. Li, and Y. Zhang. 2020a. Enhanced adsorption removal of aflatoxin B1, zearalenone and deoxynivalenol from dairy cow rumen fluid by modified nano-montmorillonite and evaluation of its mechanism. *Animal Feed Science and Technology* 259: 114366.

Zhang, Y., Y. Li, and Y. Zhang. 2020b. Preparation and intercalation structure model of halloysite-stearic acid intercalation compound. *Applied Clay Science* 187: 105451.

Zhao, H., W.F. Jaynes, and G.F. Vance. 1996. Sorption of the ionizable organic compound, dicamba (3,6-dichloro-2-methoxy benzoic acid) by organo-clays. *Chemosphere* 33: 2089–2100.

Zhao, Y., X. Gu, S. Gao, J. Geng, and X. Wang. 2012. Adsorption of tetracycline (TC) onto montmorillonite: Cations and humic acid effects. *Geoderma* 183–184: 12–18.

Zhu, L., L. Wang, and Y. Xu. 2017. Chitosan and surfactant co-modified montmorillonite: A multifunctional adsorbent for contaminant removal. *Applied Clay Science* 146: 35–42.

Zhu, R., Q. Chen, Q. Zhou, Y. Xi, J. Zhu, and H. He. 2016. Adsorbents based on montmorillonite for contaminant removal from water: A review. *Applied Clay Science* 123: 239–258.

Zhu, S., M. Xia, Y. Chu et al. 2019. Adsorption and desorption of Pb(II) on L-lysine modified montmorillonite and the simulation of interlayer structure. *Applied Clay Science* 169: 40–47.

Zohra, B., K. Aicha, S. Fatima, B. Nourredine, and D. Zoubir. 2007. Adsorption of Direct Red 2 on bentonite modified by cetyltrimethylammonium bromide. *Chemical Engineering Journal* 136: 295–305.

4 Interactions of Clay Minerals with Positively Charged Organic Compounds

4.1 INTRODUCTION

As described in Chapter 1, smectites (and vermiculites) carry a permanent negative charge as a result of isomorphous substitution within their structure. This positive charge deficiency is balanced by sorption of an equivalent amount of extraneous inorganic cations, such as sodium, potassium, calcium, and magnesium. Likewise, positively charged organic species can serve to balance the anionic clay mineral framework. Not surprisingly, many early references to the clay-organic interaction are concerned with the reactivity of clay minerals toward organic cations, or with neutralizing acid-treated clay minerals with organic bases. However, the number and variety of organic compounds that can acquire a positive charge, or act as a base, are rather limited as compared with uncharged (polar) organic molecules.

Many positively charged organic compounds contain nitrogen in their structure. Among these amines, and molecules containing amine functional groups, occupy a prominent place. As mentioned in previous chapters, adsorbed amines may be protonated when a ready source of protons is available at the silicate surface as in acid-treated montmorillonite. Cation exchange is the principal mechanism underlying the interaction of montmorillonite with positively charged pesticides, pharmaceuticals, and dyes (cf. Chapter 3). Thus, adsorption generally leads to the replacement of an equivalent amount of inorganic counterions initially present at the clay surface. Cation exchange also lies behind the formation of organoclays through intercalation of alkylammonium, quaternary ammonium cations, and cationic surfactants with hydrophobic interactions and entropy effects playing a subsidiary role.

4.2 MECHANISMS OF COMPLEX FORMATION

As early as 1916, Lloyd noted that samples of fuller's earth could remove alkaloids from aqueous solution. It was not until the 1930s, however, that the interactions of clay minerals with organic cations and bases began to be systematically investigated. For example, Smith (1934) reported that montmorillonite could take up hydrazine, ammonia, *n*-amylamine, piperidine, di-*n*-amylamine, nicotine, and strychnine from their respective aqueous hydrochloride solution with the extent of adsorption increasing in the same order. The process appeared to be one of cation exchange since the resultant complexes were stable against washing with water but treatment with a 1% NaCl solution released appreciable quantities of the adsorbed species. Subsequently, Gieseking and Jenny (1936) and Gieseking (1939) noted that some organic bases and cations could either replace the inorganic cations initially present or neutralize the hydrogen ions at the montmorillonite surface. Using H^+-montmorillonite, Hendricks (1941) also found that aliphatic and aromatic amines could neutralize the hydrogen ions up, or close to, the cation exchange capacity (CEC) of the clay mineral sample but large alkaloids (brucine, codeine) failed to do so. He postulated that the difference between the total amount of hydrogen ions and that available for reacting with the alkaloid represents the quantity 'covered' by the organic base. This 'cover-up' effect came into play when the size of the adsorbed organic molecule exceeded the area per exchange site (~ 0.8 nm^2 for montmorillonite).

DOI: 10.1201/9781003080244-4

A similar effect was reported by Chakravarti (1957) for the interaction of montmorillonite with large quaternary ammonium and pyridinium cations.

The importance of Hendricks' (1941) work is the recognition that the adsorption and intercalation of such organic bases as amines, alkaloids, purines, and nucleosides (cf. Chapter 3) involved both electrostatic (coulombic) and van der Waals forces. The basal spacing of the resultant interlayer complexes is influenced more by the thickness than by the overall dimension of the intercalated molecule. This point is illustrated in Figure 2.4 for smectites (of varying charge density) exchanged with trimethylphenylammonium (TMPA) ions.

The interlayer arrangement long-chain alkylammonium (and quaternary ammonium) ions, on the other hand, is dependent on the alkyl chain length (n_c) of the solute and the layer charge of the smectite (Intachai et al. 2021). A stable monolayer arrangement with a basal spacing of 1.37 nm is obtained for $n_c < 11$ when the area of the alkylammonium ion is smaller than the area per exchange site. When the area occupied by organic cation area exceeds that of the exchange site, a bilayer arrangement may obtain, giving a basal spacing of 1.77 nm. A pseudo-trilayer arrangement with a basal spacing of ~ 2.2 nm is formed with high charge smectites and/or large n_c values (Lagaly 1981; Hackett et al. 1998; Slade and Gates 2004; He et al. 2006a). At high solute to cation exchange capacity ratio, and with vermiculite, the paraffin-type arrangement where the alkyl chain is inclined at a high angle to the silicate surface, giving basal spacings > 2.2 nm, is preferred (Walker 1967; Williams-Daryn and Thomas 2002; He et al. 2006b; Kooli 2014). These points are illustrated by Figure 4.1. A similar arrangement also obtains for complexes of smectite with quaternary ammonium ions containing two or more alkyl chains through the imposition of layer charge density and

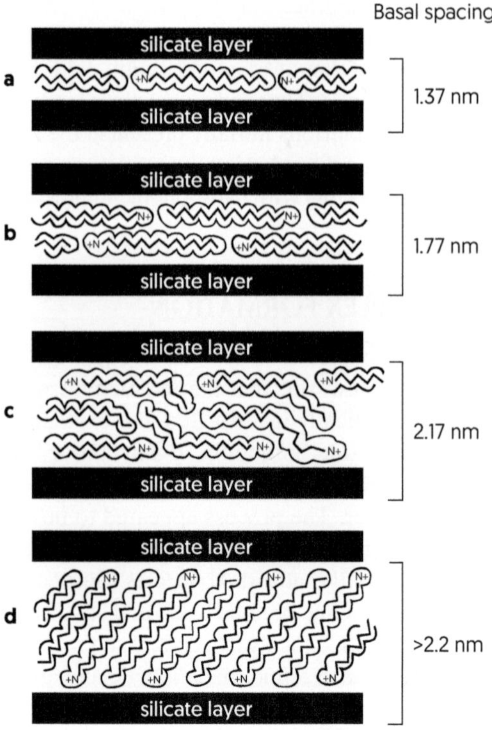

FIGURE 4.1 Possible arrangements of long-chain alkylammonium ions in the interlayer space of expanding 2:1 type layer silicates together with their respective basal spacings. (a) monolayer; (b) bilayer; (c) pseudo-trilayer; and (d) paraffin-type. (Reused with permission from *Clay Mineral Catalysis of Organic Reactions*, by B.K.G. Theng, Figure 3.6, p. 179, CRC Press, 2018.)

the requirement for close van der Waals contact between adjacent alkyl chains (Lagaly et al. 2006; Theng et al. 2008; He et al. 2010).

The influence of van der Waals forces on the adsorption of large organic cations by clay minerals was earlier indicated by Grim et al. (1947) who showed that the uptake of *n*-butyl-, *n*-dodecyl-, and ethyldimethyloctadecenyl (EDMOD)-ammonium ions by potassium-exchanged kaolinite, illite, and montmorillonite was accompanied by a stoichiometric replacement of K^+ counterions. For the relatively small *n*-butylammonium ion the maximum amount adsorbed did not exceed the cation exchange capacity (CEC) of the clay minerals even when a large excess of the organic cation was present in solution. On the other hand, the two larger organic cations were taken up in excess of the CEC. Grim et al. (1947) suggested that this excess was adsorbed by van der Waals forces. In the case of EDMOD-ammonium, the excess quantity was adsorbed as the corresponding bromide salt since there was little change in suspension pH. A similar process was postulated for the adsorption of benzyldimethyl-dodecylammonium (Tahani et al. 1999) and hexadecylammonium (Lee and Kim 2002a; Naranjo et al. 2013) by montmorillonite. For *n*-dodecylammonium, however, the equilibrium pH (~ 5.6) was appreciably less than that of the original solution (pH ~7), indicating that the excess was partly present in the form of the corresponding free amine.

Adsorption in excess of the CEC was also reported by Kurilenko and Mikhalyuk (1959) for the uptake of methyloctadecyl-, *n*-dodecyl-, tri-*n*-decyl-, and trimethyloctadecyl-ammonium ions by Na^+- and Ca^{2+}-montmorillonite. Similarly, Morel and Hénin (1956) observed excess uptake of hexamethylenediamine, dodecylpropyldiamine, and polyvinylamine cations by kaolinite, montmorillonite, and sepiolite, while the amount adsorbed beyond the CEC increased with solute molecular weight.

Barrer and Kelsey (1961) have pointed out that at least qualitatively correct interpretation of prominent physical aspects may be obtained by applying thermodynamic functions to adsorption processes in clay-organic systems. On this basis, several workers have used a thermodynamic approach to characterize the interactions of clay minerals with organic cations. To this end, Cowan and White (1958) determined the isotherms (at 20°C) for the adsorption of a homologous series of primary *n*-alkylammonium ions from ethyl-(C_2) to tetradecylammonium (C_{14}) by Na^+-montmorillonite. The isotherms belonged to the L-class (Giles et al. 1960) for which both the initial slope and the plateau adsorption increased with molecular weight. Organic cations longer than C_8 were taken up beyond the CEC of the clay sample. Similar findings were reported by Sieskind and Wey (1958).

The cation exchange process may be represented by Equations 4.2 and 4.3 (equilibria),

$$\mathrm{Na^+}_{\mathrm{clay}} + \mathrm{AH^+}_{\mathrm{soln}} \rightleftharpoons \mathrm{AH^+}_{\mathrm{clay}} + \mathrm{Na^+}_{\mathrm{soln}} \qquad \text{(Equation 4.1)}$$

$$\mathrm{AH^+}_{\mathrm{clay}} + \mathrm{AHCl}_{\mathrm{soln}} \rightleftharpoons \left(\mathrm{AH^+}_{\mathrm{clay}}\right)\left(\mathrm{AHCl}_{\mathrm{clay}}\right) \qquad \text{(Equation 4.2)}$$

and

$$\left(\mathrm{AH^+}_{\mathrm{clay}}\right)\left(\mathrm{AHCl}_{\mathrm{clay}}\right) \rightleftharpoons \mathrm{AH^+}_{\mathrm{clay}} + A_{\mathrm{clay}} + \mathrm{HCl}_{\mathrm{soln}} \qquad \text{(Equation 4.3)}$$

where A and AH^+ refer to the free amine and the amine cation (alkylammonium), respectively.

Equation 4.1 applies to all alkylamine cations used and proceeds until the sodium counterions are completely, or nearly completely, replaced. Equations 4.2 and 4.3 represent situations where adsorption exceeds the CEC of the clay mineral sample, the excess adsorbate being present either as the amine salt, or as the free amine. If Equation 4.2 predominates, the corresponding anion (in this case chloride) would also be taken up. Equation 4.3 would lead to a decrease in suspension pH and appears to be the dominant mechanism in Cowan's and White's (1958) system.

By applying the mass-action relationship to Equation 4.1, the (non-standard) free energy change of the cation exchange reaction, $-\Delta G^m$, may be derived (see later). Cowan and White (1958) found a regular increase of $-\Delta G^m$ values with molecular weight, amounting to ~ 0.63 kJ/mol per CH_2 group from C_2 to C_5 and ~ 1.67 kJ/mol from C_5 to C_{10}. This relationship between free energy change and molecular weight (or chain length) is of general applicability to the adsorption of organic compounds by montmorillonite and may be attributed to the increased contribution of van der Waals forces to the adsorption energy.

Theng et al. (1967) have extended Cowan's and White's measurements to the di-, tri-, and tetra-alkylammonium ions, ranging in length from methyl (C_1) to butyl (C_4) to clarify the effect of cation size (and shape) on relative adsorption affinity. The exchange isotherms for the ethyl (Et) derivatives with Na^+-montmorillonite are shown in Figure 4.2a where the equivalent fraction of the organic cation in solution, S_{AH^+}, is plotted against the corresponding equivalent fraction on the clay surface, Z_{AH^+}. The diagonal (broken line) of the 'square plot' denotes the line of no preference for either one of the exchanging cations. The isotherms, all of which are of the L-type, indicate that the alkylammonium ion is preferred to sodium at all values of S_{AH^+}. Moreover, the extent of preference increases with molecular weight in line with previous findings for the monoalkylammonium series (Cowan and White 1958; Sieskind and Wey 1958).

The equilibrium constant, K, of the reaction may be estimated using the simplified form of the equation developed by Gaines and Thomas (1953),

$$\ln K = \int_0^1 \ln K' dZ_{AH^+} \quad \text{(Equation 4.4)}$$

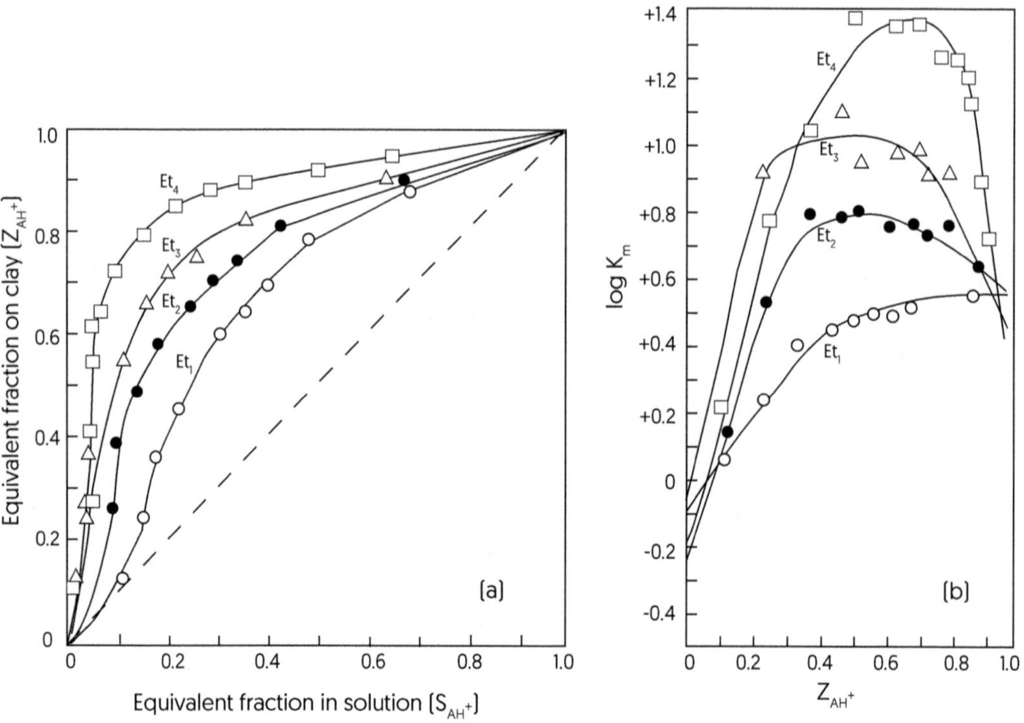

FIGURE 4.2 (a) relationship between the equivalent fraction in solution and the corresponding fraction adsorbed on montmorillonite for monoethylammonium (Et_1), diethylammonium (Et_2), triethylammonium (Et_3), and tetraethylammonium (Et_4) ions; (b) variation of the logarithm of the molar selectivity coefficient (K_m) with the equivalent fraction of adsorbed ethylammonium ions. (From Theng et al. 1967, Figure 2. Reproduced with kind permission of The Mineralogical Society of the UK and Ireland, publisher of *Clay Minerals*.)

where $K' = K_m\gamma_{Na+}/\gamma_{AH+}$. The activity coefficients of the respective cation in solution, γ_{Na+} and γ_{AH+}, may be approximated by the mean activity coefficients for the corresponding chloride salt. Since for the dilute solutions (< 0.1 M), $\gamma_{NaCl}/\gamma_{AHCl}$ does not deviate significantly from unity (Whitlow and Felsing 1944), $K' \approx K_m$, allowing K to be obtained from a plot of log K_m against Z_{AH}^+ (Figure 4.2b) by graphical integration between the limits $Z_{AH}^+ = 0$ and $Z_{AH}^+ = 1$.

It is implicitly assumed that Equation 4.1 is reversible, but as Fripiat et al. (1965) have pointed out, this type of macroscopic reversibility is seldom realized since the interlayer separation in a montmorillonite quasi-crystal varies during the exchange reaction. Basal spacing measurements (Theng et al. 1967) indicate that the replacement of sodium by alkylammonium leads to interlayer contraction and a less dispersed state. Vansant and Uytterhoeven (1972), however, have reported that the exchange of sodium for short-chain primary *n*-alkylammonium ions (C_1 to C_4) is essentially reversible.

Thus, K is best regarded as an affinity coefficient rather than as a true thermodynamic equilibrium constant. The (non-standard) free energy change of the exchange process, $-\Delta G^m$, may therefore be estimated from the relationship,

$$-\Delta G^m = RT \ln K \qquad \text{(Equation 4.5)}$$

where R is the gas constant and T is the absolute temperature. As might be predicted from the isotherms (Figure 4.2a), $-\Delta G^m$ increases regularly with molecular weight of the alkylammonium ions (Figure 4.3) although the absolute values of $-\Delta G^m$ differ from those reported by Cowan and White (1958) for the same organic cations.

Earlier, Slabaugh and Kupka (1958) compared K_m values at a single point of the isotherm, corresponding to $Z_{AH}+ = 0.7$ for the adsorption of mono-*n*-butyl-, di-*n*-butyl-, and *n*-octylammonium ions by Ca^{2+}-montmorillonite. They noted that K_m for *n*-octylammonium was greater than that for di-*n*-butylammonium, probably because the latter cation had fewer hydrogen atoms attached to nitrogen that could be H-bonded to the oxygens of the silicate surface. Since K_m varies considerably with Z_{AH}^+ (Figure 4.2b), however, such comparisons would not be valid for the whole isotherm. Slabaugh (1958) also reported thermodynamic data for the exchange adsorption of some primary *n*-alkylammonium ions (C_1, C_4, C_8, C_{10}, C_{14}, and C_{18}) by Na^+-montmorillonite. In this instance, $-\Delta G^m$ was derived from the mean K_m values determined using different Na^+/AH^+ ratios at constant Na^+ concentration, giving an improved estimate of relative affinities. Zachara et al. (1990) also noted that K_m values for some nitrogen heterocyclic compounds (e.g., pyridine) were larger than those recorded by Theng et al. (1967) for alkylammonium ions. This observation was ascribed to the close interaction between the flatly intercalated pyridine, with its delocalized positive charge, and the silicate surface.

Weiss (1963b) reported that the affinity of alkylammonium ions for montmorillonite decreased in the order $R_1NH_3^+ > R_2NH_2^+ > R_3NH^+$ where R denoted an alkyl group. On the other hand, the reverse order of affinity, as reflected by the $-\Delta G^m$ values (Figure 4.3), was observed by Theng et al. (1967), and by Vansant and Peeters (1978) in the case of Laponite. For example, for the ethyl series $-\Delta G^m$ increases in the order $Et_1 < Et_2 < Et_3$. A similar order obtains for alkylammonium ions of identical molecular weight, that is, $Et_1 < Me_2$, $Bu_1 < Et_2$ and $Pr_1 < Me_3$ where Me, Et, Pr, and Bu refer to methyl-, ethyl-, *n*-propyl-, and *n*-butyl-ammonium, respectively, while the subscripts 1, 2, 3 denote the mono-, di-, and tri-alkyl derivative, in that order.

Theng et al. (1967) have suggested that the increase in affinity with cation size is due to the enhanced contribution of van der Waals forces to the adsorption energy. If so, this contribution would offset the reduction in the number or strength of cation-to-surface hydrogen bonding in the series primary to secondary to tertiary alkylammonium ions. In line with this suggestion, Laby and Walker (1970) have reported that N–H···O–Si bonding in complexes of vermiculite with primary *n*-alkylammonium ions is rather weak. H-bonding of this type has also been indicated by applying density functional theory to complexes of montmorillonite with primary, secondary, and tertiary

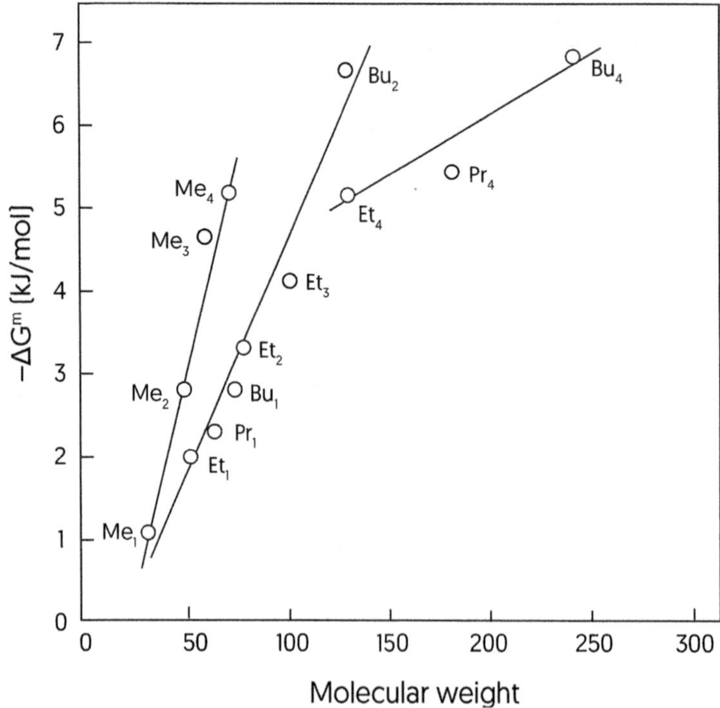

FIGURE 4.3 Relationship between molecular weight of alkylammonium ions (replacing sodium ions in montmorillonite) and the free energy change of the exchange process. Me_1, Me_2, Me_3, and Me_4 denote mono-, di-, tri-, and tetra-methylammonium ions; Et_1, Et_2, Et_3, and Et_4 denote mono-, di-, tri-, and tetra-ethylammonium ions; Pr_1 and Pr_4 denote mono- and tetra-propylammonium; Bu_1, Bu_2, and Bu_4 are mono-, di-, and tetra-butylammonium ions. (From Theng et al. 1967, Figure 3. Reproduced with kind permission of The Mineralogical Society of the UK and Ireland, publisher of *Clay Minerals*.)

n-alkylammonium ions (Scholtzová et al. 2016; Peng et al. 2018). Weak H-bonding interactions between the methyl groups of tetra-*n*-methylammonium ions and the basal oxygen atoms of the silicate layer have also been proposed by Scholtzová et al. (2013).

Similarly, molecular dynamics simulation by Heinz et al. (2007) has provided evidence for the formation of hydrogen bonds between the primary head groups of alkylammonium ions and siloxane surface oxygens (O–H distance ~ 150 pm) and between quaternary ammonium head groups (of higher lateral mobility) and interlayer surface oxygens (O–H distance ~ 290 pm). Near-IR spectroscopy of montmorillonite complexes with mono-, di-, tri-, and tetra-octylammonium ions further suggests that H-bonding between the head groups of the amine cations and basal oxygens restricts the mobility of the alkyl chains, while hydration of the complexes can cause partial dislodgement of the head groups from the ditrigonal cavities (Bizovská et al. 2018).

The data in Figure 4.3 would indicate that the shape of alkylammonium cations also influences their affinity for the montmorillonite surface. For example, the methylammonium ions (Me_1, Me_2, Me_3, Me_4) are more strongly adsorbed than would be expected from their molecular weight, while the reverse is true for Pr_4 (tetra-*n*-propylammonium) and Bu_4 (tetra-*n*-butylammonium). The shape of the organic cation apparently determines the relative extent to which van der Waals contact can be made between cation and silicate surface. Comparison between observed interlayer distances (Δ-values) and minimum molecular thicknesses indicates that the alkylammonium ions can be 'keyed' to varying extents into the ditrigonal depressions of the silicate surface (cf. Chapter 2). In this regard, the compact Me_1 (methylammonium) would be effectively keyed (Rowland and Weiss

1963), while the bulky and irregularly shaped Pr_4 and Bu_4 ions would be weakly keyed. As a result, the affinity of tetra-*n*-alkylammonium ions for montmorillonite is much less than would be expected from their respective molecular weights.

A more direct demonstration of the influence of van der Waals forces on the interaction of short-chain alkylammonium ions with montmorillonite has been provided by Fripiat et al. (1969) using infrared spectroscopy. In addition to the usual shifts in the stretching and deformation vibrations of the $-NH_3^+$ and $-NH_2^+$ groups when H-bonded to water, the intensities of the CH_2 and CH_3 deformation bands were markedly altered on exposing the clay-organic film to water vapour. These changes are particularly pronounced for the dialkylammonium cation, presumably because it is weakly keyed and hence can readily disengage from the siloxane surface. Since the frequency of the CH_2 and CH_3 bands remains sensibly constant whether water is present or absent, H-bonding between these groups and water or surface oxygens is not important. Fripiat et al. (1969) have proposed that, in the absence of water, the aliphatic chains are tightly held in the interlayer space with their axes lying parallel to silicate surface. The keying effect is reflected by the absence of the CH_3 symmetric deformation band in the methylammonium complex and of the wagging mode in the ethyl- and diethyl-ammonium complexes. Hydration of the complexes with dialkylammonium ions caused the alkyl chain to rotate but did not affect the orientation of monoalkylammonium ions.

Persuasive as these arguments are as regards the involvement of van der Waals forces in alkylammonium-exchanged montmorillonite, the thermodynamic study by Vansant and Uytterhoeven (1972) would indicate that cation hydration is equally important. By determining the isotherms for the adsorption of Me_1, Et_1, Pr_1, and Bu_1 ions by Na^+-montmorillonite at three different temperatures (4, 25, and 55°C), they were able to confirm that $-\Delta G$ increased with molecular weight but the increment of $-\Delta G$ per CH_2 group showed appreciable variations. Furthermore, the isotherms were insensitive to temperature, indicative of zero enthalpy change. The effect of increasing chain length, therefore, is to increase the entropy change (ΔS) from 10 J/K/mol (for Me_1) to 13.4 J/K/mol (for Bu_1). A large positive value for ΔS (~ 60 J/K/mol) has also been reported by Ghiaci et al. (2004) and Gamoudi et al. (2012) for the adsorption of hexadecyltrimethylammonium and hexadecylpyridinium ions, respectively.

Vansant and Uytterhoeven (1972) have suggested that variations in the hydration status of the cations together with electrostatic interactions between cation and clay surface contribute to the observed increase in $-\Delta G$ with molecular weight. In this regard, Maes et al. (1977) have proposed that the interlayer space provides a stable environment for alkylammonium ions through the intervention of hydration and van der Waals forces. Furthermore, Teppen and Aggarwal (2007) have suggested that the difference in exchange selectivity for two organic cations with the same head group is determined by the difference in their free energy of hydration. In comparing two organic cations with different head groups, the clay surface has an intrinsic preference for the molecule with the smaller head group. Nevertheless, the cation with the larger head group is selectively adsorbed because the aqueous solution has a stronger preference for the cation with a smaller head group. When the clay surface has been substantially occupied by organic cations, however, van der Waals forces would play a major role in determining cation exchange selectivity.

4.3 INTERLAYER ORGANIZATION OF ORGANIC CATIONS

Barrer and Brummer (1963) made the interesting observation that the basal spacings of (outgassed) Na^+-montmorillonite complexes with monomethyl- and tetramethyl-ammonium ions increased with the amount of organic cations adsorbed. They rationalized this finding in terms of random interstratification of sodium- and alkylammonium-rich layers within a given quasi-crystal. Using a wider range of alkylammonium ions, Theng et al. (1967) have noted a similar relationship between basal spacing and degree of organic cation saturation as did McBride and Mortland (1975) for the exchange adsorption of tetra-alkylammonium ions by Cu^{2+}-montmorillonite. Vansant and Uytterhoeven (1972) have suggested that on thermodynamic grounds this type of layer segregation is less stable than a system in which there is homogeneous mixing of the two exchanging cations in every layer. Nevertheless, layer demixing may occur when the partially exchanged complexes are dried.

Table 4.1 lists the basal spacings, measured Theng et al. (1967), of fully exchanged alkylammonium montmorillonites (after drying at 70°C). The values are greater by 0.01–0.06 nm than those reported by other workers for similar complexes (Barrer and MacLeod 1955; Barrer and Reay 1957; Diamond and Kinter 1963). The observed discrepancy may be ascribed to interlayer dehydration and/or heat-induced decomposition of the adsorbed amine cations (Chaussidon and Calvet 1965; Zhuang et al. 2019a) since the earlier basal spacing data refer to specimens that have been dried at 110°C (in air) or dehydrated by evacuation at 50°C. Furthermore, the alkyl chains become more disordered and liquid-like when the corresponding alkylammonium-montmorillonite complexes are heated (Vaia et al. 1994; Madejová et al. 2020).

If the thickness of a methyl or methylene group is taken as 0.4 nm (Pauling 1960), the Δ-values (interlayer separations) point to intercalation of a single layer of organic cations (Figure 4.4). Except perhaps for Pr_4 and Bu_4, none of the cations in Table 4.1 take up an area larger than the area per exchange site. There is no difficulty, therefore, in accommodating the full complement of the alkylammonium ions within a single layer. Table 4.1 also shows that the minimum thickness of the cations, derived from molecular ('Catalin') models, and corrected by taking 0.12 nm for the van der Waals radius of hydrogen (Pauling 1960), is always larger than the corresponding Δ-value. Since C–H···O–Si hydrogen bonding is not important in these systems (Dowdy and Mortland 1968; Mortland 1970), the apparent shortening of contact distances may be ascribed to keying of the organic cation into the siloxane oxygen surface of the montmorillonite layer.

TABLE 4.1

Basal Spacings and Δ-Values for Montmorillonite Complexes with Some Alkylammonium Ions Together with Minimum van der Waals Thicknesses of the Organic Cations Derived from 'Catalin' Molecular Models.

Alkylammonium ion	Basal spacing (nm) (1)	Δ-value[a] (nm) (2)	Minimum molecular thickness (nm) (3)	Difference (nm) (3) minus (2)
Methylammonium (Me_1)	1.25	0.29	0.37	0.08
Ethylammonium (Et_1)	1.27	0.31	0.39	0.08
n-Propylammonium (Pr_1)	1.27	0.31	0.40	0.09
n-Butylammonium (Bu_1)	1.28	0.32	0.40	0.08
Dimethylammonium (Me_2)	1.28	0.32	0.39	0.07
Diethylammonium (Et_2)	1.30	0.34	0.40	0.06
Di-*n*-butylammonium (Bu_2)	1.32	0.36	0.42	0.06
Trimethylammonium (Me_3)	1.33	0.37	0.43	0.06
Triethylammonium (Et_3)	1.32	0.36	0.47	0.11
Tetramethylammonium (Me_4)	1.38	0.42	0.53	0.11
Tetraethylammonium (Et_4)	1.40	0.44	0.55	0.11
Tetra-*n*-propylammonium (Pr_4)	1.45	0.49	N.d.	N.d.
Tetra-*n*-butylammonium (Bu_4)	1.62	0.66	N.d.	N.d.

Source: Theng et al. (1967)

[a] obtained by subtracting 0.96 nm (montmorillonite layer thickness) from the observed basal spacing

N.d. = not determined

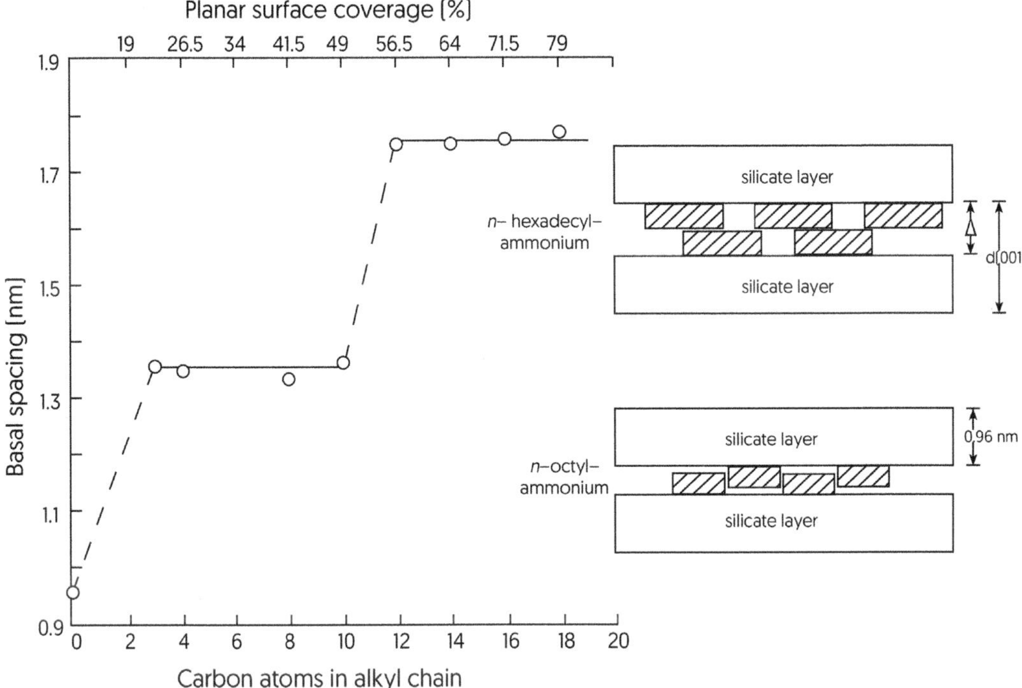

FIGURE 4.4 Variation in basal spacing with number of carbon atoms in the alkyl chain and the percentage of interlayer surface coverage by the organic cation, for complexes of montmorillonite with primary *n*-alkylammonium ions. (From Jordan, J.W., *J. Phys. Chem.*, 53, 294–306, 1949.)

Following Brindley and Hoffmann (1962), mono-, di-, and tri-alkylammonium ions apparently adopt an interlayer orientation in which the plane of the carbon zig-zag is parallel to the silicate surface, giving basal spacings of 1.25–1.33 nm (Table 4.1). The orientation of the tetra-alkylammonium ions, however, cannot be deduced with any certainty from the basal spacing of their respective complexes but the 'flattest possible' arrangement is also indicated by the X-ray diffraction data.

In this context, we would mention Jordan's (1949) early X-ray diffraction study of montmorillonite (Wyoming bentonite) complexes with primary *n*-alkylammonium ions with varying number of carbon (C) atoms in the alkyl chain (from C_3 to C_{18}). Figure 4.4 shows that for C_3 to C_{10} the basal spacings remained sensibly constant at ~ 1.36 nm, giving an interlayer separation of ~ 0.4 nm equal to the van der Waals thickness of a methyl or methylene group. For C_{12} to C_{18} the basal spacings increased by 0.4 nm to ~ 1.76 nm.

The data are consistent with the formation of single- and double-layer complexes with the alkyl chain lying parallel to the silicate surface, allowing the area covered by each amine cation to be deduced. By taking the area per interlayer exchange site as 1.65 nm^2, Jordan (1949) was able to derive the area occupied by a given organic cation as a percentage of the planar interlayer area, shown on the upper horizontal axis in Figure 4.4. If the area of the alkylammonium ion is less than half the area per exchange site, the amine cations adsorbed on one surface can fit into the gaps between the cations lying on the opposing surface, as depicted in for *n*-octylammonium. If, however, the amine cation area is greater than half the area per exchange site, such interpenetration cannot be realized. As a result, double-layer complexes are formed as exemplified by *n*-hexadecylammonium (Lagaly 1981).

Stepwise changes in basal spacing also obtain when a long-chain amine cation (e.g., *n*-octadecylammonium) is added in increasing amounts to montmorillonite, up to twice the CEC of the sample (Jordan et al. 1950). The first and second plateaux are comparable with those shown in Figure 4.4. When the solute concentration exceeds the CEC, however, the chains in the double-layer complex stand at an angle to the silicate surface. At still higher amounts adsorbed, the alkyl chains are close-packed in a near-vertical position, allowing van der Waals interactions to be established between adjacent alkyl chains.

The stepwise change in basal spacing with alkyl chain length (Figure 4.4) is indicative of montmorillonite (SWy-1) having a closely homogeneous layer charge distribution. When this distribution is heterogeneous, however, the monolayer to bilayer and bilayer to pseudo-trilayer transitions (Figure 4.1) become less discrete and more gradual as Mercier and Detellier (1994) have reported for complexes of STx-1 montmorillonite with tetra-alkylammonium ions, and Lee et al. (2005) for complexes of SAz montmorillonite with hexadecyltrimethylammonium ions. Molecular dynamics simulation further indicates that montmorillonites with a low CEC tend to give a stepwise, while those with a high CEC would show a continuous, increase in basal spacing with alkyl chain length (Heinz et al. 2007).

The arrangement and packing of primary *n*-alkylammonium ions in the interlayer space of vermiculite and related (mica-type) minerals have been the topic of many investigations (Weiss et al. 1956; Sutherland and MacEwan 1961; Garrett and Walker 1962; Weiss 1963a; Johns and Sen Gupta 1967; Walker 1967). Besides being dependent on the alkyl chain length of the cation and layer charge density of the vermiculite sample, the arrangement of intercalated *n*-alkylammonium ions is also influenced by the method of preparing the complex. Here we summarize Walker's (1967) data and conclusions with respect to the intercalation of *n*-alkylammonium ions into different vermiculites (cf. Chapter 1). The charge per formula unit (Table 1.1) of the sample from West Chester is slightly smaller than that of Kenya vermiculite.

Walker (1967) has proposed that the alkyl chains, adsorbed on opposing silicate surfaces, interpenetrate to varying degrees during complex formation. When the solute concentration is sufficiently high, complete double layers of alkylammonium ions (with free amine and probably also water and amine salt in the gaps between) are formed in the interlayer region for which the average increment per C atom is ~ 0.204 nm (Figure 4.5, line A). Assuming a *trans-trans* (ordered) conformation of the alkyl chain, and with the N–C bond normal to the silicate layer, the chain is tilted at an angle of ~ 54° to the silicate layer. Taking the radius of the terminal methyl group as 0.2 nm and assuming no chain overlap, the theoretical basal spacings may be calculated. Since these values are smaller by about 0.03 nm for even-C complexes and by 0.15 nm for odd-C complexes, it seems reasonable to suppose that terminal methyl groups overlap at the centre of the interlayers.

Here we might mention the molecular dynamics simulations study by Zhou et al. (2014) who noted that the conformation of interlayer cetyltrimethylammonium (CTMA) ions in montmorillonite is influenced by water content and CTMA loading. Interestingly, the percentage of disordered (*gauche*) conformations decreases when either the system is water saturated, or the surfactant loading is very high and steric hindrance comes into play. Near-IR and ^{13}C NMR spectroscopy of montmorillonite complexes with alkylammonium ions further indicate that *gauche* conformers are steadily replaced by ordered *all-trans* conformers as alkyl chain length and surfactant loading increase (He et al. 2004; Slaný et al. 2019; Madejová et al. 2020).

On washing and drying of the double-layer complex, the interlayer space contracts (as excess amine, salt, and water are removed), while alkyl chains adsorbed on opposite surfaces move over one another and interpenetrate to form a single layer of ions (Figure 4.5, line B). Again, assuming a *trans-trans* conformation allows the basal spacings to be calculated. In even-C complexes (Figure 4.6A) the terminal CH_3 lies at the same level as the β-methylene group of the opposing chain, while in odd-C complexes (Figure 4.6B), it lies at the level of the opposing α-methylene. The average depth of keying of the terminal CH_3 for even-C complexes is calculated at ~ 0.01 nm and for odd-C complexes at ~ 0.05 nm. A paraffin-type arrangement with an all-trans chain conformation has also been reported by Su et al. (2016) for an expanded vermiculite complex with CTMA.

Walker's (1967) general conclusions regarding the arrangement of intercalated *n*-alkylammonium ions in (Llano) vermiculite have been confirmed by Johns and Sen Gupta (1967) using

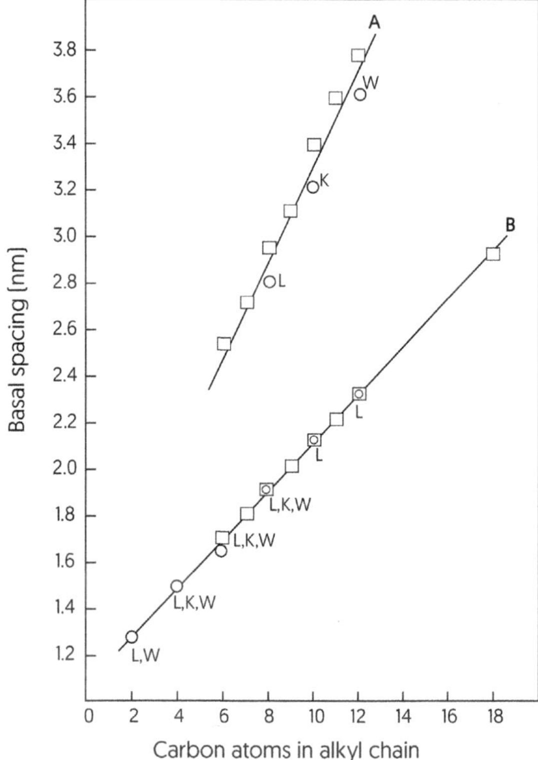

FIGURE 4.5 Relationship between basal spacing of vermiculite complexes with primary n-alkylammonium ions and alkyl chain length. Line A, double-layer complexes; line B, single-layer complexes. L = Llano vermiculite; K = Kenya vermiculite, and W = West Chester vermiculite. (Open squares are from Walker, G.F., *Clay Miner.*, 7, 129–143, 1967; open circles are from Johns, W.D. and Sen Gupta, P.K., *Am. Mineral.*, 52, 1706–1724, 1967.)

one-dimensional Fourier syntheses and by Martin-Rubi et al. (1974) using infrared spectroscopy. These workers also observed different degrees of interpenetration during complex formation but in each instance the alkyl chain is extended in the *trans-trans* conformation at an angle of ~ 55°with the silicate surface. Further, the N–C bond is normal to the silicate surface with the N atom located at 0.465 nm from the centre of the silicate layer. Only after extensive washing and drying do the basal spacings of the alkylammonium complexes reach their respective minimum values, all of which fall on a straight line coincident with that obtained by Walker (1967) for his single-layer complexes (Figure 4.5, line B).

The likelihood of hydrogen-bond formation between $-NH_3^+$ and surface oxygens has been discussed by Johns and Sen Gupta (1967) in terms of the shortening of N–O distances. By assuming that the $-NH_3^+$ group is symmetrically positioned at the centre of a ditrigonal hole, they have calculated a value of 0.30 nm for this distance. Earlier, Weiss et al. (1959) reported a minimum value of 0.285 nm for n-hexylammonium in 'batavite' (an iron-rich vermiculite), while Haase et al. (1963) gave a value of 0.282 nm for hexamethylene-diammonium in vermiculite. In comparing N–O distances for several ammonium salts, Laby and Walker (1970) have concluded that the reported distances in vermiculite complexes are not consistent with strong hydrogen bonding. They also showed that for the shorter chain cations (C_6 and less), the assumption that the N–C bond is normal to the silicate surface is only very approximately correct.

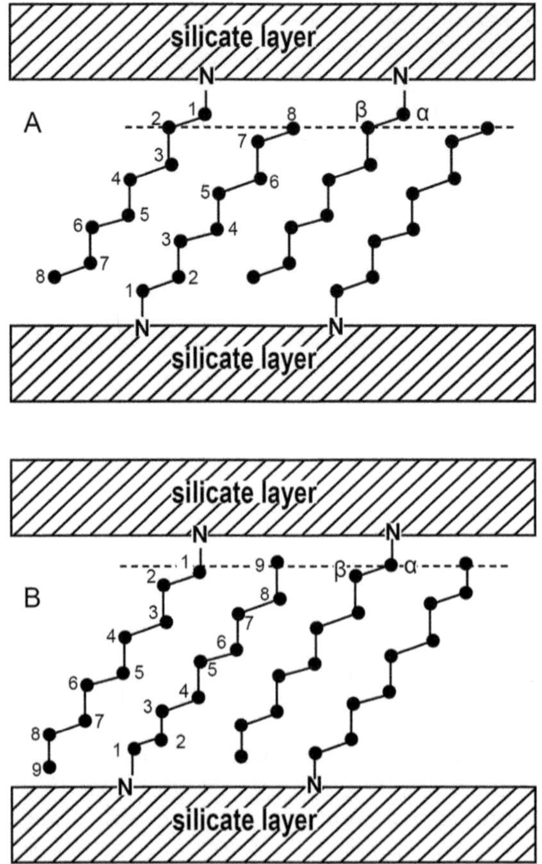

FIGURE 4.6 Diagram showing the final arrangement of interpenetrating monolayers of *n*-alkylammonium ions in the interlayer space of vermiculite. N denotes $-NH_3^+$ groups bound electrostatically to the silicate surface. (A), alkylammonium ion with an even number of carbon atoms; (B), alkylammonium ion with an odd number of carbon atoms. (From Walker 1967, Figure 1. Reproduced with kind permission of The Mineralogical Society of the UK and Ireland, publisher of *Clay Minerals*.)

Weiss (1963b) has presented X-ray data to show that the difference in interlayer arrangement between even-C and odd-C cations extends to the alkyl(α,ω)-diammonium ions. For the same batavite material (from Kropfmühl), even-C cations adopt the 'normal' *trans-trans* conformation in which the alkyl chains are inclined at an angle of ~ 56° to the silicate surface. The odd-C cations, however, apparently adopt a vertical orientation, and hence the basal spacings of the corresponding complexes are greater than those of their counterparts with even-C ions.

The basal spacings of montmorillonite complexes with 1-*n*-alkyl pyridinium ions have been reported by Greenland and Quirk (1962). Complexes containing 0.6–1.0 mmol/g of *n*-butyl-, *n*-octyl-, and *n*-dodecyl-pyridinium gave spacings of 1.40–1.43 nm. The corresponding Δ-values of 0.44–0.47 nm would indicate that the compounds were intercalated with their alkyl chains parallel to the silicate surface allowing for some degree of rotation around the N–C bond. On the other hand, complexes with methyl- and ethyl-pyridinium ions gave Δ-values that were smaller than the minimum thickness of the aliphatic chain (0.4 nm), and even the thickness of the pyridinium ring (0.37 nm). Following Greene-Kelly (1955), this finding may be explained in terms of keying of methyl and ethyl groups into the ditrigonal depressions in the silicate surface together with attractive interactions between aromatic (pyridinium) rings and interlayer surface (Praus et al. 2006; Gamoudi et al. 2013).

Dodecyl- and cetyl-pyridinium (bromides) were adsorbed in excess of the CEC of the montmorillonite sample, giving basal spacings as high as 4.2 nm (Greenland and Quirk 1962; Koh et al. 2005). This value is consistent with an interlayer arrangement in which the alkyl chains interpenetrate (Figure 4.6) when both the alkyl chain and the pyridine ring are perpendicular to the silicate surface. Molecular dynamics simulations by Schampera et al. (2015) also indicate that hexadecylammonium ions can form a disordered bilayer or a paraffin-like configuration at high amounts adsorbed (by bentonite).

4.4 FORMATION, PROPERTIES, AND SOME APPLICATIONS OF CATIONIC COMPLEXES

With reference to Equation 4.1, the adsorption of organic cations by clay minerals is influenced by, and dependent on, the valency and size of the exchangeable cations initially present at the silicate surface (Wu et al. 2015). Indeed, the values of $-\Delta G^m$ (Figure 4.3) for the replacement of Ca^{2+} ions by short-chain alkylammonium ions in montmorillonite are negative (Theng et al. 1967). That exchangeable Ca^{2+} (K^+, Mg^{2+}) ions are much more difficult to replace than Na^+ have been noted by several workers (Kurilenko and Mikhalyuk 1959; Greenland and Quirk 1962; Praus et al. 2006; Gamoudi et al. 2013). For example, McAtee (1959) found that *n*-octadecyl-, dimethyldioctadecyl(DMDO)- and dimethylbenzyllauryl(DMBL)-ammonium ions could almost quantitatively replace Na^+ ions in montmorillonite. Much less Ca^{2+} counterions, however, were exchanged by the organic cations under the same conditions. Similarly, exchangeable K^+ ions are difficult to replace by quaternary ammonium ions (Zhang et al. 1993). We might also mention that the Na^+ and Li^+ counterions in high-charge layer silicates, such as synthetic fluorotetrasilicic mica and taeniolite (cf. Chapter 1) are only partially replaceable by quaternary ammonium ions (Klapyta et al. 2001).

In examining the desorption of DMBL-ammonium ions from montmorillonite and hectorite using several primary, secondary, and tertiary amines (and DMDO ions) in an organic solvent (isooctane-isopropanol mixture), McAtee (1962, 1963) noted that the extent of desorption increased with the basicity of the replacing cation and declined as the size of the replacing cation decreased. The solvents alone, however, could extract an appreciable proportion of adsorbed quaternary ammonium ions (Jordan 1963; Flores et al. 2017). Under certain conditions, considerable replacement of one organic ion by another may occur (McAtee and Hackman 1964). This finding has an important bearing on the use of montmorillonite-organic complexes as gelling agents in grease compositions. Since the rheological properties of grease depend on the kind of clay-associated organic compound, the aforementioned process provides a means of adjusting gel consistency and thickness.

The desorption of cobaltic hexammine cations in montmorillonite by some inorganic and quaternary ammonium salts in water has been investigated by Das Kanungo et al. (1968) and Das Kanungo and Chakravarti (1971). The capacity of the various inorganic ions in desorbing $Co(NH_3)_6^{3+}$ was related to their unhydrated size, increasing in the order $Li^+ < Na^+ < K^+ < NH_4^+ < Rb^+ < Cs^+$ and $Mg^{2+} < Ca^{2+} < Ba^{2+}$, while for the organic ions the order was tetramethylammonium < tetraethylammonium < cetyltrimethylammonium < cetylpyridinium. All things being equal, the organic cations were more effective than the inorganic ones. Washing with water either fails to desorb, or displaces only a limited proportion of, the quaternary ammonium cations from their complexes with smectites (Shen 2001; Flores et al. 2017; Hue et al. 2019).

The replacement of inorganic counterions by large organic ions commonly leads to a marked reduction in water uptake by the resultant complexes. As early as 1939, Gieseking observed that the basal spacings of montmorillonite complexes with large aliphatic and aromatic amine cations did not appreciably increase on exposure to water vapour or liquid water. Similar findings were subsequently reported for complexes of montmorillonite with various organic cations and bases (Erbring and Lehmann 1944; Grim et al. 1947; Kobakhidze and Shishniahvili 1965; Theng et al. 1968; Ruan et al. 2008) although Okada et al. (2015) found that the interlayer space of benzylammonium-smectites expands when the complexes are immersed in water.

Hendricks (1941) suggested that the interlayer surface, occupied by the organic species, was no longer available to water molecules. In accord with this suggestion, Theng et al. (1968) have found that the swelling (in water) of Na^+-montmorillonite, containing increasing proportions of exchangeable alkylammonium ions, decreases with surface coverage by the organic cations. Similarly, Seki et al. (2009) have indicated that the amount of water (vapour) adsorbed increases as the size of the occupying organic cation (quaternary diammonium) decreases, and the CEC of the montmorillonite host increases. Subsequently, Pálková et al. (2015) reported that water vapour uptake decreased in the order Li^+- > TMA^+- > $CTMA^+$-montmorillonite. Shah et al. (2013) have also made the interesting observation that montmorillonite modified with quaternary ammonium ions, containing a benzyl group, is highly hydrophobic.

On the other hand, montmorillonite intercalated with small organic cations (e.g., tetramethylammonium), or compact quaternary ammonium ions, such as trimethylphenylammonium (cf. Figure 2.4), can take up appreciable amounts of polar and non-polar organic solutes from the vapour phase or aqueous solution. By occupying a fraction of the interlayer surface, small and compact organic cations act as 'pillars' propping the silicate layers apart to yield a series of microporous, shape-selective adsorbents (Barrer and MacLeod 1955; Barrer and Reay 1957; Diamond and Kinter 1963; Lee et al. 1989, 1990; Jaynes and Boyd 1990; Kukkadapu and Boyd 1995).

Because of their hydrophobic (organophilic) properties interlayer montmorillonite-alkylammonium complexes have been widely used as thickeners and gelling agents of organic liquids (Jordan et al. 1950; Granquist and McAtee 1963; Slabaugh and Hiltner 1968; Slabaugh and Hanson 1969; Lu et al. 2010). An alkyl chain of at least 10 carbon atoms is required before appreciable gelation occurs (Jordan 1949). With reference to Figure 4.4 this onset of large swelling coincides with the point of 50% surface coverage when a monolayer of close-packed, interpenetrating organic cations are attached to opposing silicate layers. The influence of alkylammonium 'dosage' on the viscosity of the resultant organo-montmorillonite gel has also been noted by Liu et al. (2010).

Besides being dependent on chain length, gelation is influenced by the polarity of the solvent. Polar (e.g., alcohols, ketones) and non-polar (e.g., benzene) liquids are relatively ineffective as swelling agents, while excellent gelation occurs in nitrobenzene and binary mixtures of polar and non-polar liquids, such as methanol/toluene. Similar results have been reported by Queiroz et al. (2010) who also pointed out the importance of agitation during the mixing step. The gelation process apparently involves solvation of the 'uncoated' or exposed portion of the silicate surface by the polar component of the swelling agent, allowing van der Waals interactions to be established between the alkyl groups of the organic cation and the non-polar component of the liquid.

Following Barrer and co-workers (Barrer and MacLeod 1955; Barrer and Reay 1957; Barrer and Brummer 1963; Barrer and Millington 1967), Slabaugh and Hiltner (1968) have determined the vapour phase isotherms for the adsorption of some alcohols, *n*-heptane, nitromethane, and ethyl acetate by montmorillonite containing *n*-alkylammonium ions with alkyl chain lengths from 4 to 18 carbon atoms. Concurrently, the basal spacing of the intercalation complexes was measured. Following adsorption of methanol and ethanol at $P/P_0 < 0.8$, the basal spacing of complexes with C_{12}, C_{14}, C_{16}, and C_{18} remain virtually constant at ~ 1.8 nm corresponding to a double layer of *n*-alkylammonium ions in the interlayer space in an arrangement as depicted in Figure 4.4 for C_{18}. Above $P/P_0 = 0.8$, however, there is an abrupt increase in basal spacing coincident with the formation of a single layer of interpenetrating alkyl chains standing perpendicular to the silicate surface.

Slabaugh and Hiltner (1968) have also suggested that swelling is a two-step process. The first step involves solute adsorption to external clay particles and to interlayer surfaces that are not already occupied by amine cations. Thus, little or no interlayer expansion is observed. The second step occurs at high adsorbate pressures when the alkyl chains are displaced from the silicate surface. As a result, new surfaces are formed to which polar molecules can be introduced while the displaced alkyl chains are solvated by alcohols, nitromethane, and ethyl acetate. Water vapour would influence, and participate in, the first step, while paraffins would only interact with the alkyl chain of amine cations. Thus, neither polar nor non-polar solutes give rise to appreciable swelling, while molecules possessing both polar and non-polar groups can form large gel volumes.

Barrer and Millington (1967) found that the basal spacings of montmorillonite and hectorite complexes with a series of *n*-alkylammonium and α,ω-alkyldiammonium ions increased following intercalation of benzene, *n*-heptane, acetonitrile, and α,ω-diaminopropane gases and vapours. Adsorption of benzene and α,ω-diaminopropane gave rise to complexes with basal spacings of ~ 1.43 and 1.50 nm, respectively, while exposure to acetonitrile yielded two series of complexes with basal spacings of ~ 1.63 and ~ 1.99 nm, respectively. The final interlayer separation was apparently controlled by the penetrant rather than by the valency or size of the resident organic cation.

When the alkyl chains of *n*-alkylammonium ions are 'detached' from the silicate surface as they are in vermiculite complexes (Figure 4.6), polar and non-polar organic liquids (phenols; esters, ethers, ketones, halogen and nitro compounds, hydrocarbons) can enter and expand the interlayer space (Weiss 1963a). Intercalation of these compounds, particularly those possessing *n*-alkyl groups in their molecules, presumably causes the interpenetrating *n*-alkylammonium chains to separate. As a result, gaps are formed between the extended chains attached to opposing silicate surfaces into which molecules of the swelling liquid can be accommodated.

Interlayer complexes of vermiculites with short-chain *n*-alkylammonium ions, notably *n*-butylammonium (Bu_1) can show extensive interlayer ('macroscopic') swelling when they are immersed in water giving interlayer spacings of tens of nanometres (Walker 1960; Garrett and Walker 1962; Walker and Garrett 1967). Table 4.2 lists the basal spacings of various vermiculites, saturated with different *n*-alkylammonium ions, after immersion in water.

Macroscopic swelling of vermiculite-Bu_1 complexes also occurs on immersion in dilute aqueous solutions of *n*-butylammonium chloride. Thus, a linear relationship obtains between the interlayer spacing (Δ-value) of the complexes and the inverse square root of Bu_1Cl concentration (Figure 4.7). At high concentrations, interlayer spacings of 0.96 nm are measured. These values do not change until the concentration is reduced to about 0.06 M when there is an abrupt 'jump' in swelling. Williams et al. (1996) have observed similarly for vermiculites exchanged with *n*-propyl- and *n*-butyl-ammonium ions, using neutron diffraction.

TABLE 4.2
Basal Spacings of Various Vermiculites Exchanged with *n*-Alkylammonium Ions Following Immersion in Water.

Alkylammonium ion	Young River vermiculite (x = 0.6)	Kenya vermiculite (x = 0.65)	West Chester vermiculite (x = 0.7)	Macon County vermiculite (x = 0.8)
Methylammonium	1.22	1.22	1.23	1.26
Ethylammonium	1.26	1.26	1.27	1.26
n-Propylammonium	E.i.s.	E.i.s.	E.i.s.	E.i.s.
n-Butylammonium	E.i.s.	E.i.s.	E.i.s.	E.i.s.
Isoamylammonium	E.i.s.	E.i.s.	E.i.s.	E.i.s.
n-Amylammonium	1.98	2.02	2.00	1.93
n-Hexylammonium	2.12	n.d.	n.d.	2.14
n-Heptylammonium	2.22	2.26	2.23	2.20

Source: Garrett, W.G and Walker, G.F., *Clays Clay Miner.*, 9, 557–567, 1962
x = charge per formula unit (cf. Table 1.1)
E.i.s. = extensive interlayer swelling (cf. Table 1.2)
n.d. = not determined

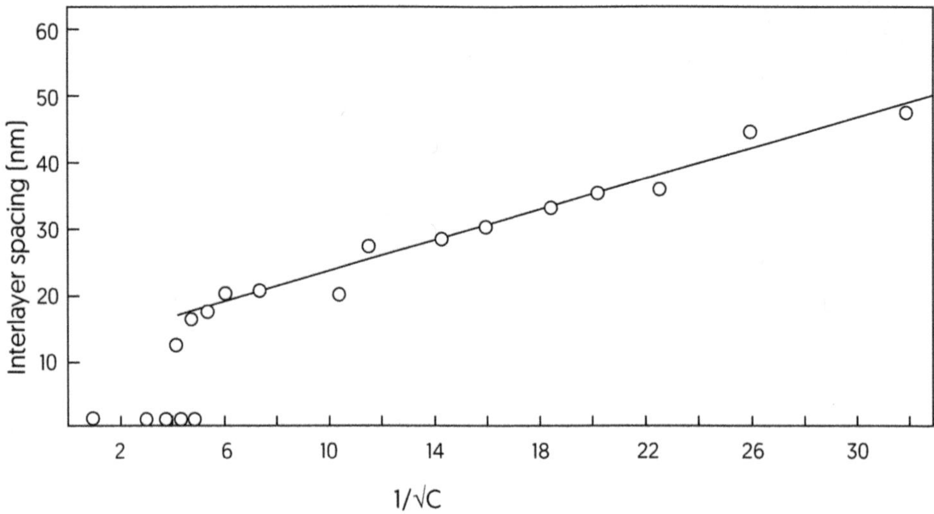

FIGURE 4.7 Variation of interlayer spacing (Δ-value) of vermiculite-butylammonium complexes with 1/√C where C is the concentration of *n*-butylammonium chloride in solution. (After Garrett and Walker 1962, Figure 1. Reproduced with kind permission of The Clay Minerals Society, publisher of *Clays and Clay Minerals*.)

Following Frank and Evans (1945), Walker and Garrett (1967) have suggested that ordered water structures ('icebergs') build around the *n*-butylammonium ion. Here, the water molecules act as hosts of the organic ions while the latter provide support for the water. The formation of such *clathrate* structures in the interlayer space evidently promotes the dissociation of Bu_1 ions from the silicate surface and the formation of diffuse double layers. Low-angle X-ray diffraction analysis of swollen Bu_1^+-vermiculite single crystals also indicates that macroscopic swelling arises from the interaction of diffuse double layers on interlayer surfaces (Rausell-Colom 1964). By measuring the basal spacing of the swollen gel as a function of an externally applied load, the swelling pressure, and hence the repulsive forces involved in expanding the complex, can be estimated. The difference between the calculated repulsion and the experimentally derived swelling pressure, at any given equilibrium interlayer separation, was found to be of the same order of magnitude as the van der Waals attraction between two opposing silicate layers.

The behaviour of Bu_1^+-vermiculite closely resembles that of the Li^+-exchanged counterpart (Norrish and Rausell-Colom 1963; Skipper et al. 1995). In both systems, there is a range of electrolyte concentrations in which only limited interlayer expansion takes place. When a given critical level of dilution is reached, however, there is a sudden transition from limited to extensive interlayer swelling, giving rise to gel formation. Unlike that of the Li^+-exchanged material, the swelling of Bu_1^+-vermiculite is very temperature sensitive (Garrett and Walker 1962). Thus, Bu_1^+-vermiculite does not show macroscopic swelling on immersion in water at > 50°C, presumably because the alkylammonium-water clathrate is not as thermally stable as the lithium-water association.

Unlike vermiculite, montmorillonite saturated with *n*-butylammonium ions does not show extensive interlayer expansion in water. Indeed, none of the alkylammonium ions of Table 4.2 shows this feature. However, when increasing amounts of short-chain primary *n*-alkylammonium ions are introduced into Ca^{2+}-montmorillonite, and the resultant complexes allowed to wet at a hydrostatic suction of 1500 N/m^2, macroscopic swelling is observed (Greenland et al. 1964; Theng et al. 1968). Figure 4.8 shows that the swelling of the resultant complexes increases to a maximum and then decreases. The water content of the complex containing approximately equal proportions of Ca^{2+} and Me_1, Et_1, Pr_1, and Bu_1 ions (~ 9 g/g) is comparable with that measured for Na^+-montmorillonite (12 g/g), while the parent Ca^{2+}-exchanged material takes up only 2 g/g at the same suction.

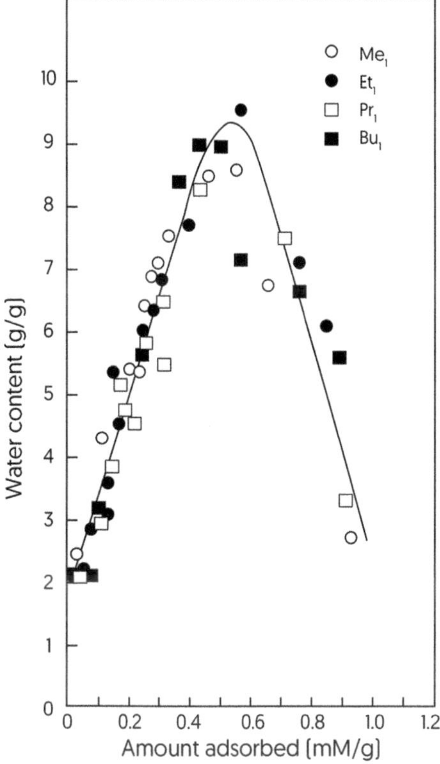

FIGURE 4.8 Variation in the water content of montmorillonite-alkylammonium complexes (determined at a hydrostatic suction of 1500 N/m^2) with the amount of organic cations adsorbed or degree of saturation of the clay mineral exchange sites by alkylammonium ions. Me_1 = monomethylammonium; Et_1 = monoethylammonium; Pr_1 = mono-*n*-propylammonium; Bu_1 = mono-*n*-butylammonium. (From Theng et al. 1968, Figure 2b. Reproduced with kind permission of The Mineralogical Society of the UK and Ireland, publisher of *Clay Minerals*.)

As already remarked on, much of the initial increase in swelling may be ascribed to interlayer expansion due to the formation of diffuse double layers. Thus, the 1.9 nm line of the parent Ca^{2+}-montmorillonite becomes progressively less sharp as the proportion of alkylammonium ions in the complex increases. When ~ 45 mmol/g has been adsorbed, this line is absent from the X-ray diffraction pattern of the swollen complex. The introduction of small *n*-alkylammonium ions disrupts the ordered water structures around the divalent cation. As a result, the entropy of the system increases while the electrostatic attraction between cation and clay surface decreases. Macroscopic swelling of this kind, however, is apparently restricted to alkylammonium ions containing less than five carbon atoms. With larger organic cations there is a progressive decrease in water uptake as the amount adsorbed increases. In this regard, we might mention that montmorillonite swelling can be inhibited by adsorption of N1,N1'-(ethane-1,2-diethyl)bis(N1-(2-aminoethyl)ethane-1,2-diamine (Yuan et al. 2022).

4.5 ORGANICALLY MODIFIED AND COVALENTLY GRAFTED CLAY MINERALS

In an early survey of the literature on clay-organic chemistry, Nahin (1963) proposed the following definition for organically modified clay minerals: 'Organoclays are reaction products of specific clay minerals with specific organic molecules involving the formation of chemical bonds'. Here we use the term *organoclays* to denote organically modified clay minerals, especially smectites, in which the inorganic counterions have been replaced by either short-chain (compact) or long-chain quaternary ammonium ions and related cationic surfactants in an aqueous medium (Scheme 4.1). The underlying mechanism here is one of cation exchange (cf. Equation 4.1), supplemented by van der Waals (hydrophobic), H-bonding, and cation-dipole interactions. Organoclays can also form by a semisolid-solid or solid-solid reaction between clay mineral and organic modifying agent (Ogawa et al. 1990, 1995; Paiva and Morales 2012; Yu et al. 2014; Gamoudi et al. 2013, 2015; Alves et al. 2016; Rigail-Cedeño et al. 2023).

In comparison with cationic surfactants, such as hexadecyltrimethylammonium (HDTMA)- or cetyltrimethylammonium (CTMA)-bromide, anionic, non-ionic, and zwitterionic surfactants together with (cationic) gemini and ionic liquids have not featured widely as organic modifying agents. Scheme 4.1 shows the chemical structures of representative compounds in each surfactant class.

Anionic surfactants, such as sodium dodecylsulphonate and *n*-lauroyl sarcosinate, commonly fail to intercalate into montmorillonite but can do so when the clay mineral has previously been modified with quaternary ammonium cations (Chen et al. 2008a; Fu et al. 2016). For example, Liao et al. (2016) were able to sequentially intercalate CTMA-bromide, sodium dodecylsulphate, and hydrogenated castor oil into Na^+-montmorillonite. Likewise, azobenzene and aminoazobenzene can co-intercalate with alkylammonium ions into smectites (Koteja et al. 2017) as does the combination of chitosan and HDTMA ions (Zhu et al. 2017a) and that of non-ionic amides and octadecyltrimethylammonium ions (Zhou et al. 2019). We might add here that montmorillonite, modified with cetylpyridinium-lauroyl sarcosinate surfactants, is reported to be bactericidal against *Streptococcus mutans* (Şahiner et al. 2022).

Because of their negative surface charge, clay minerals tend to repel, or at best show a small capacity for adsorbing, *anionic surfactants* from aqueous solutions. Furthermore, adsorption here is generally confined to external clay surfaces. In this regard, the particle edge plays an important role because its surface carries (under-coordinated) OH groups that can acquire either a positive or negative charge, depending on the pH of the ambient solution (cf. Figure 1.5). These terminal hydroxyls can also participate in ligand exchange reactions with anionic species in solution (cf. Equation 1.1; Figure 5.2). For example, Sánchez-Martín et al. (2008) reported that adsorption of sodium dodecylsulphate (SDS) to various clay minerals, including montmorillonite, did not change the basal spacing of the adsorbents, while kaolinite and sepiolite could take up more SDS (presumably by ligand exchange) than did montmorillonite since the former minerals are relatively rich in surface hydroxyl groups.

Under highly acidic conditions, the molecules of anionic surfactants would essentially be uncharged. Like uncharged organic species, both SDS and sodium stearate could intercalate into Ca^{2+}-montmorillonite at $pH \leq 3$ (Zhang et al. 2010) and interact with the interlayer surface, probably through a water-bridging mechanism (cf. Figure 5.1). On this basis, Lin and Juang (2002) suggested that SDS could intercalate into a commercial acid-treated (KSF) montmorillonite, while Zhuang et al. (2015) reported the formation by ball milling of an interlayer complex of Ca^{2+}-montmorillonite with a combined SDS/non-ionic surfactant, giving a basal spacing larger than 5 nm.

Unlike the anionic type, *non-ionic surfactants* are strongly adsorbed by, and can readily intercalate into, smectites. The underlying mechanism here is a combination of H-bonding, ion-dipole, and van der Waals interactions rather than electrostatic attraction. Thus, Shen (2001) obtained L- to H-class isotherms for the adsorption of polyethylene glycol ethers and Brij 78 by bentonite (at neutral pH), giving basal spacings of 1.44–1.78 nm. For interlayer complexes of Na^+-montmorillonite with low molecular weight polyethylene glycols, Sarier and Onder (2010) and Nourmoradi et al. (2012) measured basal spacings of 1.69–1.75 nm and ~ 1.75 nm, respectively. These values are consistent with the formation of a double-layer complex in which the surfactant molecules are apparently keyed into the ditrigonal depressions of the siloxane surface (Parfitt and Greenland 1970). Similarly, triethylene glycol monodecyl ether ($C_{10}E_3$; Scheme 4.1), can also form a double-layer

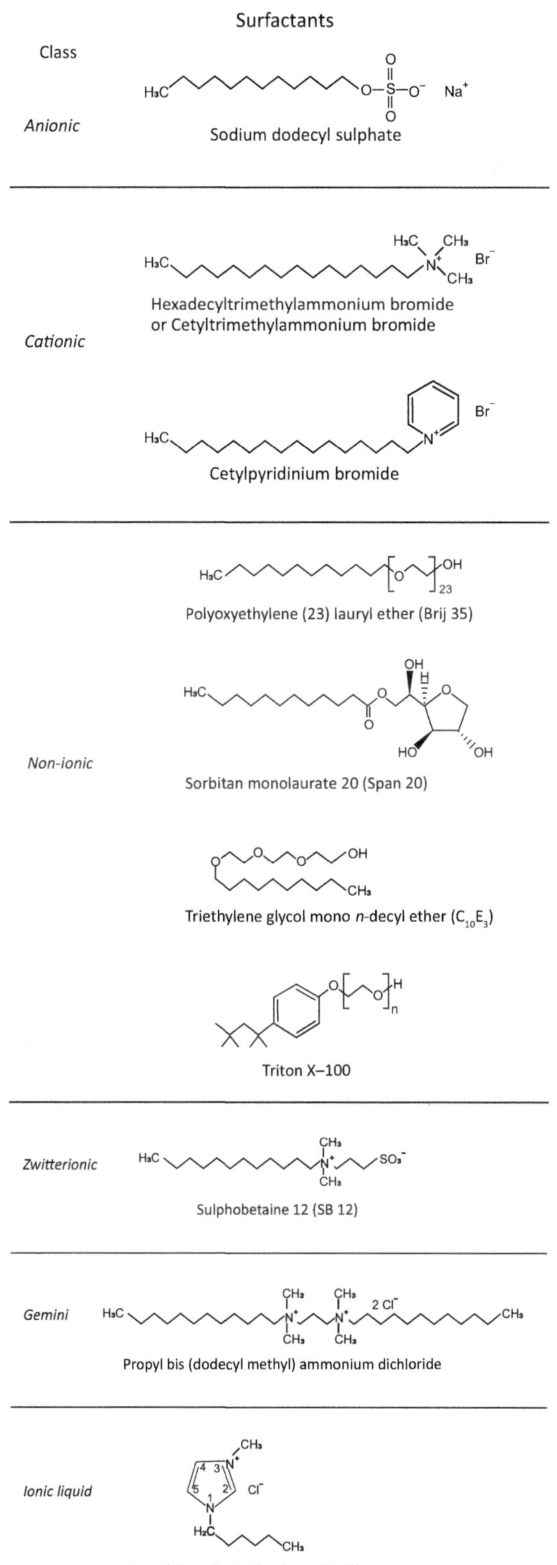

SCHEME 4.1 Molecular structures of some anionic, cationic, non-ionic, zwitterionic, and gemini surfactants and an ionic liquid.

complex with Ca^{2+}-montmorillonite (Guégan et al. 2009). However, below the critical micelle concentration $C_{10}E_3$ intercalates as a flat monolayer into Na^+-montmorillonite (Guégan et al. 2015). The extensive literature on the interactions of clay minerals with short-chain uncharged organic polymers has been reviewed by Theng (2012).

The systematic investigation into the interactions of smectites with *zwitterionic surfactants* was initiated by Zhu and co-workers (Zhu et al. 2011, 2015a, 2017b). As the term suggests, the molecules here possess both positively and negatively charged groups which, in the case of sulphobetaine (Scheme 4.1), are represented by dimethylammonium and sulphonate groups, respectively. The formation, properties, and practical applications of betaines and related zwitterionic surfactants have been reviewed by Kelleppan et al. (2021).

Zhu et al. (2011) suggested that for loadings of less than the cation exchange capacity (CEC) of montmorillonite, sulphobetaine (SB) was intercalated by replacing the Ca^{2+} counterions. At loading levels higher than the CEC, the excess surfactant would adsorb to SB that has already occupied interlayer exchange sites as well as to external particle surfaces and interparticle pores. Having neither layer charge nor CEC, talc fails to intercalate SB (Ma et al. 2015). On the other hand, the intercalation of SB into polyhydroxy-aluminium (Al_{13})-pillared montmorillonite apparently occurs through electrostatic attraction between the negatively charged group of the zwitterion and the interlayer Al_{13} cation (Ma et al. 2014). A cation-exchange mechanism has also been proposed by Lazorenko et al. (2018) for the intercalation of zwitterionic oleylamidopropyl betaine surfactant into Na^+-montmorillonite.

Using sulphobetaines of varying molecular weight (SB12, SB14, SB16), Zhu et al. (2015a) further noted that both the extent of intercalation and the basal spacing of the resultant complexes increased with the alkyl chain length and loading level of the surfactants. Accordingly, there was a corresponding increase in the capacity of the surfactant-modified clays for taking up *p*-nitrophenol and nitrobenzene from aqueous solution. More importantly, Zhu et al. (2017b) subsequently found that surfactant adsorption led to the release of only a limited amount of Ca^{2+} ions into solution. It would therefore appear that SB was intercalated through ion-dipole interactions, or more likely through water-bridging (cf. Figure 5.1), rather than a cation exchange mechanism.

As already remarked on, the capacity of montmorillonite for removing organic pollutants from water can be appreciably enhanced by prior intercalation of zwitterionic surfactants. For example, Gu et al. (2015) observed that montmorillonite modified with 3-(N,N-dimethylhexadecylammonino)-propane sulphonate could effectively take up paraquat and amitrole. Similarly, Ma et al. (2016) found that sulphobetaine 16-modified montmorillonite to be a good adsorbent of phenol and Cu(II), while Liu et al. (2016) used montmorillonite modified with octadecane-betaine to remove bisphenol A and Cd(II) from solution. More recently, Sun et al. (2020) prepared an interlayer complex of montmorillonite with a mixture of non-ionic (polyoxyethylene ether) and zwitterionic (lauramidopropyl betaine) surfactants for the efficient and simultaneous removal of aflatoxin B_1 and zearalenone.

Complexes of clay minerals with (cationic) *gemini surfactants* have attracted a great deal of attention in terms of their formation, properties, and capacity for adsorbing organic contaminants and pollutants (Shen and Gao 2019). The molecules in this surfactant class consist of a long alkyl chain, a cationic group, a spacer, a second cationic group, and another alkyl chain (Menger and Littau 1993; Menger and Keiper 2000; Sharma et al. 2017). In the case of propyl bis(dodecyl methyl) ammonium (Scheme 4.1) dodecane constitutes the alkyl chain, dimethylammonium represents the cationic group, and propane is the spacer. Such a structure provides scope for varying alkyl chain length, cationic group character, and spacer size. Thus, an interesting research topic would be to measure the adsorption of gemini surfactants with varying spacer size by layer silicates differing in layer charge density and hence negative charge separation. Taleb et al. (2018), for example, found that the level of loading by, and the spacer size/length of, the gemini surfactant affect the thermal stability of the corresponding organoclay.

Li and Rosen (2000) have reported on the modification of Na^+-montmorillonite by a series of gemini surfactants with the general formula of $[C_nH_{2n+1}N^+(CH_3)_2\text{-}CH_2CH_2]_2.2Br^-$ where $n = 10, 12,$

14, and 16. The capacity of the resultant organoclays for removing 2-naphthol and 4-chlorophenol from water was superior to that of their counterparts containing the corresponding single-chain surfactant. This finding reflects the efficiency of gemini surfactants in expanding the montmorillonite quasi-crystal to form a highly hydrophobic interlayer space into which extraneous organic solutes can be accommodated (Ni et al. 2009; Zhou et al. 2009; Ren et al. 2018). For example, Luo et al. (2015) found that uptake of methyl orange by reduced-charge montmorillonites, modified by similar gemini surfactants, increased with layer charge. Other examples of dye removal by gemini surfactant-modified montmorillonite/bentonite are listed in Table 3.2.

Yang et al. (2016) have reported on the intercalation into Na^+-montmorillonite of gemini surfactants having pyridinium as the cationic group. The basal spacings of the resultant organoclays indicate that the interlayer surfactants adopt a pseudo-trilayer and paraffin-type chain arrangement (Figure 4.1). The surfactant-modified clays are effective in taking up bisphenol A from aqueous solution through a combination of hydrophobic and π–π interactions. Electrostatic attraction ('anion exchange') is also an important mechanism controlling the adsorption of negatively charged solutes, such as perchlorate (Srinivasarao et al. 2018), bromophenol blue (Xiang et al. 2019), and alkyl xanthates (Huang et al. 2019).

Yang et al. (2016) also found that solute removal (from water) followed pseudo second-order kinetics. Furthermore, the isotherms are of the Langmuir- or Freundlich-type rather than linear indicating that the process is not merely one of solute partitioning. Similar observations have been reported for the adsorption of perchlorate (Srinivasarao et al. 2018), various phenols (Xue et al. 2013; Yang et al. 2014, 2015; Xu et al. 2018; Yu et al. 2018), and benzotriazoles (Li et al. 2019; Zhang et al. 2021) to gemini surfactant-modified montmorillonite and vermiculite. Interestingly, such organo-vermiculites are also effective in removing coastal marine algae (Wu et al. 2010) and phenols (Cao et al. 2019) as well as methyl orange and crystal violet (Cao et al. 2020).

Like gemini surfactants, *ionic liquids* are positively charged compounds. As such, their intercalation into smectites is primarily controlled by cation exchange (electrostatic attraction), at least at low levels of uptake. Hydrophobic (van der Waals) interactions come into play when adsorption exceeds the CEC of the clay samples, accompanied by interlayer dehydration (Li et al. 2014; Takahashi et al. 2014; Lv et al. 2015; Wang et al. 2015). A typical example of an ionic liquid is 1-hexyl-3-methylimidazolium chloride (Scheme 4.1). Here the positively charged group is associated with the N_3 atom in the imidazole ring while the hexyl group (C_6H_{11}) is attached to the N_1 atom with chloride as the charge-balancing anion. Thus, this imidazolium-based ionic liquid is commonly designated as C_6mimCl.

As would be expected, the intercalation of ionic liquids (ILs) is accompanied by an increase in the basal spacing of the montmorillonite quasi-crystal, the extent of which is related to the size (alkyl chain length) and interlayer conformation of the solute molecules (Reinert et al. 2012; Takahashi et al. 2012, 2013; Wu et al. 2014; Lv et al. 2015; Nasser et al. 2016; Montaño et al. 2017; Zhao et al. 2019; Xiao et al. 2020; He et al. 2022). The expansion in basal spacing ('crystalline' swelling), resulting from the interlayer uptake of ILs, should not be confused with the known ability of ILs to inhibit the extensive interlayer swelling (in water) of Na^+-montmorillonite (Ren et al. 2019; He et al. 2020, 2022; Khan et al. 2020, 2022). Here we might also mention that IL-modified montmorillonites can effectively remove pharmaceuticals and dyes from aqueous solutions (Lawal and Moodley 2015, 2018; Wu et al. 2019; Yılmazoğlu et al. 2022).

Earlier, Wu et al. (2014) used molecular dynamics modelling in conjunction with X-ray diffraction to probe the interlayer configuration of $C_n$mimCl in montmorillonite (SWy-2) as a function of chain length (n). Apparently, C_4mim^+ and C_8mim^+ intercalate as (flat) monolayers, while $C_{12}mim^+$ and $C_{16}mim^+$ adopt a flat or tilted bilayer configuration. Wu et al. (2015) also found that intercalation of C_{16}mimCl into Na^+-, K^+-, Mg^{2+}-, Ca^{2+}-, and Fe^{3+}-montmorillonites led to the replacement (desorption) of different amounts of exchangeable cations, the extent of which is related to the ionic potential of the counterions (cf. Table 2.1). We suggest that the 'cover-up' effect (Hendricks 1941) would also come into play. Thus, Reinert et al. (2012) found that only slightly more than 80% of the sodium counterions in montmorillonite could be replaced by various imidazolium- and pyridinium-based ionic liquids. Similarly, Pei et al. (2019) reported that 1-butyl-3-methylimidazolium tetrafluoroborate failed to completely replace the Na^+ ions, occupying exchange sites on the montmorillonite surface.

The formation and properties of organoclays began to be systematically investigated in the 1950s (Weiss 1963a, 1963b; Barrer 1978). Since then, the focus of research has been directed to the practical applications of organically modified smectites, especially as adsorbents of organic pollutants and contaminants (e.g., dyes) as well as carriers and delivery systems for the controlled release of drugs and pesticides (Xu et al. 1997; Theng et al. 2008; Park et al. 2011; Zhu et al. 2015b; Awad et al. 2019). These aspects of organoclay chemistry have been described in Chapters 2 and 3. Perhaps the single most widely known application of organoclays, however, is in the formulation and synthesis of polymer-clay nanocomposites (Le Baron et al. 1999; Le Pluart et al. 2002; Ruiz-Hitzky and Van Meerbeek 2006; Chen et al. 2008b; de Paiva et al. 2008; Theng 2012; Gul et al. 2016).

Montmorillonites modified with small compact quaternary ammonium cations (QAC), such as tetramethyl- and tetraethyl-ammonium (Barrer and MacLeod 1955; Barrer and Brummer 1963; McBride and Mortland 1975; Stevens and Anderson 1996; Stevens et al. 1996; Fuller et al. 2007; Ruan et al. 2008; Yoko and Ogawa 2010), 1,4-diazobicyclo(2,2,2)-octane ($DABCO^{2+}$) (Mortland and Berkheiser 1976; Shabtai et al. 1977; Van Leemput et al. 1983), trimethylphenylammonium (Jaynes and Boyd 1991a; Jaynes and Vance 1999; Sharmasarkar et al. 2000), benzylammonium (Okada et al. 2015), and benzyl tri-*n*-butylammonium (Saeed et al. 2022) have been referred to as adsorptive clays or 'type I' organoclays. As already remarked on, the cationic modifying agents here act as 'pillars' (cf. Figure 2.4), enabling hydrophobic organic solutes to occupy the 'interpillar' pore space where they interact with both the QAC and the exposed siloxane surface, yielding curvilinear adsorption isotherms. Sorption also rises with a decrease in layer charge density and QAC occupancy but decreases in the presence of water (Lee et al. 1989, 1990; Pálková et al. 2009; Okada et al. 2015; Zhu et al. 2016; cf. Figure 2.20). Evidently, the hydrophilicity of natural smectite surfaces is not an intrinsic property; rather, it is due to the presence of hydrated inorganic counterions. We might also add that cation-π interactions play an important part in the binding of aromatic compounds and hydrocarbons to cation- and tetra-alkylammonium-exchanged montmorillonite (Qu et al. 2008; Hou et al. 2015).

Likewise, ortho-phenanthroline and tris-bipyridyl complexes of Fe^{2+} and Cu^{2+} can prop the silicate layers in hectorite (Berkheiser and Mortland 1977; Traynor et al. 1978). Okada et al. (2005) also found that saponite, pillared with tris(2,2'-bipyridine)ruthenium cations, was a better adsorbent of phenols than its relatively high-charge montmorillonite counterpart. Subsequently, Herling et al. (2012) observed a striking increase in micropore volume and pore diameter for a low-charge hectorite that has been pillared with dimethyl-$DABCO^{2+}$ and rhodium-tris-2,2'-bipyridine [$(Rh(bpy)_3{}^{3+}$] cations.

Modification with long-chain QAC, notably hexadecyltrimethylammonium (HDTMA) or cetyltrimethylammonium (CTMA), gives rise to 'type II' organoclays. Since the interlayer space here is largely occupied by the QAC, these organically modified clays have a low porosity and surface area. As we have already seen (Chapter 3), the sorption of various polar organic solutes from water commonly yields linear or near-linear isotherms indicative of solute partitioning between surface and solution (Boyd et al. 1988a; Jaynes and Boyd 1991b; Groisman et al. 2004; Lee et al. 2004; He et al. 2006b; Roberts et al. 2006; Upson and Burns 2006; Rodríguez-Cruz et al. 2008; Zhu and Zhu 2008a; Shu et al. 2010; Sekrane et al. 2011; Vidal and Volzone 2012; Parolo et al. 2014; Ruan et al. 2015; Zhu et al. 2016; Abbas et al. 2017). Solute uptake also increases with smectite layer charge density and interlayer separation (cf. Figure 2.19).

In some instances, however, the isotherms for adsorption of some organic solutes to HDTMA- and related QAC-modified montmorillonites conform to the Langmuir, Freundlich, or related models rather than the linear type. This finding would indicate solute removal is influenced by other factors, such as solute-surface interaction and changes in surfactant conformation, besides simple

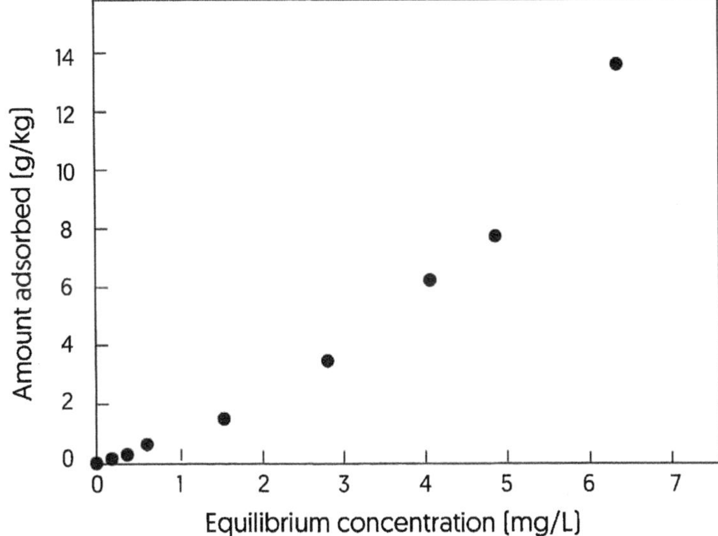

FIGURE 4.9 Isotherm for the adsorption of naphthalene from aqueous solution to montmorillonite exchanged with octadecyltrimethylammonium ions. (From Yuan, G., *J. Environ. Sci. Health Part A*, 39, 2661–2670, 2004.)

partitioning. Such is the case for the uptake of BTEX compounds (Carvalho et al. 2012), diphenylamine (Chen et al. 2014), and various phenolic compounds (Lo et al. 1998; Farkas and Dékány 2001; Juang et al. 2002; Witthuhn et al. 2005; Okada and Ogawa 2010; Zhou et al. 2008, 2011; Nguyen et al. 2013; Park et al. 2013; Zheng et al. 2013; Huang et al. 2015) by type II organoclays. Likewise, Yuan (2004) found that the isotherm for the removal of naphthalene by octadecyltrimethylammonium-montmorillonite was non-linear (Figure 4.9). Ghezali et al. (2018) observed similarly for the uptake of 2,4,6-trichlorophenol by a bentonite complex with benzyldimethyltetradecyl ammonium as did Intachai et al. (2021) for the adsorption of rhodamine 6G to some alkylammonium-exchanged bentonite. On the other hand, naphthalene adsorption to HDTMA-modified kaolinite and halloysite yielded linear isotherms (Lee and Kim 2002b). Moreover, there is evidence to show that such organic solutes as *p*-nitrophenol can displace and alter the interlayer arrangement of the surfactant molecule (Zhou et al. 2007; Frost et al. 2008). Molecular modelling further indicates that phenols can form H-bonds with the interlayer siloxane surface and the inorganic counterions of montmorillonite at low loading levels of HDTMA (Zhou et al. 2015).

Here we would interpolate that montmorillonites, exchanged with quaternary phosphonium cations (QPC), are often used (in preference to their QAC-exchanged counterparts) for the synthesis of polymer-clay nanocomposites because QPC-based organoclays have a relatively high thermal stability (Xie et al. 2002; Hedley et al. 2007; Patel et al. 2007; Calderon et al. 2008; Mittal 2012; Prahsarn et al. 2014; Suin et al. 2014; Ezquerro et al. 2015; Alves et al. 2016; Madejová et al. 2021).

Type II organoclays containing HDTMA can be prepared using different concentrations (C_s) of HDTMA-Br from 0.5 to > 2.0 times the cation exchange capacity (CEC) of the parent montmorillonites, ranging from 44 to 138 $cmol_c$/kg. Data on the attendant changes in basal spacing, collected by He et al. (2014), are summarized in Figure 4.10. For C_s = 0.5 CEC the basal spacings of the resultant organoclays remain essentially constant at ~ 1.5 nm. The spacings increase slightly from ~ 1.7 to ~ 2.2 nm for C_s = 1.0 CEC. For montmorillonites with a CEC > 80 $cmol_c$/kg, the basal spacings increase from 2.0–2.5 nm to 3.2–3.8 nm at C_s = 1.5 CEC, and to 3.5–4.0 nm at C_s = 2.0 CEC. These observations may be ascribed to the progressive intercalation of HDTMA as a flat monolayer, a bilayer, a pseudo-trilayer, and a paraffin-type arrangement (cf. Figure 4.1), while the point scatter

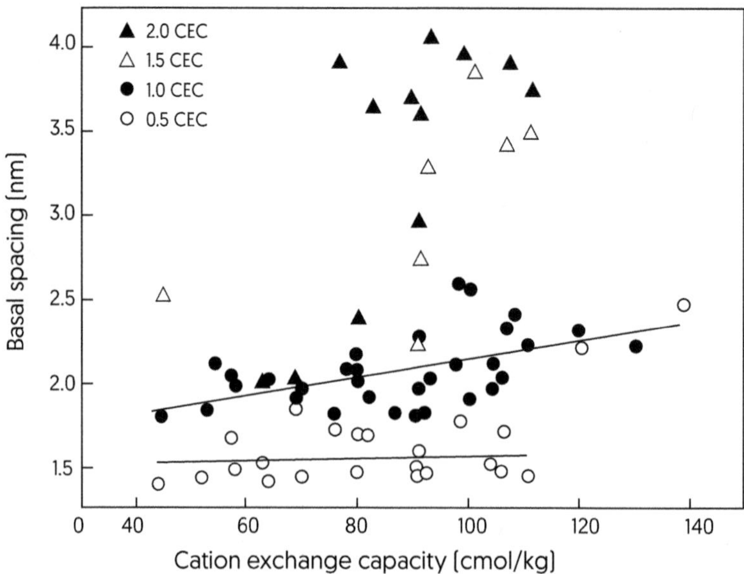

FIGURE 4.10 Relationship between the basal spacing of montmorillonite complexes with hexadecyltrimethylammonium (HDTMA) ions, prepared using different concentrations of HDTMA-Br, ranging from 0.5 to 2.0 times the cation exchange capacity (CEC) of the parent clay mineral samples. (From He, H. et al., *Appl. Clay Sci.*, 100, 22–28, 2014.)

reflects variation in layer charge density and packing density among the montmorillonite samples (Slade and Gates 2004). More recently, Qiu et al. (2020) also found that Na^+-montmorillonite of low charge density was easier to expand by intercalation of QAC than its high-charge counterpart, while the basal spacing of the interlayer complexes increased with alkyl chain length. We might also add that HDTMA, adsorbed in excess of 60% of the CEC of kaolinite, apparently assumed a bilayer conformation on external particle surfaces (Lee and Kim 2002b).

He et al. (2006a) have reported that as the HDTMA loading increases from zero (for the pristine Na^+-montmorillonite) to 2.0 CEC, the BET-N_2 surface area decreases from 55 to 1 m^2/g, and the pore volume from 0.11 to 0.01 cm^3/g although the average pore diameter increases from ~ 7 to ~ 16 nm. For a HDTMA loading of ≤ 1.0 CEC, most of the surfactant is intercalated, while for loadings of ≥ 1.5 CEC the surfactant is adsorbed to both interlayer and external particle surfaces. Zawrah et al. (2014) have suggested similarly based on X-ray diffraction, infrared spectroscopy, and scanning electron microscopy data. At fractional surfactant loadings, the removal of polar organic contaminants (e.g., *p*-nitrophenol) from aqueous solution involves both surface adsorption and partitioning, yielding non-linear isotherms. On the other hand, partitioning is the dominant mechanism controlling organic solute uptake at high surfactant loadings, giving rise to C-class (linear) isotherms (Boyd et al. 1988b).

Chiou (1989) has found that the capacity of type II organoclays for taking up non-ionic organic compounds from water is inversely related to the aqueous solubility of the solutes. Thus, the affinity of HDTMA-smectite for phenols increases in the order phenol < chlorophenol < dichlorophenol < trichlorophenol while their solubility in water decreases in the same order (Mortland et al. 1986; Lo et al. 1998). Figure 4.11 shows this relationship for a range of organic compounds with variable water solubility (Beall 2003).

Thermodynamic measurements by Zhu and Zhu (2008b) further indicate that the sorption of non-ionic organic solutes, such as naphthalene, to organoclays with a low HDTMA packing density is driven by the change in both the enthalpy ($\Delta H°$) and entropy ($-T\Delta S°$) terms. As the surfactant packing density increases, however, solute uptake is largely driven by the entropy term. This

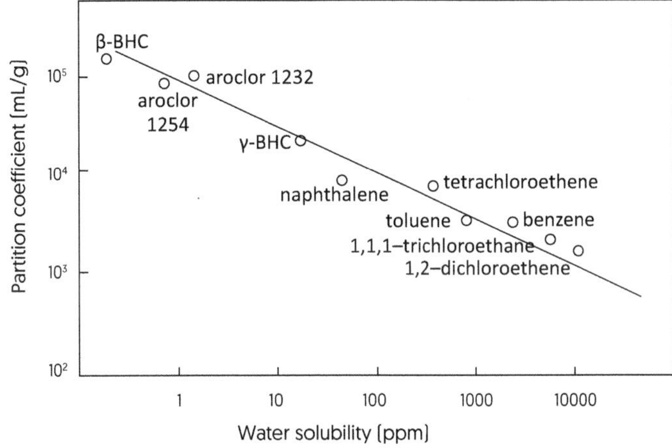

FIGURE 4.11 Relationship between the partition coefficient (ratio of concentration of a compound in water to its concentration in the organoclay) and the solubility of the compound in water for a variety of uncharged organic molecules. In this instance, the organoclay is an interlayer complex of montmorillonite with dimethyl dehydrogenated tallow. (From Beall, G.W., *Appl. Clay Sci.*, 24, 11–20, 2003.)

observation led them to suggest that the formation of organic phases, comprising surfactant aggregates, varies with packing density.

Earlier, Theng et al. (1998) found that the solid-phase intercalation of phenanthrene (Ph) into a dry complex of tetradecyltrimethylammonium (TDTMA)-montmorillonite induced the TDTMA chains to change conformation from a flat bilayer to a paraffin-like arrangement (Figure 4.1). Solid-state ^{13}C NMR spectroscopy indicated that the TDTMA chains became disordered and intimately mixed with Ph. As a result, the intercalated solute was not available to degrading bacteria and their enzymes, at least over the limited period (11 days) of incubation in the laboratory (Theng et al. 2001). The naturally occurring counterpart of this phenomenon is an interlayer clay-organic complex, separated from a surface horizon of a Spodosol in New Zealand. The intercalated organic material has a radiocarbon age of 6700 years BP because it has long been protected from microbial decomposition and contamination by modern carbon (Theng et al. 1992).

The large volume of literature on the synthesis, characterization, and applications of organoclays that has accumulated over the last few decades has periodically been reviewed (Ortego et al. 1991; Xu et al. 1997; Beall 2003; Lagaly et al. 2006; Patel et al. 2006; de Paiva et al. 2008; Ruiz-Hitzky et al. 2010; Park et al. 2011; Lee and Tiwari 2012; Sarkar et al. 2012; Yuan et al. 2013; He et al. 2014; Nafees and Waseem 2014; Zhu et al. 2016; He and Zhu 2017; Zhao et al. 2017; Awad et al. 2019; Guégan 2019; Shen and Gao 2019; Zhang et al. 2019; Zhuang et al. 2019b; Guo et al. 2020; Gil et al. 2021; Perelomov et al. 2021).

Although covalently grafted clay minerals do not feature large as adsorbents of organic compounds, we will briefly describe their synthesis and properties for the sake of completeness. Research into the grafting of organic functional groups to clay mineral surfaces dates back to the 1940s when Berger (1941) reported to have methylated the surface silanol groups of H-montmorillonite using diazomethane (CH_2N_2) according to Equation 4.6,

$$[\text{surface}]-OH + CH_2N_2 \rightarrow [\text{surface}]-OCH_3 + N_2^{\uparrow} \qquad \text{(Equation 4.6)}$$

Berger's work, however, together with the results of early attempts at preparing *organic derivatives* of clay minerals (Deuel et al. 1950; Deuel 1952; Spencer and Gieseking 1952), have been called into question (Brown et al. 1952; Greenland and Russell 1955) on the grounds that montmorillonite has a low content of exposed ≡ Si–OH groups. On the other hand, palygorskite and sepiolite can be readily methylated by diazomethane (Hermosín and Cornejo 1986) since their surfaces are rich in silanol groups (cf. Figures 1.17 and 1.18). The methylation and silylation of these minerals are further described in Chapter 7.

Nevertheless, Dai and Huang (1999) have proposed that the *outer-surface* hydroxyls of kaolinite (cf. Figure 1.4) could be esterified using diazomethane or acetyl chloride. Little doubt, however, attaches to the esterification of the *inner-surface* hydroxyls using alkanols, diols, and glycol mono-ethers when the individual silicate layers of kaolinite particles have previously been separated by intercalation of dimethyl sulphoxide or *N*-methylformamide (Tunney and Detellier 1993; Komori et al. 2000; Gardolinski and Lagaly 2005).

The grafting process or *silylation* involves the hydrolysis of an organosilane coupling agent, R–SiX_3, (Equation 4.7), followed by condensation of the hydrolysis product, R-$Si(OH)_3$, with a surface silanol group (Equation 4.8),

$$\text{R-SiX}_3 + 3\text{H}_2\text{O} \rightarrow \text{R-Si}(\text{OH})_3 + 3\text{HX} \qquad \text{(Equation 4.7)}$$

$$[\text{surface}] - \text{OH} + \text{R-Si}(\text{OH})_3 \rightarrow [\text{surface}] - \text{O} - \text{SiR}(\text{OH})_2 + \text{H}_2\text{O} \qquad \text{(Equation 4.8)}$$

where R denotes a functional group, such as $-CH_2NH_2$, γ-methacryloxypropyl, γ-aminopropyl, glycidylpropyl, octyldimethyl, phenyldimethyl, vinyldimethyl, while X represents an alkoxy group ($-OCH_3$ or $-OC_2H_5$). Among the large variety of coupling agents, γ- or 3-aminopropyl triethoxysilane (APTES) with the structural formula of $NH_2(CH_2)_3Si(OC_2H_5)_3$, has featured prominently. In common with organoclays, silylated clay minerals show an enhanced capacity for taking up organic contaminants from aqueous solution as compared to the pristine (raw) materials (Guo et al. 2020; Gil et al. 2021).

Being richly endowed with structural hydroxyl groups (cf. Figure 1.18), sepiolite can be readily grafted with organosilanes (Ruiz-Hitzky and Fripiat 1976; Undabeytia et al. 2019; Junior et al. 2020). Similarly, imogolite lends itself to being silylated as do Laponite and acid-treated chrysotile (Johnson and Pinnavaia 1990, 1991; Herrera et al. 2005; Mendelovici and Frost 2005; Wheeler et al. 2005; Zanzottera et al. 2012).

In particular, APTES has been used for the covalent grafting of kaolinite (Tonlé et al. 2007; Avila et al. 2010; Guerra et al. 2012; Yang et al. 2012; Tao et al. 2014; Zhang et al. 2015), halloysite (Yuan et al. 2008; Daitx et al. 2015; Peixoto et al. 2016), fluorohectorite (He et al. 2005), imogolite (Johnson and Pinnavaia 1990; Qi et al. 2008), Laponite (Wheeler et al. 2005; Pereira et al. 2007; Peraro et al. 2020; Thiebault et al. 2020), montmorillonite (Tonlé et al. 2003; He et al. 2005; Shanmugharaj et al. 2006; Uchida et al. 2006; Shen et al. 2007; Piscitelli et al. 2010; Su et al. 2012; Bertuoli et al. 2014; Asgari and Sundararaj 2018), palygorskite (Moreira et al. 2017), pyrophyllite (Sayılkan et al. 2004), saponite (Avila et al. 2010; Marçal et al. 2015; Tao et al. 2016), sepiolite (Moreira et al. 2017; Undabeytia et al. 2019), and vermiculite (Alves et al. 2013).

Organosilanes other than APTES, and related compounds have also been used for the silylation of montmorillonite, hectorite, sepiolite, acid-activated smectites, pillared interlayered clays, and organoclays (Song and Sandí 2001; Herrera et al. 2005; Zhu et al. 2007; Guimarães et al. 2009; Phothitontimongkol et al. 2009; Piscitelli et al. 2010; Paul et al. 2011; Ianchis et al. 2012; Sepehri et al. 2014; Romanzini et al. 2015; Horri et al. 2019; Kostenko et al. 2019; Undabeytia et al. 2019; Dumitru et al. 2022; Pinos et al. 2022).

Like quaternary phosphonium cation-exchanged smectites, silylated clay minerals have a high thermal stability (and hydrophobicity) and hence have found applications in the synthesis of polymer-clay nanocomposites (Le Pluart et al. 2002; Ruiz-Hitzky and Van Meerbeek 2006; Zhang et al. 2006; Wang et al. 2007; Choi et al. 2009; Mishra et al. 2011; Silva et al. 2011; Ianchis et al. 2012; Tao et al. 2014; Kotal and Bhowmick 2015; Khajehpour et al. 2015; Romanzini et al. 2015; Souza et al. 2016; Varadwaj et al. 2016; Asgari et al. 2017; Theng 2018; Abbassi et al. 2021).

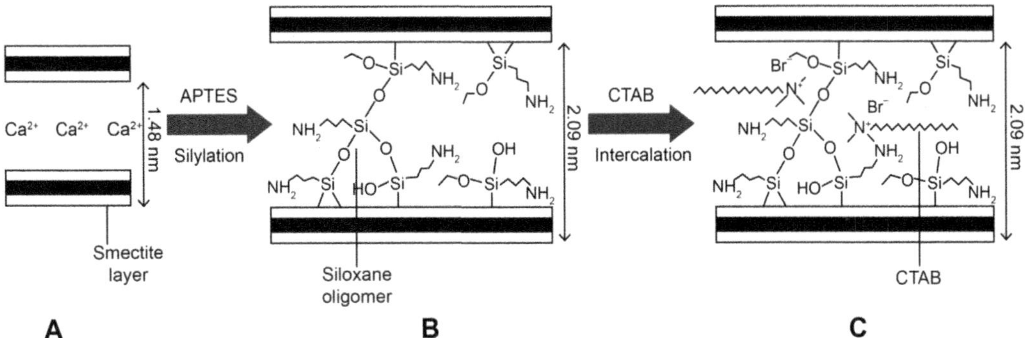

FIGURE 4.12 Diagram showing the silylation of Ca^{2+}-montmorillonite (A) using γ-aminopropyl triethoxysilane (APTES) to yield the 'locked' silylated complex (B) into which cetyltrimethylammonium bromide (CTAB) is intercalated to form a new complex (C) with the same basal spacing as (B). (After Su et al. 2012, Graphical abstract. Reused with permission from *Clay Mineral Catalysis of Organic Reactions*, by B.K.G. Theng, Figure 3.7, p. 182, CRC Press, 2018.)

In a review on clay mineral silylation, He et al. (2013) and Varadwaj et al. (2016) have pointed out that the success of silane grafting is dependent on such factors as surface hydroxyl density, functionality and structure of the silane coupling agent, solvent polarity, and reaction temperature. The use of non-polar solvents (e.g., toluene, cyclohexane) is conducive to intercalation, hydrolysis, and condensation of the silane molecules into polysiloxane oligomers (Su et al. 2013). On intercalation, these oligomers can form covalent bonds with opposing interlayer surfaces, increasing the basal spacing from 1.48 nm (for the parent Ca^{2+}-montmorillonite) to 2.09 nm (Su et al. 2012). X-ray diffraction together with infrared and NMR spectroscopy measurements by Uchida et al. (2006) indicate that APTES is intercalated into Na^{+}-montmorillonite as a siloxane cubic octamer ('cage') structure rather than as a 'random' polymer. On the other hand, 3-aminopropyl diethoxymethylsilane gives rise to poly(dimethylsiloxane) on intercalation. In either case, the individual layers in a montmorillonite quasi-crystal would be 'locked', preventing further interlayer expansion when cetyltrimethylammonium bromide (CTAB) is introduced (Figure 4.12).

The locking of contiguous silicate layers would also inhibit particle exfoliation during the preparation of polymer-clay nanocomposites and hence adversely affect the properties of the composite products. On the other hand, the highly porous organosilane-grafted material can usefully serve as sorbents and molecular sieves of organic contaminants and pollutants (He et al. 2014). In this context, we would mention the locking of Laponite layers by grafting the edge silanols with mono- and trifunctional alkoxy silanes (Herrera et al. 2005; Wheeler et al. 2005). Subsequently, Wang et al. (2007) synthesized 'dual-functionalized' Laponite by grafting a silane coupling agent followed by exchanging the inorganic counterions with quaternary ammonium ions, while Mishra et al. (2011) obtained a similar product using the reverse order of treatment.

REFERENCES

Abbas, A., A.S. Sallam, A.R.A. Usman, and M.I. Al-Wabel. 2017. Organoclay-based nanoparticles from montmorillonite and natural clay deposits: Synthesis, characteristics, and application for MTBE removal. *Applied Clay Science* 142: 21–29.

Abbassi, H., R. Abidi, and M.B. Zayani. 2021. A short review on the silylated clays-polymer nanocomposites: Synthesis, properties and applications. *Journal Marocain de Chimie Hétérocyclique* 20: 117–134.

Alves, A.P.M., M.G. Fonseca, and A.F. Wanderley. 2013. Inorganic-organic hybrids originating from organosilane anchored onto leached vermiculite. *Materials Research* 16: 1–8.

Alves, J.P., P.T.V. Rosa, and A.R. Morales. 2016. A comparative study of different routes for the modification of montmorillonite with ammonium and phosphonium salts. *Applied Clay Science* 132–133: 475–484.

Asgari, M. and U. Sundararaj. 2018. Silane functionalization of sodium montmorillonite nanoclay: The effect of dispersing media on intercalation and chemical grafting. *Applied Clay Science* 153: 228–238.

Asgari, M., A. Abouelmagd, and U. Sundararaj. 2017. Silane functionalization of sodium montmorillonite nanoclay and its effect on rheological and mechanical properties of HDPE/clay nanocomposites. *Applied Clay Science* 146: 439–448.

Avila, L.R., E.H. de Faria, K.J. Ciuffi et al. 2010. New synthesis strategies for effective functionalization of kaolinite and saponite with silylating agents. *Journal of Colloid and Interface Science* 341: 186–193.

Awad, A.M., S.M.R. Shaikh, R. Jalab et al. 2019. Adsorption of organic pollutants by natural and modified clays: A comprehensive review. *Separation and Purification Technology* 228: 115719.

Barrer, R.M. 1978. *Zeolites and Clay Minerals as Sorbents and Molecular Sieves.* London, UK: Academic Press.

Barrer, R.M. and K. Brummer. 1963. Relations between partial ion exchange and interlamellar sorption in alkylammonium montmorillonites. *Transactions of the Faraday Society* 59: 959–968.

Barrer, R.M. and K.E. Kelsey. 1961. Thermodynamics of interlamellar complexes. I. Hydrocarbons in methylammonium montmorillonites *Transactions of the Faraday Society* 57: 452–462.

Barrer, R.M. and D.M. MacLeod. 1955. Activation of montmorillonite by ion exchange and sorption complexes of tetra-alkylammonium montmorillonites. *Transactions of the Faraday Society* 51: 1290–1300.

Barrer, R.M. and A.D. Millington. 1967. Sorption and intracrystalline porosity in organo-clays. *Journal of Colloid and Interface Science* 25: 359–372.

Barrer, R.M. and J.S.S. Reay. 1957. Sorption and intercalation by methylammonium montmorillonites. *Transactions of the Faraday Society* 53: 1253–1261.

Beall, G.W. 2003. The use of organo-clays in water treatment. *Applied Clay Science* 24: 11–20.

Berger, G. 1941. The structure of montmorillonite: Preliminary communication on the ability of clays and clay minerals to be methylated. *Chemische Weekblad* 38: 42–43.

Berkheiser, V.E. and M.M. Mortland. 1977. Hectorite complexes with Cu(II) and Fe(II)-1,10-phenanthroline chelates. *Clays and Clay Minerals* 25: 105–112.

Bertuoli, P.T., D. Piazza, L.C. Scienza, and A.J. Zattera. 2014. Preparation and characterization of montmorillonite modified with 3-aminopropyltriethoxysilane. *Applied Clay Science* 87: 46–51.

Bizovská, V., L. Jankovič, and J. Madejová. 2018. Montmorillonite modified with unconventional surfactants from the series of octylammonium-based cations: Structural characterization and hydration. *Applied Clay Science* 158: 102–112.

Boyd, S.A., M.M. Mortland, and C.T. Chiou. 1988b. Sorption characteristics of organic compounds on hexadecyltrimethylammonium-smectite. *Soil Science Society of America Journal* 52: 652–657.

Boyd, S.A., S. Sun, J.-F. Lee, and M.M. Mortland. 1988a. Pentachlorophenol sorption by organo-clays. *Clays and Clay Minerals* 36: 125–130.

Brindley, G.W. and R.W. Hoffmann. 1962. Orientation and packing of aliphatic chain molecules on montmorillonite: Clay-organic studies VI. *Clays and Clay Minerals* 9: 546–556.

Brown, G., R. Greene-Kelly, and K. Norrish. 1952. Organic derivatives of montmorillonite. *Nature* 169: 756–757.

Calderon, J.U., B. Lennox, and M.R. Kamal. 2008. Thermally stable phosphonium- montmorillonite organoclays. *Applied Clay Science* 40: 90–98.

Cao, G., M. Gao, T. Shen, S. Guo, B. Zhao, and Q. Zhao. 2020. Asymmetric Gemini surfactants modified vermiculite- and silica nanosheets-based adsorbents for removing methyl orange and crystal violet. *Colloids and Surfaces A* 596: 124735.

Cao, G., M. Gao, T. Shen, B. Zhao, and H. Zeng. 2019. Comparison between asymmetric and symmetric Gemini surfactant-modified novel organo-vermiculites for removal of phenols. *Industrial & Engineering Chemistry Research* 58: 12927–12938.

Carvalho, M.N., M. da Motta, M. Benachour, D.C.S. Sales, and C.A.M. Abreu. 2012. Evaluation of BTEX and phenol removal from aqueous solution by multi-solute adsorption onto smectite organoclay. *Journal of Hazardous Materials* 239–240: 95–101.

Chakravarti, S.K. 1957. Adsorption of higher alkyl quaternary ammonium and pyridinium compounds by clay minerals. *Journal of the Indian Society of Soil Science* 5: 85–90.

Chaussidon, J. and R. Calvet. 1965. Evolution of amine cations adsorbed on montmorillonite with dehydration of the mineral. *Journal of Physical Chemistry* 69: 2265–2268.

Chen, B., J.R.G. Evans, H.C. Greenwell et al. 2008b. A critical appraisal of polymer-clay nanocomposites. *Chemical Society Reviews* 37: 568–594.

Chen, C., W. Zhou, Q. Yang, L. Zhu, and L. Zhu. 2014. Sorption characteristics of nitrosodiphenylamine (NDPhA) and diphenylamine (DPhA) onto organo-bentonite from aqueous solution. *Chemical Engineering Journal* 240: 487–493.

Chen, D., J.X. Zhu, P. Yuan, S.J. Yang, T.-H. Chen, and H.P. He. 2008a. Preparation and characterization of anion-cation surfactants modified montmorillonite. *Journal of Thermal Analysis and Calorimetry* 94: 841–848.

Chiou, C.T. 1989. Theoretical considerations of the partition uptake of nonionic organic compounds by soil organic matter. In *Reactions and Movement of Organic Chemicals in Soils*, (Eds.) B.L. Sawhney and K. Brown, pp. 1–29. Madison, WI: Soil Science Society of America.

Choi, Y.Y., S.H. Lee, and S.H. Ryu. 2009. Effect of silane functionalization of montmorillonite on epoxy/montmorillonite nanocomposite. *Polymer Bulletin* 63: 47–55.

Cowan, C.T. and D. White. 1958. The mechanisms of exchange reactions occurring between sodium montmorillonite and various *n*-primary aliphatic amine salts. *Transactions of the Faraday Society* 54: 691–697.

Dai, J.C. and J.T. Huang. 1999. Surface modification of clays and clay-rubber composite. *Applied Clay Science* 15: 51–65.

Daitx, T.S., L.N. Carli, J.S. Crespo, and R.S. Mauler. 2015. Effects of the organic modification of different clay minerals and their application in biodegradable polymer nanocomposites of PHBV. *Applied Clay Science* 115: 157–164.

Das Kanungo, J.L. and S.K. Chakravarti. 1971. Behavior of quaternary ammonium ions in the desorption of $[Coen_3]^{3+}$ from H-$Coen_3$-bentonite. *Journal of Colloid and Interface Science* 35: 295–299.

Das Kanungo, J.L., S.K. Chakravarti, and S.K. Mukherjee. 1968. Studies on the sorption and desorption of $Co(NH_3)_6\,C1_3$ on bentonite. *Journal of the Indian Chemical Society* 45: 685–689.

de Paiva, L.B., A.R. Morales, and F.R. Valenzuela Díaz. 2008. Organoclays: Properties, preparation and applications. *Applied Clay Science* 42: 8–24.

Deuel, H. 1952. Organic derivatives of clay minerals. *Clay Minerals Bulletin* 1: 205–214.

Deuel, H., G. Huber, and R. Iberg. 1950. Organische derivate von Tonmineralien. *Helvetica Chimica Acta* 33: 1229–1232.

Diamond, S. and E.B. Kinter. 1963. Characterization of montmorillonite saturated with short chain amine cations. I. Interpretation of basal spacing measurements. *Clays and Clay Minerals* 10: 163–173.

Dowdy, R.H. and M.M. Mortland. 1968. Alcohol-water interactions on montmorillonite surfaces. II. Ethylene glycol. *Soil Science* 105: 36–43.

Dumitru, M.A., T. Sandu, A.L. Ciurlica et al. 2022. Organically modified montmorillonite as pH versatile carriers for delivery of 5-aminosalicylic acid. *Applied Clay Science* 218: 106415.

Erbring, H. and H. Lehmann. 1944. Austauschreaktionen an Na-Bentoniten mit grossvolumigen organischen Kolloidionen. *Kolloid Zeitschrift* 107: 201–205.

Ezquerro, C.S., G.I. Ric, C.C. Miñana, and J.S. Bermajo. 2015. Characterization of montmorillonites modified with organic divalent phosphonium cations. *Applied Clay Science* 111: 1–9.

Farkas, A. and I. Dékány. 2001. Interlamellar adsorption of organic pollutants in hydrophobic montmorillonite. *Colloid and Polymer Science* 279: 459–467.

Flores, F.M., E.L. Loveira, F. Yarza, R. Candal, and R.M. Torres Sánchez. 2017. Benzalkonium chloride surface adsorption and release by two montmorillonites and their modified organomontmorillonites. *Water Air Soil Pollution* 228: 42. https://doi.org/10.1007/s11270-016-3223-2.

Frank, H.S. and M.W. Evans. 1945. Free volume and entropy in condensed systems. III. Entropy in binary liquid mixtures; partial molal entropy in dilute solutions; structure and thermodynamics in aqueous electrolytes. *Journal of Chemical Physics* 13: 507–532.

Fripiat, J.J., P. Cloos, and A. Poncelet. 1965. Comparison between the exchange properties of montmorillonite and of a resin for alkali and alkaline-earth cations. I. Reversibility of the process. *Bulletin de la Société Chimique de France* 208–215.

Fripiat, J.J., M. Pennequin, G. Poncelet, and P. Cloos. 1969. Influence of the van der Waals force on the infrared spectra of short aliphatic alkylammonium cations held on montmorillonite. *Clay Minerals* 8: 119–134.

Frost, R.L., Q. Zhou, H. He, and Y. Xi. 2008. An infrared study of adsorption of *para*-nitrophenol on mono-, di- and tri-alkyl surfactant intercalated organoclays. *Spectrochimica Acta Part A* 69: 239–244.

Fu, M., Z. Zhang, L. Wu et al. 2016. Investigation on the co-modification process of montmorillonite by anionic and cationic surfactants. *Applied Clay Science* 132–133: 694–701.

Fuller, M., J.A. Smith, and S.E. Burns. 2007. Sorption of nonionic organic solutes from water to tetraalkylammonium bentonites: Mechanistic considerations and application of the Polanyi-Manes potential theory. *Journal of Colloid and Interface Science* 313: 405–414.

Gaines, G.L., Jr. and H.C. Thomas. 1953. Adsorption studies on clay minerals. II. A formulation of the thermodynamics of exchange adsorption. *Journal of Chemical Physics* 21: 714–718.

Gamoudi, S., N. Frini-Srasra, and E. Srasra. 2012. Influence of exchangeable cation of smectite on HDTMA adsorption: Equilibrium, kinetic and thermodynamic studies. *Applied Clay Science* 69: 99–107.

Gamoudi, S., N. Frini-Srasra, and E. Srasra. 2013. Solid-state intercalation of cationic surfactants into Tunisian smectites. *Clay Minerals* 48: 583–593.

Gamoudi, S., N. Frini-Srasra, and E. Srasra. 2015. Influence of synthesis method in preparation of $HDTMA^{+}$ and $HDPy^{+}$-illites/smectites. *Applied Clay Science* 116–117: 78–84.

Gardolinski, J.E.F.C. and G. Lagaly. 2005. Grafted organic derivatives of kaolinite: I. Synthesis, chemical and rheological characterization. *Clay Minerals* 40: 537–546.

Garrett, W.G. and G.F. Walker. 1962. Swelling of some vermiculite-organic complexes in water. *Clays and Clay Minerals* 9: 557–567.

Ghezali, S., A. Mahdad-Benzerdjeb, M. Ameri, and A.Z. Bouyakoub. 2018. Adsorption of 2,4,6-trichlorophenol on bentonite modified with benzyldimethyltetradecylammonium chloride. *Chemistry International* 4: 24–32.

Ghiaci, M., R.J. Kalbasi, H. Khani, A. Abbaspur, and H. Shariatmadari. 2004. Free-energy of adsorption of cationic surfactant onto Na-bentonite (Iran): Inspection of adsorption layer by X-ray spectroscopy. *Journal of Chemical Thermodynamics* 36: 707–713.

Gieseking, J.E. 1939. Mechanism of cation exchange in the montmorillonite-beidellite-nontronite type of clay minerals. *Soil Science* 47: 1–14.

Gieseking, J.E. and H. Jenny. 1936. Behavior of polyvalent cations in base exchange. *Soil Science* 42: 273–280.

Gil, A., L. Santamaría, S.A. Korili et al. 2021. A review of organic-inorganic hybrid clay based adsorbents for contaminants removal: Synthesis, perspectives and applications. *Journal of Environmental Chemical Engineering* 9: 105808.

Giles, C.H., T.H. MacEwan, S.N. Nakhwa, and D. Smith. 1960. Studies in adsorption. XI. A system of classification of solution adsorption isotherms and its use in diagnosis of adsorption mechanisms and in measurement of specific surface areas of solids. *Journal of the Chemical Society* 3973–3993.

Granquist, W.T. and J.L. McAtee. 1963. The gelation of hydrocarbons by montmorillonite organic complexes: The role of the dispersant. *Journal of Colloid Science* 18: 409–420.

Greene-Kelly, R. 1955. Sorption of aromatic compounds by montmorillonite. Part 1. Orientation studies. *Transactions of the Faraday Society* 51: 412–424.

Greenland, D.J. and J.P. Quirk. 1962. Adsorption of 1-*n*-alkylpyridinium bromides by montmorillonite. *Clays and Clay Minerals* 9: 484–499.

Greenland, D.J., J.P. Quirk, and B.K.G. Theng. 1964. Influence of increasing proportions of exchangeable alkylammonium ions on the swelling of calcium montmorillonite in water. *Journal of Colloid Science* 19: 837–840.

Greenland, D.J. and E.W. Russell. 1955. Organo-clay derivatives and the origin of the negative charge on clay particles. *Transactions of the Faraday Society* 51: 1300–1307.

Grim, R.E., W.H. Allaway, and F.L. Cuthbert. 1947. Reaction of different clay minerals with some organic cations. *Journal of the American Ceramic Society* 30: 137–142.

Groisman, L., C. Rav-Acha, Z. Gerstl, and U. Mingelgrin. 2004. Sorption of organic compounds of varying hydrophobicities from water and industrial wastewater by long- and short-chain organoclays. *Applied Clay Science* 24: 159–166.

Gu, Z., M. Gao, L. Lu, Y. Liu, and S. Yang. 2015. Montmorillonite functionalized with zwitterionic surfactant as a highly efficient adsorbent of herbicides. *Industrial & Engineering Chemistry Research* 54: 4947–4955.

Guégan, R. 2019. Organoclay applications and limits in the environment. *Comptes Rendus Chimie* 22: 132–141.

Guégan, R., M. Gautier, J.-M. Beny, and F. Muller. 2009. Adsorption of a $C_{10}E_3$ non-ionic surfactant on a Ca-smectite. *Clays and Clay Minerals* 57: 502–509.

Guégan, R., M. Giovanela, F. Warmont, and M. Motelica-Heino. 2015. Nonionic organoclay: A 'Swiss army knife' for the adsorption of organic micro-pollutants? *Journal of Colloid and Interface Science* 437: 71–79.

Guerra, D.L., S.P. Oliveira, R.A.S. Silva, E.M. Silva, and A.C. Batista. 2012. Dielectric properties of organofunctionalized kaolinite clay and application in adsorption mercury cation. *Ceramic International* 38: 1687–1696.

Guimarães, A.D.M.F., V.S.T. Ciminelli, and W.L. Vasconcelos. 2009. Smectite organofunctionalized with thiol groups for adsorption of heavy metal ions. *Applied Clay Science* 42: 410–414.

Gul, S., A. Kausar, B. Muhammad, and S. Jabeen. 2016. Technical relevance of epoxy/clay nanocomposite with organically modified montmorillonite: A review. *Polymer-Plastics Technology and Engineering* 55: 1393–1415.

Guo, Y.X., J.H. Liu, W.P. Gates, and C.H. Zhou. 2020. Organo-modification of montmorillonite. *Clays and Clay Minerals* 68: 601–622.

Haase, D.J., E.J. Weiss, and H. Steinfink. 1963. The crystal structure of a hexamethylenediamine-vermiculite complex. *American Mineralogist* 48: 261–270.

Hackett, E., E. Manias, and E.P. Giannelis. 1998. Molecular dynamics simulations of organically modified silicates. *Journal of Chemical Physics* 108: 7410–7415.

He, H., J. Duchet, J. Galy, and J.-F. Gérard. 2005. Grafting of swelling clay minerals with 3-aminopropyltriethoxysilane. *Journal of Colloid and Interface Science* 288: 171–176.

He, H., R.L. Frost, T. Bostrom et al. 2006a. Changes in the morphology of organoclays with $HDTMA^+$ surfactant loading. *Applied Clay Science* 31: 262–271.

He, H., R.L. Frost, F. Deng, J. Zhu, X. Wen, and P. Yuan. 2004. Conformation of surfactant molecules in the interlayer of montmorillonite studied by ^{13}C MAS NMR. *Clays and Clay Minerals* 52: 350–356.

He, H., L. Ma, J. Zhu, R.L. Frost, B.K.G. Theng, and F. Bergaya. 2014. Synthesis of organoclays: A critical review and some unresolved issues. *Applied Clay Science* 100: 22–28.

He, H., Y. Ma, J. Zhu, P. Yuan, and Y. Qing. 2010. Organoclays prepared from montmorillonites with different cation exchange capacity and surfactant configuration. *Applied Clay Science* 48: 67–72.

He, H., Q. Tao, J. Zhu, P. Yuan, W. Shen, and S. Yang. 2013. Silylation of clay mineral surfaces. *Applied Clay Science* 71: 15–20.

He, Y., L. Zhou, S. Gou et al. 2020. Synergy of imidazolium ionic liquids and flexible anionic polymer for controlling facilely montmorillonite swelling in water. *Journal of Molecular Liquids* 317: 114261.

He, Y., L. Zhou, S. Gou et al. 2022. Imidazolium Gemini ionic liquids for regulating facilely the swelling and dispersion of montmorillonite in water. *Applied Clay Science* 219: 106457.

He, H., Q. Zhou, W.N. Martens et al. 2006b. Microstructure of $HDTMA^+$-modified montmorillonite and its influence on sorption characteristics. *Clays and Clay Minerals* 54: 689–696.

He, H. and J. Zhu. 2017. Analysis of organoclays and organic adsorption by clay minerals. In *Infrared and Raman Spectroscopies of Clay Minerals*. Developments in Clay Science, Vol. 8, (Eds.) W.P. Gates, J.T. Kloprogge, J. Madejová, and F. Bergaya, pp. 310–342. Amsterdam, The Netherlands: Elsevier.

Hedley, C.B., G. Yuan, and B.K.G. Theng. 2007. Thermal analysis of montmorillonite modified with quaternary phosphonium and ammonium surfactants. *Applied Clay Science* 35: 180–188.

Heinz, H., R.A. Vaia, R. Krishnamoorti, and B.L. Farmer. 2007. Self-assembly of alkylammonium chains on montmorillonite: Effect of chain length, head group structure, and cation exchange capacity. *Chemistry of Materials* 19: 59–68.

Hendricks, S.B. 1941. Base exchange of the clay mineral montmorillonite for organic cations and its dependence upon adsorption due to van der Waals forces. *Journal of Physical Chemistry* 45: 65–81.

Herling, M.M., H. Kalo, S. Seibt, R. Schobert, and J. Breu. 2012. Tailoring the pore sizes of microporous pillared interlayered clays through layer charge reduction. *Langmuir* 28: 14713–14719.

Hermosín, M.C. and J. Cornejo. 1986. Methylation of sepiolite and palygorskite with diazomethane. *Clays and Clay Minerals* 34: 591–596.

Herrera, N.N., J.-M. Letoffe, J.-P. Reymond, and E. Bourgeat-Lami. 2005. Silylation of laponite clay particles with monofunctional and trifunctional vinyl alkoxysilanes. *Journal of Materials Chemistry* 15: 863–871.

Horri, N., E.S. Sanz-Pérez, A. Arencibia, R. Sanz, N. Frini-Srasra, and E. Srasra. 2019. Amine grafting of acid-activated bentonite for carbon dioxide capture. *Applied Clay Science* 180: 105195.

Hou, X.-J., H. Li, P. He, S. Li, and Q. Liu. 2015. Molecular-level investigation of the adsorption mechanisms of toluene and aniline on natural and organically modified montmorillonite. *Journal of Physical Chemistry A* 119: 11199–11207.

Huang, L., Y. Zhou, X. Guo, and Z. Chen. 2015. Simultaneous removal of 2,4-dchlorophenol and Pb(II) from aqueous solution using organoclays: Isotherms, kinetics and mechanism. *Journal of Industrial and Engineering Chemistry* 22: 280–287.

Huang, Q., X, Li, S. Ren, and W. Luo. 2019. Removal of ethyl, isobutyl, and isoamyl xanthates using cationic gemini surfactant-modified montmorillonites. *Colloids and Surfaces A* 580: 123723.

Hue, K.-A.A., A.A. El-Midany, and H.E. El-Shall. 2019. Adsorption/desorption stability of TCMA-modified clay in simulated digestion environment. *International Journal of Environmental Research* 13: 879–885.

Ianchis, R., M.C. Corobea, D. Donescu et al. 2012. Advanced functionalization of organoclay nanoparticles by silylation and their polystyrene nanocomposites obtained by miniemulsion polymerization. *Journal of Nanoparticle Research* 14: 1233–1245.

Intachai, S., C. Suppaso, and N. Khaorapapong. 2021. A novel process for intercalating alkylammonium ions in a Thai bentonite and its effect on adsorption performance. *Clays and Clay Minerals* 69: 477–488.

Jaynes, W.F. and S.A. Boyd. 1990. Trimethylphenylammonium-smectite as an effective adsorbent of water-soluble aromatic hydrocarbons. *Journal of the Air & Waste Management Association* 40: 1649–1653.

Jaynes, W.F. and S.A. Boyd. 1991a. Hydrophobicity of siloxane surfaces in smectites as revealed by aromatic hydrocarbon adsorption from water. *Clays and Clay Minerals* 39: 428–436.

Jaynes, W.F. and S.A. Boyd. 1991b. Clay mineral type and organic compound sorption by hexadecyltrimethylammonium-exchanged clays. *Soil Science Society of America Journal* 55: 43–48.

Jaynes, W.F. and G.F. Vance. 1999. Sorption of benzene, toluene, ethylbenzene, and xylene (BTEX) compounds by hectorite clays exchanged with aromatic organic cations. *Clays and Clay Minerals* 47: 358–365.

Johns, W.D. and P.K. Sen Gupta. 1967. Vermiculite-alkylammonium complexes. *American Mineralogist* 52: 1706–1724.

Johnson, L.M. and T.J. Pinnavaia. 1990. Silylation of tubular aluminosilicate polymer (imogolite) by reaction with hydrolyzed (γ-aminopropyl)triethoxysilane). *Langmuir* 6: 307–311.

Johnson, L.M. and T.J. Pinnavaia. 1991. Hydrolysis of γ-aminopropyl)triethoxysilane-silylated imogolite and formation of a silylated tubular silicate-layered silicate nanocomposite. *Langmuir* 7: 2636–2641.

Jordan, J.W. 1949. Organophilic bentonites. I. Swelling in organic liquids. *Journal of Physical and Colloid Chemistry* 53: 294–306.

Jordan, J.W. 1963. Organophilic clay-base thickeners. *Clays and Clay Minerals* 10: 299–308.

Jordan, J.W., B.J. Hook, and C.M. Finlayson. 1950. Organophilic bentonites. II. Organic liquid gels. *Journal of Physical and Colloid Chemistry* 54: 1196–1208.

Juang, R.-S., S.-H. Lin, and K.-H. Tsao. 2002. Mechanism of sorption of phenols from aqueous solutions onto surfactant-modified montmorillonite. *Journal of Colloid and Interface Science* 254: 234–241.

Junior, H.B., E. da Silva, M. Saltarelli et al. 2020. Inorganic-organic hybrids based on sepiolite as efficient adsorbents of caffeine and glyphosate pollutants. *Applied Surface Science Advances* 1: 100025.

Kelleppan, V.T., J.P. King, C.S.G. Butler, A.P. Williams, K.L. Tuck, and R.F. Tabor. 2021. Heads or tails? The synthesis, self-assembly, properties and uses of betaine and betaine-like surfactants. *Advances in Colloid and Interface Science* 297: 102528.

Khajehpour, M., G.A. Gelves, and U. Sundararaj. 2015. Modification of montmorillonite with alkyl silanes and fluorosurfactant for clay/fluoroelastomer nanocomposites. *Clays and Clay Minerals* 63: 1–14.

Khan, R.A., S. Kalam, K. Norrman, M.S. Kamal, M. Mahmoud, and A. Abdulraheem. 2022. Ionic liquids as clay swelling inhibitors: Adsorption study. *Energy Fuels* 36: 3596–3605.

Khan, R.A., M. Murtaza, A. Abdulraheem, M.S. Kamal, and M. Mahmoud. 2020. Imidazolium-based ionic liquids as clay swelling inhibitors: Mechanism, performance evaluation, and effect of different anions. *ACS Omega* 5: 26682–26696.

Klapyta, Z., T. Fujita, and N. Iyi. 2001. Adsorption of dodecyl- and octadecyltrimethyl-ammonium ions on a smectite and synthetic micas. *Applied Clay Science* 19: 5–10.

Kobakhidze, E.L. and M.E. Shishniashvili. 1965. Hydrophobic clays and non-aqueous clay suspensions. *Kolloidnyi Zhurnal* 28: 54–58.

Koh, S.M., M.S. Song, and T. Takagi. 2005. Mineralogy, chemical characteristics and stabilities of cetylpyridinium-exchanged smectite. *Clay Minerals* 40: 213–222.

Komori, Y., H. Enoto, R. Takenawa, S. Hayashi, Y. Sugahara, and K. Kuroda. 2000. Modification of the interlayer surface of kaolinite with methoxy groups. *Langmuir* 16: 5506–5508.

Kooli, F. 2014. Organo-bentonites with improved cetyltrimethylammonium contents. *Clay Minerals* 49: 683–692.

Kostenko, L.S., I.I. Tomashchuk, T.V. Kovalchuk, and O.A. Zaporozhets. 2019. Bentonites with grafted aminogroups: Synthesis, protolytic properties and assessing Cu(II), Cd(II) and Pb(II) adsorption capacity. *Applied Clay Science* 172: 49–56.

Kotal, M. and A.K. Bhowmick. 2015. Polymer nanocomposites from modified clays: Recent advances and challenges. *Progress in Polymer Science* 51: 127–187.

Koteja, A., M. Szczerba, and J. Matusik. 2017. Smectites intercalated with azobenzene and aminoazobenzene: Structure changes at nanoscale induced by UV light. *Journal of Physics and Chemistry of Solids* 111: 294–303.

Kukkadapu, R.K. and S.A. Boyd. 1995. Tetramethylphosphonium- and tetramethylammonium-smectites as adsorbents of aromatic and chlorinated hydrocarbons: Effect of water on adsorption efficiency. *Clays and Clay Minerals* 43: 318–323.

Kurilenko, O.D. and R.V. Mikhalyuk. 1959. Adsorption of aliphatic amines on bentonite from aqueous solutions. *Kolloidnyi Zhurnal* 21: 181–184.

Laby, R.H. and G.F. Walker. 1970. Hydrogen bonding in primary alkylammonium-vermiculite complexes. *Journal of Physical Chemistry* 74: 2369–2373.

Lagaly, G. 1981. Characterization of clays by organic compounds. *Clay Minerals* 16: 1–21.

Lagaly, G., M. Ogawa, and I. Dékány. 2006. Clay mineral organic interactions. In *Handbook of Clay Science. Developments in Clay Science*, Vol. 1, (Eds.) F. Bergaya, B.K.G. Theng, and G. Lagaly, pp. 309–377. Amsterdam, The Netherlands: Elsevier.

Lawal, I.A. and B. Moodley. 2015. Synthesis, characterization and application of imidazolium based ionic liquid modified montmorillonite sorbents for the removal of amaranth dye. *RSC Advances* 5: 61913–61924.

Lawal, I.A. and B. Moodley. 2018. Fixed-bed and batch adsorption of pharmaceuticals from aqueous solutions on ionic liquid-modified montmorillonite. *Chemical Engineering & Technology* 41: 983–993.

Lazorenko, G., A. Kasprzhitskii, and V. Yavna. 2018. Synthesis and structural characterization of betaine- and imidazoline-based organoclays. *Chemical Physics Letters* 692: 264–270.

Le Baron, P.C., Z. Wang, and T.J. Pinnavaia. 1999. Polymer-layered silicate nanocomposites: An overview. *Applied Clay Science* 15: 11–29.

Lee, J.-F., M.M. Mortland, S.A. Boyd, and C.T. Chiou. 1989. Shape-selective adsorption of aromatic compounds from water by tetramethylammonium-smectite. *Journal of the Chemical Society Faraday Transactions 1* 85: 2953–2962.

Lee, J.-F., M.M. Mortland, C.T. Chiou, D.E. Kile, and S.A. Boyd. 1990. Adsorption of benzene, toluene, and xylene by two tetramethylammonium-smectites having different charge densities. *Clays and Clay Minerals* 38: 113–120.

Lee, S.M. and D. Tiwari. 2012. Organo and inorgano-organo-modified clays in the remediation of aqueous solutions: An overview. *Applied Clay Science* 59–60: 84–102.

Lee, S.Y., W.J. Cho, P.S. Hahn, M. Lee, Y.B. Lee, and K.J. Kim. 2005. Microstructural changes of reference montmorillonites by cationic surfactants. *Applied Clay Science* 30: 174–180.

Lee, S.Y. and S.J. Kim. 2002a. Expansion of smectite by hexadecyltrimethylammonium. *Clays and Clay Minerals* 50: 435–445.

Lee, S.Y. and S.J. Kim. 2002b. Adsorption of naphthalene by HDTMA modified kaolinite and halloysite. *Applied Clay Science* 22: 55–63.

Lee, S.Y., S.J. Kim, S.Y. Chung, and C.H. Jeong. 2004. Sorption of hydrophobic organic compounds onto organoclays. *Chemosphere* 55: 781–785.

Le Pluart, L., J. Duchet, H. Sautereau, and J.F. Gérard. 2002. Surface modifications of montmorillonite for tailored interfaces in nanocomposites. *Journal of Adhesion* 78: 645–662.

Li, F. and M.J. Rosen. 2000. Adsorption of Gemini and conventional cationic surfactants onto montmorillonite and the removal of some pollutants by the clay. *Journal of Colloid and Interface Science* 224: 265–271.

Li, P., M.A. Khan, M. Xia, W. Lei, S. Zhu, and F. Wang. 2019. Efficient preparation and molecular dynamic (MD) simulations of Gemini surfactant modified layered montmorillonite to potentially remove emerging organic contaminants from wastewater. *Ceramic International* 45: 10782–10791.

Li, Z., W.-T. Jiang, P.-H. Chang, G. Lv, and S. Xu. 2014. Modification of a Ca-montmorillonite with ionic liquids and its application for chromate removal. *Journal of Hazardous Materials* 270: 169–175.

Liao, L., G. Lv, D. Cai, and L. Wu. 2016. The sequential intercalation of three types of surfactants into sodium montmorillonite. *Applied Clay Science* 119: 82–86.

Lin, S.-H. and R.-S. Juang. 2002. Heavy metal removal from water by sorption using surfactant-modified montmorillonite. *Journal of Hazardous Materials* B92: 315–326.

Liu, C., P. Wu, Y. Zhu, and L. Tran. 2016. Simultaneous adsorption of Cd^{2+} and BPA on amphoteric surfactant activated montmorillonite. *Chemosphere* 144: 1026–1032.

Liu, Y.Q., X.J. Lu, and J. Qiu. 2010. Characterization of the alky-ammonium adsorption amount in organomontmorillonite and its effect on the viscosity of organomontmorillonite gel. *Advanced Materials Research* 158: 25–34.

Lloyd, J.U. 1916. Discovery of the alkaloidal affinities of hydrous aluminium silicate. *Journal of the American Pharmaceutical Association* 5: 381–390, 490–495.

Lo, I.M.-C., S.C.-H. Lee, and R.K.-M. Mak. 1998. Sorption of non-polar and polar organics on dicetyldimethylammonium-bentonite. *Waste Management & Research* 16: 129–138.

Lu, X.J., J. Qiu, Y.Q. Liu, and P. Chen. 2010. Effect of preparation parameters on the structure and property of montmorillonite/alkylammonium complexes. *Advanced Materials Research* 158: 97–105.

Luo, Z., M. Gao, Y. Ye, and S. Yang. 2015. Modification of reduced-charge montmorillonites by a series of Gemini surfactants: Characterization and applications in methyl orange removal. *Applied Surface Science* 324: 807–816.

Lv, G., Z. Li, W.-T. Jiang, P.-H. Chang, and L. Liao. 2015. Interlayer configuration of ionic liquids in a Ca-montmorillonite as evidenced by FTIR, TG-DTG, and XRD analyses. *Materials Chemistry and Physics* 162: 417–424.

Ma, L., Q. Chen, J. Zhu et al. 2016. Adsorption of phenol and Cu(II) onto cationic and zwitterionic surfactant modified montmorillonite in single and binary systems. *Chemical Engineering Journal* 283: 880–888.

Ma, L., J. Zhu, H. He et al. 2014. Al_{13}-pillared montmorillonite modified by cationic and zwitterionic surfactants: A comparative study. *Applied Clay Science* 101: 327–334.

Ma, L., J. Zhu, H. He et al. 2015. Thermal analysis evidence for the location of zwitterionic surfactant on clay minerals. *Applied Clay Science* 112–113: 62–67.

Madejová, J., L. Jancovič, M. Slaný, and V.Hronský. 2020. Conformation heterogeneity of alkylammonium surfactants self-assembled on montmorillonite: Effect of head-group structure and temperature. *Applied Clay Science* 503: 144125.

Madejová, J., M. Barlog, L. Jancovič, M. Slaný, and H. Pálková. 2021. Comparative study of alkylammonium- and alkylphosphonium-based analogues of organo-montmorillonites. *Applied Clay Science* 200: 105894.

Maes, A., P. Marynen, and A. Cremers. 1977. The ion exchange adsorption of alkylammonium ions: An alternative view. *Clays and Clay Minerals* 25: 309–310.

Marçal, L., E.H. de Faria, E.J. Nassar et al. 2015. Organically modified saponites: SAXS study of swelling and application in caffeine removal. *ACS Applied Materials & Interfaces* 7: 10853–10862.

Martin-Rubi, J.A., J.A. Rausell-Colom, and J.M. Serratosa. 1974. Infrared absorption and X-ray diffraction study of butylammonium complexes of phyllosilicates. *Clays and Clay Minerals* 22: 87–90.

McAtee, J.L., Jr. 1959. Inorganic-organic cation exchange on montmorillonite. *American Mineralogist* 44: 1230–1236.

McAtee, J.L., Jr. 1962. Cation exchange of organic compounds on montmorillonite in organic media. *Clays and Clay Minerals* 9: 444–450.

McAtee, J.L., Jr. 1963. Organic cation exchange on montmorillonite as observed by ultraviolet analysis. *Clays and Clay Minerals* 10: 153–162.

McAtee, J.L., Jr. and J.R. Hackman. 1964. Exchange equilibria on montmorillonite involving organic cations. *American Mineralogist* 49: 1569–1577.

McBride, M.B. and M.M. Mortland. 1975. Surface properties of mixed Cu(II)-tetraalkylammonium montmorillonites. *Clay Minerals* 10: 357–368.

Mendelovici, E. and R.L. Frost. 2005. Pioneer studies on HCl and silylation treatments of chrysotile. *Journal of Colloid and Interface Science* 289: 597–599.

Menger, F.M. and J.S. Keiper. 2000. Gemini surfactants. *Angewandte Chemie International Edition* 39: 1906–1920.

Menger, F.M. and C.A. Littau. 1993. Gemini surfactants: A new class of self-assembling molecules. *Journal of the American Chemical Society* 115: 10083–10090.

Mercier, L. and C. Detellier. 1994. Intercalation of tetraalkylammonium cations into smectites and its application to internal surface area measurements. *Clays and Clay Minerals* 42: 71–76.

Mishra, A.K., S. Chattopadhyay, P.R. Rajamohanan, and G.B. Nando. 2011. Effect of tethering on the structure-property relationship of TPU-dual modified Laponite clay nanocomposites prepared by *ex-situ* and *in-situ* techniques. *Polymer* 52: 1071–1083.

Mittal, V. 2012. Modification of montmorillonites with thermally stable phosphonium cations and comparison with alkylammonium montmorillonites. *Applied Clay Science* 56: 103–109.

Montaño, D.F., H. Casanova, W.I. Cardona, and L.F. Giraldo. 2017. Functionalization of montmorillonite with ionic liquids based on 1-alkyl-3-methylimidazolium: Effect of anion and chain length. *Materials Chemistry and Physics* 198: 386–392.

Moreira, M.A., K.J. Ciuffi, V. Rives et al. 2017. Effect of chemical modification of palygorskite and sepiolite by 3-aminopropyltriethoxysilane on adsorption of cationic and anionic dyes. *Applied Clay Science* 135: 394–404.

Morel, R. and S. Hénin. 1956. Anomalies d'adsorption de cations organiques polyvalents par les minéraux argileux. *Comptes Rendus de l'Académie des Sciences, Paris* 242: 2975–2977.

Mortland, M.M. 1970. Clay-organic complexes and interactions. *Advances in Agronomy* 22: 75–117.

Mortland, M.M. and V.E. Berkheiser. 1976. Triethylene-diamine-clay complexes as matrices for adsorption and catalytic reactions. *Clays and Clay Minerals* 24: 60–63.

Mortland, M.M., S. Shaobai, and S.A. Boyd. 1986. Clay-organic complexes as adsorbents for phenol and chlorophenols. *Clays and Clay Minerals* 34: 581–595.

Nafees, M. and A. Waseem. 2014. Organoclays as sorbent material for phenolic compounds: A review. *CLEAN – Soil Air Water* 41: 1–9.

Nahin, P.G. 1963. Perspectives in applied organo-clay chemistry. *Clays and Clay Minerals* 10: 257–271.

Naranjo, P.M., E.L. Sham, E.R. Castellón, R.M. Torres Sánchez, and E.M. Farfán Torres. 2013. Identification and quantification of the interaction mechanisms between the cationic surfactant HDTMA-Br and montmorillonite. *Clays and Clay Minerals* 61: 98–106.

Nasser, M.S., S.A. Onaizi, I.A. Hussein, M.A. Saad, and M.J. Al-Marri. 2016. Intercalation of ionic liquids into bentonite: Swelling and rheological behaviors. *Colloids and Surfaces A: Physicochemical and Engineering Aspects* 507: 141–151.

Nguyen, V.N., T.D.C. Nguyen, T.P. Dao, H.T. Tran, D.B. Nguyen, and D.H. Ahn. 2013. Synthesis of organoclays and their application for the adsorption of phenolic compounds from aqueous solution. *Journal of Industrial and Engineering Chemistry* 19: 640–644.

Ni, R., Y. Huang, and C. Yao. 2009. Thermogravimetric analysis of organoclays intercalated with the Gemini surfactants. *Journal of Thermal Analysis and Calorimetry* 96: 943–947.

Norrish, K. and J.A. Rausell-Colom. 1963. Low angle: X-ray diffraction studies of the swelling of montmorillonite and vermiculite. *Clays and Clay Minerals* 10: 123–149.

Nourmoradi, H., M. Nikaeen, and M. Khiadani (Hajian). 2012. Removal of benzene, toluene, ethylbenzene and xylene (BTEX) from aqueous solutions by montmorillonite modified with nonionic surfactant: Equilibrium, kinetic and thermodynamic study. *Chemical Engineering Journal* 191: 341–348.

Ogawa, M., A. Hagiwara, T. Handa, C. Kato, and K. Kuroda. 1995. Solid-state ion exchange reactions between homoionic-montmorillonites and organoammonium salts. *Journal of Porous Materials* 1: 85–89.

Ogawa, M., T. Handa, K. Kuroda, and C. Kato. 1990. Formation of organoammonium montmorillonites by solid-solid reactions. *Chemistry Letters* 10: 71–74.

Okada, T., T. Morita, and M. Ogawa. 2005. Tris(2,2'-bipyridine)ruthenium(II)-clays as adsorbents for phenol and chlorinated phenols from aqueous solution. *Applied Clay Science* 29: 45–53.

Okada, T. and M. Ogawa. 2010. Adsorption of 4-nonylphenol on 1,1'-dioctyl-4,4'-bipyridinium-smectite intercalation compounds from aqueous solution. *Clay Science* 14: 191–194.

Okada, T., J. Oguchi, K.-I. Yamamoto, T. Shiono, M. Fujita, and T. Iiyama. 2015. Organoclays in water cause expansion that facilitates caffeine adsorption. *Langmuir* 31: 180–187.

Ortego, J.D., M. Kowalska, and D.L. Cocke. 1991. Interactions of montmorillonite with organic compounds – adsorptive and catalytic properties. *Chemosphere* 22: 769–798.

Paiva, L.B. and A.R. Morales. 2012. Organophilic bentonites based on Argentinian and Brazilian bentonites. Part 1: Influence of intrinsic properties of sodium bentonites on the final properties of organophilic bentonites prepared by solid-liquid and semisolid reactions. *Brazilian Journal of Chemical Engineering* 29: 525–536.

Pálková, H., J. Madelová, and P. Komadel. 2009. The effect of layer charge and exchangeable cations on sorption of biphenyl on montmorillonites. *Central European Journal of Chemistry* 7: 494–504.

Pálková, H., V. Hronský, V. Bizovská, and J. Madelová. 2015. Spectroscopic study of water adsorption on Li^+, TMA^+ and $HDTMA^+$ exchanged montmorillonite. *Spectrochimica Acta Part A: Molecular and Biomolecular Spectroscopy* 149: 751–761.

Parfitt, R.L. and D.J. Greenland. 1970. Adsorption of water by montmorillonite-poly(ethylene glycol) adsorption products. *Clay Minerals* 8: 317–324.

Park, Y., G.A. Ayoko, and R.L. Frost. 2011. Application of organoclays for the adsorption of recalcitrant organic molecules from aqueous media. *Journal of Colloid and Interface Science* 354: 292–305.

Park, Y., G.A. Ayoko, E. Horváth, R. Kurdi, J. Kristof, and R.L. Frost. 2013. Structural characterisation and environmental application of organoclays for the removal of phenolic compounds. *Journal of Colloid and Interface Science* 393: 319–334.

Parolo, M.E., G.S. Pettinari, T.B. Musso, M.P. Sánchez-Izquierdo, and L.G. Fernández. 2014. Characterization of organo-modified bentonite sorbents: The effect of modification conditions on adsorption performance. *Applied Surface Science* 320: 356–363.

Patel, H.A., R.S. Somani, H.C. Bajaj, and R.V. Jasra. 2006. Nanoclays for polymer nanocomposites, paints, inks, greases and cosmetic formulations, drug delivery vehicle and waste-water treatment. *Bulletin of Materials Science* 29: 133–145.

Patel, H.A., R.S. Somani, H.C. Bajaj, and R.V. Jasra. 2007. Preparation and characterization of phosphonium montmorillonite with enhanced thermal stability. *Applied Clay Science* 35: 194–200.

Paul, B., W.N. Martens, and R.L. Frost. 2011. Organosilane grafted acid-activated beidellite clay for the removal of non-ionic alachlor and anionic imazaquin. *Applied Surface Science* 257: 5552–5558.

Pauling, L. 1960. *The Nature of the Chemical Bond*, 3rd edition. Ithaca, NY: Cornell University Press.

Pei, J., X. Xing, B. Xia, Z. Wang, and Z. Luo. 2019. Study on the adsorption behavior between an imidazolium ionic liquid and Na-montmorillonite. *Molecules* 24: 1396. https://doi.org/10.3390/molecules24071396.

Peixoto, A., A.C. Fernandes, C. Pereira, J. Pires, and C. Freire. 2016. Physicochemical characterization of organosilylated halloysite clay nanotubes. *Microporous and Mesoporous Materials* 219: 145–154.

Peng, C., Y. Zhong, and F. Min. 2018. Adsorption of alkylamine cations on montmorillonite (001) surface: A density functional theory study. *Applied Clay Science* 152: 249–258.

Peraro, G.R., E.H. Donzelli, P.F. Oliveira et al. 2020. Aminofunctionalized LAPONITE® as a versatile hybrid material for chlorohexidine digluconate incorporation: Cytotoxicity and antimicrobial activities. *Applied Clay Science* 195: 105733.

Pereira, C., S. Patrício, A.R. Silva et al. 2007. Copper acetylacetonate anchored onto amine-functionalised clays. *Journal of Colloid and Interface Science* 316: 570–579.

Perelomov, L., S. Mandzhieva, T. Minkina et al. 2021. The synthesis of organoclays based on clay minerals with different structural expansion capacities. *Minerals* 11: 707. https://doi.org/10.3390/min11070707.

Phothitontimongkol, T., N. Siebers, N. Sukpirom, and F. Unob. 2009. Preparation and characterization of novel organo-clay minerals for Hg(II) ions adsorption from aqueous solution. *Applied Clay Science* 43: 343–349.

Pinos, J.Y.D.M., L.B. de Melo, S.D. de Souza, L. Marçal, and E.H. de Faria. 2022. Bentonite functionalized with amine groups by the sol-gel route as efficient adsorbent of rhodamine-B and nickel(II). *Applied Clay Science* 223: 106494.

Piscitelli, F., P. Posocco, R. Toth et al. 2010. Sodium montmorillonite silylation: Unexpected effect of the aminosilane chain length. *Journal of Colloid and Interface Science* 351: 108–115.

Prahsarn, C., N. Roungpaisan, N. Suwannamek et al. 2014. Influence of molecular structure of quaternary phosphonium salts on Thai bentonite intercalation. *Clays and Clay Minerals* 62: 13–19.

Praus, P., M. Turicová, S. Študentová, and M. Ritz. 2006. Study of cetyltrimethylammonium and cetylpyridinium adsorption on montmorillonite. *Journal of Colloid and Interface Science* 304: 29–36.

Qi, X., H. Yoon, S.-H. Lee, J. Yoon, and S.-J. Kim. 2008. Surface-modified imogolite by 3-APS-OsO_4 complex: Synthesis, characterization and its application in the dihydroxylation of olefins. *Journal of Industrial and Engineering Chemistry* 14: 136–141.

Qiu, J., D. Liu, Y. Wang et al. 2020. Comprehensive characterization of the structure and gel property of organo-montmorillonite: Effect of layer charge density of montmorillonite and carbon chain length of alkyl ammonium. *Minerals* 10: 378. https://doi.org/10.3390/min10040378.

Qu, X., P. Liu, and D. Zhu. 2008. Enhanced sorption of polycyclic aromatic hydrocarbons to tetra-alkyl ammonium modified smectites via cation-π interactions. *Environmental Science & Technology* 42: 1109–1116.

Queiroz, M.B., S.C.G. Rodrigues, H.M. Laborde, and M.G.F. Rodrigues. 2010. Swelling of Brazilian organoclays in some solvents with application on the petroleum industry. *Materials Science Forum* 660–661: 1031–1036.

Rausell-Colom, J.A. 1964. Small-angle X-ray diffraction study of the swelling of butylammonium-vermiculite. *Transactions of the Faraday Society* 60: 190–201.

Reinert, L., K. Batouche, J.-M. Lévêque et al. 2012. Adsorption of imidazolium and pyridinium ionic liquids onto montmorillonite: Characterization and thermodynamic calculations. *Chemical Engineering Journal* 209: 13–19.

Ren, H.-P., S.-P. Tian, M. Zhu et al. 2018. Modification of montmorillonite by Gemini surfactants with different chain lengths and its adsorption behavior for methyl orange. *Applied Clay Science* 151: 29–36.

Ren, Y., H. Wang, Z. Ren et al. 2019. Adsorption of imidazolium-based ionic liquid on sodium bentonite and its effects on rheological and swelling behaviors. *Applied Clay Science* 182: 105248.

Rigail-Cedeño, A.F., M.H. Cornejo, J.A. Cáceres-Zambrano, J.S. Alava-Rosado, and G. García-Mejía. 2023. Intercalation of nontronite clays from Santa Elena, Equador, using different surfactant hydrophobicity. *Minerals* 13: 272. https://doi.org/10.3390/min13020272.

Roberts, M.G., H. Li, B.J. Teppen, and S.A. Boyd. 2006. Sorption of nitroaromatics by ammonium- and organic ammonium-exchanged smectite: Shifts from adsorption/complexation to a partition-dominated process. *Clays and Clay Minerals* 54: 426–434.

Rodríguez-Cruz, M.S., M.S. Andrades, and M.J. Sánchez-Martín. 2008. Significance of the long-chain organic cation structure in the sorption of the penconazole and metalaxyl fungicides by organoclays. *Journal of Hazardous Materials* 160: 200–207.

Romanzini, D., V. Piroli, A, Frache, A.J. Zattera, and S.C. Amico. 2015. Sodium montmorillonite modified with methacryloxy and vinylsilanes: Influence of silylation on the morphology of clay/unsaturated polyester nanocomposites. *Applied Clay Science* 114: 550–557.

Rowland, R.A. and E.J. Weiss. 1963. Bentonite-methylamine complexes. *Clays and Clay Minerals* 10: 460–468.

Ruan, X., L. Zhu, and B. Chen. 2008. Adsorptive characteristics of the siloxane surfaces of reduced-charge bentonites saturated with tetramethylammonium cation. *Environmental Science & Technology* 42: 7911–7917.

Ruan, X., L. Zhu, B. Chen, G. Qian, and R.L. Frost. 2015. Combined ^{1}H NMR and LSER study for the compound-specific interactions between organic contaminants and organobentonites. *Journal of Colloid and Interface Science* 460: 119–127.

Ruiz-Hitzky, E., P. Aranda, M. Darder, and G. Rytwo. 2010. Hybrid materials based on clays for environmental and biomedical applications. *Journal of Materials Chemistry* 20: 9306–9321.

Ruiz-Hitzky, E. and J.J. Fripiat. 1976. Organomineral derivatives obtained by reacting organochlorosilanes with the surface of silicates in organic solvents. *Clays and Clay Minerals* 24: 25–30.

Ruiz-Hitzky, E. and A. Van Meerbeek. 2006. Clay mineral- and organoclay-polymer nanocomposite. In *Handbook of Clay Science*. Developments in Clay Science, Vol. 1, (Eds.) F. Bergaya, B.K.G. Theng, and G. Lagaly, pp. 583–621. Amsterdam, The Netherlands: Elsevier.

Saeed, M., A. Riaz, A. Intisar et al. 2022. Synthesis, characterization and application of organoclays for adsorptive desulfurization of fuel oil. *Scientific Reports* 12: 7362. https://doi.org/10.1038/s41598-022-11054-6.

Şahiner, A., G. Özdemir, T.H. Bulut, and S. Yapar. 2022. Synthesis and characterization of non-leaching inorgano- and organo-montmorillonites and their bactericidal properties against *Streptococcus mutans*. *Clays and Clay Minerals* 70: 481–491.

Sánchez-Martín, M.J., M.C. Dorado, C. del Hoyo, and M.S. Rodríguez-Cruz. 2008. Influence of clay mineral structure and surfactant nature on the adsorption capacity of surfactants by clays. *Journal of Hazardous Materials* 150: 115–123.

Sarier, N. and E. Onder. 2010. Organic modification of montmorillonite with low molecular weight polyethylene glycols and its use in polyurethane nanocomposite foams. *Thermochimica Acta* 510: 113–121.

Sarkar, B., Y. Xi, M. Megharaj et al. 2012. Bioreactive organoclay: A new technology for environmental remediation. *Critical Reviews in Environmental Science and Technology* 42: 435–488.

Sayılkan, H., S. Erdemoğlu, Ş. Şener, F. Sayılkan, M. Akarsu, and M. Erdemoğlu. 2004. Surface modification of pyrophyllite with amino silane coupling agent for the removal of 4-nitrophenol from aqueous solutions. *Journal of Colloid and Interface Science* 275: 530–538.

Schampera, B., R. Solc, S.K. Woche et al. 2015. Surface structure of organoclays as examined by X-ray photoelectron spectroscopy and molecular dynamics simulations. *Clay Minerals* 50: 353–367.

Scholtzová, E., J. Madejová, L. Jankovič, and D. Tunega. 2016. Structural and spectroscopic characterization of montmorillonite intercalated with *n*-butylammonium cations (n = 1–4) – modeling and experimental study. *Clays and Clay Minerals* 64: 401–412.

Scholtzová, E., D. Tunega, J. Madejová, H. Pálková, and P. Komadel. 2013. Theoretical and experimental study of montmorillonite intercalated with tetramethylammonium cation. *Vibrational Spectroscopy* 66: 123–131.

Seki, Y., T. Okada, and M. Ogawa. 2009. Preparation and vapor adsorption properties of quaternary diammonium-montmorillonites. *Microporous and Mesoporous Materials* 124: 30–35.

Sekrane, F., Z. Bouberka, A.K. Banabbou, M. Rabiller-Baudry, and Z. Derriche. 2011. Adsorption of toluene on bentonites modified by dodecyltrimethylammonium bromide. *Chemical Engineering Communications* 198: 1093–1110.

Sepehri, S., M. Rafizadeh, M. Hemmati, and H. Bouhendi. 2014. Study of the modification of montmorillonite with monofunctional and trifunctional vinyl chlorosilane. *Applied Clay Science* 97–98: 235–240.

Shabtai, J., N. Frydman, and R. Lazar. 1977. Synthesis and catalytic properties of a 1,4-diaza-bicyclo[2,2,2] octane-montmorillonite system – a novel type of molecular sieve. In *Proceedings of the 6th International Congress of Catalysis, 1976*, (Eds.) G.C. Bond, P.B. Wells, and F.C. Tompkins, pp. 660–667. London: The Chemical Society.

Shah, K.J., M.K. Mishra, A.D. Shukla, T. Imae, and D.O. Shah. 2013. Controlling wettability and hydrophobicity of organoclays modified with quaternary ammonium surfactants. *Journal of Colloid and Interface Science* 407: 493–499.

Shanmugharaj, A.M., K.Y. Rhee, and S.H. Ryu. 2006. Influence of dispersing medium on grafting of aminopropyltriethoxysilane in swelling clay materials. *Journal of Colloid and Interface Science* 298: 854–859.

Sharma, R., A. Kamal, M. Abdinejad, R.K. Mahajan, and H.-B. Kraatz. 2017. Advances in the synthesis, molecular architectures and potential applications of Gemini surfactants. *Advances in Colloid and Interface Science* 248: 35–68.

Sharmasarkar, S., W.F. Jaynes, and G.F. Vance. 2000. BTEX sorption by montmorillonite organo-clays: TMPA, ADAM, HDTMA. *Water, Air, and Soil Pollution* 119: 257–273.

Shen, T. and M. Gao. 2019. Gemini surfactant modified organo-clays for removal of organic pollutants from water: A review. *Chemical Engineering Journal* 375: 121910.

Shen, W., H. He, J. Zhu, P. Yuan, and R.L. Frost. 2007. Grafting of montmorillonite with different functional silanes via two different reaction systems. *Journal of Colloid and Interface Science* 313: 268–273.

Shen, Y.-H. 2001. Preparation of organobentonite using nonionic surfactants. *Chemosphere* 44: 989–995.

Shu, Y., L. Li, Q. Zhang, and H. Wu. 2010. Equilibrium, kinetics and thermodynamic studies for sorption of chlorobenzenes on CTMAB modified bentonite and kaolinite. *Journal of Hazardous Materials* 173: 47–53.

Sieskind, O. and R. Wey. 1958. Influence of pH on the adsorption of amines by H-montmorillonite. *Comptes Rendus de l'Académie des Sciences, Paris* 247: 74–76.

Silva, A.A., K. Dahmouche, and B.G. Soares. 2011. Nanostructure and dynamic mechanical properties of silane-functionalized montmorillonite/epoxy nanocomposites. *Applied Clay Science* 54: 151–158.

Skipper, N.T., M.V. Smalley, G.D. Williams, A.K. Soper, and C.H. Thompson. 1995. Direct measurement of the electric double-layer structure in hydrated lithium vermiculite clays by neutron diffraction. *Journal of Physical Chemistry* 99: 14201–14204.

Slabaugh, W.H. 1958. Cation exchange properties of bentonite. *Journal of Physical Chemistry* 58: 162–165.

Slabaugh, W.H. and D.B. Hanson. 1969. Solvent selectivity by an organoclay complex. *Journal of Colloid and Interface Science* 29: 460–463.

Slabaugh, W.H. and P.A. Hiltner. 1968. The swelling of alkylammonium montmorillonites. *Journal of Physical Chemistry* 72: 4295–4298.

Slabaugh, W.H. and R. Kupka. 1958. Organic cation exchange properties of calcium montmorillonite. *Journal of Physical Chemistry* 62: 599–601.

Slade, P.G. and W.P. Gates. 2004. The swelling of HDTMA smectites as influenced by their preparation and layer charges. *Applied Clay Science* 25: 93–101.

Slaný, M., L. Jankovič, and J. Madejová. 2019. Structural characterization of organo-montmorillonites prepared from a series of primary alkylamines salts: Mid-IR and near-IR study. *Applied Clay Science* 176: 11–20.

Smith, C.R. 1934. Base exchange reactions of bentonite and salts of organic bases. *Journal of the American Chemical Society* 56: 1561–1563.

Song, K. and G. Sandí. 2001. Characterization of montmorillonite surfaces after modification by organosilane. *Clays and Clay Minerals* 49: 119–125.

Souza, M.A., N.M. Larocca, and L.A. Pessan. 2016. Highly thermal stable organoclays of ionic liquids and silane organic modifiers and effect of montmorillonite source. *Journal of Thermal Analysis and Calorimetry* 126: 499–509.

Spencer, W.F. and J.E. Gieseking. 1952. Organic derivatives of montmorillonite. *Journal of Physical Chemistry* 56: 751–753.

Srinivasarao, K., S.M. Prabhu, W. Luo, and K. Sasaki. 2018. Enhanced adsorption of perchlorate by Gemini surfactant-modified montmorillonite: Synthesis, characterization and their adsorption mechanism. *Applied Clay Science* 163: 46–55.

Stevens, J.J. and S.J. Anderson. 1996. An FTIR study of water sorption on tetra- methylammonium- and trimethylphenylammonium-saturated normal and reduced-charge Wyoming montmorillonite. *Clays and Clay Minerals* 44: 142–150.

Stevens, J.J., S.J. Anderson, and S.A. Boyd. 1996. FTIR study of competitive water-arene sorption on tetramethylammonium- and trimethylphenylammonium-montmorillonites. *Clays and Clay Minerals* 44: 88–95.

Su, L., Q. Tao, H. He, J. Zhu, and P. Yuan. 2012. Locking effect: A novel insight in the silylation of montmorillonite surfaces. *Materials Chemistry and Physics* 136: 292–295.

Su, L., Q. Tao, H. He, J. Zhu, P. Yuan, and R. Zhu. 2013. Silylation of montmorillonite surfaces: Dependence on solvent nature. *Journal of Colloid and Interface Science* 391: 16–20.

Su, X., L. Ma, J. Wei, and R. Zhu. 2016. Structure and thermal stability of organo-vermiculite. *Applied Clay Science* 132–133: 261–266.

Suin, S., N.K. Shrivastava, S. Maiti, and B.B. Khatua. 2014. Facile preparation of highly exfoliated and optically transparent polycarbonate (PC)/clay mineral nanocomposites using phosphonium modified organoclay mineral. *Applied Clay Science* 95: 182–190.

Sun, Z., C. Lian, C. Li, and S. Zheng. 2020. Investigations on organo-montmorillonites modified by binary nonionic/zwitterionic surfactant mixtures for simultaneous adsorption of aflatoxin B_1 and zearalenone. *Journal of Colloid and Interface Science* 565: 11–22.

Sutherland, H.H. and D.M.C. MacEwan. 1961. Organic complexes of vermiculite. *Clay Minerals Bulletin* 4: 229–233.

Tahani, A., M. Karroua, H. Van Damme, P. Levitz, and F. Bergaya. 1999. Adsorption of a cationic surfactant on Na-montmorillonite: Inspection of adsorption layer by X-ray and fluorescence spectroscopies. *Journal of Colloid and Interface Science* 216: 242–249.

Takahashi, C., T. Shirai, and M. Fuji. 2012. Study on intercalation of ionic liquid into montmorillonite and its property evaluation. *Materials Chemistry and Physics* 135: 681–686.

Takahashi, C., T. Shirai, and M. Fuji. 2014. Selective intercalation of ionic liquid in montmorillonite and influence of water molecules. *Solid State Ionics* 267: 16–21.

Takahashi, C., T. Shirai, Y. Hayashi, and M. Fuji. 2013. Study of intercalation compounds using ionic liquids into montmorillonite and their thermal stability. *Solid State Ionics* 241: 53–61.

Taleb, K., I. Pillin, Y. Grohens, and S. Saidi-Besbes. 2018. Gemini surfactant modified clays: Effect of surfactant loading and spacer length. *Applied Clay Science* 161: 48–56.

Tao, Q., Y. Fang, T. Li et al. 2016. Silylation of saponite with 3-aminopropyltriethoxysilane. *Applied Clay Science* 132–133: 133–139.

Tao, Q., L. Su, R.L. Frost, H. He, and B.K.G. Theng. 2014. Effect of functionalized kaolinite on the curing kinetics of cycloaliphatic epoxy/anhydride system. *Applied Clay Science* 95: 317–322.

Teppen, B.J. and V. Aggarwal. 2007. Thermodynamics of organic cation exchange selectivity in smectites. *Clays and Clay Minerals* 55: 119–130.

Theng, B.K.G. 2012. *Formation and Properties of Clay-Polymer Complexes*, 2nd edition. Amsterdam, The Netherlands: Elsevier.

Theng, B.K.G. 2018. *Clay Mineral Catalysis of Organic Reactions*. Boca Raton, FL: CRC Press.

Theng, B.K.G., J. Aislabie, and R. Fraser. 2001. Bioavailability of phenanthrene intercalated into an alkylammonium-montmorillonite clay. *Soil Biology & Biochemistry* 33: 845–848.

Theng, B.K.G., G.J. Churchman, W.P. Gates, and G. Yuan. 2008. Organically modified clays for pollution uptake and environmental protection. In *Soil Mineral-Microbe-Organic Interactions*, (Eds.) Q. Huang, P.M. Huang, and A. Violante, pp. 145–174. Berlin, Germany: Springer-Verlag.

Theng, B.K.G., D.J. Greenland, and J.P. Quirk. 1967. Adsorption of alkylammonium cations by montmorillonite. *Clay Minerals* 7: 1–17.

Theng, B.K.G., D.J. Greenland, and J.P. Quirk. 1968. The effect of exchangeable alkylammonium. ions on the swelling of montmorillonite in water. *Clay Minerals* 7: 271–293.

Theng, B.K.G., R.H. Newman, and J.S. Whitton. 1998. Characterization of an alkylammonium-montmorillonite-phenanthrene intercalation complex by carbon-13 nuclear magnetic resonance spectroscopy. *Clay Minerals* 33: 221–229.

Theng, B.K.G., K.R. Tate, and P. Becker-Heidmann. 1992. Toward establishing the age, location, and identity of the inert soil organic matter of a Spodosol. *Zeitschrift für Pflanzenernährung und Bodenkunde* 155: 181–184.

Thiebault, T., J. Brendlé, G. Augé, and L. Limousy. 2020. Cleaner synthesis of silylated clay minerals for the durable recovery of ions (Co^{2+} and Sr^{2+}) from aqueous solutions. *Industrial & Engineering Chemistry Research* 59: 2104–2112.

Tonlé, I.K., T. Diaco, E. Ngameni, and C. Detellier. 2007. Nanohybrid kaolinite-based materials obtained from the interlayer grafting of 3-aminopropyltriethoxysilane and their potential use as electrochemical sensors. *Chemistry of Materials* 19: 6629–6636.

Tonlé, I.K., E. Ngameni, D. Njopwouo, C. Carteret, and A. Walcarius. 2003. Functionalization of natural smectite-type clays by grafting with organosilanes: Physico-chemical characterization and application to mercury(II) uptake. *Physical Chemistry Chemical Physics* 5: 4951–4961.

Traynor, M.F., M.M. Mortland, and T.J. Pinnavaia. 1978. Ion exchange and intersalation reactions of hectorite with tris-bipyridyl metal complexes. *Clays and Clay Minerals* 26: 318–326.

Tunney, J.J. and C. Detellier. 1993. Interlamellar covalent grafting of organic units on kaolinite. *Chemistry of Materials* 5: 747–748.

Uchida, Y., T. Sugiyama, E. Noguchi, K. Nakamura, Y. Nakamura, and K. Matsui. 2006. Intercalation of (3-aminopropyl) triethoxysilane and (3-aminopropyl) diethoxymethylsilane into montmorillonite. *Clay Science* 12(Supplement 2): 71–75.

Undabeytia, T., F. Madrid, J. Vázquez, and J.I. Pérez-Martínez. 2019. Grafted sepiolites for the removal of pharmaceuticals in water treatment. *Clays and Clay Minerals* 67: 173–182.

Upson, R.T. and S.E. Burns. 2006. Sorption of nitroaromatic compounds to synthesized organoclays. *Journal of Colloid and Interface Science* 297: 70–76.

Vaia, R.A., R.K. Teukolsky, and E.P. Gianellis. 1994. Interlayer structure and molecular environment of alkylammonium layered silicates. *Chemistry of Materials* 6: 1017–1022.

Van Leemput, L., M.S. Stul, A. Maes, J.B. Uytterhoeven, and A. Cremers. 1983. Surface properties of smectites exchanged with mono-and biprotonated 1,4, diazobicyclo(2,2,2)-octane. *Clays and Clay Minerals* 31: 261–268.

Vansant, E.F. and G. Peeters. 1978. The exchange of alkylammonium ions on Na^{+}-Laponite. *Clays and Clay Minerals* 26: 279–284.

Vansant, E.F. and J.B. Uytterhoeven. 1972. Thermodynamics of the exchange of *n*-alkylammonium ions on Na-montmorillonite. *Clays and Clay Minerals* 20: 47–54.

Varadwaj, G.B.B., K. Parida, and V.O. Nyamori. 2016. Transforming inorganic layered montmorillonite into inorganic-organic hybrid materials for various applications: A brief overview. *Inorganic Chemistry Frontiers* 3: 1100–1111.

Vidal, N. and C. Volzone. 2012. Influence of organobentonite structure on toluene adsorption from water solution. *Materials Research* 15: 944–953.

Walker, G.F. 1960. Macroscopic swelling of vermiculite crystals in water. *Nature* 187: 312–313.

Walker, G.F. 1967. Interactions of *n*-alkylammonium ions with mica-type layer lattices. *Clay Minerals* 7: 129–143.

Walker, G.F. and W.G. Garrett. 1967. Chemical exfoliation of vermiculite and the production of colloidal dispersions. *Science* 156: 385–387.

Wang, J., P.A. Wheeler, W.L. Jarrett, and L.J. Mathias. 2007. Synthesis and characterization of dual-functionalized Laponite clay for acrylic nanocomposites. *Journal of Applied Polymer Science* 106: 1496–1506.

Wang, X., B. Liu, and P. Yu. 2015. Research on the preparation and mechanism of the organic montmorillonite and its application in drilling fluid. *Journal of Nanomaterials* 2015: 10 pages. Article ID 514604. http://dx.doi.org/10.1155/2015/514604.

Weiss, A. 1963a. Organic derivatives of mica-type layer silicates. *Angewandte Chemie International Edition in English* 2: 134–143.

Weiss, A. 1963b. Mica-type layer silicates with alkylammonium ions. *Clays and Clay Minerals* 10: 191–224.

Weiss, A., A. Mehler, and U. Hofmann. 1956. Zur Kenntnis von organophilem Vermikulit. *Zeitschrift für Naturforschung* 11B: 431–434.

Weiss, A., E. Michel, and A.L. Weiss. 1959. Über den Einfluss von Wasserstoffbrücken-bindungen auf ein- und zwei-dimensionale innerkristalline Quellungsvorgänge. In *Hydrogen Bonding*, (Ed.) D. Hadži, pp. 495–508. London, UK: Pergamon Press.

Wheeler, P.A., J. Wang, J. Baker, and L.J. Mathias. 2005. Synthesis and characterization of covalently functionalized Laponite clay. *Chemistry of Materials* 17: 3012–3018.

Whitlow, E.P. and W.A. Felsing. 1944. Heats of dilution and heat capacities of aqueous solutions of mono-, di- and trimethylamine hydrochlorides. *Journal of the American Chemical Society* 66: 2028–2033.

Williams, G.D., N.T. Skipper, M.V. Smalley, A.K. Soper, and S.M. King. 1996. Structure of alkyl ammonium solutions in vermiculite clays. *Faraday Discussions* 104: 295–306.

Williams-Daryn, S. and R.K. Thomas. 2002. The intercalation of a vermiculite by cationic surfactants and its subsequent swelling with organic solvents. *Journal of Colloid and Interface Science* 255: 303–311.

Witthuhn, B., T. Pernyeszi, P. Klauth, H. Vereecken, and E. Klumpp. 2005. Sorption study of 2,4-dichlorophenol on organoclays constructed for soil bioremediation. *Colloids and Surfaces A: Physicochemical and Engineering Aspects* 265: 81–87.

Wu, L., L. Liao, and G. Lv. 2015. Influence of interlayer cations on organic intercalation of montmorillonite. *Journal of Colloid and Interface Science* 454: 1–7.

Wu, L., Q. Wang, N. Tang, and L. Gao. 2019. Preparation of ionic liquids/montmorillonite composites and its application for diclofenac sodium removal. *Journal of Contaminant Hydrology* 220: 1–5.

Wu, L., C. Yang, L. Mei, F. Qin, L. Liao, and G. Lv. 2014. Microstructure of different chain length ionic liquids intercalated into montmorillonite: A molecular dynamics study. *Applied Clay Science* 99: 266–274.

Wu, T., X. Yan, X. Cai et al. 2010. Removal of *Chattonella marina* with clay minerals modified with a Gemini surfactant. *Applied Clay Science* 50: 604–607.

Xiang, Y., M. Gao, T. Shen, G. Cao, B. Zhao, and S. Guo. 2019. Comparative study of three novel organo-clays modified with imidazolium-based Gemini surfactant on adsorption for bromophenol blue. *Journal of Molecular Liquids* 286: 110928.

Xiao, F., B.-Q. Yan, X.-Y. Zou et al. 2020. Study on ionic liquid modified montmorillonite and molecular dynamics simulation. *Colloids and Surfaces A* 587: 1245311.

Xie, W., R. Xie, W.-P. Pan et al. 2002. Thermal stability of quaternary phosphonium modified montmorillonites. *Chemistry of Materials* 14: 4837–4845.

Xu, S., G. Sheng, and S.A. Boyd. 1997. Use of organoclays in pollution abatement. *Advances in Agronomy* 59: 25–62.

Xu, Y., M.A. Khan, F. Wang, M. Xia, and W. Lei. 2018. Novel multi amine-containing Gemini surfactant modified montmorillonite as adsorbents for removal of phenols. *Applied Clay Science* 162: 204–213.

Xue, G., M. Gao, Z. Gu, Z. Luo, and Z. Hu. 2013. The removal of *p*-nitrophenol from aqueous solutions by adsorption using Gemini surfactants modified montmorillonites. *Chemical Engineering Journal* 218: 223–231.

Yang, Q., M. Gao, Z. Luo, and S. Yang. 2016. Enhanced removal of bisphenol A from aqueous solution by organo-montmorillonites modified with novel Gemini pyridinium surfactants containing long alkyl chain. *Chemical Engineering Journal* 285: 27–38.

Yang, S., M. Gao, and Z. Luo. 2014. Adsorption of 2-naphthol on the organo-montmorillonites modified by Gemini surfactants with different spacers. *Chemical Engineering Journal* 256: 39–50.

Yang, S., M. Gao, Z. Luo, and Q. Yang. 2015. The characterization of organo-montmorillonite modified with a novel aromatic-containing Gemini surfactant and its comparative adsorption for 2-naphthol and phenol. *Chemical Engineering Journal* 268: 125–134.

Yang, S.-Q., P. Yuan, H.-P. He et al. 2012. Effect of reaction temperature on grafting of γ-aminopropyl triethoxysilane (APTES) onto kaolinite. *Applied Clay Science* 62–63: 8–14.

Yılmazoğlu, M., B. Turan, P. Demircivi, and J. Hızal. 2022. Synthesis and characterization of imidazolium based ionic liquid modified montmorillonite for the adsorption of Orange II dye: Effect of chain length. *Journal of Molecular Structure* 1249: 131628.

Yoko, S. and M. Ogawa. 2010. The removal of 2-phenylphenol from aqueous solution by adsorption onto organoclays. *Bulletin of the Chemical Society of Japan* 83: 712–715.

Yu, M., M. Gao, T. Shen, and J. Wang. 2018. Organo-vermiculites modified by low-dosage Gemini surfactants with different spacers for adsorption toward *p*-nitrophenol. *Colloids and Surfaces A* 553: 601–611.

Yu, W.H., Q.Q. Ren, D.S. Tong, C.H. Zhou, and H. Wang. 2014. Clean production of CTAB-montmorillonite: Formation mechanism and swelling behavior in xylene. *Applied Clay Science* 97–98: 222–234.

Yuan, G. 2004. Natural and modified nanomaterials as sorbents of environmental contaminants. *Journal of Environmental Science and Health, Part A* 39: 2661–2670.

Yuan, G.D., B.K.G. Theng, G.J. Churchman, and W.P. Gates. 2013. Clays and clay minerals for pollution control. In *Handbook of Clay Science*, 2nd edition. Developments in Clay Science, Vol. 5A, (Eds.) F. Bergaya and G. Lagaly, pp. 587–644. Amsterdam, The Netherlands: Elsevier.

Yuan, P., P.D. Southon, Z. Liu et al. 2008. Functionalization of halloysite clay nanotubes by grafting with γ-aminopropyltriethoxysilane. *Journal of Physical Chemistry C* 112: 15742–15751.

Yuan, S., X. Zhao, R. Cheng et al. 2022. Molecular simulation of the adsorption of the hydration inhibitor N1,N1'-(ethane-1,2-diethyl)bis(N1-(2-aminoethyl)ethane-1,2-diamine onto montmorillonite. *Clay Minerals* 57: 192–201.

Zachara, J.M., C.C. Ainsworth, and S.C. Smith. 1990. The sorption of heterocyclic compounds on reference and subsurface smectite clay isolates. *Journal of Contaminant Hydrology* 6: 281–305.

Zanzottera, C., A. Vicente, E. Celasco, C. Fernandez, E. Garrone, and B. Bonelli. 2012. Physico-chemical properties of imogolite nanotubes functionalized on both external and internal surfaces. *Journal of Physical Chemistry C* 116: 7499–7506.

Zawrah, M.F., R.M. Khattab, E.M. Saad, and R.A. Gado. 2014. Effect of surfactant types and their concentration on the structural characteristics of nanoclay. *Spectrochimica Acta Part A: Molecular and Biomolecular Spectroscopy* 122: 616–623.

Zhang, H., M. Xia, F. Wang, P. Li, and M. Shi. 2021. Adsorption properties and mechanism of montmorillonite modified by two Gemini surfactants with different chain lengths for three benzotriazole emerging contaminants: Experimental and theoretical study. *Applied Clay Science* 207: 106086.

Zhang, J., R.K. Gupta, and C.A. Wilkie. 2006. Controlled silylation of montmorillonite and its polyethylene nanocomposites. *Polymer* 47: 4537–4543.

Zhang, J., C.H. Zhou, S. Petit, and H. Zhang. 2019. Hectorite: Synthesis, modification, assembly and applications. *Applied Clay Science* 177: 114–138.

Zhang, S., Q. Liu, H. Cheng, Y. Zhang, X. Li, and R.L. Frost. 2015. Intercalation of γ-aminopropyl triethoxysilane (APTES) into kaolinite interlayer with methanol-grafted kaolinite as intermediate. *Applied Clay Science* 114: 484–490.

Zhang, Z., L. Liao, and Z. Xia. 2010. Ultrasound-assisted preparation and characterization of anionic surfactant modified montmorillonites. *Applied Clay Science* 50: 576–581.

Zhang, Z.Z., D.L. Sparks, and N.C. Scrivner. 1993. Sorption and desorption of quaternary amine cations on clays. *Environmental Science & Technology* 27: 1625–1631.

Zhao, M., L. Wei, Y. Zheng, M. Liu, J. Wang, and Y. Qiu. 2019. Structural effect of imidazolium-type ionic liquid adsorption to montmorillonite. *Science of the Total Environment* 666: 858–864.

Zhao, Q., H. Choo, A. Bhatt, S.E. Burns, and B. Bate. 2017. Review of the fundamental geochemical and physical behaviors of organoclays in barrier applications. *Applied Clay Science* 142: 2–20.

Zheng, S., Z. Sun, Y. Park, G.A. Ayoko, and R.L. Frost. 2013. Removal of bisphenol A from wastewater by Ca-montmorillonite modified with selected surfactants. *Chemical Engineering Journal* 234: 416–422.

Zhou, C.H., C.J. Li, W.P. Gates, T.T. Zhu, and W.H. Yu. 2019. Co-intercalation of organic cations/amide molecules into montmorillonite with tunable hydrophobicity and swellability. *Applied Clay Science* 179: 105157.

Zhou, L., H. Chen, X. Jiang et al. 2009. Modification of montmorillonite surfaces using a novel class of cationic Gemini surfactants. *Journal of Colloid and Interface Science* 332: 16–21.

Zhou, Q., H. He, R.L. Frost, and Y. Xi. 2007. Adsorption of *p*-nitrophenol on mono-, di-, and trialkyl surfactant-intercalated organoclays: A comparative study. *Journal of Physical Chemistry* 111: 7487–7493.

Zhou, Q., H.P. He, J.X. Zhu, W. Shen, R.L. Frost, and P. Yuan. 2008. Mechanism of *p*-nitrophenol adsorption from aqueous solution by $HDTMA^+$-pillared montmorillonite – implications for water purification. *Journal of Hazardous Materials* 154: 1025–1032.

Zhou, Q., W. Shen, J. Zhu et al. 2014. Structure and dynamic properties of water saturated CTMA-montmorillonite: Molecular dynamics simulations. *Applied Clay Science* 97–98: 62–71.

Zhou, Q., R. Zhu, S.C. Parker, J. Zhu, H. He, and M. Molinari. 2015. Modelling the effects of surfactant loading level on the sorption of organic contaminants on organoclays. *RSC Advances* 5: 47022–47030.

Zhou, Y., X.-Y. Jin, H. Lin, and Z-L. Chen. 2011. Synthesis, characterization and potential application of organobentonite in removing 2,4-DCP from industrial wastewater. *Chemical Engineering Journal* 166: 176–183.

Zhu, J., Y. Qing, L. Ma, R. Zhu, and H. He. 2015a. The structure of montmorillonites modified with zwitterionic surfactants and their sorption ability. *Mineralogy and Petrology* 109: 349–355.

Zhu, J., Y. Qing, T. Wang et al. 2011. Preparation and characterization of zwitterionic surfactant-modified montmorillonites. *Journal of Colloid and Interface Science* 360: 386–392.

Zhu, J., P. Zhang, Y. Qing et al. 2017b. Novel intercalation mechanism of zwitterionic surfactant modified montmorillonites. *Applied Clay Science* 141: 265–271.

Zhu, L., S. Tian, J. Zhu, and Y. Shi. 2007. Silylated pillared clay (SPILC): A novel bentonite-based inorgano-organo composite sorbent synthesized by integration of pillaring and silylation. *Journal of Colloid and Interface Science* 315: 191–199.

Zhu, L., L. Wang, and Y. Xu. 2017a. Chitosan and surfactant co-modified montmorillonite: A multifunctional adsorbent for contaminant removal. *Applied Clay Science* 146: 35–42.

Zhu, L. and R. Zhu. 2008a. Surface structure of $CTMA^+$ modified bentonite and their sorptive characteristics towards organic compounds. *Colloids and Surfaces A: Physicochemical and Engineering Aspects* 320: 19–24.

Zhu, L. and R. Zhu. 2008b. Thermodynamics of naphthalene sorption to organoclays: Role of surfactant packing density. *Journal of Colloid and Interface Science* 322: 27–32.

Zhu, R., Q. Chen, Q. Zhou, Y. Xi, J. Zhu, and H. He. 2016. Adsorbents based on montmorillonite for contaminant removal from water: A review. *Applied Clay Science* 123: 239–258.

Zhu, R., Q. Zhou, J. Zhu, Y, Xi, and H. He. 2015b. Organo-clays as sorbents of hydrophobic organic contaminants: Sorptive characteristics and approaches to enhancing sorption capacity. *Clays and Clay Minerals* 63: 199–221.

Zhuang, G., W. Jiang, and Z. Zhang. 2019a. Organic modifiers of organo-montmorillonite in oil system under high temperatures: Desorption or degradation? *Industrial & Engineering Chemistry Research* 58: 2644–2653.

Zhuang, G., Z. Zhang, J. Guo, L. Liao, and J. Zhao. 2015. A new ball milling method to produce organo-montmorillonite from anionic and non-ionic surfactants. *Applied Clay Science* 104: 18–26.

Zhuang, G., Z. Zhang, and M. Jaber. 2019b. Organoclays used as colloidal and rheological additives in oil-based drilling fluids: An overview. *Applied Clay Science* 177: 63–81.

5 Interactions of Clay Minerals with Negatively Charged Organic Compounds

5.1 INTRODUCTION

Complex formation of clay minerals with negatively charged organic compounds has received relatively little attention since such solutes tend to be repelled from clay mineral surfaces of like charge (Weber et al. 1965; Zhang et al. 1990; Kyzioł-Komosińska et al. 2014; Cukrowicz et al. 2022). Accordingly, 2:1 type layer silicates have a low propensity for taking up anionic organic solutes from solution (Law and Kunze 1966; Kaufherr et al. 1971; Celis et al. 1999; Kubicki et al. 1999; Lackovic et al. 2003; Gimsing et al. 2007; Yeasmin et al. 2014), even when the interlayer space has been expanded by pillaring with aluminium polyoxo cations (Xin et al. 2011). Nevertheless, appreciable adsorption has been recorded under acid pH conditions when solute ionization is suppressed, causing the organic anion to behave and adsorb as an uncharged, neutral species (Dentel et al. 1995; Zhao et al. 1996; Celis et al. 1999; Greesh et al. 2008; Wu et al. 2013; Ahmat et al. 2019).

We might interpolate here that humic substances (HS), making up the bulk of soil organic matter, are polyanionic in character because of the presence in their structure of negatively charged constituents, notably phenolic and benzene carboxylic acids. The persistence and biostability of HS in soil have been ascribed to complex formation with mineral surfaces and entrapment within clay micro-aggregates (Tate and Theng 1980; Theng and Yuan 2008; Theng 2012).

The sorption of negatively charged organic molecules to clay minerals may be facilitated, if not promoted, by raising the ionic strength of the solution (Arnarson and Keil 2000; Chefetz et al. 2011; Zhao et al. 2014). The positive effect of electrolyte addition can be quite dramatic at high concentrations. Stutzmann and Siffert (1977) and Espinasse and Siffert (1979), for example, reported a two- to five-fold increase in the uptake of partly hydrolysed polyacrylamide by Na^{+}-montmorillonite at a concentration of 100 g NaCl/L over that in distilled water. The ionic strength effect may be explained in terms of charge screening by the added electrolytes. In line with this suggestion, Valentim and Joekes (2006) noted that ionic strength does not affect the uptake of sodium dodecylsulphate by extensively washed chrysotile since the surface of this layer silicate is uncharged (cf. Table 1.1). Nor does ionic strength appreciably influence the interaction of benzoic acid with montmorillonite that has previously been exchanged with cetyltrimethylammonium ions (Yan et al. 2007).

The clay-organic anion interaction is also enhanced by the presence of polyvalent inorganic counterions at basal surfaces of the mineral (Pusino et al. 2000; Chefetz et al. 2011). This finding may be ascribed to the ability of such cations to act as a bridge between solute and surface. Two types of bridging may be distinguished (Figure 5.1). In the first type, referred to as *water-bridging*, the anionic group of the solute interacts with the exchangeable cation through a water molecule to form an outer-sphere complex. The second type, referred to as *cation-bridging*, may occur under dehydrating conditions or with weakly hydrated counterions. Here the anionic group of the solute interacts directly with the cation to form an inner-sphere complex (Theng 1976; Theng and Scharpenseel 1976; Epstein and Yariv 2003; Siebner-Freibach et al. 2004; Sposito 2008). In this context, we might mention that phenolic acids, adsorbed to Fe^{3+}-montmorillonite, can undergo oxidative transformation accompanied by ferric to ferrous ion reduction (Polubesova et al. 2010).

DOI: 10.1201/9781003080244-5

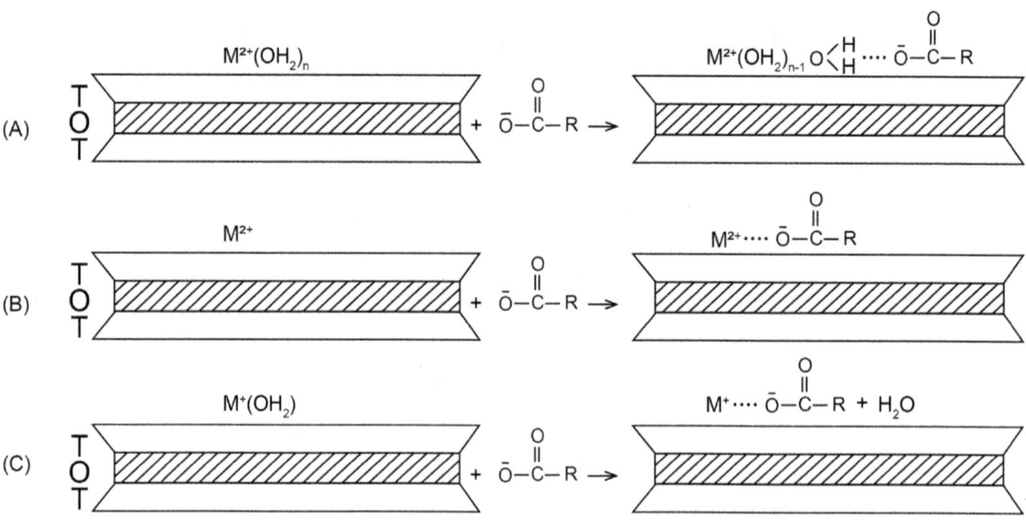

FIGURE 5.1 Diagram showing the possible types of interaction between anionic organic molecules ($RCOO^-$) and the *basal* surface of smectites. (A) water-bridging between strongly hydrated counterion and organic solute, giving rise to an outer-sphere complex; (B) cation-bridging where the counterion acts as a bridge between surface and organic solute, giving rise to an inner-sphere complex; (C) inner-sphere complex formation when the organic solute displaces a water molecule from a weakly hydrated counterion. T = tetrahedral sheet; O = octahedral sheet; M = counterion/exchangeable cation.

X-ray diffraction analysis has consistently indicated that negatively charged organic compounds do not enter the interlayer space of montmorillonite (Law and Kunze 1966; Kaufherr et al. 1971; Nagasaki et al. 2004; Xu et al. 2006; Ahmat et al. 2019). Nevertheless, Zhang et al. (2014) have proposed that perfluorooctane sulfonate could form an interlayer complex with Na^+-montmorillonite although the evidence for this is less than convincing. Intercalation, however, can and may occur under acid conditions when the negative charge on the organic anion is effectively neutralized (Greesh et al. 2008; Wu et al. 2013; Bu et al. 2019). Even humic substances are known to intercalate into montmorillonite when the solution or soil pH is less than 4 (Schnitzer and Kodama 1966; Martin Martinez and Perez Rodriguez 1969; Moinereau 1977; Theng et al. 1986; Chen et al. 2020).

5.2 FORMATION AND PROPERTIES OF COMPLEXES

Anion adsorption to clay minerals is generally confined to external particle surfaces of which the particle edge plays an important role. The participation of particle edges is especially relevant to kaolinite where the edge surface makes up a significant proportion (~ 20%) of the total external particle area. Thus, Zhang et al. (2014) found that kaolinite could take up more perfluorooctane sulfonate (through a ligand exchange mechanism) than did montmorillonite.

Under acid conditions ($pH < 6$), the kaolinite edge surface acquires a positive charge by protonation of (under-coordinated) terminal Al(OH) groups (cf. Figure 1.5), enabling anionic solutes to adsorb through electrostatic attraction (Ahmat et al. 2019). This mode of interaction has been referred to as *anion exchange* (Muljadi et al. 1966; Bidwell et al. 1970; Ferris and Jepson 1975; Sastry et al. 1995; Blockhaus et al. 1997). In this regard, Hashizume (2007) has suggested that the edge of chrysotile fibres would acquire positive charges at pH 8 and hence could adsorb aspartic and glutamic acids through anion exchange. A similar mechanism has been proposed by Wu et al. (2013) for the interaction of nalidixic acid with montmorillonite and kaolinite when the solution pH was higher than the pK_a of the solute.

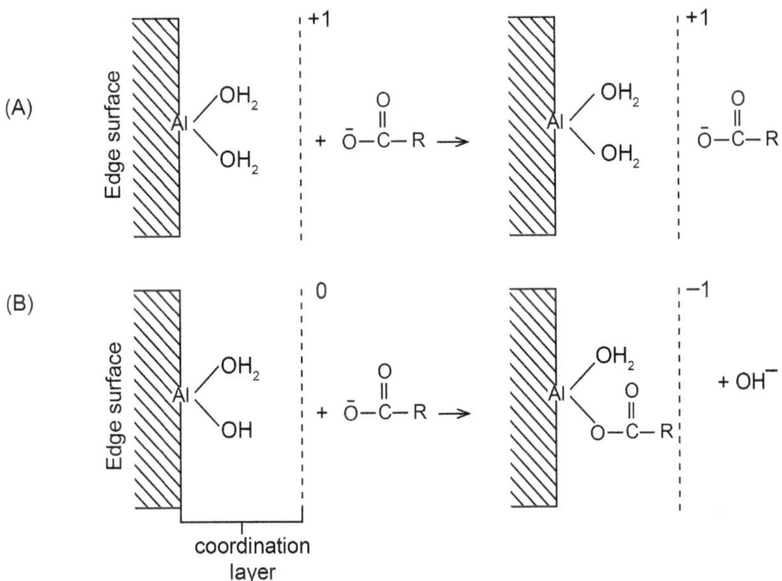

FIGURE 5.2 Diagram showing the two types of interaction between anionic organic molecules ($RCOO^-$) and the *edge* surface of clay minerals. (A) anion exchange or electrostatic attraction between the positively charged surface and negatively charged solute; (B) ligand exchange when the organic anion displaces a hydroxyl group attached to aluminium inside the coordination layer.

Negatively charged molecules can also interact with edge hydroxyl groups through *ligand exchange* (Parfitt 1978; Theng et al. 1982; Kubicki et al. 1997; Ramos and Huertas 2014). Unlike anion exchange, however, ligand exchange can operate over a wide range of pH values. As a result, the surface charge of the complex becomes less positive or more negative (Tombácz et al. 1984; Yuan et al. 2000). In addition, hydrogen-bonding may be involved in the clay-anion interaction (Kohl and Taylor 1961; Laird 1997; Aquino et al. 2003; Gautier et al. 2009; Xue et al. 2018). These modes of complex formation, involving the edge surface of clay minerals, are shown in Figure 5.2.

The edge-site hypothesis is consistent with the observation that clay minerals, in general, lose much of their capacity for taking up anionic species when their edge surface has previously been blocked or masked by treatment with phosphate (cf. Equation 1.1) or aluminium-complexing agents (Mortensen 1962; Kosiur 1977; Blockhaus et al. 1996).

Molecular dynamics simulations by Liu et al. (2017) indicate that acetate can form an inner-sphere complex at the edge surface of montmorillonite particles through a ligand exchange mechanism (Figure 5.2B). Here the organic anion enters the coordination layer and forms an inner-sphere complex by displacing a hydroxyl group coordinated to edge aluminium. Inner-sphere complexation has also been proposed by Kubicki et al. (1997) for the interaction of salicylate anions with the edge surface of illite and probably with the inner-lumen surface of acid-treated halloysite nanotubes (Spepi et al. 2016).

Such a mechanism might also underlie the adsorption of citrate by montmorillonite at low pH (Ramos and Huertas 2014) as well as the intercalation of nalidixic acid by montmorillonite at pH 4 (Wu et al. 2013). Both Specht and Frimmel (2001) and Kang and Xing (2007) have suggested similarly for the adsorption of some dicarboxylic acids by kaolinite and freeze-dried Na^+-montmorillonite, respectively. The dissolution of kaolinite, induced by citric, oxalic, and malic acids acting as organic ligands, has been described by Wang et al. (2005).

Likewise, inner-sphere complex formation underlies the interaction of adenosine-5'-phosphate with allophane at acid pH (Hashizume and Theng 2007), of *myo*-inositol hexakisphosphate with kaolinite at pH ≥ 5.5 (Hu et al. 2020), and of anionic oxytetracycline with the inner-lumen surface of halloysite (Ramanayaka et al. 2020). Density functional theory calculation further indicates that acetate can interact by H-bonding with the edge surface of pyrophyllite (Liu et al. 2008).

Early on, Siffert and Espinasse (1980) found little adsorption of oxalic and succinic acids to Na^+-montmorillonite from solution at near neutral pH as did Rabiei et al. (2016) and Ahmat et al. (2019) with respect to gallic acid. Pusino et al. (2000) observed similarly in the case of triasulfuron although appreciable uptake of this anionic herbicide was measured for Al^{3+}- and Fe^{3+}-exchanged montmorillonites. Attenuated total reflectance infrared (ATR-FTIR) spectroscopy further indicates that both citric and oxalic acids are repelled by montmorillonite (and illite) at pH ~ 6 (Yeasmin et al. 2014). Using ATR-FTIR and diffuse reflectance infrared Fourier transform (DRIFT) spectroscopy, Kang and Xing (2007) observed that adsorption of dicarboxylic acids to Na^+-exchanged montmorillonite (and kaolinite) increased with a decrease in solution pH. Moreover, uptake (at pH 4) increased with chain length in the order succinic acid < glutaric acid < adipic acid < azelaic acid, indicating the involvement of van der Waals interactions (Laird 1997; Zhao et al. 1996).

Ahmat et al. (2019) reported that gallic acid adsorption to Na^+-montmorillonite from solutions at pH 5 and 7 was limited and confined to external clay particle surfaces. They further suggested that at pH 2 this phenolic acid was intercalated in the protonated (neutral) form through electrostatic and van der Waals interactions. We would propose, however, that cation- or water-bridging is the operative mechanism here (Figure 5.1) as Greesh et al. (2008) have suggested for the intercalation of 2-acrylamido-2-methyl-1-propanesulfonic acid and related compounds into Na^+-montmorillonite. Subsequently, Ingram et al. (2016) have proposed similarly for the adsorption and (possible) intercalation of aromatic acids by acid-treated K-10 montmorillonite as did Nickels et al. (2015) for the montmorillonite-benzoic acid interaction.

In contrast to the behaviour of natural (untreated) clay minerals, organically modified clays show an appreciable propensity for adsorbing negatively charged organic solutes. Many examples of this phenomenon have already been given in Chapters 3 and 4 and listed in Table 3.1. This observation may be explained in terms of the change in the surface property of the adsorbent (from hydrophilic to hydrophobic) as well as in the surface-solute interaction (from electrostatic repulsion to one of attraction). In general, the more hydrophobic the organoclay becomes the larger is its propensity for taking up anionic organic solutes (Boyd et al. 1988; Zhao et al. 1996; An and Dultz 2007; Zhang et al. 2015). Thus, for a given organoclay uptake increases with the hydrophobicity of, and interlayer occupancy by, the modifying agent (Dentel et al. 1995; Zhao et al. 1996; Hsu et al. 2000; Rodríguez-Sarmiento and Pinzón-Bello 2001; Liao et al. 2006).

Like its thermo-exfoliated counterpart, vermiculite that has been rendered hydrophobic by treatment with polymethyl-hydro-siloxane, can efficiently remove oleic acid from water (Marcos et al. 2017). Similarly, hydrophobic dodecylamine-modified montmorillonite can effectively bind 2,4-dichlorophenoxyacetic acid, and hence reduce its phytotoxicity (Salcedo et al. 2021). Interestingly, montmorillonite modified with sodium dodecylsulphate can potentially reduce aflatoxin B1 toxicity in ruminant diets as well as decrease ruminal methanogenesis *in vitro* (Soltan et al. 2022).

The reviews by Park et al. (2011), Awad et al. (2019), and Kausor et al. (2022) have more details on the interactions of natural and modified clay minerals with organic pollutants, including negatively charged organic species. Earlier, Zhu et al. (2009) have summarized the information on regenerating organoclays that have been used as adsorbents of organic contaminants.

As might be expected, the capacity of organoclays for taking up anionic solutes increases with the degree of loading by cationic surfactants as Hsu et al. (2000) have observed for the adsorption of 2,4-dichlorophenoxy propionic acid to montmorillonite and vermiculite modified by hexadecyltrimethylammonium (HDTMA) ions. Xin et al. (2011) have reported similarly for the uptake of benzoic acid by bentonite complexes with octadecyltrimethylammonium ions as did Rodríguez-Sarmiento and Pinzón-Bello (2001) for the adsorption of sodium dodecylbenzene sulfonate by complexes of bentonite with some quaternary ammonium cations, An and Dultz (2007) for the uptake of tannic acid by montmorillonite-chitosan complexes, and Yu et al. (2010) for the interaction of Congo red with vermiculite-HDTMA complexes.

Following Zhu and Ma (2008), Yu et al. (2010) have suggested that anionic dyes adsorb by electrostatic attraction to $HDTMA^+$-exchanged bentonite (montmorillonite). Solute partitioning and hydrophobic (van der Waals) interactions may also contribute to the process (Hermosin et al. 1995; Dentel et al. 1998; Zhang et al. 2015). We might also add that the isotherms for the adsorption of negatively charged organic molecules to organically modified clay minerals (montmorillonite, palygorskite, vermiculite) generally follow the Langmuir or Freundlich model and pseudo second-order kinetics (Hermosin and Cornejo 1992; Liao et al. 2006; Yu et al. 2010; Xin et al. 2011; Bhatt et al. 2012; Zhang et al. 2015; Choi et al. 2022).

REFERENCES

Ahmat, A.M., T. Thiebault, and R. Guégan. 2019. Phenolic acids interactions with clay minerals: A spotlight on the adsorption mechanisms of gallic acid onto montmorillonite. *Applied Clay Science* 180: 105188.

An, J.-H. and S. Dultz. 2007. Adsorption of tannic acid on chitosan-montmorillonite as a function of pH and surface charge properties. *Applied Clay Science* 36: 256–264.

Aquino, A.J.A., D. Tunega, G. Haberhauer, M.H. Gerzabek, and H. Lischka. 2003. Adsorption of organic substances on broken clay surfaces: A quantum chemical study. *Journal of Computational Chemistry* 24: 1853–1863.

Arnarson, T.S. and R.G. Keil. 2000. Mechanism of pore water organic matter adsorption to montmrorillonite. *Marine Chemistry* 71: 309–320.

Awad, A.M., S.M.R. Shaikh, R. Jalab et al. 2019. Adsorption of organic pollutants by natural and modified clays: A comprehensive review. *Separation and Purification Technology* 228: 115719.

Bhatt, A.S., P.L. Sakaria, M. Vasudevan et al. 2012. Adsorption of an anionic day from aqueous medium by organoclays: Equilibrium modeling, kinetic and thermodynamic exploration. *RSC Advances* 2: 8663–8671.

Bidwell, J.I., W.B. Jepson, and G.L. Toms. 1970. The interaction of kaolinite with polyphosphate and polyacrylate in aqueous solutions – some preliminary results. *Clay Minerals* 8: 445–459.

Blockhaus, F., J.-M. Séquaris, H.D. Narres, and M.J. Schwuger. 1996. Interactions of a water-soluble polymeric detergent additive (polycarboxylate) with clay minerals from soil. *Progress in Colloid & Polymer Science* 101: 23–29.

Blockhaus, F., J.-M. Séquaris, H.D. Narres, and M.J. Schwuger. 1997. Adsorption desorption behavior of acrylic-maleic acid copolymer at clay minerals. *Journal of Colloid and Interface Science* 186: 234–247.

Boyd, S.A., S. Sun, J.-F. Lee, and M.M. Mortland. 1988. Pentachlorophenol sorption by organo-clays. *Clays and Clay Minerals* 36: 125–130.

Bu, H., D. Liu, P. Yuan, X. Zhou, H. Liu, and P. Du. 2019. Ethylene glycol monoethyl ether (EGME) adsorption by organic matter (OM)-clay complexes: Dependence on the OM type. *Applied Clay Science* 168: 340–347.

Celis, R., W.C. Koskinen, A.M. Cecchi, G.A. Bresnahan, M.J. Carrisoza, and M.A. Ulibarri. 1999. Sorption of the ionizable pesticide imazamox by organo-clays and organohydrotalcites. *Journal of Environmental Science and Health, Part B* 34: 929–941.

Chefetz, B., S. Eldad, and T. Polubesova. 2011. Interactions of aromatic acids with montmorillonite: Ca^{2+}-and Fe^{3+}-saturated clays versus Fe^{3+}-Ca^{2+}-clay system. *Geoderma* 160: 608–613.

Chen, P., Y. Wei, L. Wei et al. 2020. Quantitative assessments of organic matter uncoupling from clay surfaces in presence of salinity. *Applied Clay Science* 188: 105532.

Choi, N., Y. Son, T.-H. Kim, Y. Park, and Y. Hwang. 2022. Adsorption behaviors of modified clays prepared with structurally different surfactants for anionic dyes removal. *Environmental Engineering Research* 28: 220076. https://doi.org/10.4491/eer.2022.076.

Cukrowicz, S., P. Goj, P. Stoch, A. Bobrowski, B. Tyliszczak, and B. Grabowska. 2022. Molecular dynamic (MD) simulations of organic modified montmorillonite. *Applied Sciences* 12: 314. https://doi.org/10.3390/app12010314.

Dentel, S.K., J.Y. Bottero, K. Khatib, H. Demougeot, J.P. Duguet, and C. Anselme. 1995. Sorption of tannic acid, phenol, and 2,4,5-trichlorophenol on organoclays. *Water Research* 29: 1273–1280.

Dentel, S.K., A.I. Jamrah, and D.L. Sparks. 1998. Sorption and cosoprtion of 1,2,4-trichlorobenzene and tannic acid by organo-clays. *Water Research* 32: 3689–3697.

Epstein, M. and S. Yariv. 2003. Visible spectroscopy study of the adsorption of alizarinate by Al-montmorillonite in aqueous suspensions and in solid state. *Journal of Colloid and Interface Science* 263: 377–385.

Espinasse, P. and B. Siffert. 1979. Acetamide and polyacrylamide adsorption onto clay: Influence of the exchangeable cation and the salinity of the medium. *Clays and Clay Minerals* 27: 279–284.

Ferris, A.P. and W.B. Jepson. 1975. The exchange capacity of kaolinite and the preparation of homoionic clays. *Journal of Colloid and Interface Science* 51: 245–259.

Gautier, M., F. Muller, J.-M. Beny, L. Le Forestier, P. Alberic, and P. Baillif. 2009. Interactions of ammonium smectite with low-molecular-weight carboxylic acids. *Clay Minerals* 44: 207–219.

Gimsing, A.L., J.C. Sørensen, B.W. Strobel, and H.C.B. Hansen. 2007. Adsorption of glucosinolates to metal oxides, clay minerals and humic acid. *Applied Clay Science* 35: 212–217.

Greesh, N., P.C. Hartman, V. Cloete, and R.D. Sanderson. 2008. Adsorption of 2-acrylamido-2-methyl-1-propanesulfonic acid (AMPS) and related compounds onto montmorillonite clay. *Journal of Colloid and Interface Science* 319: 2–11.

Hashizume, H. 2007. Adsorption of some amino acids by chrysotile. *Viva Origino* 35: 60–65.

Hashizume, H. and B.K.G. Theng. 2007. Adenine, adenosine, ribose, and 5'-AMP adsorption to allophane. *Clays and Clay Minerals* 55: 599–605.

Hermosin, M.C. and J. Cornejo. 1992. Removing 2,4-D from water by organo-clays. *Chemosphere* 24: 1493–1503.

Hermosin, M.C., A. Crabb, and J. Cornejo. 1995. Sorption capacity of organo-clays for anionic and polar organic contaminants. *Fresenius Environmental Bulletin* 4: 514–519.

Hsu, Y.H., M.K. Wang, C.W. Pai, and Y.S. Wang. 2000. Sorption of 2,4-dichlorophenoxy propionic acid by organo-clay complexes. *Applied Clay Science* 16: 147–159.

Hu, Z., D.P. Jaisi, Y. Yan et al. 2020. Adsorption and precipitation of *myo*-inositol hexakisphosphate onto kaolinite. *European Journal of Soil Science* 71: 226–235.

Ingram, A.L., T.M. Nickels, D.K. Maraoulaite, and R.L. White. 2016. Thermogravimetry-mass spectrometry investigations of montmorillonite interlayer perturbations caused by aromatic acid adsorbates. *Journal of Thermal Analysis and Calorimetry* 126: 1157–1166.

Kang, S. and B. Xing. 2007. Adsorption of dicarboxylic acids by clay minerals as examined by *in situ* ATR-FTIR and *ex situ* DRIFT. *Langmuir* 23: 7024–7031.

Kaufherr, N., S. Yariv, and L. Heller. 1971. The effect of exchangeable cations on the sorption of chlorophyllin by montmorillonite. *Clays and Clay Minerals* 19: 193–200.

Kausor, M.A., S.S. Gupta, K.G. Bhattacharyya, and D. Chakrabortty. 2022. Montmorillonite and modified montmorillonite as adsorbents for removal of water soluble organic dyes: A review on current status of the art. *Inorganic Chemistry Communications* 143: 109686.

Kohl, R.A. and S.A. Taylor. 1961. Hydrogen bonding between the carbonyl group and Wyoming bentonite. *Soil Science* 91: 223–227.

Kosiur, D.R. 1977. Porphyrin adsorption by clay minerals. *Clays and Clay Minerals* 25: 365–371.

Kubicki, J.D., M.J. Itoh, L.M. Schroeter, and S.E. Apitz. 1997. Bonding mechanisms of salicylic acid adsorbed onto illite clay: An ATR-FTIR and molecular orbital study. *Environmental Science & Technology* 31: 1151–1156.

Kubicki, J.D., L.M. Schroeter, M.J. Itoh, B.N. Nguyen, and S.E. Apitz. 1999. Attenuated total reflectance Fourier-transform infrared spectroscopy of carboxylic acids adsorbed onto mineral surfaces. *Geochimica et Cosmochimica Acta* 63: 2709–2725.

Kyzioł-Komosińska, J., C. Rosik-Dulewska, M. Pająk, I. Krzyżewsla, and A. Dzieniszewska. 2014. Adsorption of anionic dyes onto natural, thermally and chemically modified smectite clays. *Polish Journal of Chemical Technology* 16: 33–40.

Lackovic, K., B.B. Johnson, M.J. Angove, and J.D. Wells. 2003. Modeling the adsorption of citric acid onto Muloorina illite and related clay minerals. *Journal of Colloid and Interface Science* 267: 49–59.

Laird, D.A. 1997. Bonding between polyacrylamide and clay mineral surfaces. *Soil Science* 162: 826–832.

Law, J.P. and G.W. Kunze. 1966. Reactions of surfactants with montmorillonite: Adsorption mechanisms. *Soil Science Society of America Journal* 30: 321–327.

Liao, C.-J., C.-P. Chen, M.-K. Wang, P.-N. Chiang, and C.-W. Pai. 2006. Sorption of chlorophenoxy propionic acids by organoclay complexes. *Environmental Toxicology* 21: 71–79.

Liu, X., X. Lu, R. Wang, H. Zhou, and S. Xu. 2008. Surface complexes of acetate on edge surfaces of 2:1 type phyllosilicates: Insights from density functional theory calculation. *Geochimica et Cosmochimica Acta* 72: 5896–5907.

Liu, X., X. Lu, Y. Zhang, C. Zhang, and R. Wang. 2017. Complexation of carboxylate on smectite surfaces. *Physical Chemistry Chemical Physics* 19: 18400–18406.

Marcos, C., R. Menéndez, and I. Rodríguez. 2017. Thermoexfoliated and hydrophobized vermiculites for oleic acid removal. *Applied Clay Science* 150: 147–152.

Martin Martinez, F. and J.L. Perez Rodriguez. 1969. Interlamellar adsorption of a black-earth humic acid on Na^+-montmorillonite. *Zeitschrift für Pflanzenernährung und Bodenkunde* 124: 52–57.

Moinereau, J. 1977. Absorption de composés humique par une montmorillonite H^+, Al^{3+}; présence de smectites à couches interfoliares organo-minérales dans les andosols. *Clay Minerals* 12: 75–82.

Mortensen, J.L. 1962. Adsorption of hydrolyzed polyacrylonitrile on kaolinite. *Clays and Clay Minerals* 9: 530–545.

Muljadi, D., A.M. Posner, and J.P. Quirk. 1966. The mechanism of phosphate adsorption on kaolinite, gibbsite and pseudo-boehmite. I. The isotherms and the effect of pH on adsorption. *Journal of Soil Science* 17: 212–231.

Nagasaki, S., Y. Nakagawa, and S. Tanaka. 2004. Sorption of nonylphenol on Na-montmorillonite. *Colloids and Surfaces A: Physicochemical and Engineering Aspects* 230: 131–139.

Nickels, T.M., A.L. Ingram, D.K. Maraoulaite, and R.L. White. 2015. Variable temperature infrared spectroscopy investigation of benzoic acid interactions with montmorillonite clay interlayer water. *Applied Spectroscopy* 69: 850–856.

Parfitt, R.L. 1978. Anion adsorption by soils and soil materials. *Advances in Agronomy* 30: 1–50.

Park, Y., G.A. Ayoko, and R.L. Frost. 2011. Application of organoclays for the adsorption of recalcitrant organic molecules from aqueous media. *Journal of Colloid and Interface Science* 354: 292–305.

Polubesova, T., S. Eldad, and B. Chefetz. 2010. Acsorption and oxidative transformation of phenolic acids by Fe(III)-montmorillonite. *Environmental Science & Technology* 44: 4203–4209.

Pusino, A., I. Braschi, and C. Gessa. 2000. Adsorption and degradation of triasulfuron on homoionic montmorillonites. *Clays and Clay Minerals* 48: 19–25.

Rabiei, M., H. Sabahi, and A.H. Rezayan. 2016. Gallic acid-loaded montmorillonite nanostructure as a new controlled release system. *Applied Clay Science* 119: 236–242.

Ramanayaka, S., B. Sarkar, A.T. Cooray, Y.S. Ok, and M. Vithanage. 2020. Halloysite nanoclay supported adsorptive removal of oxytetracycline antibiotic from aqueous media. *Journal of Hazardous Materials* 384: 121301.

Ramos, M.E. and F.J. Huertas. 2014. Adsorption of lactate and citrate on montmorillonite in aqueous solutions. *Applied Clay Science* 90: 27–34.

Rodríguez-Sarmiento, D.C. and J.A. Pinzón-Bello. 2001. Adsorption of sodium dodecylbenzene sulfonate on organophilic bentonites. *Applied Clay Science* 18: 173–181.

Salcedo, M.F., A.Y. Mansilla, S.L. Colman, M.J. Iglesias, V.A. Alvarez, and C.A. Casalongué. 2021. Efficacy of an organically modified bentonite to adsorb 2,4-dichlorophenoxyacetic acid (2,4-D) and prevent its phytotoxicity. *Journal of Environmental Management* 297: 113427.

Sastry, N.V., J.-M. Squaris, and M.J. Schwuger. 1995. Adsorption of polyacrylic acid and sodium dodecylbenzenesulfonate on kaolinite. *Journal of Colloid and Interface Science* 171: 224–233.

Schnitzer, M. and H. Kodama. 1966. Montmorillonite: Effect of pH on its adsorption of a soil humic compound. *Science* 153: 70–71.

Siebner-Freibach, H., Y. Hadar, and Y. Chen. 2004. Interaction of iron chelating agents with clay minerals. *Soil Science Society of America Journal* 68: 470–480.

Siffert, B. and P. Espinasse. 1980. Adsorption or organic diacids and sodium polyacrylate onto montmorillonite. *Clays and Clay Minerals* 28: 381–387.

Soltan, Y.A., A.S. Morsy, N.M. Hashem et al. 2022. Potential of montmorillonite modified by an organosulfur surfactant for reducing aflatoxin B_1 toxicity and ruminal methanogenesis in vitro. *BMC Veterinary Research* 18: 387. https://doi.org/10.1186/s12917-022-03476-1.

Specht, C.H. and F.H. Frimmel. 2001. An *in situ* ATR-FTIR study on the adsorption of dicarboxylic acids onto kaolinite in aqueous suspensions. *Physical Chemistry Chemical Physics* 3: 5444–5449.

Spepi, A., C. Duce, A. Pedone et al. 2016. Experimental and DFT characterization of halloysite nanotubes loaded with salicylic acid. *Journal of Physical Chemistry C* 120: 26759–26769.

Sposito, G. 2008. *The Chemistry of Soils*, 2nd edition. New York, NY: Oxford University Press.

Stutzmann, T. and B. Siffert. 1977. Contribution to the adsorption mechanism of acetamide and polyacrylamide on to clays. *Clays and Clay Minerals* 25: 392–406.

Tate, K.R. and B.K.G. Theng. 1980. Organic matter and its interactions with inorganic soil constituents. In *Soils with Variable Charge*, (Ed.) B.K.G. Theng, pp. 225–249. Lower Hutt, NZ: New Zealand Society of Soil Science.

Theng, B.K.G. 1976. Interactions between montmorillonite and fulvic acid. *Geoderma* 15: 243–251.

Theng, B.K.G. 2012. *Formation and Properties of Clay-Polymer Complexes*, 2nd edition. Amsterdam, The Netherlands: Elsevier.
Theng, B.K.G., G.J. Churchman, and R.H. Newman. 1986. The occurrence of interlayer clay-organic complexes in two New Zealand soils. *Soil Science* 142: 262–266.
Theng, B.K.G., M. Russell, G.J. Churchman, and R.L. Parfitt. 1982. Surface properties of allophane, halloysite, and imogolite. *Clays and Clay Minerals* 30: 143–149.
Theng, B.K.G. and H.W. Scharpenseel. 1976. The adsorption of ^{14}C-labelled humic acid by montmorillonite. In *Proceedings of the International Clay Conference, 1975*, (Ed.) S.W. Bailey, pp. 643–653. Wilmette, IL: Applied Publishing.
Theng, B.K.G. and G. Yuan. 2008. Nanoparticles in the soil environment. *Elements* 4: 395–399.
Tombácz, E., M. Gilde, and F. Szántó. 1984. The effects of Na-salicylate and Na-fulvate on the stability and rheological properties of Na-montmorillonite suspensions. *Acta Physica et Chimica Szeged* 30: 165–174.
Valentim, I.B. and I. Joekes. 2006. Adsorption of sodium dodecylsulfate on chrysotile. *Colloids and Surfaces A: Physicochemical and Engineering Aspects* 290: 106–111.
Wang, X., Q. Li, H. Hu, T. Zhang, and Y. Zhou. 2005. Dissolution of kaolinite induced by citric, oxalic, and malic acids. *Journal of Colloid and Interface Science* 290: 481–488.
Weber, J.B., P.W. Perry, and R.P. Upchurch. 1965. The influence of temperature and time on the adsorption of paraquat, diquat, 2,4-D, and prometone by clays, charcoal, and an anion-exchange resin. *Soil Science Society of America Proceedings* 29: 678–687.
Wu, Q., Z. Li, and H. Hong. 2013. Adsorption of the quinolone antibiotic nalidixic acid onto montmorillonite and kaolinite. *Applied Clay Science* 74: 66–73.
Xin, X., W. Si, Z. Yao et al. 2011. Adsorption of benzoic acid from aqueous solution by three kinds of modified bentonites. *Journal of Colloid and Interface Science* 359: 499–504.
Xu, S.W., J.P. Zheng, L. Tong, and K.D. Yao. 2006. Interaction of functional groups of gelatin and montmorillonite in nanocomposite. *Journal of Applied Polymer Science* 101: 1556–1561.
Xue, X., Z. Xu, I. Pedruzzi, P. Li, and J. Yu. 2018. Interaction between low molecular weight carboxylic acids and muscovite: Molecular dynamic simulation and experimental study. *Colloids and Surfaces A* 559: 8–17.
Yan, L.-G., J. Wang, H.-Q. Yu, Q. Wei, B. Du, and X.-Q. Shan. 2007. Adsorption of benzoic acid by CTAB exchanged montmorillonite. *Applied Clay Science* 37: 226–230.
Yeasmin, S., B. Sing, R.S. Kookana, M. Farrell, D.L. Sparks, and C.T. Johnston. 2014. Influence of mineral characteristics on the retention of low molecular weight organic compounds: A batch sorption-desorption and ATR-FTIR study. *Journal of Colloid and Interface Science* 432: 246–257.
Yu, X., C. Wei, L. Ke, Y.Hu, X. Xie, and H. Wu. 2010. Development of organovermiculite-based adsorbent for removing anionic dye from aqueous solution. *Journal of Hazardous Materials* 180: 499–507.
Yuan, G., B.K.G. Theng, R.L. Parfitt, and H.J. Percival. 2000. Interactions of allophane with humic acid and cations. *European Journal of Soil Science* 51: 35–41.
Zhang, L., B. Zhang, T. Wu, D. Sun, and Y. Li. 2015. Adsorption behavior and mechanism of chlorophenols onto organoclays in aqueous solution. *Colloids and Surfaces A: Physicochemical and Engineering Aspects* 484: 118–129.
Zhang, R., W. Yan, and C. Jing. 2014. Mechanistic study of PFOS adsorption on kaolinite and montmorillonite. *Colloids and Surfaces A: Physicochemical and Engineering Aspects* 462: 252–258.
Zhang, Z.Z., P.F. Low, J.H. Cushman, and C.B. Roth. 1990. Adsorption and heat of adsorption of organic compounds on montmorillonite from aqueous solutions. *Soil Science Society of America Journal* 54: 59–66.
Zhao, H., W.F. Jaynes, and G.F. Vance. 1996. Sorption of the ionizable organic compound, dicamba (3,6-dichloro-2-methoxy benzoic acid), by organo-clays. *Chemosphere* 33: 2089–2100.
Zhao, L., J. Bian, Y. Zhang, L. Zhu, and Z. Liu. 2014. Comparison of the sorption behaviors and mechanisms of perfluorosulfonates and perfluorocarboxylic acids on three kinds of clay minerals. *Chemosphere* 114: 51–58.
Zhu, L. and J. Ma. 2008. Simultaneous removal of acid dye and cationic surfactant from water by bentonite in one-step process. *Chemical Engineering Journal* 139: 503–509.
Zhu, R., J. Zhu, F. Ge, and P. Yuan. 2009. Regeneration of spent organoclays after the sorption of organic pollutants: A review. *Journal of Environmental Management* 90: 3212–3216.

6 Organic Complexes of Clay Minerals with a 1:1 Type Layer Structure

6.1 INTRODUCTION

Within the class of 1:1 type layer silicates (cf. Table 1.1), the kaolin group of minerals, represented by dickite, halloysite, kaolinite, and nacrite, can form interlayer complexes with a variety of organic compounds. Minerals in the serpentine group, of which chrysotile is a representative species, apparently fail to intercalate organic molecules. Some examples of the halloysite- and kaolinite-organic interaction have been given in the preceding chapters. Here we will describe the ability and use of the kaolin group of minerals, notably kaolinite and halloysite, to take up and remove organic compounds from solution, focusing on the processes and mechanisms underlying the intercalation of organic molecules into these minerals.

The formation, properties, and applications of kaolin-organic complexes have regularly been reviewed (Joussein et al. 2005; Lagaly et al. 2006; Cheng et al. 2012; Moyo et al. 2014; Yuan et al. 2015; Abdullayev and Lvov 2016; Tan et al. 2016; Detellier 2018; Cavallaro et al. 2019; Danyliuk et al. 2020; Neto et al. 2022). The books, edited by Pasbakhsh and Churchman (2015) and Yuan et al. (2016), contain chapters that are related and relevant to this topic. Kloprogge (2017) has described the application of Raman and infrared spectroscopies to characterizing the kaolin-organic interaction, while Zhang (2017) has described the approaches to modifying the inner surface of halloysite tubules. We would also mention the reviews by Awad et al. (2017), Hanif et al. (2016), Yu et al. (2016), Anastopoulos et al. (2018), and Saadat et al. (2022) on the use of the kaolin group of minerals in the pharmaceutical industry and water/wastewater treatment.

With respect to halloysite, the tubular forms of the mineral, commonly referred to as halloysite nanotubes (HNTs) have received the greatest attention. The mechanism underlying the formation of HNTs has been discussed by Niu (2016) and Gianni et al. (2023), while the large volume of literature on the biological, biomedical, catalytic, environmental, food packaging, health, and water treatment applications of raw and surface-modified HNTs has been summarized by Aguzzi et al. (2016), Churchman et al. (2016a), Pasbakhsh et al. (2016), Yang et al. (2017), Massaro et al. (2017, 2020a), Tharmavaram et al. (2018), Setter and Segal (2020), Fakhruddin et al. (2021), Karewicz et al. (2021), Kushwaha et al. (2021); Biddeci et al. (2022), Dube et al. (2022), Fizir et al. (2022), Li et al. (2021, 2022), Mobaraki et al. (2022), Abid et al. (2023), and Deshmukh et al. (2023).

6.2 INTERCALATION OF ORGANIC MOLECULES

6.2.1 General Aspects

As indicated in Chapter 1, the contiguous layers within a kaolinite particle are held together by O–H···O hydrogen bonding (and van der Waals forces) which, therefore, must be overcome if the organic guest molecules are to enter the interlayer space. A similar but less pronounced constraint applies to the intercalation of organic species into fully dehydrated halloysite (Churchman et al. 1984; Theng et al. 1984).

DOI: 10.1201/9781003080244-6

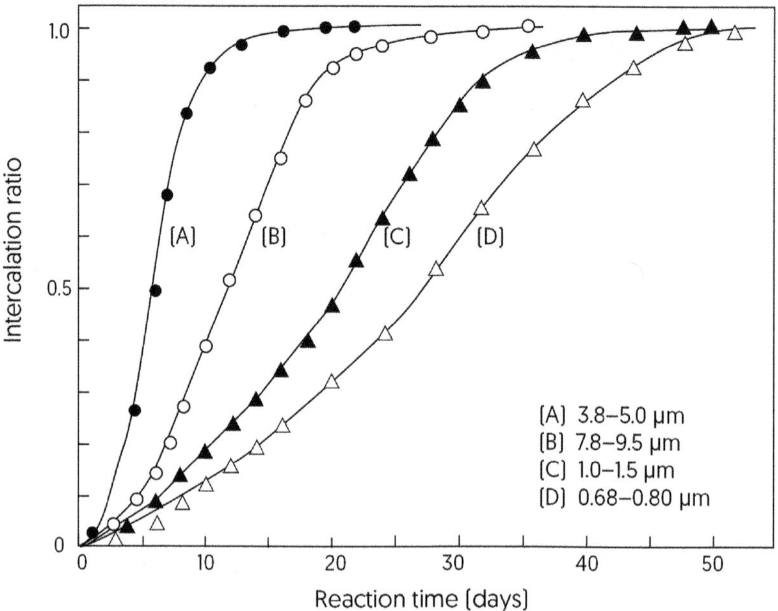

FIGURE 6.1 Relationship between intercalation ratio and reaction time for the interlayer uptake of urea by kaolinites with varying particle size. (From Weiss, A. et al. *Proc.1969 Int. Clay Conf., Tokyo*, 2, 180–184, 1970; Lagaly, G. et al., *Handbook of Clay Science*, Developments in Clay Science, Vol. 1, pp. 309–377, 2006.)

With respect to kaolinite, Weiss et al. (1970) have pointed out that larger particles are more easily intercalated than their smaller counterparts (Figure 6.1). This observation has been ascribed to the misfit between the tetrahedral and octahedral sheets making up the 1:1 layer structure (cf. Figure 1.3), giving rise to a structural stress. As the stress increases with particle size, the resistance to intercalation decreases (Detellier and Schoonheydt 2014). The finer the particle size the lower the structural stress and the slower the rate of intercalation (Uwins et al. 1993; Deng et al. 2002).

For kaolinites with a comparable particle size distribution, the ease of intercalation also decreases with an increase in layer stacking disorder (Cheng et al. 2010a; Fashina and Deng 2021). Earlier, Olejnik et al. (1968) reported that the rate of DMSO intercalation decreased in the order halloysite > kaolinite > dickite. Benincasa et al. (2002) also made the interesting observation that intercalation of Cd-cysteine complex was accompanied by a decrease in the structural order of kaolinite.

Similarly, a large particle size is conducive to intercalation into (dehydrated) halloysite as does a high degree of crystallinity and a low iron content (Churchman and Theng 1984; Joussein et al. 2005). In this context, we might mention the difficulty of intercalating a highly ordered kaolinite from Birdwood, South Australia, because its particles are apparently coated with a highly disordered kaolinite species (Frost et al. 2002). Nonetheless, complete intercalation can be achieved by grinding the Birdwood material with solid urea for 2 h (Makó et al. 2013). Intercalation (of formamide) into a high-defect kaolinite can also lead to a decrease in defect structures (Frost et al. 2000a)

The typical S-shape (sigmoidal) curve, describing the intercalation kinetics, has been interpreted in terms of the slow initial penetration of the kaolinite interlayers by organic guest molecules (here urea) from the particle edge inward. The resultant elastic deformation and curling of the silicate layers promote intercalation as well as the cooperative diffusion of guest molecules into the interlayer space. This process is reflected by the steep increase in the slope of the curve, reaching a plateau at maximum coverage by the organic solute (Lipsicas et al. 1986; Lagaly et al. 2006; Detellier and Schoonheydt 2014). In this respect, the unusual temperature dependence of N-methylformamide intercalation may be related to the adverse influence of temperature on the edge surface adsorption of the amide (Andreou et al. 2022).

We might also mention that the crystallinity of kaolinites within a geographic deposit often shows an inconsistent pattern with respect to particle size in that crystallinity tends to increase towards the fine particle size range (Lloyd and Conley 1970). It is also noteworthy that *b*-axis disordered kaolinites are less rather than more amenable to intercalating urea, hydrazine, and formamide (Weiss et al. 1963a, 1963b). Moreover, small variations in unit cell dimensions can cause marked differences in the reactivity of halloysite. Here a smaller *b*-dimension is associated with a tendency to hold interlayer water more firmly together with a preference for assuming a tubular particle morphology (Nagasawa 1969).

The following sections are primarily concerned with the formation and properties of interlayer complexes between the kaolin group of minerals and defined organic molecules. In addition, we will describe the interactions of organic compounds with external particle surfaces of kaolinite, halloysite, and chrysotile.

6.2.2 Formation and Properties of Interlayer Complexes

Organic compounds that can intercalate *directly* into kaolinite and halloysite-(7Å) fall into three groups (Lagaly et al. 2006; Tan et al. 2016). We would add here that (tubular) halloysite shows a greater capacity for taking up organic molecules than does kaolinite (Tan et al. 2015).

Molecules in group I can accept or donate protons to form hydrogen bonds with both inner-surface hydroxyls (aluminols) of one layer and siloxane oxygens of the opposite layer (Ledoux and White 1966a, 1966b; Frost et al. 1999; Horváth et al. 2010). Compounds in this group include *amides* (Carr and Chih 1971; Olejnik et al. 1971a, 1971b; Adams 1978; Churchman and Theng 1984; Churchman et al. 1984; Thompson 1985; Frost et al. 1999; Zamama and Knidiri 2000; Michalková et al. 2002; Joussein et al. 2007; Scholtzová et al. 2008; Zhang et al. 2018a, 2018b; Andreou et al. 2021), *hydrazine* (Weiss et al. 1963b; Theng et al. 1984; Johnston and Stone 1990; Gábor et al. 1995; Johnston et al. 2000; Ruiz Cruz and Franco 2000a, 2000b; Deng et al. 2002, 2003; Horváth et al. 2003), and *urea* (Weiss 1961; Frost and Kristof 1997; Tsunematsu and Tateyama 1999; Valášková et al. 2007; Makó et al. 2009; Nicolini et al. 2009; Rutkai et al. 2009; Zhu et al. 2012; Zhang et al. 2017).

Group II organic intercalates comprise molecules with a high dipole moment, such as pyridine-*N*-oxide, and particularly dimethyl sulphoxide (Weiss et al. 1966; Olejnik et al. 1970, 1971c; Weiss and Orth 1973; Adams and Waltl 1980; Thompson 1985; Raupach et al. 1987; Amara et al. 1995; Franco and Ruiz Cruz 2003; Mbey et al. 2013; Abou-El-Sherbini et al. 2017; Fafard et al. 2017; Zhang et al. 2018c). Intercalation of dimethyl sulphoxide (DMSO) into halloysite is commonly much faster than into kaolinite (Olejnik et al. 1968; Churchman et al. 1984; Costanzo and Giese 1986), presumably because the interlayers of halloysite-(7Å) have once been separated by a sheet of water molecules (cf. Figure 1.7). Dehydration is also known to alter the morphology of halloysite particles (Kohyama et al. 1978; Joussein et al. 2005; Ece and Schroeder 2007).

Intercalates in group III comprise ammonium, caesium, potassium, and sodium salts of short-chain fatty acids, such as acetic, butyric, lauric, oleic, palmitic, and propionic (Wada 1961; Weiss et al. 1966; Hofmann and Reingraber 1969; Carr et al. 1978; Sidheswaran et al. 1990; Frost and Kristof 1997; Frost et al. 2001; Lagaly et al. 2006; Kovács and Makó 2016). In this regard, the kaolinite-potassium acetate interaction has received the greatest amount of attention (Wada 1961; Churchman and Carr 1973; Frost et al. 1998b; Ruiz Cruz and Franco Duro 1999; Cheng et al. 2010a, 2011; Xu et al. 2011; Makó et al. 2014a; Zhong et al. 2017; Adamczyk et al. 2020). Strong hydrogen bonds apparently form between the inner-surface hydroxyl groups of kaolinite and the oxygen atoms of intercalated acetate and water. At the same time, the intercalants are weakly H-bonded through their hydrogen atoms to the siloxane surface oxygens (Ledoux and White 1964; Cheng et al. 2015a).

Interlayer complex formation may be achieved by immersing the clay sample in a concentrated solution of the salt ('solution intercalation'), grinding the dry clay mineral with the solid salt ('mechanochemical intercalation'), or mixing the kaolin sample and salt in the presence of water ('homogenization') using an agate mortar. Besides its simplicity, the mechanochemical method is highly efficient and causes minimal structural deformation (Makó et al. 2014a). Subsequently, Kovács and Makó (2016) found that cooling below 0°C could significantly promote ammonium acetate (AA) intercalation by the homogenization method. Makó et al. (2014b) further reported that AA could form two types of water-containing complexes with basal spacings of 1.7 and 1.4 nm. Homogenization can also reduce the shortcomings associated with the solution and mechanochemical means for intercalating amides, DMSO, and urea into kaolinite (Makó et al. 2019).

6.3 KAOLINITE

6.3.1 Interlayer Complex Formation

There is evidence to indicate that the propensity of kaolin minerals for intercalating urea was known to the ancient Chinese. Examination of historical notes and records has led Weiss (1963) to postulate that the secret of porcelain manufacture in medieval China was to mix kaolin, probably quarried from a deposit at Jingdezhen in Jiangxi province (Chen et al. 1997), with human urine in large pits. The mixture was then left to 'age' allowing the urea (from the decaying urine) to enter and expand the interlayer space. The resultant increase in plasticity and dry bending strength of the urea-modified kaolin allowed the material to be worked and shaped into the fine, egg-shell porcelain articles for which imperial China of medieval times was famous.

The formation and application of kaolin-urea intercalates have thus left a mark on human civilization. An outstanding example, in this regard, is the palygorskite-indigo dye complex known as 'Maya blue' (cf. Figure 7.1). As the name suggests, this material was used by the ancient Mayas as a turquoise-blue pigment in pre-Columbian Meso-American murals and artefacts (Gettens 1962; Van Olphen 1966; del Río et al. 2006; Tsiantos et al. 2012). The formation and properties of Maya blue are described in Chapter 7.

In support of Weiss' (1963) proposal, Tsunematsu and Tateyama (1999) reported extensive delamination of kaolinite by grinding the corresponding complex with urea. Valášková et al. (2007) also found that urea-intercalated kaolinite particles could be exfoliated by washing with water. Similarly, Zhu et al. (2012) noted that the particle size of kaolinite-urea complexes decreased as intercalation progressed, while Elhadj and Perrin (2021) observed that a high concentration of urea was conducive to kaolinite layer exfoliation. The breaking of interlayer hydrogen bonds in kaolinite through uptake of organic guest molecules, and the formation of new bonds involving the intercalated species, clearly alter the relationship between silicate layers making up a unit particle. In general, interlayer complex formation is conducive to layer displacement and particle exfoliation.

In an early paper, Weiss and Thielepape (1963) reported that kaolinite particles became progressively thinner and eventually exfoliated into very thin layers ('nanoscrolls') by grinding with ammonium acetate. More recently, Tao et al. (2014) obtained a silane-grafted kaolinite by grinding kaolinite with γ-aminopropyltriethoxysilane. A widely used method of exfoliating kaolinite particles into tubular halloysite-like 'nanoscrolls' is through intercalation of quaternary ammonium chlorides into a (partially) methoxy-grafted kaolinite, obtained by treating the respective interlayer DMSO/NMF/urea complex with methanol (Komori et al. 2000; Kuroda et al. 2011; Matusik et al. 2012b; Yuan et al. 2013; Cheng et al. 2015b; Liu et al. 2016; Makó et al. 2015, 2016; Li et al. 2015a, 2017; Wang et al. 2017a; Zuo et al. 2017; Khan and Wiley 2022). Likewise, da Silva et al. (2016) found that kaolinite, grafted with dipicolinic acid, could be exfoliated by treatment with water.

The sequence of processes is diagrammatically shown in Figure 6.2. The length-to-diameter ratio, specific surface area, and pore volume of synthetic kaolinite nanoscrolls are much larger than the respective values measured for natural tubular halloysite (Li et al. 2019).

Earlier, Tunney and Detellier (1996) were able to replace every third inner-surface hydroxyl with a methoxy group by heating a DMSO or NMF intercalate of kaolinite with methanol at 200–270°C.

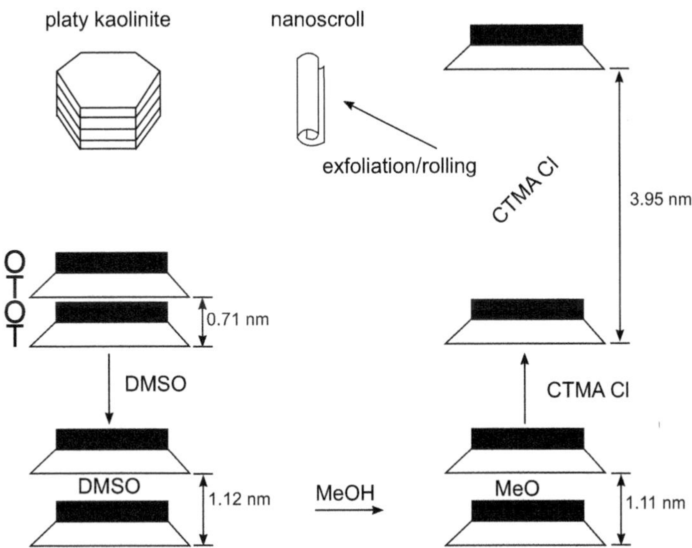

FIGURE 6.2 Diagram showing the sequence of processes involved in the transformation of platy kaolinite particles to halloysite-like nanoscrolls. The basal spacings of the parent kaolinite and the various intermediary complexes are indicated. DMSO = dimethyl sulphoxide; MeOH = methanol; MeO = methoxy; CTMA Cl = cetyltrimethylammonium chloride. (Modified from Yuan et al., *Appl. Clay Sci.* 84: 68–76, 2013.)

Similarly, ethylene glycol and glycerol could be grafted to interlayer aluminol groups by heating the compounds in the presence of an interlayer complex of kaolinite with DMSO or potassium acetate (Janek et al. 2007; Tunney and Detellier 1994). In a variant approach, Matusik et al. (2009) used 1,3-butanediol instead of methanol while substituting hexylamine for quaternary ammonium chloride. Letaief et al. (2006) used the kaolinite-urea complex as a precursor for the melt intersalation of ethylpyridinium chloride, while Makó et al. (2017) used the same precursor for the direct intercalation of cetyltrimethylammonium chloride from methanol at 100°C.

In this context, we would mention that methoxy-grafted/modified kaolinite can intercalate a variety of organic molecules, including amitrole (Tan et al. 2015), azobenzene (Koteja and Matusik 2019), 1,2- and 1,3-propane and butane-diols (Itagaki and Kuroda 2003; Murakami et al. 2004), benzylalkylammonium chlorides (Matusik and Kłapyta 2013; Matusik et al. 2013), 1-butyl-3-methylimidazolium bromine (Zhang et al. 2013a), 4-dimethylaminopyridine (Machida and Sugahara 2022), fatty acids (Zuo et al. 2017), 5-fluorouracil (Tan et al. 2014a, 2018), methyl viologen (Tchoumene et al. 2018), quaternary ammonium salts (Wang et al. 2017b), and trimethylphosphate (Machida et al. 2019). Similarly, methoxy-grafted dickite can intercalate hexylamine (Matusik et al. 2012a). Some alkylammonium salts can form interlayer complexes with kaolinite by displacing previously intercalated urea or dimethylformamide (Makó et al. 2017; Meziane et al. 2017). Even bulky organophosphonium salts can intercalate by displacing tetrabutylammonium bromide from its complex with kaolinite (Machida et al. 2021).

Using a concentrated (> 10 M) aqueous solution of urea, Weiss (1961) and Weiss et al. (1963a) observed rapid and substantial intercalation into kaolinite, yielding a complex with a basal spacing of 1.06 nm. The interlayer urea was weakly held and could be removed by washing the complex with water, decomposing into ammonia and carbon dioxide when heated at ~ 147°C. Formamide and hydrazine reacted similarly, expanding the *d*(001) spacing of kaolinite to 1.0 and ~ 1.04 nm, respectively (Weiss et al. 1963a, 1963b; Johnston et al. 2000). Thus, hydrazine intercalation can be used to differentiate kaolin minerals from chlorite (Wada 1968) as does complex formation with

DMSO (González García and Sánchez Camazano 1968). Hydrazine also intercalates at a faster rate than either urea or formamide, but the resultant complex is unstable in air.

It was not until the late 1940s, however, that the kaolinite-organic interaction began to be systematically investigated (MacEwan 1948, 1949). The apparent dearth of information on this topic is perhaps hardly surprising since, unlike smectite and vermiculite, kaolinite has no interlayer cations that can be solvated by water and polar organic molecules. Furthermore, the pairing of oxygen atoms and hydroxyl groups between contiguous kaolinite layers within a particle (cf. Figure 1.4) makes for strong interlayer H-bonding that needs to be broken for organic intercalation to occur.

Early attempts at intercalating organic species into kaolinite used compounds, such as acetamide, DMSO, formamide, hydrazine, and urea. Besides being capable of breaking interlayer hydrogen bonds, the intercalated molecules can themselves be H-bonded to the silicate surface (Adams 1978; Frost et al. 1999; Johnston et al. 2000). Infrared spectroscopy and molecular dynamics simulation by Zhang et al. (2017) indicate that the carbonyl group of urea is H-bonded to the aluminol surface while the amine group of urea interacts with the siloxane surface as Weiss et al. (1966) has deduced earlier from one-dimensional Fourier synthesis. Using KGa-1 kaolinite, Seifi et al. (2016) have suggested that urea could also interact through H-bonding between the NH group of the molecule and the (inner-surface) hydroxyl of the mineral. Infrared spectroscopy further indicates that the strength of bonding to inner-surface hydroxyl decreases in the order DMSO > N-methylformamide > formamide (Lipsicas et al. 1986).

X-ray diffractometry combined with NMR, Raman and infrared spectroscopies and molecular simulations indicate that DMSO intercalates through H-bonding of its oxygen atoms to inner-surface aluminols, while the sulphur atom of the molecule is directed toward the tetrahedral sheet with H-bonding between the sulphonyl oxygen and surface hydroxyl group being dominant. At the same time, one methyl group is keyed into the ditrigonal depression of the siloxane surface, while the other methyl group lies parallel to the interlayer surface (Johnston et al. 1984; Thompson and Cuff 1985; Duer et al. 1992; Hayashi 1997; Frost et al. 1998a; Fang et al. 2005; Michalková and Tunega 2007; Yariv and Lapides 2008; Zhang et al. 2015a, 2018c). The infrared spectra of kaolinite-DMSO complexes will be described later.

The intercalation process is commonly assessed and followed using X-ray diffractometry. Following Wiewióra and Brindley (1969), the intercalation ratio (degree of reaction), α, is defined as the ratio of the intensity (I) of the (001) reflection of the interlayer complex to the sum of the (001) intensity of the complex and that of the parent (unexpanded) kaolinite (Equation 6.1),

$$\alpha = I_{(001)\text{complex}} / (I_{(001)\text{complex}} + I_{(001)\text{kaolinite}}) \qquad \text{(Equation 6.1)}$$

Thus, α denotes the proportion of expanded layers formed at a given time of contact with the organic guest molecules. As Lagaly et al. (2006) have pointed out, however, α may differ from its effective value because of the intervention of Lorentzian, polarization, and structural factors. Using DMSO as intercalant, Mata-Arjona et al. (1970) derived an expression for α taking these factors into account. The value of α is also influenced by the liquid structure of DMSO (Figure 6.4B).

Although the intercalation ratio may ultimately approach unity, interlayer complex formation with kaolinites is often incomplete even after prolonged exposure to the guest compounds through solution and mechanochemical intercalation or by homogenization (Olejnik et al. 1968, 1970; Fernandez-Gonzales et al. 1976; Satokawa et al. 1997; Castrillo et al. 2015; Makó et al. 2009, 2019; Andreou et al. 2021; Neto et al. 2022). Moreover, the reactivity of kaolinite (and dickite) toward organic intercalants is influenced by mineral particle size and layer ordering as well as by the nature of the intercalant and the method of intercalation (Franco and Ruiz Cruz 2004). Prior dry grinding of kaolinite can also significantly enhance intercalate yields (MacKinnon et al. 1995). Earlier, Olejnik et al. (1968) reported that the rate of DMSO intercalation decreased in the order halloysite > kaolinite > dickite as did the number and strength of interlayer hydrogen bonds.

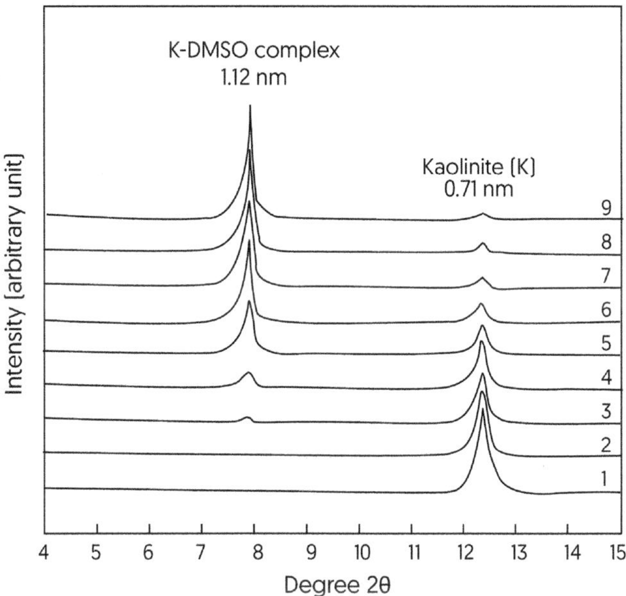

FIGURE 6.3 X-ray diffraction patterns of kaolinite and its complexes with DMSO (at 40°C) recorded at different intercalation times. (1) parent kaolinite (K), *d*(001) = 0.71 nm; (2) K-DMSO complex, 0.5 h; (3) K-DMSO complex, 5 h; (4) K-DMSO complex, 9 h; (5) K-DMSO complex, 24 h; (6) K-DMSO complex, 30 h; (7) K-DMSO complex, 48 h); (8) K-DMSO complex, 52 h; (9) K-DMSO complex, 71 h. (From Castrillo, P.D. et al., *Int. J. Miner. Proc.*, 144, 70–74, 2015.)

In the example of Figure 6.3, the basal (001) reflection for the parent kaolinite at 0.71 nm is still detectable in the X-ray diffractogram of the interlayer complex with DMSO after 71 h of exposure to the intercalant solution (Castrillo et al. 2015). For a given period of exposure, the value of α often increases with water addition, passing through a maximum with water content as Olejnik et al. (1970) have observed for complexes of (API-9) kaolinite with formamide (Figure 6.4A). Frost et al. (1999) observed close to 100% intercalation of acetamide from the melt, while Seifi et al. (2016) reported an intercalation degree of 0.69 by mixing KGa-1 kaolinite with a concentrated aqueous solution of urea.

Three classes of polar organic compounds may be distinguished based on their mode of intercalation (Weiss et al. 1963a; Olejnik et al. 1970; Lagaly et al. 2006). Class A molecules can intercalate directly from the corresponding liquid, melt, or concentrated aqueous solution. Molecules in class B enter the interlayer space by means of an 'entraining agent', while those in class C are accommodated by displacing a previously intercalated species.

Interlayer complexes with class A compounds, including DMSO, FA, NMF, and urea, are commonly prepared by homogenization in an agate mortar. The rate of intercalation may be enhanced by heating the mixture at a moderate temperature (Makó et al. 2009, 2019). Compounds in class B include urea, FA, and hydrazine. Indeed, hydrazine can apparently entrain any neutral molecule/salt that is soluble in its aqueous solution (Weiss et al. 1963a; Sidheswaran et al. 1990). The desired complex is obtained by selectively removing the entraining hydrazine by evaporation over concentrated sulphuric acid, exposure to air, or heat treatment. Table 6.1 lists some B class compounds that can intercalate through entrainment in 24% aqueous hydrazine.

We would add that heating (60°C, 8 h) a wet kaolinite-hydrazine complex yields *hydrated kaolinite* with a *d*(001) spacing of 0.84 nm (Zheng et al. 2014; Wang et al. 2017c). Earlier, Costanzo and coworkers (Costanzo et al. 1980, 1984a, 1984b; Coyne et al. 1989) described the synthesis of hydrated kaolinites with basal spacings of 0.84, 0.86, and 1.0 nm by treating the kaolinite-DMSO complex

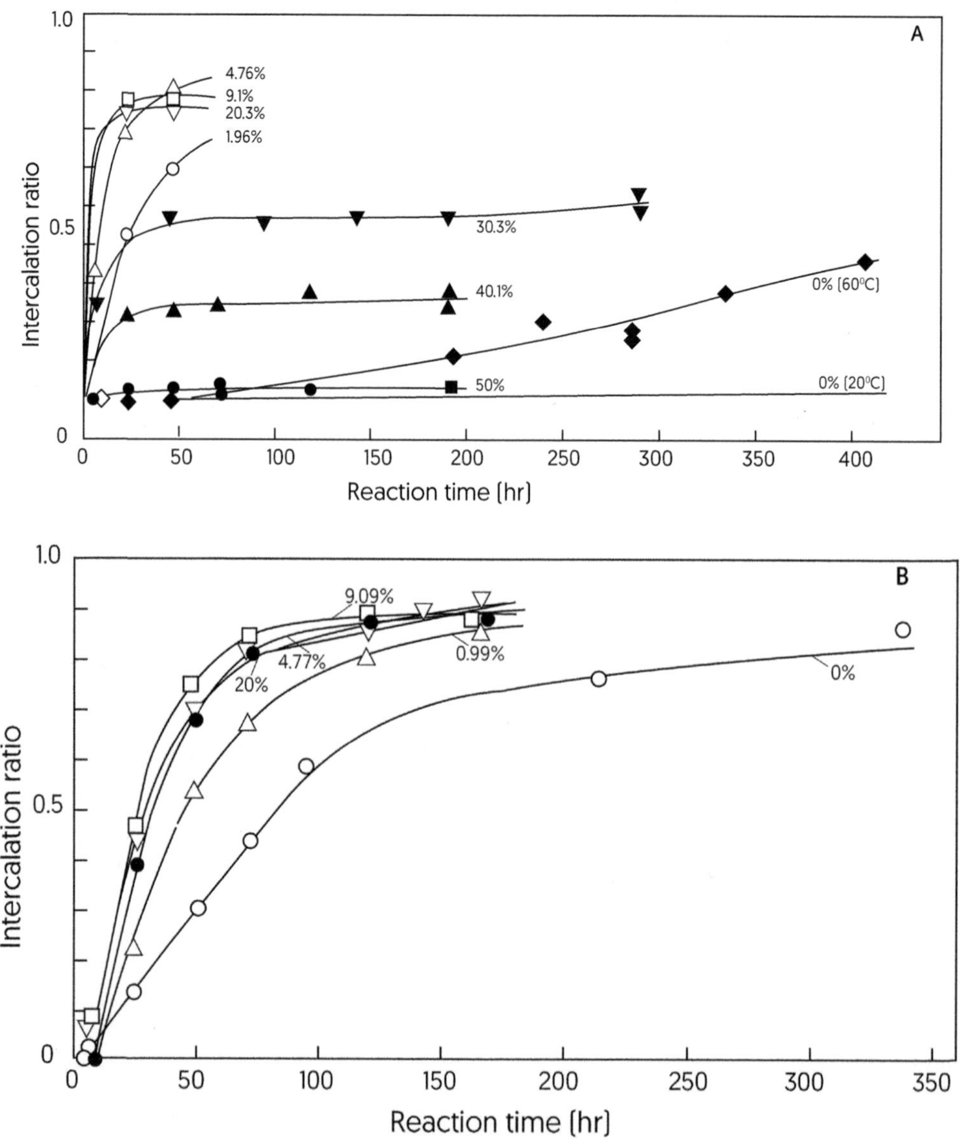

FIGURE 6.4 (A) Relationship between intercalation ratio and reaction time for the interlayer uptake (at 20°C) by kaolinite of formamide from the liquid (0% water) and from a range of aqueous formamide solutions. The percentage water content of each solution is indicated. Uptake by kaolinite of liquid formamide at 60°C is also shown; (B) as for (A) but with respect to dimethyl sulphoxide. (From Olejnik, S. et al., *Clay Miner.*, 8, 421–434, 1970; *J. Phys. Chem.*, 72, 241–249, 1968.)

with ammonium fluoride. The interlayer water in the 0.84 and 0.86 nm hydrates is keyed into the ditrigonal hole (depression) of the silicate layer, while in the 1.0 nm hydrate, as well as in halloysite-(10Å), the (associated) water is more mobile (Costanzo and Giese 1985; Lipsicas et al. 1985). Many organic molecules, including DMSO, formamide, hydrazine hydrate, 1,1-dimethylhydrazine, ethylene glycol, glycerol, and pyridine (Costanzo and Giese 1990), as well as glycine (Zheng et al. 2014) and serine (Wang et al. 2017c), can form interlayer complexes with the 0.84 nm hydrate of kaolinite. Similarly, pyridine and its mono-substituted derivatives can intercalate into a 1.0 nm

TABLE 6.1
Basal Spacings of Interlayer Kaolinite Complexes with Some Compounds Formed by Means of Entrainment in 24% Aqueous Hydrazine Solution.

	Basal spacing (nm)	
Compound	**In presence of entraining agent**	**After removal of entraining agent**
Potassium acetate	1.04	1.40
Sodium acetate	1.04	1.0
Potassium oxalate	1.04	1.02
Potassium salt of glycine	1.07	1.24 0.93[b]
Potassium salt of alanine	1.07	1.24
Potassium salt of lysine	1.06	1.48
Potassium lactate	1.06	1.11
Glycerol	1.04	1.05
n-Octylamine	3.17	3.17
Benzidine[a]	1.06	2.07[c]

Source: Weiss, A. et al. *Proc. Int. Clay Conf., Stockholm* 1, 287–305, 1963a

[a] solution of hydrazine in water-ethanol

[b] after prolonged heating at 110°C

[c] after treatment with the melt

hydrated kaolinite (Sugahara et al. 1989) as do several amino acids such as glycine, β-alanine, γ-aminobutyric acid, and δ-aminovaleric acid (Sato 1999).

Compounds in the C-class, including acetonitrile, alkylamines, amino-ethanol, ethylene glycol, nicotine, and nitrobenzene can form interlayer complexes with kaolinite that has previously been expanded by intercalation of ammonium acetate (Weiss et al. 1963a, 1966), *N*-methylformamide (Kelleher and O'Dwyer 2002; Kelleher et al. 2002; Caglar 2012), or dimethylsulphoxide (Camazano and Garcia 1966; Tunney and Detellier 1994; Gardolinski et al. 1999, 2000; Brandt et al. 2003; Tanaka et al. 2009), all of which are displaced during the process. Likewise, some alkylammonium salts can form interlayer complexes with kaolinite by displacing previously intercalated urea or dimethylformamide (Makó et al. 2017; Meziane et al. 2017). Even bulky organophosphonium salts can intercalate by displacing tetrabutylammonium bromide from its complex with kaolinite (Machida et al. 2021).

We might also mention that alkylamines, water, and ε-caprolactam can intercalate into kaolinite whose layers have previously been expanded by methanol (Gardolinski and Lagaly 2005b). In turn, the kaolinite-methanol complex may be obtained by displacing NMF, urea, or DMSO from their corresponding interlayer complexes (Figure 6.2) (Komori et al. 1999a, 1999b; Cheng et al. 2015b, 2018a). The kaolinite-methanol complex has served as a precursor in the grafting of such organic molecules as γ-aminopropyl triethoxysilane (Zhang et al. 2015b), ionic liquids (Dedzo and Detellier 2018), and pyridine-2-methanol (Caglar et al. 2013). Interestingly, kaolinite that has been modified with 1-hexyl, 3-decahexylimidazolium ionic liquid is an efficient sorbent of phenanthrene and acid red dye (Lawal and Moodley 2016). The direct grafting of (3-chloropropyl)-trimethoxy silane to HNTs, using toluene as a medium, has been reported by El-Soad et al. (2021).

Using a series of amides (FA, NMF, dimethylformamide, acetamide, *N*-methylacetamide, dimethylacetamide), pyridine-*N*-oxide (PNO), and DMSO, Olejnik et al. (1968, 1970) have been able to correlate the intercalation process with molecular properties which, in turn, are influenced by the temperature and the water content of the system. Complexes with compounds that intercalate directly from their corresponding liquid (or melt) are indefinitely stable as long as contact with the pure liquid is maintained. For complexes with class C compounds (e.g., dimethylacetamide), the intercalation ratio (Equation 6.1) is independent of the displaced organic species. They have suggested that when equilibrium is slowly attained the limitation is probably kinetic rather than thermodynamic. The latter kind of limitation is shown by water since it can remove the interlayer guest molecule without itself becoming intercalated.

The possession by the organic molecule of a large dipole moment and a small size favours its intercalation. High polarity, however, is usually associated with a large molecular size and hence these properties tend to act in opposition. The compounds used are all highly polar, having dipole moments in the range of 3.71 to 5.37 D. The dipole moments of the amides do not vary greatly with increasing methyl substitution. Because of the electron-donating inductive effect of the methyl groups, however, the basicity of the carbonyl oxygen increases with substitution, causing marked variations in the ease of intercalation. At the same time, the stability of the complexes, once formed, is enhanced by methyl substitution on the carbon atom.

As Olejnik et al. (1970) have pointed out, a large dipole moment is conducive to intercalation but would also cause extensive molecular association in the liquid state and hence depress the rate of intercalation. Thus, increasing the solute concentration would initially raise the intercalation ratio until a point is reached beyond which this ratio decreases. This finding may be ascribed to the self-association of guest molecules at high concentrations and the concomitant depletion in the population of 'free' molecules that can intercalate. Addition of water at this stage would disrupt solute molecule association and hence increase the reaction rate as Deng et al. (2003) has proposed for the kaolinite-hydrazine interaction. Further addition of water beyond a certain level, however, will cause the rate to decline for reasons already mentioned. These structural changes would account for the observation that the rate of intercalation (FA and DMSO) passes through a maximum with water content. Combined theoretical and experimental studies by Wang et al. (2021) also indicate that water plays a critical role in promoting the intercalation of DMSO into kaolinite. Raising the reaction temperature would likewise enhance the rate of intercalation by disrupting the liquid structure. These points are illustrated in Figure 6.4A.

The size and shape of molecules also influence their rate of intercalation. All things being equal, small and compact molecules can intercalate faster than their bulky counterparts. The effect of molecular shape may be illustrated by comparing the intercalation rate for dimethylformamide (DMF) with that for its isomer, *N*-methylacetamide (NMA). Having two methyl groups attached to a single atom, DMF intercalates more slowly than NMA.

Comparing the measured Δ-values (interlayer distances) with the minimum thicknesses of the intercalated compounds (Table 6.2) indicates that only a single layer of guest molecules is taken up with appreciable keying (cf. Chapter 2). In the case of PNO, the extent of contact distance shortening (0.23 nm) can be accounted for by keying of the hydrogen atoms into the siloxane surface and of the oxygen atom into the aluminol surface with the ring of the molecule being tilted at an angle of ~ 54° to the hydroxyl surface.

Similarly, the extent to which the various amides is keyed is equal to, or larger than, the value obtained with the corresponding montmorillonite complexes (cf. Chapter 2). Here again, the discrepancy between observed and calculated values is less if the plane of the molecule is allowed to tilt with respect to the silicate surface. Both Fourier analyses (Weiss et al. 1963a, 1966) and infrared spectroscopy (Ledoux and White 1964, 1966a) indicate that the tilted arrangement is more the rule than the exception. The measured Δ-values for primary, secondary, and tertiary formamides and acetamides further suggest that methyl substitution at the nitrogen atom influences the extent to which the NH group can key into the siloxane surface.

In examining the kaolin-organic interaction, classical X-ray diffractometry has been supplemented, and even supplanted, by such instrumental techniques as solid-state NMR, Raman and infrared spectroscopies, and thermal analyses (Sanz and Serratosa 2002; Langier-Kuźniarowa

TABLE 6.2
Comparing Measured Interlayer Distance (Δ-value) of Some Kaolinite-Organic Complexes with Minimum Thickness of the Corresponding Organic Molecules.

Molecules	Δ-value[a] (nm)	Molecular thickness[b] (nm)	Contact distance shortening (nm)
Pyridine-*N*-oxide (PNO)	0.53	0.76	0.23
Formamide (FA)	0.29	0.47	0.18
N-methylformamide (NMF)	0.35	0.51	0.16
Dimethylformamide (DMF)	0.49	0.59	0.10
Acetamide	0.37	0.53	0.16
N-methylacetamide (NMA)	0.41	0.53	0.12
Dimethylacetamide (DMA)	0.51	0.61	0.10

Source: Olejnik, S. et al., *Clay Miner.*, 8, 421–434, 1970

[a] Taking the individual kaolinite layer thickness as 0.72 nm

[b] For PNO the value refers to an orientation with the principal axis perpendicular to the interlayer surface. For the amides, the plane of the molecules was assumed to be perpendicular to the basal plane of kaolinite. The following van der Waals radii were used: 0.17 nm for C; 0.10 nm for aromatic ring H; and 0.35 nm for CH_3 group.

2002). Because of the general affordability and accessibility of spectrophotometers, infrared (IR) spectroscopy has been widely used to gain insight into the mechanism underlying complex formation.

The IR spectrum of kaolinite (Figure 6.5) typically shows four bands in the OH stretching region (between 3710 and 3600 cm^{-1}). The absorption bands near 3695, 3670, and 3650 cm^{-1} arise from vibrations of inner-surface (type C) hydroxyls (cf. Figure 1.4), while the band near 3620 cm^{-1} is assigned to inner (type D) hydroxyl groups (Serratosa et al. 1963; Ledoux and White 1964, 1965; Balan et al. 2014). The 3670 and 3650 cm^{-1} hydroxyls participate in interlayer hydrogen bonding, while the 3695 cm^{-1} hydroxyls are not involved.

Ledoux and White (1964, 1966a) observed that intercalation of hydrazine reduced the intensity of the 3695, 3670, and 3650 cm^{-1} bands, while additional bands of lesser intensity appeared at 3570, 3470, 3365, 3310, 3200, and 2970 cm^{-1} (Figure 6.5). With minor variations, this observation is of general validity to the intercalation of polar compounds into kaolinite. The (new) band at 3365 cm^{-1} has been attributed to the NH_2 stretching mode of intercalated hydrazine. In liquid hydrazine this mode absorbs at 3338 cm^{-1}. The broad band at 2970 cm^{-1} in the spectrum of the complex is probably due to the formation of an $O–H\cdots NH_2$ type bond.

In the case of urea, for example, intercalation principally affects the 3695 cm^{-1} band of kaolinite, causing it to weaken considerably, while additional bands appear at 3520, 3500, 3415, and 3380 cm^{-1} ascribable to NH_2 stretching vibrations. The bands at 3520 and 3415 cm^{-1} are assigned to the asymmetric and symmetric stretching modes of NH_2 groups weakly H-bonded to the inner-surface hydroxyls of kaolinite. The bands at 3500 and 3380 cm^{-1} may be attributed to NH_2 groups interacting with oxygens of the siloxane surface.

Following Cruz et al. (1969), Olejnik et al. (1971a) used IR spectroscopy to analyze the interaction of kaolinite with formamide (FA), *N*-methylformamide (NMF), and dimethylformamide (DMF). For all three compounds, interlayer complex formation causes a marked decrease in the intensity of the 3690 cm^{-1} band of kaolinite. Intercalation also led to a reduction in the intensity of the 3664 and 3648 cm^{-1} bands, the extent of which varied from one complex to another. On the basis of one-dimensional Fourier synthesis, Weiss et al. (1966) have suggested that the C=O group of intercalated NMF is slightly inclined to the aluminol surface of kaolinite to form $C{=}O\cdots H–O$ bonds.

The formation of N–H···O–Si hydrogen bonding is also likely. The results of molecular dynamics modelling by Zhang et al. (2018b) are in full accord with these suggestions.

Formamide is apparently H-bonded to the 3690 and 3664 cm^{-1} hydroxyls of kaolinite while NMF interacts with the 3690 and 3648 cm^{-1} hydroxyls. Density functional theory also indicates that the carbonyl group of FA forms a hydrogen bond with the inner-surface hydroxyl group of kaolinite, while the NH_2 group is H-bonded to the oxygen of the siloxane surface (Dawley et al. 2012; Song et al. 2013). Similarly, diffuse reflectance Fourier transform infrared (DRIFT) spectroscopy indicates the formation of H-bonding between the NH_2 group of acetamide and the siloxane oxygen of kaolinite (Frost et al. 1999).

Olejnik et al. (1971a) pointed out that their results differed from those of Cruz et al. (1969) with respect to the changes in intensity of the 3695, 3670, and 3650 cm^{-1} hydroxyl stretching bands following intercalation of FA, NMF, and DMF. Based on the magnitude of the frequency shift of perturbed hydroxyls, Cruz et al. (1969) suggested that the stability of the complexes decreased in the order DMF > NMF > formamide as did the order of electron-donating ability. On the other hand, Olejnik et al. (1971a) deduced a reversed order of stability from the changes in the IR spectra following exposure of the complexes to air or heat. Cruz et al. (1969) further suggested that DMF was H-bonded to the inner-surface hydroxyls of kaolinite through the C=O group as well as the N atom. Having two methyl groups attached to nitrogen, however, H-bonding involving the N atom would seem unlikely.

Olejnik et al. (1971b) extended their investigation to acetamide, *N*-methylacetamide (NMA), and dimethylacetamide (DMA), all of which interacted with the 3690 and 3664 cm^{-1} hydroxyls of kaolinite but not with the 3648 cm^{-1} band. These three bands correspond to the 3695, 3670, and 3650 cm^{-1} bands of Figure 6.5. That the 3690 cm^{-1} band invariably weakens on intercalation is perhaps not surprising since the 3664 and 3648 cm^{-1} hydroxyls are involved in interlayer hydrogen bonding in a kaolinite particle, while the 3690 cm^{-1} hydroxyl is relatively free. We might also mention that among the amide complexes investigated, those with formamide, acetamide, NMA, and DMA are the most stable. These compounds also react with the 3690 and 3664 cm^{-1} hydroxyls. By comparison, the NMF complex in which the guest molecule interacts with the 3690 and 3648 cm^{-1} hydroxyls of kaolinite, is less stable.

The interaction of kaolinite with DMSO has received a great deal of attention because the resultant complex can serve as a starting material in the synthesis of a variety of interlayer complexes (Letaief and Detellier 2005, 2007, 2008; Elbokl and Detellier 2008; Letaief et al. 2008; de Faria et al. 2009; de Araujo et al. 2017; Yilmaz et al. 2019). As remarked on in Chapter 4, the silylation of the inner-surface hydroxyls of kaolinite (cf. Figure 1.4) almost invariably requires prior interlayer expansion of the mineral through intercalation of small polar molecules, notably DMSO (Avila et al. 2010; He et al. 2013) The formation and applications of organically grafted kaolinite have been summarized by Dedzo and Detellier (2016).

Among the wide range and variety of organic molecules that have been grafted, using the DMSO complex as an intermediary, are alcohols (Tunney and Detellier 1993; Gardolinski and Lagaly 2005a), alkylammonium salts (Letaief and Detellier 2009), 3-aminopropyl dimethylethoxysilane (Zaharia et al. 2015), γ-aminopropyl triethoxysilane (Yang et al. 2012; Makó et al. 2023), di- and tri-ethanolamine (Letaief and Detellier 2011), Eu^{3+}- and Tb^{3+}-dipicolinates (de Araujo et al. 2017), glycol mono-ethers (Gardolinski and Lagaly 2005a), imidazolium cations (Tonle et al. 2009), polyols (Brandt et al. 2003), propane sulfone (Siegnin et al. 2022), pyridinium ionic liquids (Dedzo and Detellier 2014; Dedzo et al. 2017), and tris(hydroxymethyl) aminomethane (de Faria et al. 2010).

Olejnik et al. (1968) made the early IR spectroscopy analysis of the kaolinite-DMSO interaction, the results of which are shown in Figure 6.6. Intercalation of DMSO causes an appreciable reduction in the intensity of the 3690 cm^{-1} hydroxyl band of kaolinite together with the appearance of three additional bands at 3658, 3535 and 3499 cm^{-1}. Closely similar spectral changes have subsequently been reported by Matusik et al. (2009), Tonle et al. (2009), and Zhang et al. (2015a). By comparison,

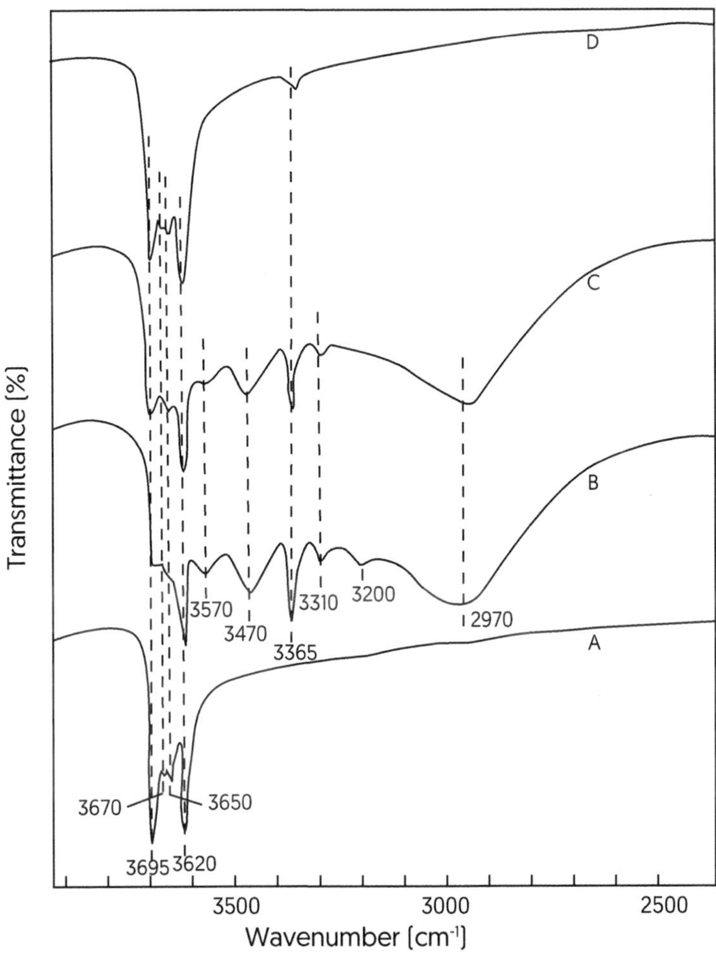

FIGURE 6.5 Infrared spectra of (A) kaolinite at 25°C; (B) kaolinite-hydrazine complex at 25°C; (C) kaolinite-hydrazine complex heated at 110°C for 2 min.; and (D) as for (C) but heated for 15 min. (From Ledoux, R.L. and White, J.L., *J. Colloid Interface Sci.*, 21, 127152, 1966a.)

intercalation of hydrazine (Figure 6.5), FA, and urea affects the 3,695, 3,670 and 3,650 cm^{-1} hydroxyls, while complex formation with amides decreases the intensity of both the 3690 cm^{-1} and either the 3670 or 3650 cm^{-1} bands of kaolinite.

As the intercalation ratio (Equation 6.1) for DMSO increases, the 3658, 3535 and 3499 cm^{-1} bands become more intense, while the 3690 cm^{-1} band further weakens. Removal of interlayer DMSO by heating the complex reduces the basal/d(001) spacing from 1.12 to 0.71 nm (cf. Figure 6.2), and restores the original hydroxyl peaks of the parent kaolinite. The measured Δ-value (~ 0.4 nm) for the interlayer complex is appreciably less than the dimension of the intercalated molecule, indicative of keying into the ditrigonal cavity of the siloxane surface (Johnston et al. 1984).

The IR spectrum of the kaolinite-DMSO complex (Figure 6.6C) further indicates that the intercalated molecule interacts with the inner-surface hydroxyls of kaolinite through either the sulphur or the oxygen atoms. The latter alternative seems more likely if the band near 1030 cm^{-1} is assigned to SO stretching, while that near 720 cm^{-1} is due to asymmetric CS stretching. The corresponding peaks in the spectrum of liquid DMSO occur at 1057 and 699 cm^{-1}, respectively (Figure 6.6A). Furthermore, the symmetric CS vibration is absent from the spectrum of the complex.

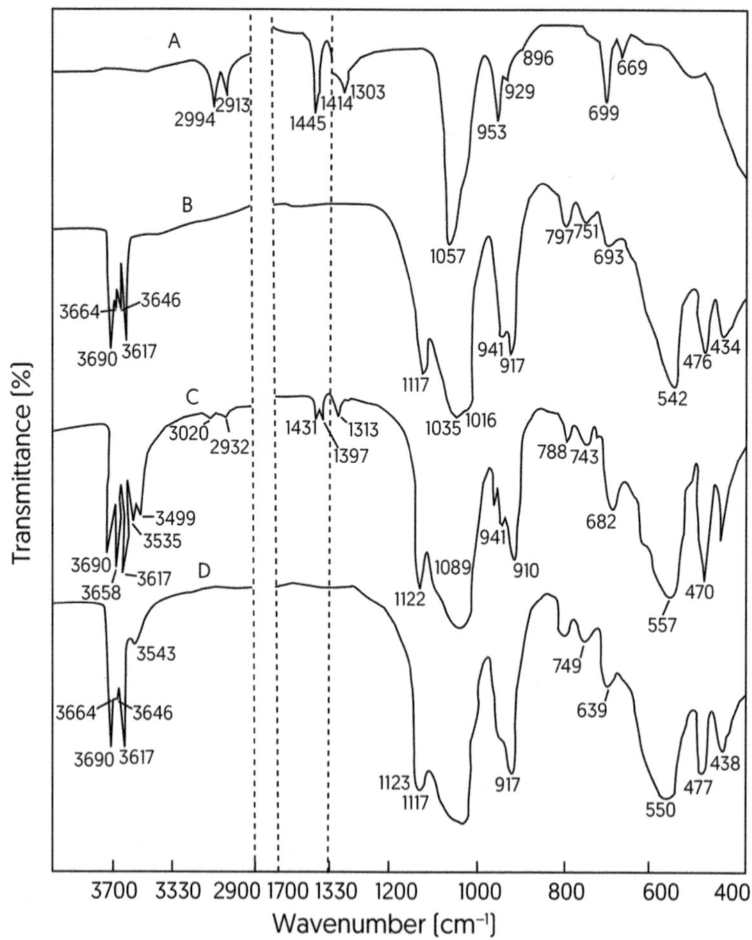

FIGURE 6.6 Infrared spectra of (A) liquid dimethyl sulphoxide (DMSO); (B) kaolinite (API-9), (C) kaolinite-DMSO complex; and (D) kaolinite-DMSO complex heated at 100°C for 15 min. (From Olejnik, S. et al., *J. Phys. Chem.*, 72, 241–249, 1968.)

Differences in interlayer H-bonding and crystallinity between the kaolinite group of minerals are also reflected in intercalation rates. In case of DMSO, these rates increased in the order of dickite < kaolinite < halloysite (Olejnik et al. 1968). The infrared spectra in the hydroxyl stretching region of these minerals would indicate that the strength and number of interlayer hydrogen bonds are greatest for dickite and least for halloysite. The intensity of the band near 3650 cm^{-1}, for example, markedly increased in the order halloysite < kaolinite < dickite. Indeed, the dickite sample failed to intercalate DMSO unless previously ground. Interestingly, Jacobs and Sterckx (1970) observed that following the removal by evacuation of DMSO from its complex with kaolinite, the IR spectrum of the hydroxyl stretching region resembled that of dickite.

Intercalation of aromatics, such as pyridine-*N*-oxide (PNO), increases the basal spacing of kaolinite from 0.71 to 1.253 nm (Weiss et al. 1966). PNO interacts through the NO group with the inner-surface hydroxyls of kaolinite (Olejnik et al. 1971c). As a result, the intensity of the 3690 and 3664 cm^{-1} bands decreases, and a single hydrogen-bonded hydroxyl band appears near 3522 cm^{-1}. The orientation of intercalated PNO, deduced from the intensity changes of certain IR bands on rotating the film of the interlayer complex, is one in which the NC_4 axis is perpendicular to the plane of the kaolinite surface as Serratosa (1966) has proposed earlier for the vermiculite-pyridinium ion complex (cf. Figure 2.2b). In this orientation, the N–O bond is pointing to the hydroxyl surface, forming N–O···H–O hydrogen bonds.

The difference in interlayer configuration between PNO and DMSO may be explained in terms of molecular shape in that PNO is planar, while DMSO has a trigonal pyramidal shape with the sulphur atom at the apex (Thomas et al. 1966). Solid-state NMR spectroscopy would indicate that the two methyl groups of DMSO at the base of the pyramid are non-equivalent. Below 27°C one methyl is keyed into the ditrigonal cavity while the other is parallel to the aluminol surface (Hayashi 1997; Fafard et al. 2017).

6.3.2 Surface Complex Formation

Many organic compounds fail to intercalate either directly or indirectly into kaolinite but can nevertheless adsorb to external particle surfaces of the mineral. In this sense, the kaolinite-organic interaction has attracted much interest because of its potential in agricultural and environmental applications. Furthermore, the process is cost-effective since kaolinite is an inexpensive raw material (Doğan et al. 2009; Mouni et al. 2018; Yang et al. 2022). We might also add that kaolinite makes up a major part of the clay fraction in variable-charge soils of the tropics (Theng 1980; Hughes et al. 2009). These soils are also rich in nano-size metal oxides (Theng and Yuan 2008) which, however, are generally less reactive toward organic molecules than is kaolinite (Harris et al. 2001). On the other hand, the adsorption capacity with respect to organophosphates increases in the order kaolinite < goethite << gibbsite (Amadou et al. 2022), presumably because phosphates are known to interact through ligand exchange with hydroxyl groups exposed at particle edge surfaces (cf. Equation 1.1).

In comparison with montmorillonite (cf. Chapter 3) the capacity of kaolinite in taking up organic compounds is appreciably smaller since its interlayer space is inaccessible to extraneous guest molecules. Thus, the maximum uptake of antibiotics by kaolinite is much less than that measured for montmorillonite under comparable experimental conditions (Figueroa et al. 2004; Essington et al. 2010; Li et al. 2010; Zhao et al. 2011, 2012, 2015). Baker and Luh (1971) reported similarly for pyridine as did Angioi et al. (2005) for chloroanilines, Polati et al. (2006) for some pesticides, Behera et al. (2010) for triclosan, Wu et al. (2013a) for nalidixic acid, Zhang et al. (2013b) for tylosin, Bhattacharyya et al. (2014) for Rhodamine B dye, Liu et al. (2019a) for microcystin-LR, and Septian et al. (2019) for ciprofloxacin and sulfadiazine. By the same token, antibiotics desorb more readily from kaolinite than montmorillonite (Wu et al. 2013b).

By contrast, Peng et al. (2009) reported that kaolinite was more effective than montmorillonite in sorbing endrin. In this instance, however, adsorption was confined to external particle and pore surfaces of both minerals. Having a larger specific surface area and pore volume, the kaolinite sample was superior in taking this uncharged organochlorine pesticide. By the same token, by increasing the surface area and pore volume of the adsorbent, prior treatment of kaolinite with strong acids (sulphuric, hydrochloric) can markedly enhance its capacity for taking up organic dyes including reactive blue 221 (Karaoğlu et al. 2010), methylene blue (Gao et al. 2016), and direct orange 34 dyes (Chargui et al. 2018), as well as 3-methoxybenzaldehyde (Koyuncu et al. 2007) and aniline (Koyuncu and Kul 2019).

All the same, the mechanism underlying the kaolinite-organic interaction closely resembles that shown by montmorillonite as described in Chapter 3. Like tetracycline, oxytetracycline is taken up by cation exchange at acid pH, and by cation complexation under near neutral pH conditions, involving the amide, carbonyl and dimethyl amino groups of the molecule (Dolui et al. 2018). Also as in montmorillonite, the adsorption to kaolinite of antibiotics (Essington et al. 2010; Zhao et al. 2015) and some cationic dyes (Karaoğlu et al. 2009; Sargin and Ünlü 2013) decreases with an increase in ionic strength of the ambient solution. Likewise, the uptake of chloroanilines by both kaolinite and montmorillonite increases with the hydrophobicity of the adsorbates (Angioi et al. 2005). We might also mention that kaolinite can apparently act as a concentrating surface (and catalyst) in the formation of adenosine from adenine and ribose (Hashizume et al. 2019).

Positively charged organic compounds are evidently adsorbed (on basal siloxane surfaces) by electrostatic attraction (cation exchange), supplemented by hydrophobic and H-bonding interactions (Harris et al. 2006; Chen et al. 2017, 2020). Thus, the uptake of ciprofloxacin and amitriptyline by kaolinite (under ambient pH conditions) is accompanied by an equivalent desorption of exchangeable cations (Li et al. 2011; Lv et al. 2013). Similarly, cation exchange is the principal mechanism underlying the interaction of kaolinite with imidazolium-type ionic liquids (Mrozik et al. 2008).

In the case of long-chain quaternary ammonium cations, both cation exchange and hydrophobic interactions are involved in controlling adsorption, while the contribution of the latter mechanism increases with chain length (Li and Gallus 2005, 2007). Malek and Ramli (2015) observed that kaolinite loaded with cetylpyridinium bromide above its critical micelle concentration was an effective antibacterial agent. Similarly, Spasojević et al. (2021) found that the capacity of kaolinite-octadecyldimethylbenzyl ammonium complexes for adsorbing ochratoxin A and zearalenone increased with surfactant surface coverage. Molecular dynamic simulations by Grančič and Tunega (2021) further suggest that phosphatidylcholine interacts with kaolinite through either its zwitterionic heads or its hydrophobic aliphatic tails.

As in smectite, the uptake of dyes by kaolinite (and halloysite) is influenced by solute concentration, pH, and temperature. The process commonly follows pseudo second-order kinetics and conforms to the Langmuir or Freundlich isotherm (Nandi et al. 2009a, 2009b; Vimonses et al. 2009; Auta and Hameed 2012; Khan et al. 2012; Rida et al. 2013; Sargin and Ünlü 2013; Anastopoulos et al. 2018; Boukhemkhem et al. 2022). Interestingly, Koyuncu and Kul (2019) and Song et al. (2019) have observed similarly for the adsorption to kaolinite of aniline and oxytetracycline, respectively.

On the other hand, uncharged organic molecules primarily adsorb by H-bonding to basal particle surfaces (Wang et al. 2013a; Guo et al. 2021). Density functional theory (DFT) further indicates that the energy for the adsorption of β-D-glucose, cellobiose, and 2,4-dinitrotoluene to the aluminol surface is appreciably larger than to the siloxane surface (Lee et al. 2013; Wang et al. 2013a). Using atomistic and quantum chemical calculations, based on DFT, Song et al. (2013) and Awad et al. (2019) have made the same suggestion with respect to the adsorption of formamide and 5-aminosalicylic acid, respectively. The theoretical study by Scott et al. (2014) and Hounfodji et al. (2021) has indicated similarly for the adsorption of some nitroaromatic and pharmaceutical compounds. Experimental investigation and DFT calculation by Liu et al. (2019b) indicate that dodecylamine and oleic acid are more strongly adsorbed to the aluminol than the siloxane surface. More recently, Kasprzhitskii et al. (2022) used DFT to deduce that the adsorption energy of aliphatic amino acids with respect to the aluminol surface is nearly three times larger than that to the siloxane surface. The siloxane surface is also involved in the H-bonding and molecular interactions of kaolinite with tributyltin (Hoch and Bandara 2005) and naproxen (Yu and Bi 2015).

The edge surface of kaolinite can play an important role in the sorption of organic compounds (Conley and Lloyd 1971), especially of negatively charged (anionic) species (Essington and Anderson 2008). Besides making up a significant proportion (up to ⅓) of the total particle surface area, the edge surface is amphoteric having a point of zero charge (PZC) of 6–7.5. Thus, the anionic surfactant sodium bis(2-ethylhexyl) sulfosuccinate interacts with the positively charged edge surface of kaolinite at acid pH, while adsorption is not detectable under alkaline conditions (Suzzoni et al. 2018). Anion exchange has also been invoked by Wu et al. (2013a) for the uptake of nalidixic acid when the solution pH is larger than the pK_a of the antibiotic. On the other hand, molecular dynamics modelling by Ke et al. (2023) suggests that electrostatic attraction drives the interaction of terminal sulfonate group of *per-* and poly-fluoroalkyl substances with the aluminol surface of kaolinite. A similar mechanism has been suggested by Zhang et al. (2014) for the interaction of kaolinite with perfluorooctane sulfonate (PFOS). We would suggest, however, that PFOS is preferentially adsorbed to particle edges since more was taken up by kaolinite than montmorillonite.

Experimental measurements and molecular dynamic simulation by Lv et al. (2014) also indicate that chlorpheniramine dominantly interacts with the edge surface of kaolinite. Ikhsan et al. (2004) have suggested similarly for aspartic acid as did Castellini et al. (2008) for malachite green.

Molecular dynamic simulation by Zeitler et al. (2017) suggests that the decahydro-2-naphtoic acid anion interacts with the (negatively charged) edge surface of kaolinite probably through a cation-bridging mechanism. DRIFT spectroscopy indicates that salicylic acid adsorbs to kaolinite (from a hexane solution) as the carboxylate (Thomas and Kelley 2009). Kim and Hyun (2014) have also raised the possibility of coordination between 2,4-dichlorophenoxyacetate and K^+-, Ca^{2+}, or Al^{3+} counterions in kaolinite. Like 2-ketogluconate (Essington and Anderson 2008), we would suggest that sodium tetraborate adsorbs by ligand exchange to the edge surface of kaolinite. Since the surface charge becomes more negative as a result (cf. Figure 5.2B), the capacity of the tetraborate-modified kaolinite for adsorbing aniline blue is much larger than that shown by the raw material (Unuabonah et al. 2008).

Here we would mention that the kaolinite complex with Ag-chlorhexidine showed a high antibacterial activity (Jou and Malek 2016). Hui et al. (2019) reported similarly for a kaolinite-hexadecyltrimethylammonium (HDTMA)-gentamycin complex. HDTMA-modified kaolinite is also a good adsorbent of *o*-xylene and phenol (Ahmed 2009) as well as chlorobenzenes (Shu et al. 2010). Lee and Kim (2002) have proposed that naphthalene is partitioned into the tails of HDTMA cations, adsorbed to kaolinite and halloysite, giving rise to linear isotherms. Earlier, Seto et al. (1978) reported the formation of complexes with ethylene glycol, 1,6-diaminohexane, and a quaternary amine using a kaolinite-ammonium propionate intercalate as an intermediary.

6.4 HALLOYSITE

6.4.1 Interlayer Complex Formation

MacEwan (1948, 1949) did much of the early research into the halloysite-organic interaction. Apart from a brief note by Caillère et al. (1950) describing the expansion of halloysite by propanol-water (basal spacing = 1.4 nm), and saturated barium hydroxide-ethanol (basal spacing = 1.7 nm), MacEwan's data stood alone for a long time. Whereas montmorillonite can intercalate two or more layers of organic molecules, halloysite seldom takes up more than a single layer (Mehdi et al. 2019). In explanation, MacEwan (1962) suggested that the halloysite layer was amphoteric in that one side consisted of a network of oxygen ions carrying negative charges, while the other side could be likened to a positively charged α-hydroxide sheet.

Like kaolinite, dehydrated halloysite can intercalate a variety of highly polar organic molecules and commonly at a faster rate and more completely than does kaolinite (Theng et al. 1984). On this basis, Churchman et al. (1984) have developed a simple formamide intercalation test for differentiating between halloysite and kaolinite in mixtures. Furthermore, the halloysite content in the mixture may be estimated by spraying the sample with formamide, examining the treated material immediately by X-ray diffractometry, and applying a modified form of Equation 6.1, namely, $I_{10}/(I_{10} + I_7)$ where I_{10} and I_7 refer to the intensity of the basal reflections at 10 and 7Å, respectively. As Joussein et al. (2007) have pointed out, however, the formamide test greatly underestimates the concentration of halloysite in samples that have been dehydrated during storage.

Early on, Weiss and Russow (1963) reported that formamide, hydrazine, and urea were intercalated as a monolayer into halloysite. Subsequently, Camazano and Garcia (1966) and Olejnik et al. (1968) found that dimethyl sulphoxide (DMSO) formed a monolayer complex with halloysite, giving a basal spacing of ~ 1.12 nm, while Churchman and Theng (1984) reported spacings between 1.04 and 1.24 nm for a series of amides. Comparable or slightly larger spacings were reported by Garrett and Walker (1959), Wada (1959a, 1959b, 1961), Frost et al. (2000b), Mellouk et al. (2009), and Cheng et al. (2010b) for complexes with ammonium, potassium, and sodium salts. For example, Frost et al. (2000c) reported a basal spacing of 1.397 nm for the interlayer complex of halloysite with potassium acetate. Raman spectroscopy further indicated that intercalated urea was H-bonded to the siloxane surface, while potassium acetate interacted with both the inner-surface hydroxyls and the siloxane oxygens of the adjacent layer (Frost and Kristof 1997).

A substantial amount of valuable information came to light when Carr and Chih (1971) published the results of their investigations on the formation and properties of halloysite complexes with a large range and variety of organic molecules. The usual criterion for intercalation, and one which was implicitly adopted by MacEwan (1962), was used – namely, that the distance separating successive layers within the complex (Δ-*value*) should exceed that of the original, fully hydrated mineral (halloysite-(10Å)). In other words, $\Delta_{complex} > \Delta_{H2O}$ (~ 0.29 nm).

As Churchman et al. (1972) have pointed out, however, halloysite-(10Å) and its fully dehydrated form, halloysite-(7Å), represent the end members of a series with variable hydration status and basal reflections that are reasonably sharp and symmetrical in profile. Since the partially hydrated forms are essentially randomly interstratified structures, the corresponding X-ray diffraction (XRD) peaks tend to be broad and asymmetric, extending across the region between the 1.0 and 0.72 nm peaks, showing no high-order reflections. Since the XRD patterns of halloysite-organic complexes might be expected to show a parallel behaviour, Carr and Chih (1971) have proposed that a fully complexed sample would show a sharp, symmetrical (001) line together with higher (rational) orders of basal reflections. Furthermore, conformity to the rule of $\Delta_{complex} > \Delta_{H2O}$ only indicates a tendency for complex formation, while intercalation cannot be unequivocally demonstrated when $\Delta_{complex} \leq \Delta_{H2O}$.

Table 6.3 summarizes Carr's and Chih's (1971) results for Te Puke halloysite from New Zealand with a tabular particle morphology using the same organic compounds as those chosen by MacEwan (1962). There is reasonably good agreement between the two sets of data, but Carr and Chih (1971) failed to observe complex formation with ethanol, acetone, acetaldehyde, and acetonitrile and recorded generally lower Δ-values. This discrepancy was ascribed to a difference in the method of preparing the respective complexes in that MacEwan exposed his halloysite sample (from Hungary) to the organic liquid for just 1 hour whereas Carr and Chih allowed a contact time of at least 3 days.

The reactivity of halloysite toward organic compounds is also influenced by particle shape. Thus, Kauri halloysite (New Zealand) with a smaller *b*-dimension and a tubular morphology, forms a complex more slowly and reaches equilibrium at a smaller Δ-value in comparison with Te Puke halloysite. Kinetic factors and differences in particle morphology may also lie behind the failure by Carr and Chih (1971) to intercalate DMSO (into Te Puke halloysite) while other workers (Camazano and Garcia 1966; Olejnik et al. 1968; Anton and Rouxhet 1977; Costanzo and Giese 1986; Churchman 1990) could obtain interlayer halloysite-DMSO with relative ease. More recently, Zhang et al. (2019, 2020a) were able to intercalate stearic acid into halloysite that had previously been expanded with DMSO, yielding an interlayer complex with a basal spacing of ~ 4 nm.

We might interpolate here that the surface composition and flow characteristics of halloysites as well as the slope behaviour and sensitivity of halloysite-rich soils are related to mineral particle shape (Smalley et al. 1980; Soma et al. 1992; Theng and Wells 1995; Moon 2016). Another point of interest is that the tubular form of halloysites, commonly referred to as halloysite nanotubes and abbreviated to HNTs, have occupied the limelight in terms of the halloysite-organic interaction.

In an attempt to correlate complex formation with molecular properties, Carr and Chih (1971) tested 127 organic compounds for their propensity for intercalating into (Te Puke) halloysite. Using the criterion of $\Delta_{complex} > \Delta_{H2O}$, only a limited number were intercalated (Table 6.3). On the basis that (hydrated) halloysites, obtained by displacing pre-intercalated potassium acetate with water (Wada 1961), appear to be less stable than the naturally occurring minerals, they also examined complex formation using rehydrated halloysite. Intercalation proceeds more slowly with rehydrated than with pristine halloysite-(10Å), indicating that structural factors influence the halloysite-organic interaction. Subsequent measurements by Franco and Ruiz Cruz (2004) indicate that the reactivity of kaolin minerals is influenced by many factors including mineral particle size and structural order, nature of organic species, intercalation method, and the presence of impurities.

Interestingly, no simple relationship was noted by Carr and Chih (1971) between the ease of complex formation and either the dielectric constant or dipole moment of the organic species. On the other hand, infrared spectroscopy clearly shows that intercalation is accompanied by the formation of hydrogen bonds between organic molecule and silicate surface. Thus, the hydroxyl stretching

TABLE 6.3
Comparing Observed Δ-value (Interlayer Distance) of Some Halloysite-Organic Complexes with the Calculated Minimum Thickness of the Intercalated Organic Compounds.

Compound (in alphabetical order)	X-ray reflections	Δ-value (nm)	Minimum molecular thickness (nm)
Acetamide	001, 003	0.39	0.40
Acetic acid	001, 003	0.44	0.40
Acetyl acetone	001, 003	0.40	0.40
N-acetylethanolamine	001, 003	0.40	0.40
β-Alanine (in water)	001, 003	0.45	0.40
Aniline HCl	001 (dd)	0.78	0.36
Betaine HCl (in water)	001 (dd)	0.59	0.57
2-Butyne-1,4-diol (in diethylether)	001, 003	0.35	0.40
Crotyl alcohol	001, 003	0.43	0.40
1,2-Cyclohexanedione (in acetone)	0.01, 0.02 (d), 0.03 (d)		
Diethylene triamine	001, 003	0.38	0.45
Ethanol	0.01 (dd)	0.35	0.40
Ethanolamine	001, 003	0.35	0.40
Ethylamine HCl (in water)	001, 003	0.55	0.36
Formaldehyde	001, 003	0.31	0.30
Formamide	001, 003	0.30	0.30
Glycine (in water)	001, 003	0.32	0.40
Hydrazine hydrate	001, 003	0.33	0.39
Hydroxylamine HCl	001, 003	0.30	0.36
Isopropanolamine	001, 003	0.38	0.45
Lactic acid	001, 003	0.46	0.42
Malonic acid (in water)	001, 003	0.35	0.39
Methylamine HCl (in water)	001, 003	0.45	0.36
1,2-Propanediol	001, 002, 003	0.44	0.42
Propionic acid	001, 003	0.49	0.40
Sarcosine (in water)	001 (dd)	0.48	0.40
Semicarbazide HCl (in water)	001, 003	0.32	0.36
Triethanolamine	001, 003	0.40	0.49
Urea	001, 003	0.38	0.30

Source: Carr, R.M. and Chih, H., *Clay Miner.*, 9, 153–166, 1971.
d = diffuse; dd = very diffuse

band at 3690 cm^{-1} in the infrared spectrum of Te Puke halloysite weakens following organic intercalation. In some instances, the observed Δ-values are smaller than the corresponding minimum calculated thicknesses of the intercalated species. This observation, however, does not necessarily indicate bond shortening because of uncertainties in assigning van der Waals radii and interlayer molecular orientation. Some doubt also attaches to the involvement of C–H···O type bonding in these systems.

What seems unambiguous is that changes in H-bonding take place during interlayer complex formation. With reference to Figure 6.6, the infrared spectrum of a halloysite-DMSO complex shows two broad bands near 3498 and 3530 cm^{-1} together with two shoulders at 3690 and 3648 cm^{-1}

(Olejnik et al. 1968). The first two bands increase in intensity while the 3690 cm^{-1} peak weakens as intercalation progresses suggesting that DMSO is H-bonded to the 3690 cm^{-1} hydroxyls via its oxygen atom.

The following general conclusions regarding the halloysite-organic interaction may be drawn from Carr's and Chih's (1971) data: (a) particle morphology influences the kinetics of complex formation; (b) the intercalated compounds are polar, commonly contain two functional groups, and rarely include cyclic and aromatic compounds; and (c) complex formation involves the formation of multiple hydrogen bonds between the organic molecule and the mineral surface.

Particle shape can also influence the capacity of halloysite for taking up uncharged organic molecules, such as amides (Churchman and Theng 1984) and *p*-nitrophenol (Theng 1995) as well as phosphate (Theng et al. 1982; Gray-Wannell et al. 2020). With reference to Figure 2.11, the maximum/plateau adsorption of *p*-nitrophenol (*p*NP) to some New Zealand halloysites increases in the following order: Matauri Bay (long tubules) < Te Puke (thick plates) < Te Akatea (slender tubules) < Opotiki (spheroids) < Hamilton (short tubules). Assuming that the phenyl ring of *p*NP is inclined at 30° to the mineral surface, the specific surface areas of the various halloysites, derived from the plateau adsorption of *p*NP, are in excellent agreement with the corresponding BET-N_2 areas (cf. Figure 2.12).

As remarked on earlier, much attention has been directed to the surface reactivity and diverse applications of halloysite nanotubes or HNTs. As shown in Figure 1.8, an individual nanotube comprises an external (negatively charged) siloxane surface, an internal (positively charged) aluminol surface, and an edge surface exposing aluminol and silanol groups with a pH-dependent charge (Theng et al. 1982; Churchman et al. 2016b; Lvov et al. 2016; Gray-Wannell et al. 2020). Besides being naturally occurring, biocompatible, non-toxic, and of low cost, HNTs (and their surface-modified forms) can accommodate a variety of organic and bioorganic molecules mostly within the lumen cavity (Lvov et al. 2016, 2019; Yang et al. 2017; Saif et al. 2018). Thus, (tubular) halloysite is a better adsorbent than (platy) kaolinite as Tan et al. (2015) have reported for amitrole, Deng et al. (2017) for benzene, and Spepi et al. (2016) in the case of salicylic acid. Ghezzi et al. (2018) further observed that the bound salicylic acid could be released in a controlled manner to act as an antibacterial agent.

Halloysite and HNTs are also efficient adsorbents of various dyes, including brilliant green (Ukkund et al. 2021a), crystal violet (Belkassa et al. 2013; Krasilin et al. 2019; Ngulube et al. 2019), direct blue 15 (Ukkund et al. 2021b), direct orange 26 (Kuśmierek et al. 2020), eosin Y (Lee et al. 2020), indigo (Zhuang et al. 2019), malachite green (Kiani et al. 2011; Altun and Ecevit 2022), methylene blue (Zhao and Liu 2008; Ngulube et al. 2019; Filice et al. 2021), methyl green (Vargas-Rodríguez et al. 2021), methyl orange (Filice et al. 2021), methyl violet (Liu et al. 2011; Altun and Ecevit 2022; Sadiku et al. 2022), neutral red (Luo et al. 2010), and safranin O (Shaik et al. 2023).

We might add that magnetic HNT, formed by incorporation of Fe_3O_4 nanoparticles, shows a reduced capacity for taking up methylene blue and neutral red (Xie et al. 2011). More recently, Mahrez et al. (2020) have suggested that the adsorption of crystal violet to a halloysite-DMSO intercalate is mediated by H-bonding between the sulphoxide oxygen and the methyl hydrogen of the dye. By analogy with kaolinite, however, it seems more likely that crystal violet is intercalated by displacing DMSO from its complex with halloysite. Besides being influenced by solute concentration, pH, and temperature, dye uptake is controlled by varied mechanisms, notably electrostatic, cation exchange and H-bonding interactions. In common with kaolinite, the adsorption process generally follows pseudo second-order kinetics and often conforms to the Langmuir isotherm model.

HNTs have also featured as adsorbents and carriers of antibiotics, pesticides, pharmaceuticals and their precursors, including 5-aminosalicylic acid (Borrego-Sánchez et al. 2018); atrazine (Gianni et al. 2022); chloroanilines (Szczepanik et al. 2014, 2017); chlorpheniramine, ciprofloxacin, diphenhydramine (Jiang et al. 2016; Cheng et al. 2018b); curcumin and its derivatives (Riela et al. 2014; Sabahi et al. 2018; Massaro et al. 2020b; Pereira et al. 2021); diuron (Gianni et al. 2022); gentamicin (Stodolak-Zych et al. 2023); ibuprofen (Tan et al. 2013); iodopropynyl butylcarbamate

(Jin et al. 2019; Zhang et al. 2020b); irinotecan (Gianni et al. 2019); isoniazid (Carazo et al. 2017); melanin (Cruz-Hernández et al. 2021); ofloxacin (Wang et al. 2014); prometryn (Grabka et al. 2015); and quercetin (Hári et al. 2016; Liu et al. 2021). We might add that quercetin can apparently form a coordination complex with the external aluminol group of kaolinite (Diar-Bakerly et al. 2014). Also interesting is the finding that the anti-inflammatory/antibacterial activity of Opotiki halloysite (New Zealand) may be related to the presence in the mineral of a nano-iron(hydr)oxide phase (Cervini-Silva et al. 2016).

Prior modification ('functionalization') of HNTs through either adsorption of, or covalent grafting with, selected organic species (Yuan et al. 2008; Peixoto et al. 2016; Massaro et al. 2014, 2017, 2018a) can markedly enhance the capacity of halloysite for taking up and slowing down the release of various drugs, dyes, pharmaceuticals, and proteins, including aspirin (Li et al. 2016), bovine serum albumin (Díaz et al. 2017), chlorogenic and salicylic acids (Kurczewska et al. 2018), clotrimazole (Massaro et al. 2018b), curcumin (Massaro and Riela 2018), diltiazem HCl and propranolol HCl (Levis and Deasy 2003), diphenhydramine HCl (Hemmatpour et al. 2015; Zargarian et al. 2015), ibuprofen and naproxen (Tan et al. 2014b; Li et al. 2015b; Kurczewska et al. 2018, 2020), insulin (Massaro et al. 2018c), ketoprofen (Sid et al. 2022), orange II (Yuan et al. 2012), and quercetin (Liu et al. 2021). Prior incorporation of multivalent metal cations can appreciably increase the adsorption of uncharged organic molecules, such as benzotriazole (Pham et al. 2022). Experimental and theoretical studies indicate that the organic guest molecules are adsorbed by electrostatic, H-bonding, and van der Waals interactions (Ferrante et al. 2017; Borrego-Sánchez et al. 2021; Liu et al. 2021; Sid et al. 2021). More recently, Fahimizadeh et al. (2023) reported that HNTs that have been loaded with yeast extract could sustainably release nutrients to support bacterial growth.

Likewise, the capacity of halloysite for taking up extraneous organic molecules (drugs, dyes) and removing organic pollutants from aqueous solutions may be greatly enhanced by prior treatment of the clay sample with mineral acids (Zhang et al. 2012; Belkassa et al. 2013; Szczepanik et al. 2014; Grabka et al. 2015; Li et al. 2015b; Spepi et al. 2016; Bediako et al. 2018; Kuśmierek et al. 2020; Lee et al. 2020; Pereira et al. 2021; Setter et al. 2022; Wierzbicka et al. 2022; Khelifa et al. 2023). This process, referred to as 'acid activation' involves a series of reactions leading to the (partial) dissolution of the aluminol sheet (Komadel and Madejová 2006; White et al. 2012; Theng 2018). As a result, there is a marked increase in the pore volume, specific surface area, and lumen diameter of HNTs, but the tubular particle shape is apparently preserved (Abdullayev et al. 2012; Garcia-Garcia et al. 2017; Li et al. 2015b; Yang et al. 2016; Deng et al. 2019; Jauković et al. 2021). Earlier, Kuroda and Kato (1979) were able to trimethylsilylate the silanol groups of halloysite following acid decomposition of the octahedral sheet.

Similarly, calcination (up to 1000°C) does not appear to alter the morphology of halloysite tubules, although appreciable structural dehydroxylation takes place together with the formation of short-range order silica and alumina (Deng et al. 2019; Jia and Fan 2019; Aytekin and Hoşgün 2020). Acid treatment of calcined halloysite leads to the formation of micropores into which guest organic molecules, such as benzene, can be accommodated (Deng et al. 2019). Dehydroxylation (at 600°C), and dealumination (by acid treatment) of halloysite lead to the partial transformation of silanols into siloxanes together with a substantial increase in specific surface area and adsorption capacity for Congo red, crystal violet, and malachite green (Bessaha et al. 2016, 2019; Belkassa et al. 2021). Jia and Fan (2019) have further suggested that the silanol groups, formed on the external tubule surface of calcined halloysite, can be grafted (silylated) with γ-aminopropyltriethoxysilane.

Similarly, alkali activation using solutions of NaOH can increase the surface area and porosity of halloysite through the dissolution of the inner-surface aluminol sheet and the progressive thinning of the tubules (White et al. 2012; Yang et al. 2016). As a result, alkali-treated HNTs show an enhanced capacity for adsorbing or removing organic molecules, such as ofloxacin (Wang et al. 2013b) and ethylene (Gaikwad et al. 2018). Density functional theory calculations and experimental measurements by Ferrante et al. (2023) further suggest that silanol groups are formed on external tubule surfaces when halloysite is treated with a concentrated (4 M) solution of NaOH.

Like montmorillonite (cf. Chapter 4), halloysite can intercalate quaternary ammonium cations, such as hexadecyltrimethylammonium (HDTMA). Mehdi et al. (2019), for example, obtained an interlayer complex with a basal spacing of 2.6 nm by mixing halloysite-(7Å) with a solution of HDTMA-bromide, equivalent to six times the CEC of the clay sample, for 14 days. At this point, the intercalation ratio (Equation 6.1) reached 42% and the basal spacing of the complex was 2.6 nm, indicating that the HDTMA chains adopted a paraffin-like arrangement in the interlayer space as shown in Figure 4.1d with respect to montmorillonite.

Subsequently, Salaa et al. (2020) obtained an intercalation ratio of 75% by displacing pre-intercalated DMSO with HDTMA. The interlayer halloysite-HDTMA that formed was effective in removing diclofenac from solution, but the resultant isotherm was non-linear. Thus, the process was not one of simple solute partitioning but rather involved both electrostatic and hydrophobic interactions. Similarly, the adsorption of phenol and chlorophenols (Słomkiewicz et al. 2020; Kurdziel et al. 2021), and that of methyl orange (Lacin and Aroguz 2020), to HDTMA-halloysite complexes gives rise to non-linear isotherms. On the other hand, naphthalene uptake by HDTMA-modified halloysite (and kaolinite) yielded linear isotherms (Lee and Kim 2002).

6.5 CHRYSOTILE

In terms of the clay-organic interaction, chrysotile does not feature as widely as kaolinite and halloysite. The small volume of literature on this topic is summarized as follows. Early research into the adsorption of polyaromatic hydrocarbons by chrysotile (Gorski and Stettler 1975; Menard et al. 1986; Fournier and Pezerat 1986) was undertaken within the framework that complex formation could markedly enhance the toxicity and carcinogenic potential of the organic solutes. In the case of phenanthrene, Fournier et al. (1989) reported that the adsorbed molecules adopted a vertical orientation, forming multilayers on the chrysotile surface. As a result, the fibres became hydrophobic and could serve as a carrier of organic species. Likewise, fluorene and dimetyl-7,12-benzanthracene formed bidimensional condensed layers (Fournier and Pezerat 1986), while pyrene and pyrene-TEMPO were adsorbed in the form of liquid-like multilayers (Ottaviani et al. 2001).

Chrysotile also shows a high affinity for various carboxylic and amino acids. Using infrared spectroscopy, Berkheiser (1982) deduced that stearic acid formed a surface Mg-stearate complex. Likewise, the carboxyl group of benzoic acid can condense with a surface hydroxyl group of chrysotile, while the phenyl group of the molecule adopts an upright orientation (Bonneau et al. 1986). The interaction of sodium dodecylbenzenesulfonate with chrysotile, in the presence of free oxygen, may even be one of catalysis rather than adsorption (Fachini and Joekes (2002).

On the other hand, the adsorption of aspartic and glutamic acids from water (at pH ~ 8) to a synthetic chrysotile primarily involves electrostatic interactions (Hashizume 2007) since the fibre surface would then be positively charged (Bales and Morgan 1985; Yang et al. 2014). The high affinity of doxylstearic acid for chrysotile (Ottaviani and Venturi 1996) may be explained in similar terms. Likewise, Hawtrey et al. (2002) invoked electrostatic interactions for the strong binding of (tritiated) deoxy-thymidine-5'-triphosphate and deoxy-adenosine-5'-triphosphate to chrysotile. In this instance, however, ligand exchange between the phosphate group of the solutes and surface hydroxyl of chrysotile (cf. Equation 1.1) is a more plausible mechanism as Hashizume and Theng (2007) have proposed for the adsorption of adenosine-5'-phosphate to allophane.

Earlier, Valerio et al. (1979) made the interesting observation that a chrysotile (from Rhodesia) could strongly adsorb basic (cationic) dyes such as acridine orange and methylene blue from aqueous solution. Indeed, Gengeç (2017) recorded a maximum capacity of 169 mg/g (pH 7; 298 K) for the adsorption of astrazon blue FGRL to a natural Turkish chrysotile. In explanation, Valerio et al. (1979) suggested the presence on the chrysotile surface of (unspecified) negatively charged sites since dye adsorption increased following magnesium extraction (depletion). It seems likely, however, that the negative surface charge originated from associated organic matter or layer silicate impurities.

Chrysotile can also form stable complexes with various fluorescent dyes, such as erythrosine B, fluorescein, rhodamine B, eosin, uranine, and calcein (Soma et al. 1993). The fluorescence spectra of the complexes indicated that the dye molecules were adsorbed as the corresponding carboxylate or phenolate species.

We might also mention that chrysotile can be silylated using organochlorosilanes to yield 'grafted' organic derivatives (Edwards 1970; Ruiz-Hitzky and Fripiat 1976). The classic procedure involves acid (HCl) dissolution of the (outer) brucite-like sheet of chrysotile, yielding silanols that can form stable Si–O–Si–C covalent bonds with the organosilyl groups of the silylating agents (Mendelovici and Frost 2005; Frost and Mendelovici 2006; Wang et al. 2009). Likewise, phenylphosphonate groups could be (covalently) grafted to surface silanols of acid-leached chrysotile (Wypych et al. 2003).

Earlier, Fonseca et al. (2001) used 3-aminopropyltrimethoxysilane and *N*-[3-(trimethoxysilyl) propyl]ethylenediamine as silylating agents. Besides showing a high capacity for adsorbing cobalt and copper cations from aqueous solutions, the resultant organo-grafted chrysotile could form C=N bonds with 2-pyridine and 2-thiophene carbaldehydes (da Fonseca and Airoldi 1999). Similarly, Liu et al. (2013) reported that chrysotile became an effective adsorbent of Cu(II) ions after silylation with γ-aminopropyltriethoxysilane.

REFERENCES

Abdullayev, E., A. Joshi, W. Wei, Y. Zhao, and Y. Lvov. 2012. Enlargement of halloysite clay nanotube lumen by selective etching of aluminum oxide. *ACS Nano* 6: 7216–7226.

Abdullayev, E. and Y. Lvov. 2016. Halloysite for controllable loading and release. In *Nanosized Tubular Clay Minerals: Halloysite and Imogolite*. Developments in Clay Science, Vol. 7, (Eds.) P. Yuan, A. Thill, and F. Bergaya, pp. 554–605. Amsterdam, The Netherlands: Elsevier.

Abid, M., A. Ben Haj Amara, an M. Bechelany. 2023. Halloysite-TiO_2 nanocomposites for water treatment: A review. *Nanomaterials* 13: 1578. https://doi.org/10.3390/nano13091578.

Abou-El-Sherbini, K.S., E.A.M. Elzahany, M.A. Wahba, S.A. Drweesh, and N.S. Youssef. 2017. Evaluation of some intercalation methods of dimethylsulphoxide onto HCl-treated and untreated Egyptian kaolinite. *Applied Clay Science* 137: 33–42.

Adamczyk, M., K. Małycha, K. Kułacz, M. Pocheć, and K. Orzechowski. 2020. Halloysite intercalated by potassium acetate. *Physicochemical problems of Mineral Processing* 56: 235–243.

Adams, J.M. 1978. Unifying features relating to the 3D structures of some intercalates of kaolinite. *Clays and Clay Minerals* 26: 291–295.

Adams, J.M. and G. Waltl. 1980. Thermal decomposition of a kaolinite: Dimethylsulfoxide intercalate. *Clays and Clay Minerals* 28: 130–134.

Aguzzi, C., G. Sandri, P. Cerezo, E. Carazo, and C. Viseras. 2016. Health and medical applications of tubular clay minerals. In *Nanosized Tubular Clay Minerals*. Developments in Clay Science. Vol. 7, (Eds.) P. Yuan, A. Thill, and F. Bergaya, pp. 708–725. Amsterdam, The Netherlands: Elsevier.

Ahmed, S.A.S. 2009. Removal of toxic pollutants from aqueous solutions by adsorption onto organo-kaolin. *Carbon Letters* 10: 305–313.

Altun, T. and H. Ecevit. 2022. Adsorption of malachite green and methyl violet 2B by halloysite nanotube: Batch adsorption experiments and Box-Behnken experimental design. *Materials Chemistry and Physics* 291: 126612.

Amadou, I., M.-P. Faucon, and D. Houben. 2022. New insights into sorption and desorption of organic phosphorus on goethite, gibbsite, kaolinite and montmorillonite. *Applied Geochemistry* 143: 105378.

Amara, A.B.H., J.B. Brahim, G. Besson, and C.H. Pons. 1995. Étude d'une nacrite intercalée par du dimethylsulfoxide et *n*-methylacetamide. *Clay Minerals* 30: 295–306.

Anastopoulos, J., A. Mittal, M. Usman et al. 2018. A review on halloysite-based adsorbents to remove pollutants in water and wastewater. *Journal of Molecular Liquids* 269: 855–868.

Andreou, F.T., B. Barylska, Z. Ciesielska et al. 2021. Intercalation of *N*-methylformamide in kaolinite: In situ monitoring by near-infrared spectroscopy and X-ray diffraction. *Applied Clay Science* 212: 106209.

Andreou, F.T., E. Siranidi, A. Derkowski, and G.D. Chryssikos. 2022. On the unusual temperature dependence of kaolinite intercalation capacity for N-methylformamide. *Clays and Clay Minerals* 7: 796–807.

Angioi, S., S. Polati, M. Roz, C. Rinaudo, V. Gianotti, and M.C. Gennaro. 2005. Sorption studies of chloroanilines on kaolinite and montmorillonite. *Environmental Pollution* 134: 35–43.

Anton, O. and P.G. Rouxhet. 1977. Note on the intercalation of kaolinite, dickite and halloysite by dimethyl sulfoxide. *Clays and Clay Minerals* 25: 259–263.

Auta, M. and B.H. Hameed. 2012. Modified mesoporous clay adsorbent for adsorption isotherm and kinetics of methylene blue. *Chemical Engineering Journal* 198–199: 219–227.

Avila, L.R., E.H. de Faria, K.J. Ciuffi et al. 2010. New synthesis strategies for effective functionalization of kaolinite and saponite with silylating agents. *Journal of Colloid and Interface Science* 341: 186–193.

Awad, M.E., E. Escamilla-Roa, A. Borrego-Sánchez, C. Viseras, A. Hernández-Laguna, and C.I. Sainz-Díaz. 2019. Adsorption of 5-aminosalicylic acid on kaolinite surfaces at a molecular level. *Clay Minerals* 54: 49–56.

Awad, M.E., A. López-Galindo, M. Setti, M.M. El-Rahmany, and C.V. Iborra. 2017. Kaolinite in pharmaceutics and biomedicine. *International Journal of Pharmaceutics* 533: 34–48.

Aytekin, M.T. and H.L. Hoşgün. 2020. Characterization studies of heat-treated halloysite nanotubes. *Chemical Papers* 74: 4547–4557.

Baker, R.A. and M.-D. Luh. 1971. Pyridine sorption from aqueous solution by montmorillonite and kaolinite. *Water Research* 5: 839–848.

Balan, E., G. Calas, and D.L. Bish. 2014. Kaolin-group minerals: From hydrogen-bonded layers to environmental recorders. *Elements* 10: 183–188.

Bales, R.C. and J.J. Morgan. 1985. Surface charge and adsorption properties of chrysotile asbestos in natural waters. *Environmental Science & Technology* 19: 1213–1219.

Bediako, E.G., E. Nyankson, D. Dodoo-Arhin et al. 2018. Modified halloysite nanoclay as a vehicle for sustained drug delivery. *Heliyon* 4: e00689. https://doi.org/10.1016/j.heliyon.2018.

Behera, S.K., S.-Y. Oh, and H.-S. Park. 2010. Sorption of triclosan onto activated carbon, kaolinite and montmorillonite: Effects of pH, ionic strength, and humic acid. *Journal of Hazardous Materials* 179: 684–691.

Belkassa, K., F. Bessaha, K. Marouf-Khelifa, I. Batonneau-Gener, J.-D. Comparot, and A. Khelifa. 2013. Physicochemical and adsorptive properties of a heat-treated and acid-leached Algerian halloysite. *Colloids and Surfaces A: Physicochemical and Engineering Aspects* 421: 26–33.

Belkassa, K., M. Khelifa, I. Batonneau-Gener, K. Marouf-Khelifa, and A. Khelifa. 2021. Understanding of the mechanism of crystal violet adsorption on modified halloysite: Hydrophobicity, performance, and interaction. *Journal of Hazardous Materials* 415: 125656.

Benincasa, E., M.F. Brigatti, D. Malferrari, L. Medici, and L. Poppi. 2002. Sorption of Cd-cysteine complexes by kaolinite. *Applied Clay Science* 21: 191–201.

Berkheiser, V.E. 1982. Adsorption of stearic acid by chrysotile. *Clays and Clay Minerals* 30: 91–96.

Bessaha, F., N. Mahrez, K. Marouf-Khelifa, A. Çoruh, and A. Khelifa. 2019. Removal of Congo red by thermally and chemically modified halloysite: Equilibrium, FTIR spectroscopy, and mechanism studies. *International Journal of Environmental Science and Technology* 16: 4253–4260.

Bessaha, F., K. Marouf-Khelifa, I. Batonneau-Gener, and A. Khelifa. 2016. Characterization and application of heat-treated and acid-leached halloysites in the removal of malachite green: Adsorption, desorption, and regeneration studies. *Desalination and Water Treatment* 57: 14609–14621.

Bhattacharyya, K.G., S. SenGupta, and G.K. Sarma. 2014. Interactions of the dye, Rhodamine B with kaolinite and montmorillonite. *Applied Clay Science* 99: 7–17.

Biddeci, G., G. Spinelli, P. Colomba, and F. Di Blasi. 2022. Nanomaterials: A review about halloysite nanotubes, properties, and application in the biological field. *International Journal of Molecular Sciences* 23: 11518. https://doi.org/10.3390/ijms231911518.

Bonneau, L., H. Suquet, C. Malard, and H. Pezerat. 1986. Studies on surface properties of asbestos. I. Active sites on surface of chrysotile and amphiboles. *Environmental Research* 41: 251–267.

Borrego-Sánchez, A., M.E. Awad, and C.I. Sainz-Díaz. 2018. Molecular modelling of adsorption of 5-aminosalicylic acid in the halloysite nanotube. *Minerals* 8: 61. https://doi.org/10.3390/min8020061.

Borrego-Sánchez, A., C.I. Sainz-Díaz, L. Perioli, and C. Viseras. 2021. Theoretical study of retinol, niacinamide and glycolic acid with halloysite clay mineral as active ingredients for topical skin care formulations. *Molecules* 26: 4392. https://doi.org/10.3390/molecules26154392.26.

Boukhemkhem, A., B. Aissa-Ouaissi-Sekkouti, J. Berdia, C. Belver, and C.B. Molina. 2022. Adsorption of crystal violet on kaolinite clay: Kinetic and equilibrium study using non-linear models. *Clay Minerals* 57: 41–50.

Brandt, K.B., T.A. Elbokl, and C. Detellier. 2003. Intercalation and interlamellar grafting of polyols in layered aluminosilicates. D-Sorbitol and adonitol derivatives of kaolinite. *Journal of Materials Chemistry* 13: 2566–2572.

Caglar, B. 2012. Structural characterization of kaolinite-nicotinamide intercalation composite. *Journal of Molecular Structure* 1020: 48–55.
Caglar, B., C. Çirak, A. Tabak, B. Afsin, and E. Eren. 2013. Covalent grafting of pyridine-2-methanol into kaolinite layers. *Journal of Molecular Structure* 1030: 12–22.
Caillère, S., R. Glaeser, J. Esquevin, and S. Hénin. 1950. Preparation of 14Å- and 17Å-halloysite. *Comptes Rendus de l'Académie des Sciences* 230: 308–310.
Camazano, M.S. and S.G. Garcia. 1966. Complejos interlaminares de caolinita y halloisita con liquidos polares. *Anales de Edafología y Agrobiología* 25: 9–25.
Carazo, E., A. Borrego-Sánchez, F. García-Villén et al. 2017. Assessment of halloysite nanotubes as vehicles of isoniazid. *Colloids and Surfaces B: Biointerfaces* 160: 337–344.
Carr, R.M., N. Chaikum, and N. Patterson. 1978. Intercalation of salts in halloysite. *Clays and Clay Minerals* 26: 144–152.
Carr, R.M. and H. Chih. 1971. Complexes of halloysite with organic compounds. *Clay Minerals* 9: 153–166.
Castellini, E., R. Andreoli, G. Malavasi, and A. Pedone. 2008. Deflocculant effects on the surface properties of kaolinite investigated through malachite green adsorption. *Colloids and Surfaces A: Physicochemical and Engineering Aspects* 329: 31–37.
Castrillo, P.D., D. Olmos, and J. González-Benito. 2015. Kinetic study of the intercalation process of dimethylsulfoxide in kaolinite. *International Journal of Mineral Processing* 144: 70–74.
Cavallaro, G., G. Lazzara, E. Rozhina et al. 2019. Organic-nanoclay composite materials as removal agents for environmental decontamination. *RSC Advances* 9: 40553–40564.
Cervini-Silva, J., A. Nieto-Camacho, E. Palacios et al. 2016. Anti-inflammatory, antibacterial, and cytotoxic activity by natural matrices of nano-iron(hydr)oxide/halloysite. *Applied Clay Science* 120: 101–110.
Chargui, H., W. Hajjaji, J. Wouters, J. Yans, and F. Jamoussi. 2018. Direct orange 34 dye fixation by modified kaolin. *Clay Minerals* 53: 271–287.
Chen, J., F.-F. Min, L. Liu, C. Liu, and F. Lu. 2017. Experimental investigation and DFT calculation of different amine/ammonium salts adsorption on kaolinite. *Applied Surface Science* 419: 241–251.
Chen, J., F.-F. Min, L.-Y. Liu, and F.-F. Jia. 2020. Adsorption of methylamine cations on kaolinite basal surfaces: A DFT study. *Physicochemical Problems of Mineral Processing* 56: 337–348.
Chen, P.-Y., M.-L. Lin, and Z. Zheng. 1997. On the origin of the name kaolin and the kaolin deposits of the Kauling and Dazhou areas, Kiangsi, China. *Applied Clay Science* 12: 1–25.
Cheng, H., Q. Liu, J. Yang, X. Du, and R.L. Frost. 2010a. Influencing factors on kaolinite-potassium acetate intercalation complexes. *Applied Clay Science* 50: 476–480.
Cheng, H., Q. Liu, J. Yang, J. Zhang, and R.L. Frost. 2010b. Thermal analysis and infrared emission spectroscopic study of halloysite-potassium acetate intercalation compound. *Thermochimica Acta* 511: 124–128.
Cheng, H., Q. Liu, J. Yang, S. Ma, and R.L. Frost. 2012. The thermal behavior of kaolinite intercalation complexes – a review. *Thermochimica Acta* 545: 1–13.
Cheng, H., Q. Liu, J. Yang, J. Zhang, R.L. Frost, and X. Du. 2011. Infrared spectroscopic study of halloysite-potassium acetate intercalation complex. *Journal of Molecular Structure* 990: 21–25.
Cheng, H., X. Hou, Q. Liu, X. Li, and R.L. Frost. 2015b. New insights into the molecular structure of kaolinite-methanol intercalation complexes. *Applied Clay Science* 109–110: 55–63.
Cheng, H., Q. Liu, P. Xu, and R. Hao. 2018a. A comparison of molecular structure and de-intercalation kinetics of kaolinite/quaternary ammonium salt and alkylamine intercalation compounds. *Journal of Solid State Chemistry* 268: 36–44.
Cheng, H., S. Zhang, Q. Liu, X. Li, and R.L. Frost. 2015a. The molecular structure of kaolinite-potassium acetate intercalation complexes: A combined experimental and molecular dynamic simulation study. *Applied Clay Science* 116–117: 273–280.
Cheng, R., H. Li, Z. Liu, and C. Du. 2018b. Halloysite nanotubes as an effective and recyclable adsorbent for removal of low-concentration antibiotics ciprofloxacin. *Minerals* 8: 387. https://doi.org/10.3390/min8090387.
Churchman, G.J. 1990. Relevance of different intercalation tests for distinguishing halloysite from kaolinite in soils. *Clays and Clay Minerals* 38: 591–599.
Churchman, G.J., L.P. Aldridge, and R.M. Carr. 1972. The relationship between the hydrated and dehydrated states of an halloysite. *Clays and Clay Minerals* 20: 241–246.
Churchman, G.J. and R.M. Carr. 1973. Dehydration of the potassium acetate complex of halloysite. *Clays and Clay Minerals* 21: 423–424.
Churchman, G.J., P. Pasbakhsh, and S. Hillier. 2016a. The rise and rise of halloysite. *Clays and Clay Minerals* 51: 303–308.

Churchman, G.J., P. Pasbakhsh, D.J. Lowe, and B.K.G. Theng. 2016b. Unique but diverse: Some observations on the formation, structure and morphology of halloysite. *Clay Minerals* 51: 395–416.
Churchman, G.J. and B.K.G. Theng. 1984. Interactions of halloysites with amides: Mineralogical factors affecting complex formation. *Clay Minerals* 19: 161–175.
Churchman, G.J., J.S. Whitton, G.G.C. Claridge, and B.K.G. Theng. 1984. Intercalation methods using formamide for differentiating halloysite from kaolinite. *Clays and Clay Minerals* 32: 241–248.
Conley, R.F. and M.K. Lloyd. 1971. Adsorption studies on kaolinites II. Adsorption of amines. *Clays and Clay Minerals* 19: 273–282.
Costanzo, P.M., C.V. Clemency, and R.F. Giese, Jr. 1980. Low temperature synthesis of a 10 Å hydrate of kaolinite using dimethylsulfoxide and ammonium fluoride. *Clays and Clay Minerals* 28: 155–156.
Costanzo, P.M. and R.F. Giese, Jr. 1985. Dehydration of synthetic hydrated kaolinites: A model for the dehydration of halloysite(10Å). *Clays and Clay Minerals* 33: 415–423.
Costanzo, P.M. and R.F. Giese, Jr. 1986. Ordered halloysite: Dimethylsulfoxide intercalate. *Clays and Clay Minerals* 34: 105–107.
Costanzo, P.M. and R.F. Giese, Jr. 1990. Ordered and disordered organic intercalates of 8.4-Å, synthetically hydrated kaolinite. *Clays and Clay Minerals* 38: 160–170.
Costanzo, P.M., R.F. Giese, Jr., and C.V. Clemency. 1984a. Synthesis of a 10-Å hydrated kaolinite. *Clays and Clay Minerals* 32: 29–35.
Costanzo, P.M., R.F. Giese, Jr., and M. Lipsicas. 1984b. Static and dynamic structure of water in hydrated kaolinites. I. The static structure. *Clays and Clay Minerals* 32: 419–428.
Coyne, L.M., P.M. Costanzo, and B.K.G. Theng. 1989. Luminescence and ESR studies of relationships between O^- -centres and structural iron in natural and synthetically hydrated kaolinites. *Clay Minerals* 24: 671–693.
Cruz, M., A. Laycock, and J.L. White. 1969. Perturbation of OH groups in intercalated kaolinite donor-acceptor complexes. I. Formamide-, methylformamide-, and dimethylformamide-kaolinite complexes. *Proceedings of the International Clay Conference, Tokyo* 1: 775–789.
Cruz-Hernández, M., F.D. Velázquez-Herrera, N. Villa-Ruano, M. Giovanela, J.D.S. Crespo, and G. Fetter. 2021. High-performance antifungal nanohybrid materials composed of melanin-clays. *Applied Clay Science* 211: 106201.
da Fonseca, M.G. and C. Airoldi. 1999. Action of silylating agents on a chrysotile surface and subsequent reactions with 2-pyridine and 2-thiophene carbaldehydes. *Journal of Materials Chemistry* 9: 1375–1380.
Danyliuk, N., J. Tomaszewska, and T. Tatarchuk. 2020. Halloysite nanotubes and halloysite-based composites for environmental and biomedical applications. *Journal of Molecular Liquids* 309: 113077.
da Silva, A.C., K.J. Ciuffi, M.J. dos Reis, P.S. Calefi, and E.H. de Faria. 2016. Influence of physical/chemical treatments to delamination of nanohybrid kaolinite-dipicolinate. *Applied Clay Science* 126: 251–258.
Dawley, M.M., A.M. Scott, F.C. Hill, J. Leszczynski, and T.M. Orlando. 2012. Adsorption of formamide on kaolinite surfaces: A combined infrared experimental and theoretical study. *Journal of Physical Chemistry C* 116: 23981–23991.
de Araujo, D.T., K.J. Ciuffi, E.J. Nassar et al. 2017. Eu^{3+}- and Tb^{3+}-dipicolinate complexes covalently grafted into kaolinite as luminescence-functionalized clay hybrid materials. *Journal of Physical Chemistry C* 121: 5081–5088.
Dedzo, G.K. and C. Detellier. 2014. Intercalation of two phenolic acids in an ionic liquid-kaolinite nanohybrid material and desorption studies. *Applied Clay Science* 97–98: 153–159.
Dedzo, G.K. and C. Detellier. 2016. Functional nanohybrid materials derived from kaolinite. *Applied Clay Science* 130: 33–39.
Dedzo, G.K. and C. Detellier. 2018. Clay minerals – ionic liquids, nanoarchitectures, and applications. *Advanced Functional Materials* 28: 1703845.
Dedzo, G.K., B.B. Nguelo, I.T. Kenfack, E. Ngameni, and C. Detellier. 2017. Molecular control of the functional and spatial interlayer environment of kaolinite by the grafting of selected pyridinium ionic liquids. *Applied Clay Science* 143: 445–451.
de Faria, E.H., K.J. Ciuffi, E.J. Nassar, M.A. Vicente, R. Trujillano, and P.S. Calefi. 2010. Novel reactive amino-compound: Tris(hydroxymethyl) aminomethane covalently grafted on kaolinite. *Applied Clay Science* 48: 516–521.
de Faria, E.H., O.J. Lima, K.J. Ciuffi et al. 2009. Hybrid materials prepared by interlayer functionalization of kaolinite with pyridine-carboxylic acids. *Journal of Colloid and Interface Science* 335: 210–215.
del Río, M.S., P. Matinetto, C. Reyes-Valerio, E. Dooryhée, and M. Suárez. 2006. Synthesis and acid resistance of Maya blue pigment. *Archaeometry* 48: 115–130.

Deng, L., P. Yuan, D. Liu et al. 2017. Effects of microstructure of clay minerals, montmorillonite, kaolinite and halloysite, on their benzene adsorption behaviors. *Applied Clay Science* 143: 184–191.
Deng, L., P. Yuan, D. Liu et al. 2019. Effects of calcination and acid treatment on improving benzene adsorption performance of halloysite. *Applied Clay Science* 181: 105240.
Deng, Y, J.B. Dixon, and G.N. White. 2003. Molecular configurations and orientations of hydrazine between structural layers of kaolinite. *Journal of Colloid and Interface Science* 257: 208–227.
Deng, Y., G.N. White, and J.B. Dixon. 2002. Effect of structural stress on the intercalation rate of kaolinite. *Journal of Colloid and Interface Science* 250: 379–393.
Deshmukh, R.K., L. Kumar, and K.K. Gaikwad. 2023. Halloysite nanotubes for food packaging application: A review. *Applied Clay Science* 234: 106856.
Detellier, C. 2018. Functional kaolinite. *Chemical Record* 18: 868–877.
Detellier, C. and R.A. Schoonheydt. 2014. From platy kaolinite to nanorolls. *Elements* 10: 201–206.
Diar-Bakerly, B., D. Hirsemann, H. Kalo, R. Schobert, and J. Breu. 2014. Modification of kaolinite by grafting of siderophilic ligands to the external octahedral surface. *Applied Clay Science* 90: 67–72.
Díaz, M., M.A. Villa-García, R. Duarte-Silva, and M. Rendueles. 2017. Preparation of organo-modified kaolinite sorbents: The effect of surface functionalization on protein adsorption performance. *Colloids and Surfaces A* 530: 181–190.
Doğan, M., M.H. Karaoğlu, and M. Alkan. 2009. Adsorption kinetics of maxilon yellow 4GL and maxilon red GRL dyes on kaolinite. *Journal of Hazardous Materials* 165: 1142–1151.
Dolui, M., S. Rakshit, M.E. Essington, and G. Lefèvre. 2018. Probing oxytetracycline sorption mechanism on kaolinite in a single ion and binary mixtures with phosphate using in situ ATR-FTIR spectroscopy. *Soil Science Society of America Journal* 82: 826–838.
Dube, S., D. Rawtani, N. Khatri, and G. Parikh. 2022. A deep delve into the chemistry and biocompatibility of halloysite nanotubes: A new perspective on an idiosyncratic nanocarrier for delivering drugs and biologics. *Advances in Colloid and Interface Science* 309: 102776.
Duer, M.J., J. Rocha, and J. Klinowski. 1992. Solid-state NMR studies of the molecular motion in the kaolinite: DMSO intercalate. *Journal of the American Chemical Society* 114: 6867–6874.
Ece, Ö.I. and P.A. Schroeder. 2007. Clay mineralogy and chemistry of halloysite and alunite deposits in the Turplu area, Balıkesir, Turkey. *Clays and Clay Minerals* 55: 18–35.
Edwards, H. 1970. Study of the reactions of surface hydroxyl groups of a chrysotile asbestos with organic silanes by means of infra-red spectroscopy. *Journal of Applied Chemistry* 20: 76–79.
Elbokl, T.A. and C. Detellier. 2008. Intercalation of cyclic amides in kaolinite. *Journal of Colloid and Interface Science* 323: 338–348.
Elhadj, M.-S.Y. and F.X. Perrin. 2021. Influencing parameters of mechanochemical intercalation of kaolinite with urea. *Applied Clay Science* 213: 106250.
El-Soad, A.M.A., G. Lazzara, A.V. Pestov et al. 2021. Grafting of (3-chloropropyl)-trimethoxy silane on halloysite nanotubes surface. *Applied Sciences* 11: 5534. https://doi.org/10.3390/app11125534.
Essington, M.E. and R.M. Anderson. 2008. Competitive adsorption of 2-ketogluconate and inorganic ligands onto gibbsite and kaolinite. *Soil Science Society of America Journal* 72: 595–604.
Essington, M.E., J. Lee, and Y. Seo. 2010. Adsorption of antibiotics by montmorillonite and kaolinite. *Soil Science Society of America Journal* 74: 1577–1588.
Fachini, A. and I. Joekes. 2002. Interaction of sodium dodecylbenzenesulfonate with chrysotile fibers: Adsorption or catalysis? *Colloids and Surfaces A: Physicochemical and Engineering Aspects* 201: 151–160.
Fafard, J., V. Terskikh, and C. Detellier. 2017. Solid-state ^{1}H and ^{27}Al NMR studies of DMSO-kaolinite intercalates. *Clays and Clay Minerals* 65: 206–219.
Fahimizadeh, M., P. Pasbakhsh, L.S. Mae, J.B.L. Tan, and R.K. Singh Raman. 2023. Sustained-release of nutrients by yeast extract-loaded halloysite nanotubes supports bacterial growth. *Applied Clay Science* 240: 106979.
Fakhruddin, K., R. Hassan, M.U.A. Khan et al. 2021. Halloysite nanotubes and halloysite-based composites for biomedical applications. *Arabian Journal of Chemistry* 14: 103294.
Fang, Q., S. Huang, and W. Wang. 2005. Intercalation of dimethyl sulfoxide in kaolinite: Molecular dynamics simulation study. *Chemical Physics Letters* 411: 233–237.
Fashina, B. and Y. Deng. 2021. Stacking disorder and reactivity of kaolinites. *Clays and Clay Minerals* 69: 354–365.
Fernandez-Gonzales, M., A. Weiss, and G. Lagaly. 1976. Über das Verhalten nordwest-spanischer Kaoline bei der Bildung von Einlagerungsverbindungen. *Keramische Zeitschrift* 28: 55–58.

Ferrante, F., N. Armata, G. Cavallaro, and G. Lazzara. 2017. Adsorption studies of molecules on the halloysite surfaces: A computational and experimental investigation. *Journal of Physical Chemistry C* 121: 2951–2958.

Ferrante, F., M. Bertini, C. Ferlito, L. Lisuzzo, G. Lazzara, and D. Duca. 2023. A computational and experimental investigation of halloysite silicic surface modifications after alkaline treatment. *Applied Clay Science* 232: 106813.

Figueroa, R.A., A. Leonard, and A.A. MacKay. 2004. Modeling tetracycline antibiotic sorption to clays. *Environmental Science & Technology* 38: 476–483.

Filice, S., C. Bongiorno, S. Libertino et al. 2021. Structural characterization and adsorption properties of Dunino raw halloysite mineral for dye removal from water. *Materials* 14: 3676. https://doi.org/10.3390/ma14133676.

Fizir, M., W. Liu, X. Tang, F. Wang, and Y. Benmokadem. 2022. Design approaches, functionalization, and environmental and analytical applications of magnetic halloysite nanotubes: A review. *Clays and Clay Minerals* 70: 660–694.

Fonseca, M.G., A.S. Oliveira, and C. Airoldi. 2001. Silylating agents grafted onto silica derived from leached chrysotile. *Journal of Colloid and Interface Science* 240: 533–538.

Fournier, J., B. Fubini, V. Bolis, and H. Pezerat. 1989. Thermodynamic aspects in the adsorption of polynuclear aromatic hydrocarbons on chrysotile and silica – possible relation to synergistic effects in lung toxicity. *Canadian Journal of Chemistry* 67: 289–296.

Fournier, J. and H. Pezerat. 1986. Studies on surface properties of asbestos. III. Interactions between asbestos and polynuclear aromatic hydrocarbons. *Environmental Research* 41: 276–295.

Franco, F. and M.D. Ruiz Cruz. 2003. Thermal behaviour of dickite-dimethylsulfoxide intercalation complex. *Journal of Thermal Analysis and Calorimetry* 73: 151–165.

Franco, F. and M.D. Ruiz Cruz. 2004. Factors influencing the intercalation degree ('reactivity') of kaolin minerals with potassium acetate, formamide, dimethylsulphoxide and hydrazine. *Clay Minerals* 39: 193–205.

Frost, R.L. and J. Kristof. 1997. Intercalation of halloysite: A Raman spectroscopic study. *Clays and Clay Minerals* 45: 551–563.

Frost, R.L., J. Kristof, E. Horvath, and J.T. Kloprogge. 2000a. Vibrational spectroscopy of formamide-intercalated kaolinites. *Spectrochimica Acta Part A* 56: 1191–1204.

Frost, R.L., J. Kristof, E. Mako, and J.T. Kloprogge. 2000b. Modification of the hydroxyl surface of potassium acetate intercalated halloysite between 25 and 300 °C. *American Mineralogist* 85: 1735–1743.

Frost, R.L., J. Kristof, G.N. Paroz, and J.T. Kloprogge. 1998a. Molecular structure of dimethyl sulfoxide intercalated kaolinites. *Journal of Physical Chemistry B* 102: 8519–8532.

Frost, R.L., J. Kristof, G.N. Paroz, and J.T. Kloprogge. 1999. Intercalation of kaolinite with acetamide. *Physics and Chemistry of Minerals* 26: 257–263.

Frost, R.L., J. Kristof, and T.H. Tran. 1998b. Kinetics of deintercalation of potassium acetate from kaolinite – a Raman spectroscopic study. *Clay Minerals* 33: 605–617.

Frost, R.L., J. Kristof, E. Horvath, and J.T. Kloprogge. 2000c. Rehydration and phase changes of potassium acetate-intercalated halloysite at 298 K. *Journal of Colloid and Interface Science* 226: 318–327.

Frost, R.L., J. Kristof, J.T. Kloprogge, and E. Horvath. 2001. Modification of the hydroxyl surface in cesium acetate intercalated kaolinite. *Langmuir* 17: 4067–4073.

Frost, R.L. and E. Mendelovici. 2006. Modification of fibrous silicates surfaces with organic derivatives: An infrared spectroscopic study. *Journal of Colloid and Interface Science* 294: 47–52.

Frost, R.L., S.J. Van der Gaast, M. Zbik, J.T. Kloprogge, and G.N. Paroz. 2002. Birdwood kaolinite: A highly ordered kaolinite that is difficult to intercalate – an XRD, SEM and Raman spectroscopic study. *Applied Clay Science* 20: 177–187.

Gábor, M., M. Tóth, J. Kristóf, and G. Komáromi-Hiller. 1995. Thermal behavior and decomposition of intercalated kaolinite. *Clays and Clay Minerals* 43: 223–228.

Gaikwad, K.K., S. Singh, and Y.S. Lee. 2018. High adsorption of ethylene by alkali-treated halloysite nanotubes for food-packaging applications. *Environmental Chemistry Letters* 16: 1055–1062.

Gao, W., S. Zhao, H. Wu, W. Deligeer, and S. Asuha. 2016. Direct acid activation of kaolinite and its effects on the adsorption of methylene blue. *Applied Clay Science* 126: 98–106.

Garcia-Garcia, D., J.M. Ferri, L. Ripoll, M. Hidalgo, J. Lopez-Martinez, and R. Balart. 2017. Characterization of selectively etched halloysite nanotubes by acid treatment. *Applied Surface Science* 422: 616–625.

Gardolinski, J.E.F.C. and G. Lagaly. 2005a. Grafted organic derivatives of kaolinite: I. Synthesis, chemical and rheological characterization. *Clay Minerals* 40: 537–546.

Gardolinski, J.E.F.C. and G. Lagaly. 2005b. Grafted organic derivatives of kaolinite: II. Intercalation of primary n-alkylamines and delamination. *Clay Minerals* 40: 547–556.

Gardolinski, J.F., P. Peralta-Zamora, and F. Wypych. 1999. Preparation and characterization of a kaolinite-1-methyl-2-pyrrolidone intercalation compound. *Journal of Colloid and Interface Science* 211: 137–141.

Gardolinski, J.F., L.P. Ramos, G.P. de Souza, and F. Wypych. 2000. Intercalation of benzamide into kaolinite. *Journal of Colloid and Interface Science* 221: 284–290.

Garrett, W.G. and G.F. Walker. 1959. The cation exchange capacity of hydrated halloysite and the formation of halloysite-salt complexes. *Clay Minerals Bulletin* 4: 75–80.

Gengeç, E. 2017. The adsorption of basic dye (astrazon blue FGRL) from aqueous solutions onto two different clays: Talc and chrysotile. *Karaelmas Fen ve Mühendislik Dergisi* 7: 438–448.

Gettens, R.J. 1962. Maya blue: An unsolved problem in ancient pigments. *American Antiquity* 7: 557–564.

Ghezzi, L., A. Spepi, M. Agnolucci et al. 2018. Kinetics of release and antibacterial activity of salicylic acid loaded into halloysite nanotubes. *Applied Clay Science* 160: 88–94.

Gianni, E., K. Avgoustakis, M. Pšenička, M. Pospíšil, and D. Papoulis. 2019. Halloysite nanotubes as carriers for irinotecan: Synthesis and characterization by experimental and molecular simulation methods. *Journal of Drug Delivery Science and Technology* 52: 568–576.

Gianni, E., D. Moreno-Rodríguez, L. Janković, E. Scholtzová, and M. Pospíšil. 2022. How herbicides like atrazine and diuron interact with the spiral halloysite structure. *Journal of Environmental Chemical Engineering* 10: 108785.

Gianni, E., M. Pšenička, K. Macková et al. 2023. New detail insight into halloysite structure: Mechanism behind nanotubular morphology described by density functional theory and molecular dynamics supported by experiments. *Journal of Molecular Structure* 1287: 135639.

González García, S. and M. Sánchez Camazano. 1968. Differentiation of kaolinite from chlorite by treatment with dimethyl-sulphoxide. *Clay Minerals* 7: 447–450.

Gorski, C.H. and L.E. Stettler. 1975. The adsorption of water and benzene on amosite and chrysotile asbestos. *American Industrial Hygiene Association Journal* 36: 292–298.

Grabka, D., M. Raczyńska-Żak, K. Czech, P.M. Słomkiewicz, and M.A. Jóźwiak. 2015. Modified halloysite as an adsorbent for prometryn from aqueous solutions. *Applied Clay Science* 114: 321–329.

Grančič, P. and D. Tunega. 2021. Hydrophobicity and charge distribution effects in the formation of bioorgano-clays. *Minerals* 11: 1102. https://doi.org/10.3390/min11101102.

Gray-Wannell, N., P.J. Holliman, H.C. Greenwell, E. Delbos, and S. Hillier. 2020. Adsorption of phosphate by halloysite (7 Å) nanotubes (HNTs). *Clay Minerals* 55: 184–193.

Guo, F., M. Zhou, J. Xu et al. 2021. Glyphosate adsorption onto kaolinite and kaolinite-humic acid composites: Experimental and molecular dynamics studies. *Chemosphere* 263: 127979.

Hanif, M., F. Jabbar, S. Sharif, G. Abbas, A. Farooq, and M. Aziz. 2016. Halloysite nanotubes as a new drug-delivery system: A review. *Clay Minerals* 51: 469–477.

Hári, J., P. Polyák, D. Mester et al. 2016. Adsorption of an active molecule on the surface of halloysite for controlled release application: Interaction, orientation, consequences. *Applied Clay Science* 132–133: 167–174.

Harris, R.G., J.D. Wells, M.J. Angove, and B.B. Johnson. 2006. Modeling the adsorption of organic dye molecules to kaolinite. *Clays and Clay Minerals* 54: 456–465.

Harris, R.G., J.D. Wells, and B.B. Johnson. 2001. Selective adsorption of dyes and other organic molecules to kaolinite and oxide surfaces. *Colloids and Surfaces A: Physicochemical and Engineering Aspects* 180: 131–140.

Hashizume, H. 2007. Adsorption of some amino acids by chrysotile. *Viva Origino* 35: 60–65.

Hashizume, H. and B.K.G. Theng. 2007. Adenine, adenosine, ribose and 5'AMP adsorption to allophane. *Clays and Clay Minerals* 55: 599–605.

Hashizume, H., B.K.G. Theng, S. van der Gaast, and K. Fujii. 2019. Formation of adenosine from adenine and ribose under conditions of repeated wetting and drying in the presence of clay minerals. *Geochimica et Cosmochimica Acta* 265: 495–504.

Hawtrey, A.O., A. Pieterse, J.M. van Zyl, A. Bester, and J.S. Harington. 2002. Binding of deoxy-thymidine-5'-triphosphate and deoxyadenosine-5'-triphosphate to chrysotile asbestos. *South African Journal of Science* 98: 557–559.

Hayashi, S. 1997. NMR study of dynamics and evolution of guest molecules in kaolinite/dimethyl sulfoxide intercalation compound. *Clays and Clay Minerals* 45: 724–732.

He, H., Q. Tao, J. Zhu, P. Yuan, W. Shen, and S. Yang. 2013. Silylation of clay mineral surfaces. *Applied Clay Science* 71: 15–20.

Hemmatpour, H., V. Haddadi-Asl, and H. Roghani-Mamaqani. 2015. Synthesis of pH-sensitive poly(N,N-dimethylaminoethyl methacrylate)-grafted halloysite nanotubes for adsorption and controlled release of DPH and DS drugs. *Polymer* 65: 143–153.

Hoch, M. and A. Bandara. 2005. Determination of the adsorption process of tributyltin (TBT) and monobutyltin (MBT) onto kaolinite surface using Fourier transform infrared (FTIR) spectroscopy. *Colloids and Surfaces A: Physicochemical and Engineering Aspects* 253: 117–124.

Hofmann, U. and R. Reingraber. 1969. Einlagerungsverbindungen in wasserarmem Halloysit. *Zeitschrift für anorganische und allgemeine Chemie* 369: 208–211.

Horváth, E., J. Kristóf, and R.L. Frost. 2010. Vibrational spectroscopy of intercalated kaolinites. Part I. *Applied Spectroscopy Reviews* 45: 130–147.

Horváth, E., J. Kristóf, R.L. Frost, A. Rédey, V. Vágvölgyi, and T. Cseh. 2003. Hydrazine-hydrate intercalated halloysite under controlled-rate thermal analysis conditions. *Journal of Thermal Analysis and Calorimetry* 71: 707–714.

Hounfodji, J.W., W.G. Kanhounnon, G. Kpotin et al. 2021. Molecular insights on the adsorption of some pharmaceutical residues from wastewater on kaolinite surfaces. *Chemical Engineering Journal* 407: 127176.

Hughes, J.C., R.J. Gilkes, and R.D. Hart. 2009. Intercalation of reference and soil kaolins in relation to physicochemical and structural properties. *Applied Clay Science* 45: 24–35.

Hui, L.C., N.A.N.N. Malek, M.Z.A. Awaluddin, M.H. Asraf, S.N. Ishak, and A.A. Hadi. 2019. Adsorption of gentamicin on surfactant-kaolinite and its antibacterial activity. *Malaysian Journal of Fundamental and Applied Sciences* 15: 686–689.

Ikhsan, J., B.B. Johnson, J.D. Wells, and M.J. Angove. 2004. Adsorption of aspartic acid on kaolinite. *Journal of Colloid and Interface Science* 273: 1–5.

Itagaki, T. and K. Kuroda. 2003. Organic modification of the interlayer surface of kaolinite with propanediols by transesterification. *Journal of Materials Chemistry* 13: 1064–1068.

Jacobs, H. and M. Sterckx. 1970. Contribution a l'étude de l'intercalation du dimethylsulfoxide dans le réseau de la kaolinite. *Reunion Hispano-Belga de Minerales de la Arcilla, Madrid* 154–160.

Janek, M., K. Emmerich, S. Heissler, and R. Nüesch. 2007. Thermally induced grafting reactions of ethylene glycol and glycerol intercalates of kaolinite. *Chemistry of Materials* 19: 684–693.

Jauković, V., D. Krajišnik, A. Daković et al. 2021. Influence of selective acid-etching on functionality of halloysite-chitosan nanocontainers for sustained drug release. *Materials Science & Engineering C* 123: 112029.

Jia, S. and M. Fan. 2019. Silanization of heat-treated halloysite nanotubes using -aminopropyltriethoxysilane. *Applied Clay Science* 180: 105204.

Jiang, W.-T., P.-H. Chang, Y. Tsai, and Z. Li. 2016. Halloysite nanotubes as a carrier for the uptake of selected pharmaceuticals. *Microporous and Mesoporous Materials* 220: 298–307.

Jin, X., R. Zhang, M. Su et al. 2019. Functionalization of halloysite nanotubes by enlargement and layer-by-layer assembly for controlled release of the fungicide iodopropynyl butylcarbamate. *RSC Advances* 9: 42062–42070.

Johnston, C.T., D.L. Bish, J. Eckert, and L.A. Brown. 2000. Infrared and inelastic neutron scattering study of the 1.03- and 0.95-nm kaolinite-hydrazine intercalation complexes. *Journal of Physical Chemistry B* 104: 8080–8088.

Johnston, C.T., G. Sposito, D.F. Bocian, and R.R. Birge. 1984. Vibrational spectroscopic study of the interlamellar kaolinite-dimethyl sulfoxide complex. *Journal of Physical Chemistry* 88: 5959–5964.

Johnston, C.T. and D.A. Stone. 1990. Influence of hydrazine on the vibrational modes of kaolinite. *Clays and Clay Minerals* 38: 121–128.

Jou, S.K. and N.A.N.N. Malek. 2016. Characterization and antibacterial activity of chlorhexidine loaded silver-kaolinite. *Applied Clay Science* 127–128: 1–9.

Joussein, E., S. Petit, J. Churchman, B. Theng, D. Righi, and B. Delvaux. 2005. Halloysite clay minerals – a review. *Clay Minerals* 40: 383–426.

Joussein, E., S. Petit, and B. Delvaux. 2007. Behavior of halloysite clay under formamide treatment. *Applied Clay Science* 35: 17–24.

Karaoğlu, M.H., M. Doğan, and M. Alkan. 2009. Removal of cationic dyes by kaolinite. *Microporous and Mesoporous Materials* 122: 20–27.

Karaoğlu, M.H., M. Doğan, and M. Alkan. 2010. Removal of reactive blue 221 by kaolinite from aqueous solutions. *Industrial & Engineering Chemistry Research* 49: 1534–1540.

Karewicz, A., A. Machowska, M. Kasprzyk, and G. Ledwójcik. 2021. Application of halloysite nanotubes in cancer therapy – a review. *Materials* 14: 2943. https://doi.org/10.3390/ma14112943.

Kasprzhitskii, A., G. Lazorenko, D.S. Kharytonnau, M.A. Osipenko, A.A. Kasach, and I.I. Kurilo. 2022. Adsorption mechanism of aliphatic amino acids on kaolinite surfaces. *Applied Clay Science* 226: 106566.

Ke, Z.-W., S.-J. Wei, P. Shen, Y.-M. Chen, and Y.-C. Li. 2023. Mechanism for the adsorption of *per-* and polyfluoroalkyl substances on kaolinite: Molecular dynamics modeling. *Applied Clay Science* 232: 106804.

Kelleher, B.P. and T.F. O'Dwyer. 2002. Intercalation of benzamide into expanded kaolinite under ambient environmental conditions. *Clays and Clay Minerals* 50: 331–335.

Kelleher, B.P., D. Sutton, and T.F. O'Dwyer. 2002. The effect of kaolinite intercalation on the structural arrangements of *N*-methylformamide and 1-methyl-2-pyrrolidone. *Journal of Colloid and Interface Science* 255: 219–224.

Khan, M.S.I. and J.B. Wiley. 2022. Rapid synthesis of kaolinite nanoscrolls through microwave processing. *Nanomaterials* 12: 3141. https://doi.org/103.390/nano12183141.

Khan, T.A., S. Dahiya, and I. Ali. 2012. Use of kaolinite as adsorbent: Equilibrium, dynamics and thermodynamic studies on the adsorption of Rhodamine B from aqueous solution. *Applied Clay Science* 69: 58–66.

Khelifa, M., S. Mellouk, G.L. Lecomte-Nana, I. Batonneau-Gener, K. Marouf-Khelifa, and A. Khelifa. 2023. Methodological approach to the chloramphenicol adsorption by acid-leached halloysites: Preparation, characterization, performance and mechanism. *Microporous and Mesoporous Materials* 348: 112412.

Kiani, G., M. Dostali, A. Rostami, and A.R. Khataee. 2011. Adsorption studies on the removal of malachite green from aqueous solutions onto halloysite nanotubes. *Applied Clay Science* 54: 34–39.

Kim, M. and S. Hyun. 2014. Effect of surface coordination on 2,4-D sorption by kaolinite from methanol/water mixture. *Chemosphere* 103: 329–335.

Kloprogge, J.T. 2017. Raman and infrared spectroscopies of intercalated kaolinite groups minerals. In *Infrared and Raman Spectroscopies of Clay Minerals*. Developments in Clay Science, Vol. 8, (Eds.) W.P. Gates, J.T. Kloprogge, J. Madejová, and F. Bergaya, pp. 343–410. Amsterdam, The Netherlands: Elsevier.

Kohyama, N., K. Fukushima, and A. Fukami. 1978. Observation of the hydrated form of tubular halloysite by an electron microscope equipped with an environmental cell. *Clays and Clay Minerals* 26: 25–40.

Komadel, P. and J. Madejová. 2006. Acid activation of clay minerals. In *Handbook of Clay Science*. Developments in Clay Science, Vol. 1, (Eds.) F. Bergaya, B.K.G. Theng, and G. Lagaly, pp. 263–287. Amsterdam, The Netherlands: Elsevier.

Komori, Y., H. Enoto, R. Takenawa, S. Hayashi, Y. Sugahara, and K. Kuroda. 2000. Modification of the interlayer surface of kaolinite with methoxy groups. *Langmuir* 16: 5506–5508.

Komori, Y., A. Matsumura, T. Itagaki, Y. Sugahara, and K. Kuroda. 1999b. Preparation of a kaolinite-ε-caprolactam intercalation compound. *Clay Science* 11: 47–55.

Komori, Y., Y. Sugahara, and K. Kuroda. 1999a. Intercalation of alkylamines and water into kaolinite with methanol kaolinite as an intermediate. *Applied Clay Science* 15: 241–252.

Koteja, A. and J. Matusik. 2019. Preparation of azobenzene-intercalated kaolinite and monitoring of the photo-induced activity. *Clay Minerals* 54: 57–66.

Kovács, A. and E. Makó. 2016. Cooling as the key parameter in formation of kaolinite-ammonium acetate and halloysite-ammonium acetate complexes using homogenization method. *Colloids and Surfaces A: Physicochemical and Engineering Aspects* 508: 70–78.

Koyuncu, H. and A.R. Kul. 2019. Removal of aniline from aqueous solution by activated kaolinite: Kinetic, equilibrium and thermodynamic studies. *Colloids and Surfaces A* 569: 59–66.

Koyuncu, H., A.R. Kul, N. Yıldız, A. Çalımlı, and H. Ceylan. 2007. Equilibrium and kinetic studies for the sorption of 3-methoxybenzaldehyde on activated kaolinites. *Journal of Hazardous Materials* 141: 128–139.

Krasilin, A.A., D.P. Danilovich, E.B. Yudina, S. Bruyere, J. Ghanbaja, and V.K. Ivanov. 2019. Crystal violet adsorption by oppositely twisted heat-treated halloysite and pecoraite nanoscrolls. *Applied Clay Science* 173: 1–11.

Kurczewska, J., M. Cegłowski, B. Messyasz, and G. Schroeder. 2018. Dendrimer-functionalized halloysite nanotubes for effective drug delivery. *Applied Clay Science* 153: 134–143.

Kurczewska, J., M. Cegłowski, and G. Schroeder. 2020. PAMAM-halloysite Dunino hybrid as an effective adsorbent of ibuprofen and naproxen from aqueous solutions. *Applied Clay Science* 190: 105603.

Kurdziel, K., M. Raczyńska-Żak, and A. Kolbus. 2021. Removal of phenol and monochlorophenols pollution from aqueous solutions with HDTMA-modified halloysite. *Solvent Extraction and Ion Exchange* 39: 399–423.

Kuroda, K. and C. Kato. 1979. Synthesis of the trimethylsilylation derivative of halloysite. *Clays and Clay Minerals* 27: 53–56.

Kuroda, Y., K. Ito, K. Itabashi, and K. Kuroda. 2011. One-step exfoliation of kaolinites and their transformation into nanoscrolls. *Langmuir* 27: 2028–2035.

Kushwaha, S.K.S., N. Kushwaha, P. Pandey, and B. Fatma. 2021. Halloysite nanotubes for nanomedicine: Prospects, challenges and applications. *BioNanoScience* 11: 200–208.

Kuśmierek, K., A. Świątkowski, E. Wierzbicka, and I. Legocka. 2020. Enhanced adsorption of direct orange 26 dye in aqueous solutions by modified halloysite. *Physicochemical Problems of Mineral Processing* 56: 693–701.

Lacin, D. and A.Z. Aroguz. 2020. Kinetic studies on adsorption behavior of methyl orange using modified halloysite, as an eco-friendly adsorbent. *SN Applied Sciences* 2: 2091. https://doi.org/10.1007/s42452-020-03799-4.

Lagaly, G., M. Ogawa, and I. Dékány. 2006. Clay mineral organic interactions. In *Handbook of Clay Science*. Developments in Clay Science, Vol. 1, (Eds.) F. Bergaya, B.K.G. Theng, and G. Lagaly, pp. 309–377. Amsterdam, The Netherlands: Elsevier.

Langier-Kuźniarowa, A. 2002. Thermal analysis of organo-clay complexes. In *Organo-Clay Complexes and Interactions*, (Eds.) S. Yariv and H. Cross, pp. 273–344. New York: Marcel Dekker.

Lawal, I.A. and B. Moodley. 2016. Column, kinetic and isotherm studies of PAH (phenanthrene) and dye (acid red) on kaolin modified with 1-hexyl, 3-decahexyl imidazolium ionic liquid. *Journal of Environmental Chemical Engineering* 4: 2774–2784.

Ledoux, R.L. and J.L. White. 1964. Infrared study of the OH groups in expanded kaolinite. *Science* 143: 244–246.

Ledoux, R.L. and J.L. White. 1965. Infrared studies of the hydroxyl groups in intercalated kaolinite complexes. *Clays and Clay Minerals* 13: 289–315.

Ledoux, R.L. and J.L. White. 1966a. Infrared studies of hydrogen bonding between kaolinite surfaces and intercalated potassium acetate, hydrazine, formamide and urea. *Journal of Colloid and Interface Science* 21: 127–152.

Ledoux, R.L. and J.L. White. 1966b. Infrared studies of hydrogen bonding of organic compounds on oxygen and hydroxyl surfaces of layer lattice silicates. *Proceedings of the International Clay Conference, Jerusalem* 1: 361–374.

Lee, H., K. Lee, J. Park, J. Lee, and J. Noh. 2020. Highly efficient adsorption of anionic dye on acid-treated halloysite nanotubes. *Journal of Nanoscience and Nanotechnology* 20: 5024–5027.

Lee, S.G., J.I. Choi, W. Koh, and S.S. Jang. 2013. Adsorption of β-D-glucose and cellobiose on kaolinite surfaces: Density functional theory (DFT) approach. *Applied Clay Science* 71: 73–81.

Lee, S.Y. and S.J. Kim. 2002. Adsorption of naphthalene by HDTMA modified kaolinite and halloysite. *Applied Clay Science* 22: 55–63.

Letaief, S. and C. Detellier. 2005. Reactivity of kaolinite in ionic liquids: Preparation and characterization of a 1-ethyl pyridinium chloride-kaolinite intercalate. *Journal of Materials Chemistry* 15: 4734–4740.

Letaief, S. and C. Detellier. 2007. Nanohybrid materials from the intercalation of imidazolium ionic liquids in kaolinite. *Journal of Materials Chemistry* 17: 1476–1484.

Letaief, S. and C. Detellier. 2008. Ionic liquids-kaolinite nanostructured materials: Intercalation of pyrrolidinium salts. *Clays and Clay Minerals* 56: 82–89.

Letaief, S. and C. Detellier. 2009. Functionalization of the interlayer surfaces of kaolinite by alkylammonium groups from ionic liquids. *Clays and Clay Minerals* 57: 638–648.

Letaief, S. and C. Detellier. 2011. Application of thermal analysis for the characterization of intercalated and grafted organo-kaolinite nanohybrid materials. *Journal of Thermal Analysis and Calorimetry* 104: 831–839.

Letaief, S., T. Diaco, W. Pell, S.I. Gorelsky, and C. Detellier. 2008. Ionic conductivity of nanostructured hybrid materials designed from imidazolium ionic liquids and kaolinite. *Chemistry of Materials* 20: 7136–7142.

Letaief, S., T.A. Elbokl, and C. Detellier. 2006. Reactivity of ionic liquids with kaolinite: Melt intersalation of ethylpyridinium chloride in an urea-kaolinite pre-intercalate. *Journal of Colloid and Interface Science* 302: 254–258.

Levis, S.R. and P.B. Deasy. 2003. Use of coated microtubular halloysite for the sustained release of diltiazem hydrochloride and propranolol hydrochloride. *International Journal of Pharmaceutics* 253: 145–157.

Li, H., X. Zhu, H. Zhou, and S. Zhong. 2015b. Functionalization of halloysite nanotubes by enlargement and hydrophobicity for sustained release of analgesic. *Colloids and Surfaces A: Physicochemical and Engineering Aspects* 487: 154–161.

Li, Q., T. Ren, P. Perkins, X. Hu, and X. Wang. 2021. Applications of halloysite nanotubes in food packaging for improving film performance and food preservation. *Food Control* 124: 107876.

Li, X., Q. Liu, H. Cheng, S. Zhang, and R.L. Frost. 2015a. Mechanism of kaolinite sheets curling via the intercalation and delamination process. *Journal of Colloid and Interface Science* 444: 74–80.

Li, X., X. Cui, S. Wang et al. 2017. Methoxy-grafted kaolinite preparation by intercalation of methanol: Mechanism of its structural variability. *Applied Clay Science* 137: 241–248.

Li, X., J. Ouyang, H. Yang, and S. Chang. 2016. Chitosan modified halloysite nanotubes as emerging porous microspheres for drug carrier. *Applied Clay Science* 126: 306–312.

Li, X., D. Wang, Q. Liu, and S. Komarneni. 2019. A comparative study of synthetic tubular kaolinite nanoscrolls and natural halloysite nanotubes. *Applied Clay Science* 168: 421–427.

Li, Y., X. Yuan, L. Jiang et al. 2022. Manipulation of the halloysite clay nanotube lumen for environmental remediation: A review. *Environmental Science: Nano* 9: 841–866.

Li, Z. and L. Gallus. 2005. Surface configuration of sorbed hexadecyltrimethylammonium on kaolinite as indicated by surfactant and counterion sorption, cation desorption, and FTIR. *Colloids and Surfaces A: Physicochemical and Engineering Aspects* 264: 61–67.

Li, Z. and L. Gallus. 2007. Adsorption of dodecyl trimethylammonium and hexadecyl trimethylammonium onto kaolinite – competitive adsorption and chain length. *Applied Clay Science* 35: 250–257.

Li, Z., H. Hong, L. Liao et al. 2011. A mechanistic study of ciprofloxacin removal by kaolinite. *Colloids and Surfaces B: Biointerfaces* 88: 339–344.

Li, Z., L. Schulz, C. Ackley, and N. Fenske. 2010. Adsorption of tetracycline on kaolinite with pH-dependent surface charges. *Journal of Colloid and Interface Science* 351: 254–260.

Lipsicas, M., R. Raythatha, R.F. Giese, Jr., and P.M. Costanzo. 1986. Molecular motions, surface interactions, and stacking disorder in kaolinite intercalates. *Clays and Clay Minerals* 34: 635–644.

Lipsicas, M., C. Straley, P.M. Costanzo, and R.F. Giese, Jr. 1985. Static and dynamic structure of water in hydrated kaolinites. II. The dynamic structure. *Journal of Colloid and Interface Science* 107: 221–230.

Liu, K, B. Zhu, Q. Feng et al. 2013. Adsorption of Cu(II) ions from aqueous solutions on modified chrysotile: Thermodynamic and kinetic studies. *Applied Clay Science* 80–81: 38–45.

Liu, L., F. Min, J. Chen, F. Lu, and L. Shen. 2019b. The adsorption of dodecylamine and oleic acid on kaolinite surfaces: Insights from DFT calculation and experimental investigation. *Applied Surface Science* 470: 27–35.

Liu, Q., X. Li, and H. Cheng. 2016. Insight into the self-adaptive deformation of kaolinite layers into nanoscrolls. *Applied Clay Science* 124–125: 175–182.

Liu, R., B. Zhang, D. Mei, H. Zhang, and J. Liu. 2011. Adsorption of methyl violet from aqueous solution by halloysite nanotubes. *Desalination* 268: 111–116.

Liu, S.-T., X.-G. Chen, S.-L. Zhang, X.-M. Liu, and J.-J. Zhang. 2021. Preparation and characterization of halloysite-based carriers for quercetin loading and release. *Clays and Clay Minerals* 69: 94–104.

Liu, Y.-L., H.W. Walker, and J.J. Lenhart. 2019a. Adsorption of microcystin-LR onto kaolinite, illite and montmorillonite. *Chemosphere* 220: 696–705.

Lloyd, M.K. and R.E. Conley. 1970. Adsorption studies on kaolinites. *Clays and Clay Minerals* 18: 37–46.

Luo, P., Y. Zhao, B. Zhang, J. Liu, Y. Yang, and J. Liu. 2010. Study on the adsorption of neutral red from aqueous solution onto halloysite nanotubes. *Water Research* 44: 1489–1497.

Lv, G., C. Stockwell, J. Niles, S. Minegar, Z. Li, and W-T. Jiang. 2013. Uptake and retention of amitriptyline by kaolinite. *Journal of Colloid and Interface Science* 411: 198–203.

Lv, G., L. Wu, Z. Li, L. Liao, and M. Liu. 2014. Binding sites of chlorpheniramine on 1:1 layered kaolinite from aqueous solution. *Journal of Colloid and Interface Science* 424: 16–21.

Lvov, Y., A. Panchal, Y. Fu et al. 2019. Interfacial self-assembly in halloysite nanotube composites. *Langmuir* 35: 8646–8657.

Lvov, Y., W. Wang, L. Zhang, and R. Fakhrullin. 2016. Halloysite clay nanotubes for loading and sustained release of functional compounds. *Advanced Materials* 28: 1227–1250.

MacEwan, D.M.C. 1948. Complexes of clays with organic compounds. I. Complex formation between montmorillonite and halloysite and certain organic liquids. *Transactions of the Faraday Society* 44: 349–367.

MacEwan, D.M.C. 1949. Clay mineral complexes with organic liquids. *Clay Minerals Bulletin* 3: 44–46.

MacEwan, D.M.C. 1962. Interlamellar reactions of clays and other substances. *Clays and Clay Minerals* 9: 431–443.

Machida, S., R. Guégan, and Y. Sugahara. 2021. A kaolinite-tetrabutylphosphonium bromide intercalation compound as an effective intermediate for intercalation of bulky organophosphonium salts. *Applied Clay Science* 206: 106038.

Machida, S., N. Idota, and Y. Sugahara. 2019. Interlayer grafting of kaolinite using trimethylphosphate. *Dalton Transactions* 48: 11663–11673.

Machida, S. and Y. Sugahara. 2022. An orientation of 4-dimethylaminopyridine in kaolinite interlayers. *Clay Science* 26: 9–15.

MacKinnon, I.D.R., P.J.R. Uwins, A.J.E. Yago, and J.G. Thomson. 1995. Grinding of kaolinite improves intercalate yield. In *Clays: Controlling the Environment. Proceedings of the 10th International Clay Conference, 1993, Adelaide*, (Eds.), G.J. Churchman, R.W. Fitzpatrick, and R.A. Eggleton, pp. 196–200. Melbourne, Australia: CSIRO Publishing.

Mahrez, N., F. Bessaha, K. Mrouf-Khelifa, A. Çoruh, and A. Khelifa. 2020. Performance and mechanism of interaction of crystal violet with organohalloysite. *Desalination and Water Treatment* 207: 410–419.

Makó, E., A. Kovács, V. Antal, and T. Kristóf. 2017. One-pot exfoliation of kaolinite by solvothermal cointercalation. *Applied Clay Science* 146: 131–139.

Makó, E., A. Kovács, Z. Ható, and J. Kristóf. 2015. Simulation assisted characterization of kaolinite-methanol intercalation complexes synthesized using cost-efficient homogenization method. *Applied Surface Science* 357: 626–634.

Makó, E., A. Kovács, Z. Ható, B. Zsirka, and T. Kristóf. 2014b. Characterization of kaolinite-ammonium acetate complexes prepared by one-step homogenization method. *Journal of Colloid and Interface Science* 431: 125–131.

Makó, E, A. Kovács, E. Horváth, and J. Kristóf. 2014a. Kaolinite-potassium acetate and halloysite-potassium acetate complexes prepared by mechanochemical, solution and homogenization techniques: A comparative study. *Clay Minerals* 49: 457–471.

Makó, E., A. Kovács, R. Katona, and T. Kristóf. 2016. Characterization of kaolinite-cetyltrimethylammonium chloride intercalation complex synthesized through eco-friend kaolinite-urea pre-intercalation complex. *Colloids and Surfaces A: Physicochemical and Engineering Aspects* 508: 265–271.

Makó, E., A. Kovács, and T. Kristóf. 2019. Influencing parameters of direct homogenization intercalation of kaolinite with urea, dimethyl sulfoxide, formamide, and *N*-methylformamide. *Applied Clay Science* 182: 105287.

Makó, E., J. Kristóf, E. Horváth, and V. Vágvölgyi. 2009. Kaolinite-urea complexes obtained by mechanochemical and aqueous suspension techniques – a comparative study. *Journal of Colloid and Interface Science* 330: 367–373.

Makó, E., J. Kristóf, E. Horváth, and V. Vágvölgyi. 2013. Mechanochemical intercalation of low reactivity kaolinite. *Applied Clay Science* 83–84: 24–31.

Makó, E., Z. Sarkadi, Z. Ható, and T. Kristóf. 2023. Characterization of kaolinite-3-aminopropyltriethoxysilane intercalation complexes. *Applied Clay Science* 231: 106753.

Malek, N.A.N.N. and N.I. Ramli. 2015. Characterization and antibacterial activity of cetylpyridinium bromide (CPB) immobilized on kaolinite with different CPB loadings. *Applied Clay Science* 109–110: 8–14.

Massaro, M., A. Campofelice, C.G. Colletti, G. Lazzara, R. Noto, and S. Riela. 2018b. Functionalized halloysite nanotubes: Efficient carrier systems for antifungine drugs. *Applied Clay Science* 160: 186–192.

Massaro, M., G. Cavallaro, C.G. Colletti et al. 2018a. Chemical modification of halloysite nanotubes for controlled loading and release. *Journal of Materials Chemistry B* 6: 3415–3433.

Massaro, M., G. Cavallaro, C.G. Colletti et al. 2018c. Halloysite nanotubes for efficient loading, stabilization and controlled release of insulin. *Journal of Colloid and Interface Science* 524: 156–164.

Massaro, M., G. Lazzara, S. Milioto, R. Noto, and S. Riela. 2017. Covalently modified halloysite clay nanotubes: Synthesis, properties, biological and medical applications. *Journal of Materials Chemistry B* 5: 2867–2882.

Massaro, M., R. Noto, and S. Riela. 2020a. Past, present and future perspectives on halloysite clay minerals. *Molecules* 25: 4863. https://doi.org/10.3390/molecules25204863.

Massaro, M., P. Poma, C.G. Colletti et al. 2020b. Chemical and biological evaluation of cross-linked halloysite-curcumin derivatives. *Applied Clay Science* 184: 105400.

Massaro, M. and S. Riela. 2018. Organoclay nanomaterials based on halloysite and cyclodextrin as carriers for polyphenolic compounds. *Journal of Functional Biomaterials* 9: 61. https://doi.org/10.3390/jfb9040061.

Massaro, M., S. Riela, P. Lo Meo et al. 2014. Functionalized halloysite multivalent glycocluster as a new drug delivery system. *Journal of Materials Chemistry B* 2: 7732–7738.

Mata-Arjona, A., A. Ruiz-Amil, and E. Inaraja-Martin. 1970. Cinetica del proceso de sorcion del dimetilsulfoxido en caolinita: Estudio por difraccion de rayos X. *Reunion Hispano-Belga de Minerales de la Arcilla, Madrid*, 115–120.

Matusik, J., A. Gaweł, E. Bielańska, W. Osuch, and K. Bahranowski. 2009. The effect of structural order on nanotubes derived from kaolin-group minerals. *Clays and Clay Minerals* 57: 452–464.

Matusik, J., A. Gawel, and K. Bahranowski. 2012a. Grafting of methanol in dickite and intercalation of hexylamine. *Applied Clay Science* 56: 63–67.

Matusik, J. and Z. Kłapyta. 2013. Characterization of kaolinite intercalation compounds with benzalkylammonium chlorides using XRD, TGA/DTA and CHNS elemental analysis. *Applied Clay Science* 83–84: 433–440.

Matusik, J., Z. Kłapyta, and Z. Olejniczak. 2013. NMR and IR study of kaolinite intercalation compounds with benzylalkylammonium chlorides. *Applied Clay Science* 83–84: 426–432.

Matusik, J., E. Scholtzova, and D. Tunega. 2012b. Influence of synthesis conditions on the formation of a kaolinite-methanol complex and simulation of its vibrational spectra. *Clays and Clay Minerals* 60: 227–239.

Mbey, J.A., F. Thomas, C.J. Ngally Sabouang, Liboum, and D. Njopwouo. 2013. An insight on the weakening of the interlayer bonds in a Cameroonian kaolinite through DMSO intercalation. *Applied Clay Science* 83–84: 327–335.

Mehdi, K., S. Bendenia, G.L. Lecomte-Nana et al. 2019. A new approach about the intercalation of hexadecyltrimethylammonium into halloysite: Preparation, characterization, and mechanism. *Chemical Papers* 73: 131–139.

Mellouk, S., S. Cherifi, M. Sassi et al. 2009. Intercalation of halloysite from Djebel Debagh (Algeria) and adsorption of copper ions. *Applied Clay Science* 44: 230–236.

Menard, H., L. Noel, J. Khorami, J.-L. Jouve, and J. Dunnigan. 1986. The adsorption of polyaromatic hydrocarbons on natural and chemically modified asbestos fibers. *Environmental Research* 40: 84–91.

Mendelovici, E. and R.L. Frost. 2005. Pioneer studies on HCl and silylation treatments of chrysotile. *Journal of Colloid and Interface Science* 289: 597–599.

Meziane, O., A. Bensedira, M. Guessoum, and N. Haddaoui. 2017. Preparation and characterization of intercalated kaolinite with: Urea, dimethylformamide and an alkylammonium salt using guest displacement reaction. *Journal of Materials and Environmental Sciences* 8: 3625–3636.

Michalková, A. and D. Tunega. 2007. Kaolinite-dimethylsulfoxide intercalate – A theoretical study. *Journal of Physical Chemistry C* 111: 11259–11266.

Michalková, A., D. Tunega, and L.T. Nagy. 2002. Theoretical study of interactions of dickite and kaolinite with small organic molecules. *Journal of Molecular Structure (Theochem)* 581: 37–49.

Mobaraki, M., S. Karnik, Y. Li, and D.K. Mills. 2022. Therapeutic applications of halloysite. *Applied Sciences* 12: 87. https://doi.org/10.3390/app12010087.

Moon, V. 2016. Halloysite behaving badly: Geomechanics and slope behaviour of halloysite-rich soils. *Clay Minerals* 51: 517–528.

Mouni, L., L. Belkhiri, J.-C. Bollinger et al. 2018. Removal of methylene blue from aqueous solutions by adsorption on kaolin: Kinetic and equilibrium studies. *Applied Clay Science* 153: 38–45.

Moyo, F., R. Tandlich, B.S. Wilhelmi, and S. Balaz. 2014. Sorption of hydrophobic organic compounds on natural sorbents and organoclays from aqueous and non-aqueous solutions: A mini-review. *International Journal of Environmental Research and Public Health* 11: 5020–5048.

Mrozik, W., C. Jungnickel, M. Skup, P. Urbaszek, and P. Stepnowski. 2008. Determination of the adsorption mechanism of imidazolium-type ionic liquids onto kaolinite: Implications for their fate and transport in the soil environment. *Environmental Chemistry* 5: 299–306.

Murakami, J., T. Itagaki, and K. Kuroda. 2004. Synthesis of kaolinite-organic nanohybrids with butanediols. *Solid State Ionics* 172: 279–282.

Nagasawa, K. 1969. Kaolin minerals in Cenozoic sediments of central Japan. *Proceedings of the International Clay Conference, Tokyo* 1: 15–30.

Nandi, B.K., A. Goswami, and M.K. Purkait. 2009a. Adsorption characteristics of brilliant green dye on kaolin. *Journal of Hazardous Materials* 161: 387–395.

Nandi, B.K., A. Goswami, and M.K. Purkait. 2009b. Removal of cationic dyes from aqueous solutions by kaolin: Kinetic and equilibrium studies. *Applied Clay Science* 42: 583–590.

Neto, J.C.M., N.R. Nascimento, R.H. Bello et al. 2022. Kaolinite review: Intercalation and production of polymer nanocomposites. *Engineered Science* 17: 28–44.

Ngulube, T., J.R. Gumbo, V. Masindi, and A. Maity. 2019. Evaluation of the efficacy of halloysite nanotubes in the removal of acidic and basic dyes from aqueous solution. *Clay Minerals* 54: 197–207.

Nicolini, K.P., C.R.B. Fukamachi, F. Wypych, and A.S. Mangrich. 2009. Dehydrated halloysite intercalated mechanochemically with urea: Thermal behavior and structural aspects. *Journal of Colloid and Interface Science* 338: 474–479.

Niu, J. 2016. Formation mechanisms of tubular structure of halloysite. In *Nanosized Tubular Clay Minerals*. Developments in Clay Science, Vol. 7, (Eds.) P. Yuan, A. Thill, and F. Bergaya, pp. 387–407. Amsterdam, The Netherlands: Elsevier.

Olejnik, S., L.A.G. Aylmore, A.M. Posner, and J.P. Quirk. 1968. Infrared spectra of kaolin mineral-dimethyl sulfoxide complexes. *Journal of Physical Chemistry* 72: 241–249.

Olejnik, S., A.M. Posner, and J.P. Quirk. 1970. The intercalation of polar organic compounds into kaolinite. *Clay Minerals* 8: 421–434.

Olejnik, S., A.M. Posner, and J.P. Quirk. 1971a. The I.R. spectra of interlamellar kaolinite-amide complexes. I. The complexes of formamide, *N*-methylformamide and dimethylformamide. *Clays and Clay Minerals* 19: 83–94.

Olejnik, S., A.M. Posner, and J.P. Quirk. 1971b. The infrared spectra of interlamellar kaolinite-amide complexes. II. Acetamide, *N*-methylacetamide and dimethylacetamide. *Journal of Colloid and Interface Science* 37: 536–547.

Olejnik, S., A.M. Posner, and J.P. Quirk. 1971c. Infrared spectrum of the kaolinite-pyridine-*N*-oxide complex. *Spectrochimica Acta* 27A: 2005–2009.
Ottaviani, M.F. and F. Venturi. 1996. Physicochemical study on the adsorption properties of asbestos. 1. EPR study on the adsorption of organic radicals. *Journal of Physical Chemistry* 100: 265–273.
Ottaviani, M.F., F. Venturi, M.R. Pokhrel, T. Schmutz, and S.H. Bossmann. 2001. Physicochemical studies on the adsorption properties of asbestos. *Journal of Colloid and Interface Science* 238: 371–380.
Pasbakhsh, P. and G.J. Churchman (Eds.). 2015. *Natural Mineral Nanotubes*. Oakville, ON: Apple Academic Press.
Pasbakhsh, P., R. de Silva, V. Vahedi, and G.J. Churchman. 2016. Halloysite nanotubes: Prospects and challenges of their use as additives and carriers – a focused review. *Clay Minerals* 51: 479–487.
Peixoto, A., A.C. Fernandes, C. Pereira, J. Pires, and C. Freire. 2016. Physicochemical characterization of organosilylated halloysite clay nanotubes. *Microporous and Mesoporous Materials* 219: 145–154.
Peng, X., J. Wang, B. Fan, and Z. Luan. 2009. Sorption of endrin to montmorillonite and kaolinite clays. *Journal of Hazardous Materials* 168: 210–214.
Pereira, V.D.A., I. dos S. Paz, A.L. Gomes, L.A. Leite, P.B.A. Fechine, and M.D.S. Filho. 2021. Effects of acid activation on the halloysite nanotubes for curcumin incorporation and release. *Applied Clay Science* 200: 105953.
Pham, T.H., T.C. Nguyen, A.K. Le, P.K. Le, and D.N. Son. 2022. Insights into effects of metal cations on the adsorption of benzotriazole on halloysite nanotubes: An experimental and DFT study. *Journal of Physical Chemistry C* 126: 2920–2929.
Polati, S., S. Angioi, V. Gianotti, F. Gosetti, and M.C. Gennaro. 2006. Sorption of pesticides on kaolinite and montmorillonite as a function of hydrophilicity. *Journal of Environmental Science and Health, Part B* 41: 333–344.
Raupach, M., P.F. Barron, and J.G. Thompson. 1987. Nuclear magnetic resonance, infrared and X-ray diffraction study of dimethylsulfoxide and dimethylselenoxide intercalates with kaolinite. *Clays and Clay Minerals* 35: 208–219.
Rida, K., S. Bouraoui, and S. Hadnine. 2013. Adsorption of methylene blue from aqueous solution by kaolin and zeolite. *Applied Clay Science* 83–84: 99–105.
Riela, S., M. Massaro, C.G. Colletti et al. 2014. Development and characterization of co-loaded curcumin/triazole-halloysite systems and evaluation of their potential anticancer activity. *International Journal of Pharmaceutics* 475: 613–623.
Ruiz Cruz, M.D. and F. Franco. 2000a. Thermal behavior of the kaolinite-hydrazine intercalation complex. *Clays and Clay Minerals* 48: 63–67.
Ruiz Cruz, M.D. and F. Franco. 2000b. Thermal decomposition of a dickite-hydrazine intercalation complex. *Clays and Clay Minerals* 48: 586–592.
Ruiz Cruz, M.D. and F.I. Franco Duro. 1999. New data on the kaolinite-potassium acetate complex. *Clay Minerals* 34: 565–577.
Ruiz-Hitzky, E. and J.J. Fripiat. 1976. Organomineral derivatives obtained by reacting organochlorosilanes with the surface of silicates in organic solvents. *Clays and Clay Minerals* 24: 25–30.
Rutkai, G., E. Makó, and T. Kristóf. 2009. Simulation and experimental study of intercalation of urea in kaolinite. *Journal of Colloid and Interface Science* 334: 65–69.
Saadat, S., D. Rawtani, and G. Parikh. 2022. Clay minerals-based drug delivery systems for anti-tuberculosis drugs. *Journal of Drug Delivery Science and Technology* 76: 103755.
Sabahi, H., M. Khorami, A.H. Rezayan, Y. Jafari, and M.H. Karami. 2018. Surface functionalization of halloysite nanotubes via curcumin inclusion. *Colloids and Surfaces A* 538: 834–840.
Sadiku, M., T. Selimi, A. Berisha et al. 2022. Removal of methyl violet from aqueous solution by adsorption onto halloysite nanoclay: Experiment and theory. *Toxics* 10: 445. https://doi.org/10.3390/toxics10080445.
Saif, M.J., H.M. Asif, and M. Naveed. 2018. Properties and modification methods of halloysite nanotubes: A state-of-the-art review. *Journal of the Chilean Chemical Society* 63: 3. http://dx.doi.org/10.4067/s0717-97072018000304109.
Salaa, F., S. Bendenia, G.L. Lecomte-Nana, and A. Khelifa. 2020. Enhanced removal of diclofenac by an organohalloysite intercalated via a novel route: Performance and mechanism. *Chemical Engineering Journal* 396: 125226.
Sanz, J. and J.M. Serratosa. 2002. Nuclear magnetic resonance spectroscopy of organo-clay complexes. In *Organo-Clay Complexes and Interactions*, (Eds.) S. Yariv and H. Cross, pp. 223–272. New York: Marcel Dekker.
Sargin, I. and N. Ünlü. 2013. Insights into cationic methyl violet 6B dye-kaolinite interactions: Kinetic, equilibrium and thermodynamic studies. *Clay Minerals* 48: 85–95.

Sato, M. 1999. Preparation of kaolinite-amino acid intercalates derived from hydrated kaolinite. *Clays and Clay Minerals* 47: 793–802.

Satokawa, S., R. Miyawaki, S. Tomura, and Y. Shibasaki. 1997. DMSO-intercalation of synthetic kaolinites. *Clay Science* 10: 231–239.

Scholtzová, E., L. Benco, and D. Tunega. 2008. A model study of dickite intercalated with formamide and *N*-methylformamide. *Physics and Chemistry of Minerals* 35: 299–309.

Scott, A.M., E.A. Burns, and F.C. Hill. 2014. Theoretical study of adsorption of nitrogen-containing environmental contaminants on kaolinite surfaces. *Journal of Molecular Modeling* 20: 2373.

Seifi, S., M.T. Diatta-Dieme, P. Blanchart, G.L. Lecomte-Nana, D. Kobor, and S. Petit. 2016. Kaolin intercalated by urea: Ceramic applications. *Construction and Building Materials* 113: 579–585.

Septian, A., S. Oh, and W.S. Shin. 2019. Sorption of antibiotics onto montmorillonite and kaolinite: Competition modelling. *Environmental Technology* 40: 2940–2953.

Serratosa, J.M. 1966. Infrared analysis of the orientation of pyridine molecules in clay complexes. *Clays and Clay Minerals* 14: 385–391.

Serratosa, J.M., A. Hidalgo, and J.M. Vinas. 1963. Infrared study of the OH groups in kaolin minerals. *Proceedings of the International Clay Conference, Stockholm* 1: 17–26.

Seto, H., M.I. Cruz-Cumplido, and J.J. Fripiat. 1978. Reactivity of a long-spacing ammonium propionate-kaolinite intercalate toward diol, diamines and quaternary ammonium salts. *Clay Minerals* 13: 309–323.

Setter, O.P., L. Dahan, H.A. Hamad, and E. Segal. 2022. Acid-etched halloysite nanotubes as superior carriers for ciprofloxacin. *Applied Clay Science* 228: 106629.

Setter, O.P. and E. Segal. 2020. Halloysite nanotubes – the nano-bio interface. *Nanoscale* 12: 23444–23460.

Shaik, S.A., U. Roy, S. Sengupta, and A. Goswami. 2023. Adsorption of Safranin O on halloysite nanotubes: A mechanistic case study for efficient wastewater remediation. *International Journal of Environmental Science and Technology* 20: 5405–5426.

Shu, Y., L. Li, Q. Zhang, and H. Wu. 2010. Equilibrium, kinetics and thermodynamic studies for sorption of chlorobenzenes on CTMAB modified bentonite and kaolinite. *Journal of Hazardous Materials* 1173: 47–53.

Sid, D., M. Baitiche, L. Arrar et al. 2022. Improved biological performance of ketoprofen using novel modified halloysite clay nanotubes. *Applied Clay Science* 216: 106341.

Sid, D., M. Baitiche, R. Bourzami et al. 2021.Experimental and theoretical studies of the interaction of ketoprofen in halloysite nanotubes. *Colloids and Surfaces A: Physicochemical and Engineering Aspects* 627: 127136.

Sidheswaran, P., A.N. Bhat, and P. Ganguli. 1990. Intercalation of salts of fatty acids into kaolinite. *Clays and Clay Minerals* 38: 29–32.

Siegnin, R., G.K. Dedzo, and E. Ngameni. 2022. Sulfonation of the interlayer surface of kaolinite. *Applied Clay Science* 226: 106570.

Słomkiewicz, B. Szczepanik, and M. Czaplicka. 2020. Adsorption of phenol and chlorophenols by HDTMA modified halloysite nanotubes. *Materials* 13: 3309. https://doi.org/10.3390/ma13153309.

Smalley, I.J., C.W. Ross, and J.S. Whitton. 1980. Clays from New Zealand support the inactive particle theory of soil sensitivity. *Nature* 288: 576–577.

Soma, M., G.J. Churchman, and B.K.G. Theng. 1992. X-ray photoelectron spectroscopic analysis of halloysites with different composition and particle morphology. *Clay Minerals* 27: 413–421.

Soma, Y., H. Seyama, and M. Soma. 1993. Adsorption of fluorescent dyes to chrysotile asbestos. *Clay Science* 9: 9–20.

Song, K.-H., X. Wang, P. Qian, C. Zhang, and Q. Zhang. 2013. Theoretical study of interaction of formamide with kaolinite. *Computational and Theoretical Chemistry* 1020: 72–80.

Song, Y., E.A. Sackey, H. Wang, and H. Wang. 2019. Adsorption of oxytetracycline on kaolinite. *PLoS One* 14(11): e0225335. https://doi.org/10.1371/journal.pone.0225335.

Spasojević, M., A. Daković, G.E. Rottinghaus et al. 2021. Influence of surface coverage of kaolin with surfactant ions on adsorption of ochratoxin A and zearalenone. *Applied Clay Science* 205: 106040.

Spepi, A., C. Duce, A. Pedone et al. 2016. Experimental and DFT characterization of halloysite nanotubes loaded with salicylic acid. *Journal of Physical Chemistry C* 120: 26759–26769.

Stodolak-Zych, E., A. Rapacz-Kmita, M. Gajek, A. Różycka, M. Dudek, and S. Kluska. 2023. Functionalized halloysite nanotubes as potential drug carriers. *Journal of Functional Biomaterials* 14: 167. https://doi.org/10.3390/jfb14030167.

Sugahara, Y., S. Satokawa, K.-I. Yoshioka, K. Kuroda, and C. Kato. 1989. Kaolinite-pyridine intercalation compound derived from hydrated kaolinite. *Clays and Clay Minerals* 37: 143–150.

Suzzoni, A., L. Barre, E. Kohler, P. Levitz, L.J. Michot, and J. M'Hamdi. 2018. Interactions between kaolinite clay and AOT. *Colloids and Surfaces A* 556: 309–315.

Szczepanik, B., P. Słomkiewicz, M. Garnuszek, and K. Czech. 2014. Adsorption of chloroanilines from aqueous solutions on the modified halloysite. *Applied Clay Science* 101: 260–264.

Szczepanik, B., P. Słomkiewicz, M. Garnuszek et al. 2017. Effect of temperature on halloysite acid treatment for efficient chloroaniline removal from aqueous solutions. *Clays and Clay Minerals* 65: 155–167.

Tan, D., P. Yuan, F. Annabi-Bergaya, F. Dong, D. Liu, and H. He. 2015. A comparative study of tubular halloysite and platy kaolinite as carriers for the loading and release of the herbicide amitrole. *Applied Clay Science* 114: 190–196.

Tan, D., P. Yuan, F. Annabi-Bergaya, D. Liu, and H. He. 2014a. High-capacity loading of 5-fluorouracil on the methoxy-modified kaolinite. *Applied Clay Science* 100: 60–65.

Tan, D., P. Yuan, F. Annabi-Bergaya et al. 2013. Natural halloysite nanotubes as mesoporous carriers for the loading of ibuprofen. *Microporous and Mesoporous Materials* 179: 89–98.

Tan, D., P. Yuan, F. Annabi-Bergaya et al. 2014b. Loading and *in vitro* release of ibuprofen in tubular halloysite. *Applied Clay Science* 96: 50–55.

Tan, D., P. Yuan, F. Dong, H. He, S. Sun, and Z. Liu. 2018. Selective loading of 5-fluorouracil in the interlayer space of methoxy-modified kaolinite for controlled release. *Applied Clay Science* 159: 102–106.

Tan, D., P. Yuan, D. Liu, and P. Du. 2016. Surface modifications of halloysite. In *Nanosized Tubular Clay Minerals: Halloysite and Imogolite*. Developments in Clay Science, Vol. 7, (Eds.) P. Yuan, A. Thill, and F. Bergaya, pp. 167–201. Amsterdam, The Netherlands: Elsevier.

Tanaka, Y., T. Okada, and M. Ogawa. 2009. Adsorption of tetrakis (*p*-sulfonatophenyl) porphyrin on kaolinite. *Journal of Porous Materials* 16: 623–629.

Tao, Q., L. Su, R.L. Frost et al. 2014. Silylation of mechanically ground kaolinite. *Clay Minerals* 49: 559–568.

Tchoumene, R., G.K. Dedzo, and E. Ngameni. 2018. Preparation of methyl viologen-kaolinite intercalation compound: Controlled release and electrochemical applications. *ACS Applied Materials & Interfaces* 10: 34534–34542.

Tharmavaram, M., G. Pandey, and D. Rawtani. 2018. Surface modified halloysite nanotubes: A flexible interface for biological, environmental and catalytic applications. *Advances in Colloid and Interface Science* 261: 82–101.

Theng, B.K.G. (Ed.). 1980. *Soils with Variable Charge*. Lower Hutt, NZ: New Zealand Society of Soil Science.

Theng, B.K.G. 1995. On measuring the specific surface area of clays and soils by adsorption of *para*-nitrophenol: Use and limitations. In *Clays: Controlling the Environment. Proceedings of the 10th International Clay Conference, 1993, Adelaide*, (Eds.) G.J. Churchman, R.W. Fitzpatrick, and R.A. Eggleton, pp. 304–310. Melbourne, Australia: CSIRO Publishing.

Theng, B.K.G. 2018. *Clay Mineral Catalysis of Organic Reactions*. Boca Raton, FL: CRC Press.

Theng, B.K.G., G.J. Churchman, J.S. Whitton, and G.G.C. Claridge. 1984. Comparison of intercalation methods for differentiating halloysite from kaolinite. *Clays and Clay Minerals* 32: 249–258.

Theng, B.K.G., M. Russell, G.J. Churchman, and R.L. Parfitt. 1982. Surface properties of allophane, halloysite, and imogolite. *Clays and Clay Minerals* 30: 143–149.

Theng, B.K.G. and N. Wells. 1995. The flow characteristics of halloysite suspensions. *Clay Minerals* 30: 99–106.

Theng, B.K.G. and G. Yuan. 2008. Nanoparticles in the soil environment. *Elements* 4: 395–399.

Thomas, J.E. and M.J. Kelley. 2009. The adsorption of salicylic acid onto γ-alumina and kaolinite from solution in hexane studied using diffuse reflectance infrared Fourier transform spectroscopy (DRIFT). *Journal of Colloid and Interface Science* 338: 389–394.

Thomas, R., C.B. Shoemaker, and K. Eriks. 1966. The molecular and crystal structure of dimethyl sulfoxide, Me_2SO. *Acta Crystallographica* 21: 12–20.

Thompson, J.G. 1985. Interpretation of solid state ^{13}C and ^{29}Si nuclear magnetic resonance spectra of kaolinite intercalates. *Clays and Clay Minerals* 33: 173–180.

Thompson, J.G. and C. Cuff. 1985. Crystal structure of kaolinite: Dimethylsulfoxide intercalate. *Clays and Clay Minerals* 33: 490–500.

Tonle, I.K., S. Letaief, E. Ngameni, and C. Detellier. 2009. Nanohybrid materials from the grafting of imidazolium cations on the interlayer surfaces of kaolinite: Application as electrode modifier. *Journal of Materials Chemistry* 19: 5996–6003.

Tsiantos, C., M. Tsampodimou, G.H. Kacandes, M.S del Río, V. Gionis, and G.D. Chryssikos. 2012. Vibrational investigation of indigo-palygorskite association(s) in synthetic Maya blue. *Journal of Materials Science* 47: 3415–3428.

Tsunematsu, K. and H. Tateyama. 1999. Delamination of urea-kaolinite complex by using intercalation procedures. *Journal of the American Ceramic Society* 82: 1589–1591.
Tunney, J.J. and C. Detellier. 1993. Interlamellar covalent grafting of organic units on kaolinite. *Chemistry of Materials* 5: 747–748.
Tunney, J.J. and C. Detellier. 1994. Preparation and characterization of two distinct ethylene glycol derivatives of kaolinite. *Clays and Clay Minerals* 42: 552–560.
Tunney, J.J. and C. Detellier. 1996. Chemically modified kaolinite: Grafting of methoxy groups on the interlamellar aluminol surface of kaolinite. *Journal of Materials Chemistry* 6: 1679–1685.
Ukkund, S.J., P. Puthiyillam, H.M. Alshehri et al. 2021a. Adsorption method for the remediation of brilliant green dye using halloysite nanotube: Isotherm, kinetic and modeling studies. *Applied Sciences* 11: 8088. https://doi.org/10.3390/app11178088.
Ukkund, S.J., P. Puthiyillam, A.E. Anqi et al. 2021b. A recent study on remediation of direct blue 15 dye using halloysite nanotubes. *Applied Sciences* 11: 8196. https://doi.org/10.3390/app11178196.
Unuabonah, E.I., K.O. Adebowale, and F.A. Dawodu. 2008. Equilibrium, kinetic and Sorber design studies on the adsorption of aniline blue dye by sodium tetraborate-modified kaolinite clay adsorbent. *Journal of Hazardous Materials* 157: 397–409.
Uwins, P.J.R., I.D.R. Mackinnon, J.G. Thompson, and A.J.E. Yago. 1993. Kaolinite: NMF intercalates. *Clays and Clay Minerals* 41: 707–717.
Valášková, M., M. Rieder, V. Matějka, P. Čapková, and A. Slíva. 2007. Exfoliation/delamination of kaolinite by low-temperature washing of kaolinite-urea intercalates. *Applied Clay Science* 35: 108–118.
Valerio, F., R. Puntoni, and L. Santi. 1979. Adsorption properties of U.I.C.C. Rhodesian chrysotile and crocidolite in aqueous solution – effects of cation depletion. *American Industrial Hygiene Association Journal* 40: 781–788.
Van Olphen, H. 1966. Maya blue: A clay-organic pigment? *Science* 154: 645–646.
Vargas-Rodríguez, Y.M., A. Obaya, J.F. García-Petronilo et al. 2021. Adsorption studies of aqueous solutions of methyl green for halloysite nanotubes: Kinetics, isotherms, and thermodynamic parameters. *American Journal of Nanomaterials* 9: 1–11.
Vimonses, V., S. Lei, B. Jin, C.W.K. Chow, and C. Saint. 2009. Adsorption of Congo red by three Australian kaolins. *Applied Clay Science* 43: 465–472.
Wada, K. 1959a. Oriented penetration of ionic compounds between the silicate layers of halloysite. *American Mineralogist* 44: 153–165.
Wada, K. 1959b. An interlayer complex of halloysite with ammonium chloride. *American Mineralogist* 44: 1237–1247.
Wada, K. 1961. Lattice expansion of kaolin minerals by treatment with potassium acetate. *American Mineralogist* 46: 78–91.
Wada, K. 1968. Hydrazine intercalation – intersalation for differentiation of kaolin minerals from chlorites. *American Mineralogist* 53: 334–339.
Wang, D., Q. Liu, H. Cheng, S. Zhang, and X. Zuo. 2017a. Effect of reaction temperature on intercalation of octyltrimethylammonium chloride into kaolinite. *Journal of Thermal Analysis and Calorimetry* 128: 1555–1564.
Wang, D., Q. Liu, D. Hou, H. Cheng, and R.L. Frost. 2017b. The formation mechanism of organoammonium-kaolinite by solid-solid reaction. *Applied Clay Science* 146: 195–200.
Wang, J., L. Fu, H. Yang, X. Zuo, and D. Wu. 2021. Energetics, interlayer molecular structures, and hydration mechanisms of dimethyl sulfoxide (DMSO)-kaolinite nanoclay guest-host interactions. *Journal of Physical Chemistry Letters* 12: 9973–9981.
Wang, L., A. Lu, Z. Xiao, J. Ma, and Y. Li. 2009. Modification of nano-fibriform silica by dimethyldichlorosilane. *Applied Surface Science* 255: 7542–7546.
Wang, Q., J. Zhang, and A. Wang. 2013b. Alkali activation of halloysite for adsorption and release of ofloxacin. *Applied Surface Science* 287: 54–61.
Wang, Q., J. Zhang, Y. Zheng, and A. Wang. 2014. Adsorption and release of ofloxacin from acid- and heat-treated halloysite. *Colloids and Surfaces B: Biointerfaces* 113: 51–58.
Wang, X., P. Qian, K. Song, C. Zhang, and J. Dong. 2013a. The DFT study of adsorption of 2,4-dinitrotoluene on kaolinite surfaces. *Computational and Theoretical Chemistry* 1025: 16–23.
Wang, Z., W. Zheng, Z. Zhang et al. 2017c. Formation of a kaolinite-serine intercalation compound via exchange of the pre-intercalated transition molecules in kaolinite with serine. *Applied Clay Science* 135: 378–385.
Weiss, A. 1961. Eine Schichteinschlussverbindung von Kaolinit mit Harnstoff. *Angewandte Chemie* 73: 736.
Weiss, A. 1963. A secret of Chinese porcelain manufacture. *Angewandte Chemie International Edition in English* 2: 697–703.

Weiss, A., H.O. Becker, H. Orth, G. Mai, H. Lechner, and K.J. Range. 1970. Particle size effect and reaction mechanism of the intercalation into kaolinite. *Proceedings of the 1969 International Clay Conference, Tokyo* 2: 180–184.

Weiss, A. and H. Orth. 1973. Zur Kenntnis der Intercalationsverbindungen von Kaolinit, Nakrit, Dickit und Halloysit mit Pyridin-N-oxid und Picolin-N-oxid. *Zeitschrift für Naturforschung B* 28: 252–254.

Weiss, A. and J. Russow. 1963. Über das Einrollen von Kaolinitkristallen zu Halloysitähnlichen Röhren und einen Unterschied zwischen Halloysit und röhrchenformigem Kaolinit. *Proceedings of the International Clay Conference, Stockholm* 2: 69–74.

Weiss, A. and W. Thielepape. 1963. Über eine Verbesserung der technischen Eigenschaften von Kaolinit durch die Herstellung von Einlagerungsverbindungcn. *Proceedings of the International Clay Conference, Stockholm* 2: 427–429.

Weiss, A., W. Thielepape, G. Göring, W. Ritter, and H. Schäfer. 1963a. Kaolinit-Einlagerungs-Verbindungen. *Proceedings of the International Clay Conference, Stockholm* 1: 287–305.

Weiss, A., W. Thielepape, and H. Orth. 1966. Neue Kaolinit-Einlagerungsverbindungen. *Proceedings of the International Clay Conference, Jerusalem* 1: 277–293.

Weiss, A., W. Thielepape, W. Ritter, H. Schäfer, and G. Göring. 1963b. Zur Kenntnis von Hydrazin-Kaolinit. *Zeitschrift für anorganische und allgemeine Chemie* 320: 183–204.

White, G.N., D.V. Bavykin, and F.C. Walsh. 2012. The stability of halloysite nanotubes in acidic and alkaline aqueous suspensions. *Nanotechnology* 23: 065705. https://doi.org/10.1088/0957-4484/23/6/065705.

Wierzbicka, E., K. Kuśmierek, A. Światkowski, and I. Legocka. 2022. Efficient rhodamine B dye removal from water by acid- and organo-modified halloysite. *Minerals* 12: 350. https://doi.org/10.3390/min12030350.

Wiewióra, A. and G.W. Brindley. 1969. Potassium acetate intercalation in kaolinites and its removal: Effect of material characteristics. *Proceedings of the International Clay Conference, Tokyo* 1: 723–733.

Wu, Q., Z. Li, and H. Hong. 2013a. Adsorption of the quinolone antibiotic nalidixic acid onto montmorillonite and kaolinite. *Applied Clay Science* 74: 66–73.

Wu, Q., Z. Li, H. Hong, R. Li, and W.-T. Jiang. 2013b. Desorption of ciprofloxacin from clay mineral surfaces. *Water Research* 47: 259–268.

Wypych, F., W.H. Schreiner, N. Mattoso, D.H. Mosca, R. Marangoni, and C.A. da S. Bento. 2003. Covalent grafting of phenylphosphonate groups onto layered silica derived from *in situ*-leached chrysotile fibers. *Journal of Materials Chemistry* 13: 304–307.

Xie, Y., D. Qian, D. Wu, and X. Ma. 2011. Magnetic halloysite nanotubes/iron oxide composites for the adsorption of dyes. *Chemical Engineering Journal* 168: 959–963.

Xu, H., M. Wang, Q. Liu et al. 2011. Stability of the compounds obtained by intercalating potassium acetate molecules in kaolinite from coal measures. *Journal of Physics and Chemistry of Solids* 72: 24–28.

Yang, S.-Q., P. Yuan, H.-P. He et al. 2012. Effect of reaction temperature on grafting of γ-aminopropyl triethoxysilane (APTES) onto kaolinite. *Applied Clay Science* 62–63: 8–14.

Yang, D., L. Xie, E. Bobicki, Z. Xu, Q. Liu, and H. Zeng. 2014. Probing anisotropic surface properties and interaction forces of chrysotile rods by atomic force microscopy and rheology. *Langmuir* 30: 10809–10817.

Yang, H., Y. Zhang, and J. Ouyang. 2016. Physicochemical properties of halloysite. In *Nanosized Tubular Clay Minerals*. Developments in Clay Science, Vol. 7, (Eds.) P. Yuan, A. Thill, and F. Bergaya, pp. 67–91. Amsterdam, The Netherlands: Elsevier.

Yang, Y., Y. Chen, F. Leng, L. Huang, Z. Wang, and W. Tian. 2017. Recent advances on surface modification of halloysite nanotubes for multifunctional applications. *Applied Sciences* 7: 1215. https://doi.org/10.3390/app7121215.

Yang, Y., Z. Zhong, J. Li, H. Du, and Z. Li. 2022. Efficient with low-cost removal and adsorption mechanisms of norfloxacin, ciprofloxacin and ofloxacin on modified thermal kaolin: Experimental and theoretical studies. *Journal of Hazardous Materials* 430: 128500.

Yariv, S. and I. Lapides. 2008. Thermo-infrared spectroscopy analysis of dimethyl-sulfoxide-kaolinite intercalation complexes. *Journal of Thermal Analysis and Calorimetry* 94: 433–440.

Yilmaz, B., E.T. Irmak, Y. Turhan, S. Doğan, M. Doğan, and O. Turhan. 2019. Synthesis, characterization and biological properties of intercalated kaolinite nanoclays: Intercalation and biocompatibility. *Advances in Materials Science* 19: 83–99.

Yu, C. and E. Bi. 2015. Roles of functional groups of naproxen in its sorption to kaolinite. *Chemosphere* 138: 335–339.

Yu, L., H. Wang, Y. Zhang, B. Zhang, and J. Liu. 2016. Recent advances in halloysite nanotube derived composites for water treatment. *Environmental Science: Nano* 3: 28–44.

Yuan, P., P.D. Southon, Z. Liu, and C.J. Kepert. 2012. Organosilane functionalization of halloysite nanotubes for enhanced loading and controlled release. *Nanotechnology* 23: 375705.

Yuan, P., P.D. Southon, Z. Liu et al. 2008. Functionalization of halloysite clay nanotubes by grafting with γ-aminopropyltriethoxysilane. *Journal of Physical Chemistry C* 112: 15742–15751.

Yuan, P., D. Tan, and F. Annabi-Bergaya. 2015. Properties and applications of halloysite nanotubes: Recent research advances and future prospects. *Applied Clay Science* 112–113: 75–93.

Yuan, P., D. Tan, F. Annabi-Bergaya, W. Yan, D. Liu, and Z. Liu. 2013. From platy kaolinite to aluminosilicate nanoroll via one-step delamination of kaolinite: Effect of the temperature of intercalation. *Applied Clay Science* 83–84: 68–76.

Yuan, P., A. Thill, and F. Bergaya (Eds.). 2016. *Nanosized Tubular Clay Minerals: Halloysite and Imogolite*. Developments in Clay Science, Vol. 7. Amsterdam, The Netherlands: Elsevier.

Zaharia, A., F.-X. Perrin, M. Teodorescu et al. 2015. New organophilic kaolin clays based on single-point grafted 3-aminopropyl dimethylethoxysilane. *Physical Chemistry Chemical Physics* 17: 24908–24916.

Zamama, M. and M. Knidiri. 2000. IR study of dickite-formamide intercalate, $Al_2Si_2O_5(OH)_4$-H_2NCOH. *Spectrochimica Acta Part A* 56: 1139–1147.

Zargarian, S.S., V. Haddadi-Asl, and H. Hematpour. 2015. Carboxylic acid functionalization of halloysite nanotubes for sustained release of diphenhydramine hydrochloride. *Journal of Nanoparticle Research* 17: 218. https://doi.org/10.1007/s11051-015-3032-3.

Zeitler, T.R., J.A. Greathouse, R.T. Cygan, J.T. Fredrich, and G.R. Jerauld. 2017. Molecular dynamics simulation of resin adsorption at kaolinite edge sites: Effect of surface deprotonation on interfacial structure. *Journal of Physical Chemistry C* 121: 22787–22796.

Zhang, A.-B., L. Pan, H.-Y. Zhang et al. 2012. Effects of acid treatment on the physico-chemical and pore characteristics of halloysite. *Colloids and Surfaces A: Physicochemical and Engineering Aspects* 396: 182–188.

Zhang, H. 2017. Selective modification of inner surface of halloysite nanotubes: A review. *Nanotechnology Reviews* 6: 573–581.

Zhang, L., C.-J. Wang, Z.-F. Yan et al. 2013a. Kaolinite nanomaterial: Intercalation of 1-butyl-3-methylimidazolium bromine in a methanol-kaolinite pre-intercalate. *Applied Clay Science* 86: 106–110.

Zhang, Q., C. Yang, W. Huang, Z. Dang, and X. Shu. 2013b. Sorption of tylosin on clay minerals. *Chemosphere* 93: 2180–2186.

Zhang, R., W. Yan, and C. Jing. 2014. Mechanistic study of PFOS adsorption on kaolinite and montmorillonite. *Colloids and Surfaces A: Physicochemical and Engineering Aspects* 462: 252–258.

Zhang, S., Q. Liu, H. Cheng, F. Gao, C. Liu, and B.J. Teppen. 2018c. Mechanism responsible for intercalation of dimethyl sulfoxide in kaolinite: Molecular dynamics simulations. *Applied Clay Science* 151: 46–53.

Zhang, S., Q. Liu, H. Cheng, and F. Zeng. 2015a. Combined experimental and theoretical investigation of interactions between kaolinite inner surface and intercalated dimethyl sulfoxide. *Applied Surface Science* 331: 234–240.

Zhang, S., Q. Liu, H. Cheng, Y. Zhang, X. Li, and R.L. Frost. 2015b. Intercalation of γ-aminopropyl triethoxysilane (APTES) into kaolinite interlayer with methanol-grafted kaolinite as intermediate. *Applied Clay Science* 114: 484–490.

Zhang, S., Q. Liu, F. Gao, R. Ma, Z. Wu, and B.J. Teppen. 2018b. Interfacial structure and interaction of kaolinite intercalated with *N*-methylformamide insight from molecular dynamics modeling. *Applied Clay Science* 158: 204–210.

Zhang, S., Q. Liu, F. Gao, and B.J. Teppen. 2018a. Molecular dynamics simulation of basal spacing, energetics, and structure evolution of a kaolinite-formamide intercalation complex and their interaction. *Journal of Physical Chemistry C* 122: 3341–3349.

Zhang, S., Q. Liu, F. Gao et al. 2017. Mechanism associated with kaolinite intercalation with urea: Combination of infrared spectroscopy and molecular dynamics simulation studies. *Journal of Physical Chemistry C* 121: 402–409.

Zhang, R., Y. Li, Y. He, and D. Qin. 2020b. Preparation of iodopropynyl butylcarbamate loaded halloysite and its anti-mildew activity. *Journal of Materials Research and Technology* 9: 10148–10156.

Zhang, Y., Y. Li, Y. Zhang. 2020a. Preparation and intercalation structure model of halloysite-stearic acid intercalation compound. *Applied Clay Science* 187: 105451.

Zhang, Y., Y. Li, Y. Zhang et al. 2019. Thermal behavior and kinetic analysis of halloysite-stearic acid complex. *Journal of Thermal Analysis and Calorimetry* 135: 2429–2436.

Zhao, M. and P. Liu. 2008. Adsorption behavior of methylene blue on halloysite nanotubes. *Microporous and Mesoporous Materials* 112: 419–424.

Zhao, Y., J. Geng, X. Wang, X. Gu, and S. Gao. 2011. Tetracycline adsorption on kaolinite: pH, metal cations and humic acid effects. *Ecotoxicology* 20: 1141–1147.

Zhao, Y., X. Gu, S. Gao, J. Geng, and X. Wang. 2012. Adsorption of tetracycline (TC) onto montmorillonite: Cations and humic acid effects. *Geoderma* 183–184: 12–18.

Zhao, Y., X. Gu, S. Li, R. Han, and G. Wang. 2015. Insights into tetracycline adsorption onto kaolinite and montmorillonite: Experimental and modeling. *Environmental Science and Pollution Research* 22: 17031–17040.

Zheng, W., J. Zhou, Z. Zhang et al. 2014. Formation of intercalation compound of kaolinite-glycine via displacing guest water by glycine. *Journal of Colloid and Interface Science* 432: 278–284.

Zhong, X.-H., Y. Liu, T. Xu, and W.-Y. Liu. 2017. Studies on the thermal behavior and decomposition mechanism of dickite-potassium acetate complexes. *Journal of Thermal Analysis and Calorimetry* 129: 1095–1102.

Zhu, X., C. Yan, and J. Chen. 2012. Application of urea-intercalated kaolinite for paper coating. *Applied Clay Science* 55: 114–119.

Zhuang, G., M. Jaber, F. Rodrigues, B. Rigaud, P. Walter, and Z. Zhang. 2019. A new durable pigment with hydrophobic surface based on natural nanotubes and indigo: Interactions and stability. *Journal of Colloid and Interface Science* 552: 204–217.

Zuo, X., D. Wang, S. Zhang, Q. Liu, and H. Yang. 2017. Effect of interaction agents on morphology of exfoliated kaolinite. *Minerals* 7: 249. https://doi.org/10.3390/min7120249.

7 Organic Complexes of Clay Minerals with Layer-Ribbon and Short-Range Order Structures

7.1 INTRODUCTION

In comparison with the 2:1 and 1:1 layer silicates, the fibrous clay minerals with a layer-ribbon structure, represented by palygorskite and sepiolite (cf. Table 1.1), have attracted less attention with respect to their interactions with organic compounds. The same comment could be made regarding clay minerals with short-range order structure, exemplified by allophane and imogolite. Furthermore, all these minerals are incapable of intercalating organic species, precluding X-ray diffraction from assessing organic complex formation.

Nevertheless, palygorskite and sepiolite (together with their organically modified forms) have found biomedical, cosmetic, industrial, and pharmaceutical applications as nanocarriers of drugs and nanofillers of polymers because of their abundance, biocompatibility, low cost, and non-toxicity (Chang et al. 2009; Eusepi et al. 2020; Cao et al. 2022; Saadat et al. 2022; Yang and Wang 2022; Biddeci et al. 2023; Ruiz et al. 2023). On the other hand, the allophane-organic interaction has attracted attention because allophane-rich soils have a large propensity for adsorbing and stabilizing organic matter (Theng 1980; Saggar et al. 1994; Parfitt et al. 1997; Dahlgren et al. 2004; Bruun et al. 2010; Chevallier et al. 2010; Calabi-Floody et al. 2011; Ding et al. 2019), while imogolite-organic complex formation lies behind the actual and potential applications of this nanotubular mineral in biology, catalysis, gas storage, and especially in the synthesis of polymer nanocomposites (Ohashi et al. 2004; Yah et al. 2010; Ma et al. 2012a; Yuan and Wada 2012; Paineau 2018; Theng 2018; Govan et al. 2021).

7.2 PALYGORSKITE AND SEPIOLITE

Both palygorskite and sepiolite can take up a variety of organic compounds into their channel (tunnel) structure without causing layer-ribbon expansion or any visible change in their X-ray diffraction patterns (Tartaglione et al. 2008; Chang et al. 2009; Xi et al. 2010; Bakhtiary et al. 2013; Tsai et al. 2016; Fitaroni et al. 2019; Hernández et al. 2023). Steric factors, however, are known to influence the adsorption of organic molecules and their access to the fibre interior (Leal et al. 2017). Thus, straight-chain hydrocarbons are more easily adsorbed than their branched counterparts (Nederbragt 1949). Similarly, small and compact organic molecules, such as methanol and ethanol, can access the channels of palygorskite whereas longer chain alcohols apparently fail to do so (Serna and Vanscoyoc 1979; Serratosa 1979; Kuang and Detellier 2005). Likewise, the channel structure of both palygorskite and sepiolite is accessible to butylamine and pyridine (Shuali et al. 1990, 1991). The drug praziquantel, however, can only enter the (larger) sepiolite channel (Borrego-Sánchez et al. 2018, 2020), while aflatoxin apparently fails to access the channels of either natural or heated palygorskite as well as those of sepiolite (Akbar et al. 2022).

In measuring the surface area of sepiolite by sorption of nitrogen, oxygen, methylamine, and ethylamine vapours, Dandy (1968) found evidence to indicate that these molecules could penetrate

DOI: 10.1201/9781003080244-7

the channel structure, at least partially. Earlier, Barrer and Mackenzie (1954) proposed similarly for the sorption of methanol and ethanol by palygorskite. Interestingly, palygorskite also showed selective uptake of hydrocarbons in the order *n*-heptane > isooctane and *n*-pentane > iso-pentane > neopentane (Barrer et al. 1954). Surface area measurements by Inagaki et al. (1990), using a variety of adsorbates, also indicate that the channels of sepiolite are accessible to ethanol, pyridine, and benzene but not to ethylbenzene, 1,3,5-trimethylbenzene, and *tert*-butylbenzene. Inagaki et al. (1995) further found that isoprene, adsorbed to intrachannel (and external particle) surfaces of sepiolite, could be polymerized through a Brønsted acid catalyzed mechanism (Theng 2018).

Likewise, such compounds as amitriptyline, crystal violet (CV), desferrioxamine-B, fluoroquinolone, methylene blue (MB), and spiramycin are excluded because of their large molecular dimensions (Rytwo et al. 1998; Shariatmadari et al. 1999; Peng et al. 2010; Shirvani and Nourbakhsh 2010; Chen et al. 2011; Sturini et al. 2016; Tsai et al. 2016; Habibi et al. 2018; Youcef et al. 2019). Nevertheless, partial channel penetration by indigo (Ovarlez et al. 2009; Giustetto et al. 2011), methylene blue (Zhang et al. 2015b; Moreira et al. 2017), quaternary ammonium cations (Gök et al. 2008; Karataş et al. 2013), phenanthrene (González-Santamaría et al. 2017), and Basic Red 46 (Chen et al. 2019) is possible, if not plausible.

Even chlorophyll-*a* and β-carotene can apparently penetrate the channels of sepiolite that has been 'acid-activated' by treatment with nitric acid (Sabah 2007; Sabah et al. 2007). By increasing specific surface area and porosity, acid activation (Rusmin et al. 2016; Habibi et al. 2018; Zhu et al. 2018) and ion beam bombardment (Zhang et al. 2013) can enhance the capacity of palygorskite for adsorbing organic compounds. Thus, acid-activated and thermally treated sepiolite, modified with quaternary ammonium ions, is an effective adsorbent of BTEX (benzene, toluene, ethylbenzene, xylene) compounds (Varela et al. 2021).

Prior freeze-drying of palygorskite can markedly improve its performance as a substrate for the sustained release of antibiotics, such as ofloxacin (Wang et al. 2014). Earlier, Sugiura and Fukumoto (1992) reported that sepiolite complexed with 2-aminobenzoic acid was superior to the raw mineral in adsorbing acetaldehyde. Likewise, sepiolite modified with ethylenediamine tetraacetic acid (EDTA) and hexadecyltrimethylammonium can effectively remove phenol and chlorophenol from ethanolic solutions (Yildiz and Gür 2007).

For organic molecules that are not sterically constrained, those with strongly polar groups are generally preferred to weakly polar species. Uncharged polar compounds, including 2-aminopyrimidine, isoniazid, nicotinamide, pyrene, pyridine, pyridine derivatives, quinoline are taken up by displacing the zeolitic water in the channels (cf. Figures 1.17 and 1.18), forming H-bonds to (external) silanol groups and water molecules coordinated to Mg(II) ions at the ribbon edge, or by direct coordination to terminal Mg(II) cations (Barrer 1978; Serna and Vanscoyoc 1979; Fernandez Hernandez and Fernandez Alvarez 1983; Mikhail et al. 1983; Singer 1989; Sabah and Çelik 2002; Akyüz and Akyüz 2003, 2005; Kuang and Detellier 2004; Kuang et al. 2003, 2006; Vico and Acebal 2006; Ovarlez et al. 2009; Akyüz et al. 2001, 2010; Cobas et al. 2014; Sabah and Ouki 2017; Carazo et al. 2018). We might also add that adenine and guanine are stable under gamma irradiation when adsorbed to palygorskite (Meléndez-López et al. 2023).

On the other hand, many non-polar compounds are excluded from, or can only gain limited entry into, the channel structure (Barrer 1978; Serratosa 1979; Singer 1989). Since such compounds would interact only weakly through (non-specific) van der Waals forces, the extent of their sorption is much less than that shown by polar molecules of comparable size (Haden and Schwint 1967).

Gerstl and Yaron (1981) also made the interesting observation that parathion was isomerized on adsorption to Ca^{2+}-palygorskite that had been preheated at temperatures up to 850°C. In line with the structural changes that occur during dehydration and dehydroxylation of palygorskite (cf. Chapter 1), the phosphate group of adsorbed parathion was apparently distorted, making the molecule conducive to isomerization. Earlier, Gerstl and Mingelgrin (1979) reported that Ca^{2+}-palygorskite could take up more paraquat from water than its tetramethylammonium (TMA)-exchanged counterpart. As described in Chapter 2, it seems likely that in the former instance paraquat interacts with exchangeable Ca^{2+} by means of a water bridge. In the latter case, on the other hand, the pesticide molecule would be adsorbed to the siloxane surface exposed between TMA-occupied exchange sites. The high uptake of paraquat by hexadecyltrimethylammonium (HDTMA)-montmorillonite may be ascribed to solute partitioning between organoclay and water (Roberts et al. 2006). A similar mechanism might be operative in the uptake of naphthalene by dodecyltrimethylammonium-modified sepiolite (Gök et al. 2008) and in the adsorption of oxytetracycline by sepiolite modified with HDTMA and sodium dodecyl benzene sulphonate (Wu et al. 2020).

The interactions of surfactants, dyes, and herbicides with palygorskite and sepiolite have been summarized by Shuali et al. (2011). The review by Mu and Wang (2016) focuses on methods of modifying palygorskite to enhance its capacity for adsorbing dyes, while that by Kausar et al. (2018) is concerned with the removal of dyes from textile wastewater by different clay minerals, including palygorskite and sepiolite. The formation, properties, and functional applications of surface-modified palygorskite and sepiolite have been summarized by Wang and Wang (2019b) and Tian et al. (2019), respectively.

In the case of monovalent cationic dyes, such as CV, MB, astrazon blue FGRL, maxilon blue 5G and GRL, carbinol form of methyl green, and thioflavine-T, the maximum amount adsorbed can far exceed the cation exchange capacity (CEC) of the clay mineral samples (Aznar et al. 1992; Rytwo et al. 1998, 2000; Shariatmadari et al. 1999; Casal et al. 2001; Doğan et al. 2006; Özdemir et al. 2006; Karagozoglu et al. 2007; Alkan et al. 2008; Ongen et al. 2012; Mu and Wang 2016; He et al. 2018; Yang et al. 2018). In this regard, MB is less 'reactive' because of its tendency to form aggregates in solution (Al-Futaisi et al. 2007; Youcef et al. 2019). However, prior dry grinding, heating, and treatment with chloroacetic acid, sodium hexametaphosphate, and sodium humate as well as plasma activation can markedly enhance the capacity of palygorskite for adsorbing MB (Chen et al. 2011; Liu et al. 2012; Zhang et al. 2015a, 2021; He et al. 2018; Sousa et al. 2019). Similarly, prior treatment of sepiolite with low molecular organic acids (acetic, citric, oxalic) has a positive effect on the adsorption of quinclorac (Yang et al. 2021).

Kinetic studies suggest that the rate-limiting step is intra-particle diffusion and commonly follows pseudo second-order kinetics (Alkan et al. 2007; Doğan et al. 2006, 2007; Cobas et al. 2014; Han et al. 2014; Duman et al. 2015). On the other hand, Santos and Boaventura (2016) reported a maximum uptake close to the CEC for the cationic Basic Red 46 azo dye using an impure sample of sepiolite. The amount adsorbed over and above the CEC is presumably associated with (neutral) silanol groups on external particle/fibre surfaces (Casal et al. 2001), possibly in the form of multilayers (Al-Futaisi et al. 2007). Sepiolite complexed with the cationic dye, thioflavine-T, can stabilize the photo-labile trifluralin herbicide, while sepiolite modified with crystal violet is effective in removing dispersed organic contaminants from winery effluents (Rytwo et al. 2011).

A two-stage process has also been proposed by Sevim et al. (2011) for the adsorption of the large, multivalent cationic cobalt porphyrazine to sepiolite. The first stage is one of cation exchange while the second stage is characterized by π–π and electrostatic interactions. In the case of quaternary ammonium cations, the amount adsorbed increases with chain length (Karataş et al. 2013). Adsorption here can also exceed the CEC (of sepiolite) with the excess forming multilayers on the fibre surface which becomes hydrophobic as a result (Lemić et al. 2005; Zhuang et al. 2018a). More recently, Yang et al. (2023) prepared hydrophobic sepiolite by attaching saturated fatty acids to the mineral surface and found that the complex with octadecanoic acid was efficient in removing diesel oil from water.

On the other hand, divalent organic cations (e.g., diquat, paraquat and methyl green) do not adsorb much above the CEC (of sepiolite), indicative of a cation exchange process (Rytwo et al. 2002). As such, their uptake is little influenced by solution pH whereas solute interactions with silanol groups are pH dependent (Hajjaji et al. 2006; Eren and Afsin 2007). Similarly, the adsorption of ciprofloxacin (Chang et al. 2016) and amitriptyline (Tsai et al. 2016) to palygorskite at acidic to neutral pH is one of cation exchange whereas uptake at alkaline pH apparently occurs through cation-bridging and complexation (Sposito et al. 1999; Theng 2012) Electrostatic attraction, supplemented by H-bonding, has also been proposed by Meirelles et al. (2019) for the palygroskite-ethambutol interaction, while Hernández et al. (2023) have suggested that van der Waals and hydrophobic interactions are involved in the uptake of sulfamethoxazole by palygorskite at acid pH.

Although solution pH affects the uptake of cationic dyes by sepiolite (Özdemir et al. 2006), the pH-dependency of adsorption is especially pronounced with anionic dyes, such as Acid Red 57 (Alkan et al. 2004); Acid Blue 25, 62, 193, and 294 (Alkan et al. 2005b; Özcan et al. 2005, 2006; Han et al. 2014); Reactive Blue 15, 62, and 221 (Alkan et al. 2005b; Tabak et al. 2009); and Brilliant Yellow (Bingol et al. 2010). Here uptake markedly increases or reaches a maximum at acidic pH when the edge surface of the fibres is positively charged, allowing the dye molecules to adsorb through electrostatic interactions. Hydrogen-bonding of the anionic group of methyl orange to zeolitic and coordinated water in palygorskite as well as cation-bridging (Chen et al. 2012; Yang et al. 2018) have also been proposed. More recently, de Souza et al. (2021) measured the adsorption to a Brazilian palygorskite of several dyes (Nile blue, methylene blue, dithizone) under neutral to basic pH conditions. In this regard, we might mention that palygorskite coated with iron nanoparticles and then exchanged with cetyltrimethylammonium (CTA) cations can effectively decolourize and degrade Acid orange 7 dye (Quan et al. 2018). Similarly, an Algerian palygorskite that has been modified with iron oxide can effectively remove linuron from wastewater (Belaroui et al. 2018).

The interactions of palygorskite (and sepiolite) with indigo dye have received a great deal of attention because the resultant clay-indigo complex ('nanohybrid') is identifiable with 'Maya blue'. This pigment was used by the Mayas in mural paintings (Figure 7.1), pottery, and sculptures during the first millennium AD in Mesoamerica (Gómez-Romero and Sanchez 2005; Arnold et al. 2008, 2012; Sánchez del Rio et al. 2011). Maya blue is formed through a solid-solid reaction between clay and dye followed by mild heating (120–150°C) of the complex (van Olphen 1966). A similar approach was used by Wang et al. (2021) to incorporate fluorescent yellow X-10GFF dye molecules.

FIGURE 7.1 Painting of a warrior with a Maya blue background. (From Gómez-Romero, P. and Sanchez, C., *New J. Chem.*, 29, 57–58, 2005.)

Thermal analysis and solid-state NMR spectroscopy indicate that the carbonyl and amino groups of indigo form hydrogen bonds with silanol groups at fibre edges (Hubbard et al. 2003). As a result, the channel openings are blocked preventing entry of dye molecules into intra-particle pores. Synchrotron powder diffraction and Raman spectroscopy by del Rio et al. (2009) confirm channel blocking by indigo molecules but also raise the possibility of partial channel penetration. Molecular modelling (Chiari et al. 2003) and total energy density functional theory (DFT) calculations (Sánchez-Ochoa et al. 2017) together with thermal and infrared measurements (Giustetto et al. 2011; Ovarlez et al. 2006, 2011) indicate that indigo can be H-bonded not only to edge silanols but also to structural water and octahedrally coordinated Mg^{2+} (cf. Figures 1.17 and 1.18). The role of coordinated water in stabilizing the indigo-sepiolite hybrid system has been described by Li et al. (2023). Thus, the channel structure of palygorskite and sepiolite is, at least partially, accessible to indigo (Reinen et al. 2004), although Fois et al. (2003) have argued for full accessibility based on molecular dynamics simulations. In this regard, Yin et al. (2022) have pointed out that the capacity of sepiolite in taking up cationic dyes is not only influenced by molecular size but also dependent on the type of functional group.

Zhang et al. (2015c, 2015d) were able to synthesize a Maya blue-like pigment using methylene blue (MB) or methyl violet (MV) as the dye component. They first attached the dye to palygorskite (PAL) and then coated the PAL-MB or PAL-MV complex with a layer of silica through polycondensation of tetraethoxysilane in the presence of acetic acid or CTA bromide. The resultant PAL-MB/MV-SiO_2 pigment was resistant to chemical elution, heating, and UV radiation. Similarly, Ouellet-Plamondon et al. (2015) encapsulated methylene blue (and methyl red) into sepiolite to obtain a coloured pigment resembling Maya blue, while Giustetto et al. (2017) obtained a palygorskite-methyl red composite by grinding the mineral with the dye at 140°C under acidic conditions.

Palygorskite and sepiolite are also good adsorbents of (long-chain) quaternary ammonium cations and related surfactants, such as benzyldimethyl octadecylammonium (Zhuang et al. 2018b), cetylpyridinium (Bakhtiary et al. 2013), didodecyl dimethylammonium (Ghrab et al. 2018), dihexadecyl dimethylammonium (Rodríguez-Cruz et al. 2008), dioctadecyl dimethylammonium (Sarkar et al. 2011, 2012), dodecyltrimethylammonium (Özcan and Gök 2012), hexadecyltrimethylammonium (Casal et al. 2001; Li et al. 2003; Lei et al. 2009; Liu et al. 2014; de Gois da Silva et al. 2014; Zhao et al. 2014; Middea et al. 2017; Karataş et al. 2018; Weng et al. 2018; Fitaroni et al. 2019; Silva et al. 2021), octadecyltrimethylammonium (Huang et al. 2008; Sarkar et al. 2011), and stearyl dimethylbenzylammonium (Kovacević et al. 2011), while Zhao et al. (2016, 2017) chose a gemini surfactant (N1-dodecyl-N1, N1, N2, N2-tetramethyl-N2-octylethane-1,2-diaminium bromide, 12–2–12) as the modifying surfactant.

Up to the CEC of these fibrous minerals, the surfactants are adsorbed by cation-exchange (electrostatic attraction), while solute uptake in excess of the CEC largely involves hydrophobic and van der Waals interactions (Yuan et al. 2013; Karataş et al. 2018). As a result, the (external) fibre surface becomes essentially hydrophobic and may even acquire positive charges to which anionic solutes are attracted (Li et al. 2003; Sarkar et al. 2010; Middea et al. 2017). More recently, Lobato-Aguilar et al. (2023) incorporated benzalkonium chloride into palygorskite (and montmorillonite) up to twice the CEC of the minerals. After an initial burst of release during the first 5 h of testing time, the adsorbed antibacterial drug was progressively released over the following 19 h period.

Organically modified palygorskite and sepiolite are organophilic and show a high capacity for adsorbing and removing a variety of organic solutes, including pesticides (Akçay and Yurdakoç 2000; Casal et al. 2001; Akçay et al. 2005; Sanchez-Martin et al. 2006; Rodríguez-Cruz et al. 2008; Xi et al. 2010; Kovacević et al. 2011; Bakhtiary et al. 2013), polycyclic aromatics (Gök et al. 2008; Özcan and Gök 2012; Zhao et al. 2016, 2017), tannins (Huang et al. 2008), terpenic compounds (Ghrab et al. 2018), and anionic dyes (Özcan and Özcan 2005; Huang et al. 2007b; Chen and Zhao 2009; Sarkar et al. 2011; Middea et al. 2017; Karataş et al. 2018) as well as organic contaminants and pollutants in wastewater, industrial effluents, and soil (Huang et al. 2007a, 2008; Sarkar et al. 2010, 2012; Liu et al. 2014). The adsorption process commonly conforms to the Langmuir isotherm and follows pseudo second-order kinetics (Özcan and Özcan 2005; Huang et al. 2007a; Gök et al. 2008; Chen and Zhao 2009; Sarkar et al. 2012; Zhao et al. 2014; Zhang et al. 2018).

Here we might also mention that the complex of sepiolite with Ethomeen T/15 (surfactant) can function as slow-release formulation of the herbicide mesotrione (Galán-Jiménez et al. 2020). In developing a controlled release formulation, Maqueda et al. (2009) used ultrasound to entrap the herbicide metribuzin in a sepiolite gel, while Zhang et al. (2016) used a one-pot hydrothermal process to modify palygorskite with various organic acids. The resultant complexes with chloroacetic and citric acids were much more effective than the parent mineral in removing several dyes (methylene blue, methyl violet, malachite green) from solution. Subsequently, Weng et al. (2018) and Zhuang et al. (2019) reported that surfactant-modified palygorskite and sepiolite can suitably serve as additives in oil-based drilling fluids.

Palygorskite and sepiolite can be rendered organophilic by methylation (of surface silanol groups) with diazomethane (Hermosín and Cornejo 1986) or more commonly, by silylation with organochlorosilanes following acid treatment of the minerals (Ruiz-Hitzky and Fripiat 1976; Alkan et al. 2005c; Undabeytia et al. 2019) or alkoxysilanes (Mendelovici and Portillo 1976; He et al. 2013; Liu et al. 2018). Among the alkoxysilanes, γ-aminopropyl triethoxysilane (APTES) has been the most widely used silylating agent (Demirbaş et al. 2007; Cui et al. 2013; Mu and Wang 2016; Moreira et al. 2017). We refer to Chapter 4 for more details.

Silylated palygorskite and sepiolite are essentially organophilic and hence show a high propensity for adsorbing organic solutes, including dyes (Xue et al. 2010, 2011; Moreira et al. 2017; Zhang et al. 2018) and pharmaceuticals (Undabeytia et al. 2019). These materials are also good adsorbents of heavy metals (Celis et al. 2000; Doğan et al. 2008; Cui et al. 2013; Liang et al. 2013; Beyli et al. 2016; Xu et al. 2021) and effective in decolourizing palm oil (Tian et al. 2014).

7.3 ALLOPHANE AND IMOGOLITE

Because of their large specific surface area, microporosity, and variable charge characteristics, allophane might be expected to show an interesting behaviour towards organic compounds. Research into this interaction, however, is hampered by the variation in composition, and lack of long-range order, of the mineral component (Levard et al. 2012). Furthermore, the separation of allophane nanoparticles from the bulk soil is laborious and time consuming (Hashizume and Theng 1999; Calabi-Floody et al. 2011). We might also mention that natural allophanes (from Ecuador, Japan, and New Zealand) have anti-inflammatory properties (Cervini-Silva et al. 2015).

Research into the allophane-organic interaction is also hampered by the lack of large mineable deposits of raw allophane except for the massive (> 5 m thick) layer covering an area larger than 4000 km^2 in Ecuador (Kaufhold et al. 2009). For these reasons, the reactivity of allophane towards organic species has often been assessed using synthetic samples of the mineral. Although similar limitations apply to using imogolite, the imogolite-organic interaction has attracted appreciable attention because of its potential environmental and industrial applications (Yuan and Wada 2012; Bonelli 2016; Ma et al. 2012a, 2016; Paineau and Launois 2019).

The reactivity of allophane and imogolite towards organic compounds is largely associated with the presence of $(OH)Al(OH_2)$ groups exposed at spherule wall perforations in allophane (cf. Figure 1.20) and at tubule ends in imogolite (cf. Figure 1.22). Since these groups can gain protons under acidic conditions and lose protons at alkaline pH, the allophane- and imogolite-organic interactions are sensitive to, and dependent on, solution/suspension pH. The PZC of allophane ranges from pH 5 to 6 (Theng et al. 1982; Clark and McBride 1984; Hashizume and Theng 2007).

Early on, Horikawa (1975) reported that allophane (and imogolite) were effective in taking up anionic surfactants from acid solutions (pH 4) when the mineral surface was positively charged. Subsequently, Inoue et al. (1978) observed a steep decline in the adsorption of (linear and branched) dodecylbenzene sulfonates by allophanic soils as solution pH was raised, while the surface charge on allophane becomes increasingly negative. Hanudin et al. (2000, 2002) have reported similarly for the interaction of allophane with some low molecular weight organic acids, indicative of dehydroxylation involving a ligand exchange process (cf. Equation 1.2). More recently, Remlinger et al. (2023) found that adsorption of glyphosate to (synthetic) allophane at pH 5 was more than double that at pH 7.

A similar pH-dependency was observed by Okamura (1990) with respect to 2-hydroxy-3-naphthoic acid and naphthochrome green, by Cea et al. (2007) for 2,4-dichlorophenol and pentachlorophenol, and by Nishikiori et al. (2009) for benzoic acid, phthalic acid, and benzaldehyde. The picture, however, was not so clear-cut for salicylic acid because of its tendency to form a surface chelate with allophane. Likewise, oxalate could form a bidentate (chelation) complex (Hanudin et al. 2002), while the formation of monodentate inner- and outer-sphere complexes of chlorophenol with allophane was suggested by Cea et al. (2005). We might also mention that chlorophenols, adsorbed from the gas phase to allophane, can be transformed into various benzene derivatives, including benzoquinone (Soma and Soma 1989).

Inoue et al. (1978) further proposed that the anionic surfactants, previously mentioned, interacted with allophane through ligand exchange between their sulfonate groups and surface hydroxyls of the mineral. This process seems more probable than electrostatic interaction (anion exchange) that Nishikiori et al. (2010) have suggested for the adsorption of linear alkylbenzene sulfonate micelles to allophane (at pH 6). Indeed, ligand exchange is the dominant mechanism involved in the interactions of allophane with anionic organic species, containing carboxylate and phosphate side chains (Parfitt et al. 1999; Yuan et al. 2000; Huang et al. 2016; Remlinger et al. 2023). In this regard, the strong adsorption of adenosine-5'-phosphate to allophane is ascribable to ligand exchange between the phosphate side chain of 5'-AMP and the hydroxyl of $(OH)Al(OH_2)$ groups of the mineral, yielding monodentate and bidentate (inner-sphere) complexes (Hashizume and Theng 2007; Iyoda et al. 2012). Even at pH 8 when the surface charge on allophane is negative, the amount of 5'-AMP adsorbed was more than an order of magnitude larger than that measured for adenine, adenosine, and ribose (Hashizume and Theng 2007).

Ligand exchange can occur even when the adsorbent surface is negatively charged (Hashizume and Theng 2007). Another distinguishing feature of ligand exchange is that the negative surface charge increases as adsorption progresses as Hanudin et al. (2002) have observed during the uptake of some organic acids (acetate, citrate, oxalate) by allophane (cf. Equation 1.2). Similarly, the adsorption of chlorophenols to an allophanic soil through ligand exchange leads to a downward shift in the soil PZC (Cea et al. 2005, 2010) as Hingston et al. (1967) have described. Figure 7.2 illustrates this feature for the ligand exchange reaction between the surface hydroxyl groups of allophane and the carboxylate groups of humic acid (Yuan et al. 2000).

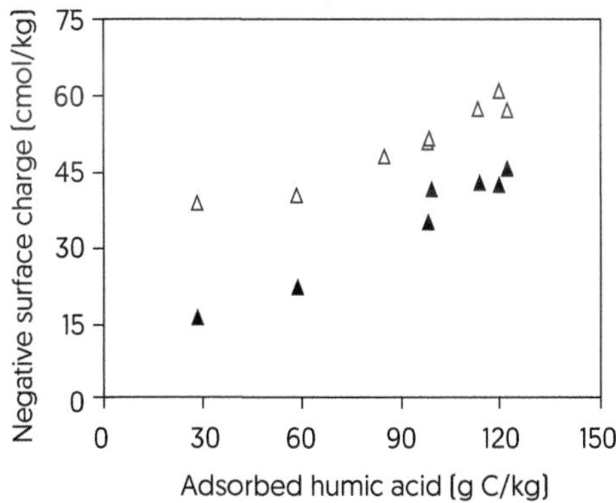

FIGURE 7.2 Relationship between negative surface charge of allophane-humic acid complexes and amount of humic acid adsorbed (from a 2mM solution of $CaCl_2$). Open triangles: adjusted to pH 7; closed triangles: unadjusted pH. (From Yuan, G. et al., *European J. Soil Sci.*, 51, 35–41, 2000.)

Hashizume and Theng (2007) have reported that the adsorption of (uncharged) adenine and adenosine by allophane decreases in the order pH 8 > pH 6 > pH 4. If the underlying mechanism is largely one of electrostatic attraction, little uptake would occur at pH 4 since both the organic solutes and mineral surface are positively charged. They have therefore suggested that other attractive forces, such as van der Waals and H-bonding, are operative. Based on the pH-dependent charge characteristics of allophane, electrostatic interactions between solute and surface would increase with solution pH in line with the experimental data.

In the case of cationic organic species, however, uptake would be favoured at alkaline pH when the surface charge on allophane is negative. Abidin et al. (2013), for example, reported that (soil) allophanes could adsorb more ammonia and pyridine than montmorillonite. Although appreciable adsorption of organic cations may still occur near the isoelectric point (pH ~ 6) of allophane, they are not strongly held as Knight and Tomlinson (1967) have observed for paraquat.

This point is well illustrated by the adsorption and (reversible) desorption of ammonium and *n*-alkylammonium ions to (soil) allophane near pH 6 (Theng 1972). Figure 7.3 shows that adsorption increases linearly with concentration up to a certain value before turning into a long flat plateau. Linear ('C-class') isotherms are indicative of solute partitioning between the bulk and interstitial (pore) water (Giles et al. 1960). The finite intercepts of the isotherms suggest the occurrence of some uptake by electrostatic and H-bonding interactions. These cations can apparently penetrate intra-aggregate pores and spread over interspherule surfaces, thereby creating fresh adsorption sites until a given 'saturation' point is reached. Since the initial slope of the isotherms is a measure of solute-surface interaction (during partition), the affinity of *n*-alkylammonium ions for allophane decreases with molecular weight (number of carbon atoms in the alkyl chain). As the size of the

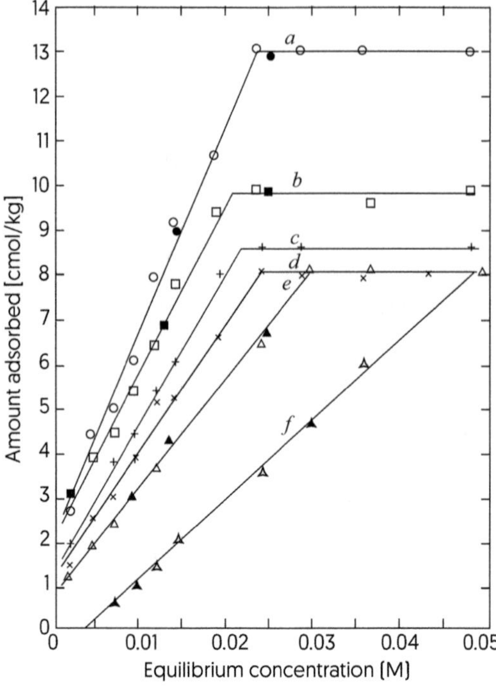

FIGURE 7.3 Isotherms for the adsorption of ammonium and a series of primary *n*-alkylammonium ions by allophane at 20°C. Open symbols: adsorption; closed symbols: desorption. (a) ammonium; (b) methylammonium; (c) ethylammonium; (d) *n*-propylammonium; (e) *n*-butylammonium; (f) *n*-hexylammonium. (From Theng, B.K.G., *Nature* 238, 150–151, 1972.)

alkylammonium cations increases, their entry into intra-aggregate pores becomes more difficult as does access to the available pore space.

The influence of solution pH on amino acid uptake by allophane is less straightforward than that shown by either organic anions or cations. This finding is related to the pH-dependent charge characteristics of the solute. Hashizume and Theng (1999), for example, found that the adsorption of DL-alanine by (soil) allophane decreased in the order pH 6 > pH 8 > pH 4. Thus, adsorption was highest at or near the isoelectric point (pI) of the solute and declined on either side of the pI (pH ~6). Since alanine exists as a zwitterion at pH ~ 6, while the (net) surface charge on allophane is then close to zero, it seems plausible that the charge-charge separation between COO^- and NH^{3+} groups on the solute molecule 'matches' that between $(OH_2)^+Al(H_2O)$ and $(OH)Al(OH)^-$ groups on the allophane surface. Furthermore, adsorption increased linearly up to a certain concentration (region I), levelled off to a flat plateau (region II), followed by a second linear increase (region III). They suggested that region I described adsorption to inter-aggregate surfaces, while region II denoted full coverage of this surface, and region III indicated intra-aggregate penetration by alanine followed by adsorption to interspherule surfaces.

Subsequently, Hashizume et al. (2002) investigated the adsorption of D- and L-alanine together with their respective dimers by three soil allophanes with different Al/Si ratios. Although the allophane samples differed in their capacity for adsorbing alanine, none showed a clear preference for either the D- or the L-form of alanine. On the other hand, the Te Kuiti allophane from New Zealand (Al/Si = 1.6) preferred L-alanyl-L-alanine to the D-enantiomer. The size, intra-molecular charge separation, and surface orientation of the zwitterionic L-alanine dimer apparently combined to confer 'structural chirality' to the allophane-organic complex, leading to the preferential uptake of the L-alanyl-alanine (Yamagishi 1987). Although this point needs to be checked, the apparent ability of allophane to discriminate for the L-enantiomer of alanyl-alanine has implications for the origin of biochirality.

Early publications on the imogolite-organic interaction are concerned with the adsorption of humic substances by natural (soil-derived) imogolite. Following Greenland (1971), Inoue and Wada (1971a, 1971b) suggested that humic acid adsorbed to imogolite through ligand exchange between its carboxylate groups and surface hydroxyls of the mineral (cf. Equation 1.2), supplemented by hydrogen-bonding and van der Waals interactions. Parfitt et al. (1977) have proposed similarly for the imogolite-fulvic acid interaction.

Wada and Henmi (1972) used the retention of quaternary ammonium chlorides to probe the microporosity of (natural) imogolite. Both inter- and intra-tubule pores would be accessible to $(C_4H_9)_4N^+$ with a diameter of 0.9–1.1 nm but only the inter-tubule pores could be penetrated by $(CH_3)_2(C_6H_5)(C_6H_5CH_2)N^+$. By the same token, acridine (diameter ~ 0.8 nm) could move freely within the intra-tubule pores (Harsh et al. 1992). Electron paramagnetic resonance (EPR) spectroscopy further indicated that the bisglycine-Cu(II) complex adsorbed to two different sites on imogolite (Goodman et al. 1984). Force field calculations by Moreno-Rodríguez et al. (2023) suggest that the herbicides, diuron and atrazine, are preferentially adsorbed to external and edge surfaces of imogolite through H-bonding and van der Waals interactions.

More recent investigations on the imogolite-organic interaction have been conducted using synthetic imogolite and its surface-modified forms (Yamamoto et al. 2001; Ma et al. 2012b; Yuan et al. 2016). A common method of imogolite synthesis is to hydrolyze a mixture of tetraethyl orthosilicate (TEOS) with aluminium-*sec*-butoxide in a solution of perchloric acid (Farmer and Fraser 1979; Farmer et al. 1983). Substituting triethoxymethylsilane (TEMS) for TEOS yields methyl-imogolite (Me-IMO) with a unit formula composition of $(OH)_3Al_2O_3SiCH_3$.

Having a larger intra-tubule pore and specific surface area than the unmodified mineral, Me-IMO shows an enhanced capacity for adsorbing methane (Bottero et al.2011; Park et al. 2014). Subsequently, Picot et al. (2019) have described the partial replacement of the methyl group in Me-IMO by various functional groups, such as 3-bromopropyl, chloromethyl, 3-aminopropyl, and

3-mercaptopropyl. As a result, the polarity of the intra-tubule walls is markedly altered as indicated by the optical response of adsorbed Nile red.

Hongo et al. (2013) have reported that imogolite, synthesized from rice husk ash, could adsorb 0.88 cmol/kg acetaldehyde (in 60 min), presumably through H-bonding to surface aluminol and silanol groups. Earlier, Bonelli et al. (2009) found that the inner (intra-tubule) silanols of imogolite (after dehydration at 300°C) were accessible to carbon monoxide, ammonia, and methanol but only partially to phenol. Interestingly, 2-methyl-4-chlorophenoxyacetic acid is mainly adsorbed to external tubule surfaces of Me-IMO although the intra-tubule pore diameter of the material is of the order of 2 nm (Bonelli et al. 2013). On the other hand, dichloromethane can apparently access the intra-tubule pores of Me-IMO if only partially (Nasi et al. 2020).

In common with allophane, ligand exchange appears to be the primary mechanism underlying the interaction of imogolite with organic molecules that contain phosphate and phosphonate groups, such as octadecylphosphonic acid (Yamamoto et al. 2001), 2-acidphosphoxyethyl methacrylate (Yamamoto et al. 2005), alkyl phosphate (Ma et al. 2011), DNA (Jiravanichanun et al. 2012), ammonium dodecyl phosphate (Ma et al. 2012b), and decylphosphonic acid (Lange et al. 2019).

Likewise, glutamic acid (Glu) showed the strongest affinity for imogolite in comparison with other amino acids (Bonini et al. 2017). On that basis, these workers reacted imogolite with a synthetic surfactant, having a Glu polar head and a C_{12} alkyl tail, and used the resultant complex ('hybrid') as an adsorbent of rhodamine B isothiocyanate. In this connection, we might mention the synthesis of hydrophobic imogolite nanotubes by 'complexation' of lauric acid to external AlOH groups (Thomas et al. 2012). Ligand exchange together with electrostatic attraction and H-bonding have also been proposed by Shafia et al. (2016) for the adsorption of Acid Orange 7 to iron-doped imogolite.

The presence of aluminol groups on its outer tubule surface, and that of silanol groups on its inner tubule surface, make imogolite a good candidate for silylation. Johnson and Pinnavaia (1990) used APTES (in aqueous acetic acid) to silylate a synthetic imogolite (cf. Equations 4.7 and 4.8). ^{29}Si NMR spectroscopy indicated that APTES was mostly bound to external tubule surfaces although they subsequently reported that the internal SiOH surface was partially accessible to the reagent (Johnson and Pinnavaia 1991). Indeed, up to 35% of the intra-tubule silanols may be 'functionalized' in this manner (Kang et al. 2011). In the case of Me-IMO, however, APTES can only be grafted to external tubule surfaces since the intra-tubule hydroxyls have previously been replaced by methyl groups. As a result, the outer surface of silylated Me-IMO is enriched with amine groups (Zanzottera et al. 2012).

Earlier, Qi et al. (2008) were able to attach osmium tetroxide to the amine groups of silylated imogolite and used the resultant complex as a catalyst in the dehydroxylation of olefins. Similarly, Guerra et al. (2010, 2011) grafted 3-chloropropyltriethoxysilane to synthetic imogolite and then reacted the silylated mineral with 2-mercaptothiazoline (MTZ) and 2-mercaptobenzimidazole (MBI). The capacity of imogolite-MTZ for taking up methylene blue was superior to that of the parent silylated mineral, while the complexes of imogolite with both MTZ and MBI could efficiently remove rubidium from aqueous solution. Subsequently, Shikinaka et al. (2015) reported that the aluminol group of imogolite could form a covalent bond with the aldehyde group of benzaldehyde, yielding a hydrophobic material for the synthesis of polyester nanohybrid films. Similarly, Picot et al. (2019) prepared hybrid imogolites with a bifunctional internal cavity by combining methyltrimethoxysilane with some organosilanes.

REFERENCES

Abidin, Z., N. Matsue, and T. Henmi. 2013. Adsorption of amines on nano-ball allophane and its molecular orbital analysis. *Clay Science* 17: 67–73.

Akbar, S., M.S. Akhtar, A. Khan, G. Jilani, B. Fashina, and Y. Deng. 2022. Aflatoxin adsorption by natural and heated sepiolite and palygorskite in comparison with adsorption by smectite. *Clays and Clay Minerals* 70: 733–752.

Akçay, G., M. Akçay, and K. Yurdakoç. 2005. Removal of 2,4-dichlorophenoxyacetic acid from aqueous solutions by partially characterized organophilic sepiolite: Thermodynamic and kinetic calculations. *Journal of Colloid and Interface Science* 281: 27–32.

Akçay, G. and K. Yurdakoç. 2000. Removal of various phenoxyalkanoic acid herbicides from water by organo-clays. *Acta Hydrochimica et Hydrobiologica* 28: 300–304.

Akyüz, S. and T. Akyüz. 2003. FT-IR spectroscopic investigations of surface and intercalated 2-aminopyrimidine adsorbed on sepiolite and montmorillonite from Anatolia. *Journal of Molecular Structure* 651–653: 205–210.

Akyüz, S. and T. Akyüz. 2005. Study on the interaction of nicotinamide with sepiolite, loughlinite and palygorskite by IR spectroscopy. *Journal of Molecular Structure* 744–747: 47–52.

Akyüz, S., T. Akyüz, and E. Akalin. 2010. Adsorption of isoniazid onto sepiolite-palygorskite group of clays: An IR study. *Spectrochimica Acta Part A: Molecular and Biomolecular Spectroscopy* 75: 1304–1307.

Akyüz, S., T. Akyüz, and A.E. Yakar. 2001. FT-IR spectroscopic investigation of adsorption of 3-aminopyridine on sepiolite and montmorillonite from Anatolia. *Journal of Molecular Structure* 565–566: 487–491.

Al-Futaisi, A., A. Jamrah, and R. Al-Hanai. 2007. Aspects of cationic dye molecule adsorption to palygorskite. *Desalination* 214: 327–342.

Alkan, M., S. Çelikçapa, Ö. Demirbaş, and M. Doğan. 2005b. Removal of reactive blue 221 and acid blue 62 anionic dyes from aqueous solutions by sepiolite. *Dyes and Pigments* 65: 251–259.

Alkan, M., Ö. Demirbaş, S. Çelikçapa, and M. Doğan. 2004. Sorption of acid red 57 from aqueous solution onto sepiolite. *Journal of Hazardous Materials* B116: 135–145.

Alkan, M., Ö. Demirbaş, and M. Doğan. 2007. Adsorption kinetics and thermodynamics of an anionic dye onto sepiolite. *Microporous and Mesoporous Materials* 101: 388–396.

Alkan, M., M. Doğan, G.Turhan, Ö. Demirbaş, and P. Turan. 2008. Adsorption kinetics and mechanism of maxilon blue 5G dye on sepiolite from aqueous solutions. *Chemical Engineering Journal* 139: 213–223.

Alkan, M., G. Tekin, and H. Namli. 2005c. FTIR and zeta potential measurements of sepiolite treated with some organosilanes. *Microporous and Mesoporous Materials* 84: 75–83.

Arnold, D.E., B.F. Bohor, H. Neff et al. 2012. The first evidence of pre-Columbian sources of palygorskite for Maya blue. *Journal of Archaeological Science* 39: 2252–2260.

Arnold, D.E., J.R. Branden, P.R. Williams, G.M. Feinman, and J.P. Brown. 2008. The first direct evidence for the production of Maya blue: Rediscovery of a technology. *Antiquity* 82: 151–164.

Aznar, J.A., B. Casal, E. Ruiz-Hitzky et al. 1992. Adsorption of methylene blue on sepiolite gels: Spectroscopic and rheological studies. *Clay Minerals* 27: 101–108.

Bakhtiary, S., M. Shirvani, and H. Shariatmadari. 2013. Characterization and 2,4-D adsorption of sepiolite nanofibers modified by N-cetylpyridinium cations. *Microporous and Mesoporous Materials* 168: 30–36.

Barrer, R.M. 1978. *Zeolites and Clay Minerals as Sorbents and Molecular Sieves*. London, UK: Academic Press.

Barrer, R.M. and N. Mackenzie. 1954. Sorption by attapulgite. Part I. Availability of intracrystalline channels. *Journal of Physical Chemistry* 58: 560–568.

Barrer, R.M., N. Mackenzie, and D.M. MacLeod. 1954. Sorption by attapulgite. Part II. Selectivity shown by attapulgite, sepiolite and montmorillonite for *n*-paraffins. *Journal of Physical Chemistry* 58: 568–572.

Belaroui, L.S., A. Ouali, A. Bengueddach, A.L. Galindo, and A. Peña. 2018. Adsorption of linuron by an Algerian palygorskite modified with magnetic iron. *Applied Clay Science* 164: 26–33.

Beyli, P.T., M. Doğan, M. Alkan, A. Türkyilmaz, Y. Turhan, and Ö. Demirbaş. 2016. Characterization, adsorption, and electrokinetic properties of modified sepiolite. *Desalination and Water Treatment* 57: 19248–19261.

Biddeci, G., G. Spinelli, P. Colomba, and F. Di Blasi. 2023. Halloysite nanotubes and sepiolite for health applications. *International Journal of Molecular Sciences* 24: 4801. https://doi.org/10.3390/ijms24054801.

Bingol, D., N. Tekin, and M. Alkan. 2010. Brilliant yellow dye adsorption onto sepiolite using a full factorial design. *Applied Clay Science* 50: 315–321.

Bonelli, B. 2016. Surface chemical modifications of imogolite. In *Nanosized Tubular Clay Minerals: Halloysite and Imogolite.* Developments in Clay Science, Vol. 7, (Eds.) P. Yuan, A. Thill, and F. Bergaya, pp. 279–307. Amsterdam, The Netherlands: Elsevier.

Bonelli, B., M. Armandi, and E. Garrone. 2013. Surface properties of alumino-silicate single-walled nanotubes of the imogolite type. *Physical Chemistry Chemical Physics* 15: 13381–13390.

Bonelli, B., I Bottero, N. Ballarini, S. Passeri, F. Cavani, and E. Garrone. 2009. IR spectroscopic and catalytic characterization of the acidity of imogolite-based systems. *Journal of Catalysis* 264: 15–30.

Bonini, M., A. Gabbani, S. Del Buffa et al. 2017. Adsorption of amino acids and glutamic acid-based surfactants on imogolite clays. *Langmuir* 33: 2411–2419.

Borrego-Sánchez, A., E. Carazo, C. Aguzzi, and C. Viseras. 2018. Biopharmaceutical improvement of praziquantel by interaction with montmorillonite and sepiolite. *Applied Clay Science* 160: 173–179.
Borrego-Sánchez, A., C. Viseras, and C.I. Sainz-Díaz. 2020. Molecular interactions of praziquantel drug with nanosurfaces of sepiolite and montmorillonite. *Applied Clay Science* 197: 105774.
Bottero, I., B. Bonelli, S.E. Ashbrook et al. 2011. Synthesis and characterization of hybrid organic/inorganic nanotubes of the imogolite type and their behaviour towards methane adsorption. *Physical Chemistry Chemical Physics* 13: 744–750.
Bruun, T.B., B. Elberling, and B.T. Christensen. 2010. Lability of soil organic carbon in tropical soils with different clay minerals. *Soil Biology & Biochemistry* 42: 888–895.
Calabi-Floody, M., J.S. Bendall, A.A. Jara et al. 2011. Nanoclays from an Andisol: Extraction, properties and carbon stabilization. *Geoderma* 161: 159–167.
Cao, L., W. Xie, H. Cui et al. 2022. Fibrous clays in dermopharmaceutical and cosmetic applications: Traditional and emerging perspectives. *International Journal of Pharmaceutics* 625: 122097.
Carazo, E., A. Borrego-Sánchez, F. García-Villén et al. 2018. Adsorption and characterization of palygorskite-isoniazid nanohybrids. *Applied Clay Science* 160: 180–185.
Casal, B., J. Merino, J.-M. Serratosa, and E. Ruiz-Hitzky. 2001. Sepiolite-based materials for the photo- and thermal-stabilization of pesticides. *Applied Clay Science* 18: 245–254.
Cea, M., J.C. Seaman, A.A. Jara, B. Fuentes, M.L. Mora, and M.C. Diez. 2007. Adsorption behavior of 2,4-dichlorophenol and pentachlorophenol in an allophanic soil. *Chemosphere* 67: 1354–1360.
Cea, M., J.C. Seaman, A.A. Jara, M.L. Mora, and M.C. Diez. 2005. Describing chlorophenol sorption on variable-charge soil using the triple-layer model. *Journal of Colloid and Interface Science* 292: 171–178.
Cea, M., J.C. Seaman, A.A. Jara, M.L. Mora, and M.C. Diez. 2010. Kinetic and thermodynamic study of chlorophenol sorption in an allophanic soil. *Chemosphere* 78: 86–91.
Celis, R., M.C. Hermosín, and J. Cornejo. 2000. Heavy metal adsorption by functionalized clays. *Environmental Science & Technology* 34: 4593–4599.
Cervini-Silva, J., A. Nieto-Camacho, V. Gómez-Vidales, S. Kaufhold, and B.K.G. Theng. 2015. The anti-inflammatory activity of natural allophane. *Applied Clay Science* 105–106: 48–51.
Chang, P.-H., W.-T. Jiang, Z. Li, C.-Y. Kuo, Q. Wu, J.-S. Jean, and G. Lv. 2016. Interaction of ciprofloxacin and probe compounds with palygorskite PFl-1. *Journal of Hazardous Materials* 303: 55–63.
Chang, P.-H, Z. Li, T.-L. Yu et al. 2009. Sorptive removal of tetracycline from water by palygorskite. *Journal of Hazardous Materials* 165: 148–155.
Chen, H. and J. Zhao. 2009. Adsorption study for removal of Congo red anionic dye using organo-attapulgite. *Adsorption* 15: 381–389.
Chen, H., Z. Zhang, G. Zhuang, and R. Jiang. 2019. A new method to prepare 'Maya red' pigment from sepiolite and basic red 46. *Applied Clay Science* 174: 38–46.
Chen, H., J. Zhao, A. Zhong, and Y. Jin. 2011. Removal capacity and adsorption mechanism of heat-treated palygorskite clay for methylene blue. *Chemical Engineering Journal* 174: 143–150.
Chen, H., A. Zhong, J. Wu, J. Zhao, and H. Yan. 2012. Adsorption behaviors and mechanisms of methyl orange on heat-treated palygorskite clays. *Industrial and Engineering Chemistry Research* 51: 14026–14036.
Chevallier, T., T. Woignier, J. Toucet, and E. Blanchart. 2010. Organic carbon stabilization in the fractal pore structure of andosols. *Geoderma* 159: 182–188.
Chiari, G, R. Giustetto, and G. Ricciardi. 2003. Crystal structure refinements of palygorskite and Maya blue from molecular modelling and powder synchrotron diffraction. *European Journal of Mineralogy* 15: 21–33.
Clark, C.J. and M.B. McBride. 1984. Cation and anion retention by natural and synthetic allophane and imogolite. *Clays and Clay Minerals* 32: 291–299.
Cobas, M., L. Ferreira, M.A. Sanromán, and M. Pazos. 2014. Assessment of sepiolite as a low-cost adsorbent for phenanthrene and pyrene removal: Kinetic and equilibrium studies. *Ecological Engineering* 70: 287–294.
Cui, H., Y. Qian, Q. Li, Z. Wei, and J. Zhai. 2013. Fast removal of Hg(II) ions from aqueous solution by amine-modified attapulgite. *Applied Clay Science* 72: 84–90.
Dahlgren, R.A., M. Saigusa, and F.C. Ugolini. 2004. The nature, properties and management of volcanic soils. *Advances in Agronomy* 82: 113–182.
Dandy, A.J. 1968. Sorption of vapors by sepiolite. *Journal of Physical Chemistry* 72: 334–339.
de Gois da Silva, M.L., A.C. Fortes, M.E.R. Oliveira et al. 2014. Palygorskite organophilic for dermopharmaceutical application. *Journal of Thermal Analysis and Calorimetry* 115: 2287–2294.
del Rio, M.S., E. Boccaleri, M. Milanesio et al. 2009. A combined synchrotron diffraction and vibrational study of the thermal treatment of palygorskite-indigo to produce Maya blue. *Journal of Materials Science* 44: 5524–5536.

Demirbaş, Ö, M. Alkan, M. Doğan, Y. Turhan, H. Namli, and P. Turan. 2007. Electrokinetic and adsorption properties of sepiolite modified with 3-aminopropyltriethoxysilane. *Journal of Hazardous Materials* 149: 650–656.

de Souza, C.G., T.C.L. de Jesus, R.C. dos Santos et al. 2021. Characterization of Brazilian palygorskite (Guadalupe region) and adsorptive behaviour for solvatochromic dyes. *Clay Minerals* 56: 55–64.

Ding, Y., Y. Lu, P. Liao et al. 2019. Molecular fractionation and sub-nanoscale distribution of dissolved organic matter on allophane. *Environmental Science: Nano* 6: 2037–2048.

Doğan, M., M. Alkan, Ö. Demirbaş, Y. Özdemir, and C. Özmetin. 2006. Adsorption kinetics of maxilon blue GRL onto sepiolite from aqueous solutions. *Chemical Engineering Journal* 124: 89–101.

Doğan, M., Y. Özdemir, and M. Alkan. 2007. Adsorption kinetics and mechanism of cationic methyl violet and methylene blue dyes onto sepiolite. *Dyes and Pigments* 75: 701–713.

Doğan, M., Y. Turhan, M. Alkan, H. Namli, P. Turan, and Ö. Demirbaş. 2008. Functionalized sepiolite for heavy metal ions adsorption. *Desalination* 230: 248–268.

Duman, O., S. Tunç, and T.G. Polat. 2015. Adsorptive removal of triarylmethane dye (basic red 9) from aqueous solution by sepiolite as effective and low-cost adsorbent. *Microporous and Mesoporous Materials* 210: 176–184.

Eren, E. and B. Afsin. 2007. Investigation of a basic dye adsorption from aqueous solution onto raw and pretreated sepiolite surfaces. *Dyes and Pigments* 73: 162–167.

Eusepi, P., L. Marinelli, A. Borrego-Sánchez et al. 2020. Nano-delivery systems based on carvacrol prodrugs and fibrous clays. *Journal of Drug Delivery Science and Technology* 58: 101815.

Farmer, V.C., M.J. Adams, A.R. Fraser, and F. Palmieri. 1983. Synthetic imogolite: Properties, synthesis, and possible applications. *Clay Minerals* 18: 459–472.

Farmer, V.C. and A.R. Fraser. 1979. Synthetic imogolite, a tubular hydroxyaluminium silicate. *Developments in Sedimentology* 27: 547–553.

Fernandez Hernandez, M.N. and T. Fernandez Alvarez. 1983. Effect of dehydration on the adsorption properties of sepiolite and palygorskite. II. Adsorption of organic substances. *Anales de Química, Serie B* 79: 342–347.

Fitaroni, L.B., T. Venâncio, F.H. Tanaka, J.C.F. Gimenez, J.A.S. Costa, and S.A. Cruz. 2019. Organically modified sepiolite: Thermal treatment and chemical and morphological properties. *Applied Clay Science* 179: 105149.

Fois, E., A. Gamba, and A. Tilocca. 2003. On the unusual stability of Maya blue paint: Molecular dynamics simulations. *Microporous and Mesoporous Materials* 57: 263–272.

Galán-Jiménez, M.D.C., E. Morillo, F. Bonnemoy, C. Mallet, and T. Undabeytia. 2020. A sepiolite-based formulation for slow release of the herbicide mesotrione. *Applied Clay Science* 189: 105503.

Gerstl, Z. and U. Mingelgrin. 1979. A note on the adsorption of organic molecules on clays. *Clays and Clay Minerals* 27: 285–290.

Gerstl, Z. and B. Yaron. 1981. Stability of parathion on attapulgite as affected by structural and hydration changes. *Clays and Clay Minerals* 29: 53–59.

Ghrab, S., M. Eloussaief, S. Lambert, S. Bouaziz, and M. Benzina. 2018. Adsorption of terpenic compounds onto organo-palygorskite. *Environmental Sustainability and Pollution Prevention* 25: 18251–18262.

Giles, C.H., T.H. MacEwan, S.N. Nakhwa, and D. Smith. 1960. Studies in adsorption. XI. A system of classification of solution adsorption isotherms and its use in diagnosis of adsorption mechanisms and in measurement of specific surface areas of solids. *Journal of the Chemical Society* 3973–3993.

Giustetto, R., A. Idone, and E. Diana. 2017. FT-Raman and surface-enhanced Raman scattering (SERS) spectroscopic study of a methyl red-palygorskite hybrid nanocomposite: Isomerization and protonation of the guest dye. *Journal of Raman Spectroscopy* 48: 507–517.

Giustetto, R., O. Wahyudi, I. Corazzari, and F. Turci. 2011. Chemical stability and dehydration behavior of a sepiolite/indigo Maya blue pigment. *Applied Clay Science* 52: 41–50.

Gök, Ö, A.S. Özcan, and A. Özcan. 2008. Adsorption kinetics of naphthalene onto organo-sepiolite from aqueous solutions. *Desalination* 220: 96–107.

Gómez-Romero, P. and C. Sanchez. 2005. Hybrid materials: Functional properties. From Maya Blue to 21st century materials. *New Journal of Chemistry* 29: 57–58.

González-Santamaría, D.E., E. López, A. Ruiz, R. Fernández, A. Ortega, and J. Cuevas. 2017. Adsorption of phenanthrene by stevensite and sepiolite. *Clay Minerals* 52: 341–350.

Goodman, B.A., H.L. Green, and D.B. McPhail. 1984. An electron paramagnetic resonance (EPR) study of the adsorption of copper complexes on montmorillonite and imogolite. *Geochimica et Cosmochimica Acta* 48: 2143–2150.

Govan, J., N. Arancibia-Miranda, M. Escudey, B. Bonelli, and F. Tasca. 2021. Imogolite: A nanotubular aluminosilicate: Synthesis, derivatives, analogues, and general and biological applications. *Materials Chemistry Frontiers* 5: 6779–6802.

Greenland, D.J. 1971. Interactions between humic and fulvic acids and clays. *Soil Science* 111: 34–41.

Guerra, D.L., A.C. Batista, R.R. Viana, and C. Airoldi. 2010. Adsorption of methylene blue on raw and MTZ/imogolite hybrid surfaces: Effect of concentration and calorimetric investigation. *Journal of Hazardous Materials* 183: 81–86.

Guerra, D.L., A.C. Batista, R.R. Viana, and C. Airoldi. 2011. Adsorption of rubidium on raw and MTZ- and MBI-imogolite hybrid surfaces: An evidence of the chelate effect. *Desalination* 275: 107–117.

Habibi, A., L.S. Belaroui, A. Bengueddach, A.L. Galindo, C.I. Sainz Díaz, and A. Peña. 2018. Adsorption of metronidazole and spiramycin by an Algerian palygorskite. Effect of modification with tin. *Microporous and Mesoporous Materials* 268: 293–302.

Haden, W.L., Jr. and I.A. Schwint. 1967. Attapulgite – its properties and applications. *Industrial & Engineering Chemistry* 59: 59–69.

Hajjaji, M., A. Alami, and A. Ell Bouadili. 2006. Removal of methylene blue from aqueous solution by fibrous clay minerals. *Journal of Hazardous Materials* B135: 188–192.

Han, Z.-X., Z. Zhu, D.-D. Wu, J. Wu, and Y.-R. Liu. 2014. Adsorption kinetics and thermodynamics of acid blue 25 and methylene blue dye solutions on natural sepiolite. *Synthesis and Reactivity in Inorganic, Metal-Organic, and Nano-Metal Chemistry* 44: 140–147.

Hanudin, E., N. Matsue, and T. Henmi. 2000. Change in charge characteristics of allophane with adsorption of low molecular weight organic acids. *Clay Science* 11: 243–255.

Hanudin, E., N. Matsue, and T. Henmi. 2002. Reactions of some short-range ordered aluminosilicates with selected organic ligands. *Developments in Soil Science* 28A: 319–332.

Harsh, J.B., S.J. Traina, J. Boyle, and Y. Yang. 1992. Adsorption of cations on imogolite and their effect on surface charge characteristics. *Clays and Clay Minerals* 40: 700–706.

Hashizume, H. and B.K.G. Theng. 1999. Adsorption of DL-alanine by allophane: Effect of pH and unit particle aggregation. *Clay Minerals* 34: 233–238.

Hashizume, H. and B.K.G. Theng. 2007. Adenine, adenosine, ribose and 5'AMP adsorption to allophane. *Clays and Clay Minerals* 55: 599–605.

Hashizume, H., B.K.G. Theng, and A. Yamagishi. 2002. Adsorption and discrimination of alanine and alanyl-alanine enantiomers by allophane. *Clay Minerals* 37: 551–557.

He, D., H. Huang, W. Xu, F. Qin, and S. Liu. 2018. Adsorption properties and mechanism of purified palygorskite on methylene blue. *Arabian Journal of Geosciences* 11: 658. https://doi.org/10.1007/s12517-018-4015-3.

He, H., Q. Tao, J. Zhu, P. Yuan, W. Shen, and S. Yang. 2013. Silylation of clay mineral surfaces. *Applied Clay Science* 71: 15–20.

Hermosín, M.C. and J. Cornejo. 1986. Methylation of sepiolite and palygorskite with diazomethane. *Clays and Clay Minerals* 34: 591–596.

Hernández, D., L. Quiñones, L. Lazo et al. 2023. Removal of an emergent contaminant by a palygorskite from Pontezuela/Cuban region. *Journal of Porous Materials* 30: 1149–1161.

Hingston, F.J., R.J. Atkinson, A.M. Posner, and J.P. Quirk. 1967. Specific adsorption of anions. *Nature* 215: 1459–1461.

Hongo, T., J. Sugiyama, A. Yamazaki, and A. Yamasaki. 2013. Synthesis of imogolite from rice husk ash and evaluation of its acetaldehyde adsorption ability. *Industrial & Engineering Chemistry Research* 52: 2111–2115.

Horikawa, Y. 1975. Effects of ionic surface-active agents on the floatability and electrophoretic mobility of selected clays. *Clay Science* 4: 281–290.

Huang, J., Y. Liu, Q. Jin, X. Wang, and J. Yang. 2007b. Adsorption studies of a water-soluble dye, reactive red MF-3B, using sonication-surfactant-modified attapulgite clay. *Journal of Hazardous Materials* 143: 541–548.

Huang, J., Y. Liu, and X. Wang. 2008. Selective adsorption of tannin from flavonoids by organically modified attapulgite clay. *Journal of Hazardous Materials* 160: 382–387.

Huang, J., X. Wang, Q. Jin, Y. Liu, and Y. Wang. 2007a. Removal of phenol from aqueous solution by adsorption onto OTMAC-modified attapulgite. *Journal of Environmental Management* 84: 229–236.

Huang, Y.-T., D.J. Lowe, G.J. Churchman et al. 2016. DNA adsorption by nanocrystalline allophane spherules and nanoaggregates, and implications for carbon sequestration in andisols. *Applied Clay Science* 120: 40–50.

Hubbard, B., W. Kuang, A. Moser, G.A. Facey, and C. Detellier. 2003. Structural study of Maya blue: Textural and solid-state multinuclear magnetic resonance characterization of the palygorskite-indigo and sepiolite-indigo adducts. *Clays and Clay Minerals* 51: 318–326.

Inagaki, S., Y. Fukushima, H. Doi, and O. Kamigaito. 1990. Pore size distribution and adsorption selectivity of sepiolite. *Clay Minerals* 25: 99–105.

Inagaki, S., Y. Fukushima, and M. Miyata. 1995. Inclusion polymerization of isoprene in the channels of sepiolite. *Research on Chemical Intermediates* 21: 167–180.

Inoue, K., K. Kaneko, and M. Yoshida. 1978. Adsorption of dodecylbenzenesulfonates by soil colloids and their degradation. *Soil Science and Plant Nutrition* 24: 91–102.

Inoue, T. and K. Wada. 1971a. Reactions between humified clover extracts and imogolite as a model of humus clay interaction: Part I. *Clay Science* 4: 61–70.

Inoue, T. and K. Wada. 1971b. Reactions between humified clover extracts and imogolite as a model of humus clay interaction: Part II. *Clay Science* 4: 71–80.

Iyoda, F., S. Hayashi, S. Arakawa et al. 2012. Synthesis and adsorption characteristics of hollow spherical allophane nano-particles. *Applied Clay Science* 56: 77–83.

Jiravanichanun, N., K. Yamamoto, K. Kato et al. 2012. Preparation and characterization of imogolite/DNA hybrid hydrogels. *Biomacromolecules* 13: 276–281.

Johnson, L.M. and T.J. Pinnavaia. 1990. Silylation of a tubular aluminosilicate polymer (imogolite) by reaction with hydrolyzed (γ-aminopropyl)triethoxysilane. *Langmuir* 6: 307–311.

Johnson, L.M. and T.J. Pinnavaia. 1991. Hydrolysis of (γ-aminopropyl)triethoxysilane-silylated imogolite and formation of a silylated tubular silicate-layered silicate nanocomposite. *Langmuir* 7: 2636–2641.

Kang, D.-Y., J. Zang, C.W. Jones, and S. Nair. 2011. Single-walled aluminosilicate nanotubes with organic-modified interiors. *Journal of Physical Chemistry C* 115: 7676–7685.

Karagozoglu, B., M. Tasdemir, E. Demirbas, and M. Kobya. 2007. The adsorption of basic dye (Astrazon Blue FGRL) from aqueous solutions onto sepiolite, fly ash and apricot shell activated carbon: Kinetic and equilibrium studies. *Journal of Hazardous Materials* 147: 297–396.

Karataş, D., D. Senol-Arslan, and O. Ozdemir. 2018. Experimental and atomic modeling of the adsorption of acid azo dye 57 to sepiolite. *Clays and Clay Minerals* 66: 427–438.

Karataş, D., A. Tekin, and M.S. Çelik. 2013. Adsorption of quaternary amine surfactants and their penetration into the intracrystalline cavities of sepiolite. *New Journal of Chemistry* 37: 3936–3948.

Kaufhold, S., A. Kaufhold, R. Jahn et al. 2009. A new massive deposit of allophane raw material in equador. *Clays and Clay Minerals* 57: 72–81.

Kausar, A., M. Iqbal, A. Javed et al. 2018. Dyes adsorption using clay and modified clay: A review. *Journal of Molecular Liquids* 256: 395–407.

Knight, B.A.G. and T.E. Tomlinson. 1967. The interaction of paraquat (1:1'-dimethyl 4:4'-dipyridylium dichloride) with mineral soils. *Journal of Soil Science* 18: 233–243.

Kovacević, D., J. Lemić, M. Damjanović, R. Petroijević, D. Janaćković, and T. Stanić. 2011. Fenitrothion adsorption – desorption on organo-minerals. *Applied Clay Science* 52: 109–114.

Kuang, W. and C. Detellier. 2004. Insertion of acetone molecules in the nanostructured tunnels of palygorskite. *Canadian Journal of Chemistry* 82: 1527–1535.

Kuang, W. and C. Detellier. 2005. Structuration of organo-minerals: Nanohybrid materials resulting from the incorporation of alcohols in the tunnels of palygorskite. *Studies in Surface Science and Catalysis* 156: 451–456.

Kuang, W., G.A. Facey, and C. Detellier. 2006. Organo-mineral nanohybrids. Incorporation, coordination and structuration role of acetone molecules in the tunnels of sepiolite. *Journal of Materials Chemistry* 16: 179–185.

Kuang, W., G.A. Facey, C. Detellier, B. Casal, J.M. Serratosa, and E. Ruiz-Hitzky. 2003. Nanostructured hybrid materials formed by sequestration of pyridine molecules in the tunnels of sepiolite. *Chemistry of Materials* 15: 4956–4967.

Lange, T., T. Charpentier, F. Gobeaux, S. Charton, F. Testard, and A. Thill. 2019. Partial transformation of imogolite by decylphosphonic acid yields an interface active composite material. *Langmuir* 35: 4068–4076.

Leal, M., V. Martínez-Hernández, R. Meffe, J. Lillo, and I. de Bustamante. 2017. Clinoptilolite and palygorskite as sorbents of neutral emerging organic contaminants in treated wastewater: Sorption-desorption studies. *Chemosphere* 175: 534–542.

Lei, Z., Q. Yang, S. Wu, and X. Song. 2009. Reinforcement of polyurethane/epoxy interpenetrating network nanocomposites with an organically modified palygorskite. *Journal of Applied Polymer Science* 111: 3150–3162.

Lemić, J., M. Tomašević-Canović, M. Djuričić, and T. Stanić. 2005. Surface modification of sepiolite with quaternary amines. *Journal of Colloid and Interface Science* 292: 11–19.

Levard, C., E. Doelsch, I. Basile-Doelsch et al. 2012. Structure and distribution of allophanes, imogolite and proto-imogolite in volcanic soils. *Geoderma* 183–184: 100–108.

Li, L., G. Zhuang, and P. Yuan. 2023. Coordinated water stabilizes indigo-sepiolite hybrid pigments: Evidence from acid modification. *Applied Clay Science* 236: 106877.

Li, Z., C.A. Willms, and K. Kniola. 2003. Removal of anionic contaminants using surfactant-modified palygorskite and sepiolite. *Clays and Clay Minerals* 51: 445–451.

Liang, X., Y. Xu, X. Tan et al. 2013. Heavy metal adsorbents mercapto and amino functionalized palygorskite: Preparation and characterization. *Colloids and Surfaces A: Physicochemical and Engineering Aspects* 426: 98–105.

Liu, P., H. Wang, and C. Pan. 2018. Surface organo-functionalization of palygorskite nanorods with γ-mercaptopropyltrimethoxysilane. *Applied Clay Science* 159: 37–41.

Liu, Y., W. Wang, and A. Wang. 2012. Effect of dry grinding on the microstructure of palygorskite and adsorption efficiency for methylene blue. *Powder Technology* 225: 124–129.

Liu, Z., G. Zhang, and M.K. Wang. 2014. Immobilization of polychlorinated biphenyls in contaminated soils using organoclays. *Applied Clay Science* 101: 297–303.

Lobato-Aguilar, H.A., W.A. Herrera-Kao, S. Duarte-Aranda et al. 2023.Characterization and drug release of benzalkonium chloride-loaded organo-palygorskite or organo-montmorillonite. *Clay Minerals* 58: 102–112.

Ma, W., J. Kim, H. Otsuka, and A. Takahara. 2011. Surface modification of individual imogolite nanotubes with alkyl phosphate from aqueous solution. *Chemistry Letters* 40: 159–161.

Ma, W., W.O. Yah, H. Otsuka, and A. Takahara. 2012a. Application of imogolite clay nanotubes in organic-inorganic nanohybrid materials. *Journal of Materials Chemistry* 22: 11887–11892.

Ma, W., W.O. Yah, H. Otsuka, and A. Takahara. 2012b. Surface functionalization of aluminosilicate nanotubes with organic molecules. *Beilstein Journal of Nanotechnology* 3: 82–100.

Ma, W., Y. Higaki, and A. Takahara. 2016. *Nanosized Tubular Clay Minerals: Halloysite and Imogolite.* Developments in Clay Science, Vol. 7, (Eds.) P. Yuan, A. Thill, and F. Bergaya, pp. 628–671. Amsterdam, The Netherlands: Elsevier.

Maqueda, C., P. Partal, J. Villaverde, and J.L. Perez-Rodriguez. 2009. Characterization of sepiolite-gel-based formulations for controlled release of pesticides. *Applied Clay Science* 46: 289–295.

Meirelles, L., E. Carazo, A. Borrego-Sánchez et al. 2019. Design and characterization of a tuberculostatic hybrid based on interaction of ethambutol with a raw palygorskite. *Applied Clay Science* 181: 105213.

Meléndez-López, A., J. Cruz-Castañeda, A. Negrón-Mendoza et al. 2023. Gamma irradiation of adenine and guanine adsorbed into hectorite and attapulgite. *Heliyon* 9: e16071.

Mendelovici, E. and D.C. Portillo. 1976. Organic derivatives of attapulgite–I. Infrared spectroscopy and X-ray diffraction studies. *Clays and Clay Minerals* 24: 177–182.

Middea, A., L.S. Spinelli, F.G. Souza, Jr., R. Neumann, T.L.A.P. Fernandes, and O da F.M. Gomes. 2017. Preparation and characterization of an organo-palygorskite-Fe_3O_4 nanomaterial for removal of anionic dyes from wastewater. *Applied Clay Science* 139: 45–53.

Mikhail, R.S., N.M. Guindy, and S. Hanafi. 1983. Surface properties of attapulgite. *Journal of Vacuum Science & Technology A* 1: 267–270.

Moreira, M.A., K.J. Ciuffi, V. Rives et al. 2017. Effect of chemical modification of palygorskite and sepiolite by 3-aminopropyltriethoxysilane on adsorption of cationic and anionic dyes. *Applied Clay Science* 135: 394–404.

Moreno-Rodríguez, D., E. Gianni, M. Pospíšil, and E. Scholtzová. 2023. Is imogolite a suitable adsorbent agent for the herbicides like diuron and atrazine? *Journal of Molecular Liquids* 380: 121732.

Mu, B. and A. Wang. 2016. Adsorption of dyes onto palygorskite and its composites: A review. *Journal of Environmental Chemical Engineering* 4: 1274–1294.

Nasi, R., F. Sannino, P. Picot et al. 2020. Hybrid organic-inorganic nanotubes effectively adsorb some organic pollutants in aqueous phase. *Applied Clay Science* 186: 105449.

Nederbragt, G.W. 1949. Separation of long-chain and compact molecules by adsorption to attapulgite-containing clays. *Clay Minerals Bulletin* 3: 72–75.

Nishikiori, H., K. Kobayashi, S. Kubota, N. Tanaka, and T. Fujii. 2010. Removal of detergents and fats from wastewater using allophane. *Applied Clay Science* 47: 325–329.

Nishikiori, H., J. Shindoh, N. Takahashi, T. Takagi, N. Tanaka, and T. Fujii. 2009. Adsorption of benzene derivatives on allophane. *Applied Clay Science* 43: 160–163.

Ohashi, F., S. Tomura, K. Akaku, S. Hayashi, and S.-I. Wada. 2004. Characterization of synthetic imogolite nanotubes as gas storage. *Journal of Materials Science* 39: 1799–1801.

Okamura, Y. 1990. Consecutive adsorption of salicylic acid and its structural analogues on allophanic clay. *Clay Science* 7: 325–335.

Ongen, A., H.K. Ozcan, E.E. Ozbas, and N. Balkaya. 2012. Adsorption of astrazon blue FGRL onto sepiolite from aqueous solutions. *Desalination and Water Treatment* 40: 129–136.

Ouellet-Plamondon, C., P. Aranda, A. Favier, G. Habert, H. van Damme, and E. Ruiz-Hitzky. 2015. The Maya blue nanostructured material concept applied to colouring geopolymers. *RSC Advances* 5: 98834–98841.

Ovarlez, S., A.-M. Chaze, F. Giulieri, and F. Delamare. 2006. Indigo chemisorption in sepiolite. Application to Maya blue formation. *Comptes Rendus Chimie* 9: 1243–1248.

Ovarlez, S., F. Giulieri, A-M. Chaze, F. Delamare, J. Raya, and J. Hirschinger. 2009. The incorporation of indigo molecules in sepiolite tunnels. *Chemistry: A European Journal* 15: 11326–11332.

Ovarlez, S., F. Giulieri, F. Delamare, N. Sbirrazzuoli, and A.-M. Chaze. 2011. Indigo-sepiolite nanohybrids: Temperature-dependent synthesis of two complexes and comparison with indigo-palygorskite systems. *Microporous and Mesoporous Materials* 142: 371–380.

Özcan, A.S. and Ö. Gök. 2012. Structural characterization of dodecyltrimethylammonium (DTMA) bromide modified sepiolite and its adsorption isotherm studies. *Journal of Molecular Structure* 1007: 36–44.

Özcan, A. and A.S. Özcan. 2005. Adsorption of acid red 57 from aqueous solutions onto surfactant-modified sepiolite. *Journal of Hazardous Materials* B125: 252–259.

Özcan, A., E.M. Öncü, and A.S. Özcan. 2006. Kinetics, isotherm and thermodynamic studies of adsorption of acid blue 193 from aqueous solutions onto natural sepiolite. *Colloids and Surfaces A: Physicochemical Engineering Aspects* 277: 90–97.

Özcan, A.S., S. Tetik, and A. Özcan. 2005. Adsorption of acid dyes from aqueous solutions onto sepiolite. *Separation Science and Technology* 39: 301–320.

Özdemir, Y., M. Doğan, and M. Alkan. 2006. Adsorption of cationic dyes from aqueous solutions by sepiolite. *Microporous and Mesoporous Materials* 96: 419–427.

Paineau, E. 2018. Imogolite nanotubes: A flexible nanoplatform with multipurpose applications. *Applied Sciences* 8: 1921. https://doi.org/10.3390/app8101921.

Paineau, E. and P. Launois. 2019. Nanomaterials from imogolite: Structure, properties and functional materials. In *Nanomaterials from Clay Minerals: A New Approach to Green Functional Materials*, (Eds.) A. Wang and W. Wang, pp. 257–284. Amsterdam, The Netherlands: Elsevier.

Parfitt, R.L., A.R. Fraser, and V.C. Farmer. 1977. Adsorption on hydrous oxides. III. Fulvic acid and humic acid on goethite, gibbsite and imogolite. *Journal of Soil Science* 28: 289–296.

Parfitt, R.L., B.K.G. Theng, J.S. Whitton, and T.G. Shepherd. 1997. Effects of clay minerals and land use on organic matter pools. *Geoderma* 75: 1–12.

Parfitt, R.L., G. Yuan, and B.K.G. Theng. 1999. A ^{13}C-NMR study of the interactions of soil organic matter with aluminium and allophane in podzols. *European Journal of Soil Science* 50: 695–700.

Park, G., H. Lee, S.U. Lee, and D. Sohn. 2014. Strain energy and structural property of methyl substituted imogolite. *Molecular Crystals and Liquid Crystals* 599: 68–71.

Peng, S., S. Wang, T. Chen, S. Jiang, and C. Huang. 2010. Adsorption kinetics of methylene blue from aqueous solutions onto palygorskite. *Acta Geologica Sinica* 80: 236–242.

Picot, P., F. Gobeaux, T. Coradin, and A. Thill. 2019. Dual internal functionalization of imogolite nanotubes as evidence by optical properties of Nile red. *Applied Clay Science* 178: 105133.

Qi, X., H. Yoon, S-H. Lee, J. Yoon, and S-J. Kim. 2008. Surface-modified imogolite by 3-APS-OsO_4 complex: Synthesis, characterization and its application in the dihydroxylation of olefins. *Journal of Industrial and Engineering Chemistry* 14: 136–141.

Quan, G., L. Kong, Y. Lan, J. Yan, and B. Gao. 2018. Removal of acid orange 7 by surfactant-modified iron nanoparticle supported on palygorskite: Reactivity and mechanism. *Applied Clay Science* 152: 173–182.

Reinen, D., P. Köhl, and C. Müller. 2004. The nature of the colour centres in 'Maya blue' – the incorporation of organic pigment molecules into the palygorskite lattice. *Zeitschrift für anorganische und allgemeine Chemie* 630: 97–103.

Remlinger, V.I., K.R. Lenhardt, M.V. Rechberger et al. 2023. Glyphosate adsorption on synthetic allophanes and halloysite: Effects of pH and mineral properties. *Journal of Plant Nutrition and Soil Science* 186: 568–579.

Roberts, M.G., H. Li, B.J. Teppen, and S.A. Boyd. 2006. Sorption of nitroaromatics by ammonium- and organic ammonium-exchanged smectite: Shifts from adsorption complexation to a partition-dominated process. *Clays and Clay Minerals* 54: 426–434.

Rodríguez-Cruz, M.S., M.S. Andrades, and M.J. Sánchez-Martín. 2008. Significance of the long-chain organic cation structure in the sorption of the penconazole and metalaxyl fungicides by organo clays. *Journal of Hazardous Materials* 160: 200–207.

Ruiz, A.I., C. Ruiz-Garcia, and E. Ruiz-Hitzky. 2023. From old to new inorganic materials for advanced applications: The paradigmatic example of the sepiolite clay mineral. *Applied Clay Science* 235: 106874.

Ruiz-Hitzky, E. and J.J. Fripiat. 1976. Organomineral derivatives obtained by reacting organochlorosilanes with the surface of silicates in organic solvents. *Clays and Clay Minerals* 24: 25–30.

Rusmin, R., B. Sarkar, B. Biswas, J. Churchman, Y. Liu, and R. Naidu. 2016. Structural, electrokinetic and surface properties of activated palygorskite for environmental application. *Applied Clay Science* 134: 95–102.

Rytwo, G., S. Nir, M. Crespin, and L. Margulies. 2000. Adsorption and interactions of methyl green with montmorillonite and sepiolite. *Journal of Colloid and Interface Science* 222: 12–19.

Rytwo, G., S. Nir, L. Margulies et al. 1998. Adsorption of monovalent organic cations on sepiolite: Experimental results and model calculations. *Clays and Clay Minerals* 46: 340–348.

Rytwo, G., A. Rettig, and Y. Gonen. 2011. Organo-sepiolite particles for efficient pretreatment of organic wastewater: Application to winery effluents. *Applied Clay Science* 51: 390–394.

Rytwo, G., D. Tropp, and C. Serban. 2002. Adsorption of diquat, paraquat and methyl green on sepiolite: Experimental results and model calculations. *Applied Clay Science* 20: 273–282.

Saadat, S., D. Rawtani, and G. Parikh. 2022. Clay minerals-based drug delivery systems for anti-tuberculosis drugs. *Journal of Drug Delivery Science and Technology* 76: 103755.

Sabah, E. 2007. Decolorization of vegetable oils: Chlorophyll-*a* adsorption by acid-activated sepiolite. *Journal of Colloid and Interface Science* 310: 1–7.

Sabah, E. and M.S. Çelik. 2002. Interaction of pyridine derivatives with sepiolite. *Journal of Colloid and Interface Science* 251: 33–38.

Sabah, E., M. Çinar, and M.S. Çelik. 2007. Decolorization of vegetable oils: Adsorption mechanism of β-carotene on acid-activated sepiolite. *Food Chemistry* 100: 1661–1668.

Sabah, E. and S. Ouki. 2017. Adsorption of pyrene from aqueous solutions onto sepiolite. *Clays and Clay Minerals* 65: 14–26.

Saggar, S., K.R. Tate, C.W. Feltham, C.W. Childs, and A. Parshotam. 1994. Carbon turnover in a range of allophanic soils amended with ^{14}C-labelled glucose. *Soil Biology & Biochemistry* 26: 1263–1271.

Sánchez del Rio, M., A. Doménech, M.T. Doménech-Carbó, M.L.V. de Agredos Pascual, M. Suárez, and E. García-Romero. 2011. The Maya blue pigment. *Developments in Clay Science* 3: 453–481.

Sanchez-Martin, M.J., M.S. Rodriguez-Cruz, M.S. Andrades, and M. Sanchez-Camazano. 2006. Efficiency of different clay minerals modified with a cationic surfactant in the adsorption of pesticides: Influence of clay type and pesticide hydrophobicity. *Applied Clay Science* 31: 216–228.

Sánchez-Ochoa, F., G.H. Cocoletzi, and G. Canto. 2017. Trapping and diffusion of organic dyes inside the palygorskite clay: The ancient Maya blue pigment. *Microporous and Mesoporous Materials* 249: 111–117.

Santos, S.C.R. and R.A.R. Boaventura. 2016. Adsorption of cationic and anionic azo dyes on sepiolite clay: Equilibrium and kinetic studies in batch mode. *Journal of Environmental Chemical Engineering* 4: 1473–1483.

Sarkar, B., M. Megharaj, Y. Xi, and R. Naidu. 2012. Surface charge characteristics of organo-palygorskites and adsorption of *p*-nitrophenol in flow-through reactor system. *Chemical Engineering Journal* 185–186: 35–43.

Sarkar, B., Y. Xi, M. Megharaj, G.S.R. Krishnamurti, and R. Naidu. 2010. Synthesis and characterization of novel organopalygorskites for removal of *p*-nitrophenol from aqueous solution: Isothermal studies. *Journal of Colloid and Interface Science* 350: 295–304.

Sarkar, B., Y. Xi, M. Megharaj, and R. Naidu. 2011. Orange II adsorption on palygorskite modified with alkyl trimethylammonium and dialkyl dimethylammonium bromide – an isothermal and kinetic study. *Applied Clay Science* 51: 370–374.

Serna, C.J. and G.E. Vanscoyoc. 1979. Infrared study of sepiolite and palygorskite surfaces. In *International Clay Conference 1978*. Developments in Sedimentology, Vol. 27, (Eds.) M.M. Mortland and V.C. Farmer, pp. 197–206. Amsterdam, The Netherlands: Elsevier.

Serratosa, J.M. 1979. Surface properties of fibrous clay minerals (palygorskite and sepiolite). In *International Clay Conference, 1978*. Developments in Sedimentology, Vol. 27, (Eds.) M.M. Mortland and V.C. Farmer, pp. 99–109. Amsterdam, The Netherlands: Elsevier.

Sevim, A.M., R. Hojiyev, A. Gül, and M.S. Çelik. 2011. An investigation of the kinetics and thermodynamics of the adsorption of a cationic cobalt porphyrazine onto sepiolite. *Dyes and Pigments* 88: 25–38.

Shafia, E., S. Esposito, M Armandi, E. Bahadori, E. Garrone, and B. Bonelli. 2016. Reactivity of bare and Fe-doped alumino-silicate nanotubes imogolite with H_2O_2 and the azo-dye acid orange 7. *Catalysis Today* 277: 89–96.

Shariatmadari, H., A.R. Mermut, and M.B. Benke. 1999. Sorption of selected cationic and neutral organic molecules on palygorskite and sepiolite. *Clays and Clay Minerals* 47: 44–53.

Shikinaka, K., A. Abe, and K. Shigehara. 2015. Nanohybrid film consisted of hydrophobized imogolite and various aliphatic polyesters. *Polymer* 68: 279283.

Shirvani, M. and F. Nourbakhsh. 2010. Desferrioxamine-B adsorption to and iron dissolution from palygorskite and sepiolite. *Applied Clay Science* 48: 393–397.

Shuali, U., S. Nir, and G. Rytwo. 2011. Adsorption of surfactants, dyes and cationic herbicides on sepiolite and palygorskite: Modifications, applications and modelling. In *Developments in Palygorskite-Sepiolite Research*. Developments in Clay Science, Vol. 3, (Eds.) F. Galan and A. Singer, pp. 351–374. Amsterdam, The Netherlands: Elsevier.

Shuali, U., M. Steinberg, S. Yariv, M. Müller-Vonmoos, G. Kahr, and A. Rub. 1990. Thermal analysis of sepiolite and palygorskite treated with butylamine. *Clay Minerals* 25: 107–119.

Shuali, U., S. Yariv, M. Steinberg, M. Müller-Vonmoos, G. Kahr, and A. Rub. 1991. Thermal analysis of pyridine-treated sepiolite and palygorskite. *Clay Minerals* 26: 497–506.

Silva, R.P., A.G.B. Gois, M.O. Ramme, T.N.C. Dantas, J.L.M. Barillas, and V.C. Santanna. 2021. Adsorption of cetyltrimethyl ammonium bromide surfactant for organophilization of palygorskite clay. *Clay Minerals* 56: 140–147.

Singer, A. 1989. Palygorskite and sepiolite group minerals. In *Minerals in Soil Environments*, 2nd edition, (Eds.) J.B. Dixon and S.B. Weed, pp. 829–872. Madison, WI: Soil Science Society of America.

Soma, Y. and M. Soma. 1989. Formation of hydroxydibenzofurans from chlorophenols adsorbed on Fe-ion exchanged montmorillonite. *Chemosphere* 18: 1895–1902.

Sousa, H.R., L.S. Silva, P.A.A. Sousa et al. 2019. Evaluation of methylene blue removal by plasma activated palygorskites. *Journal of Materials Research and Technology* 8: 5432–5442.

Sposito, G., N.T. Skipper, R. Sutton, S.-H. Park, A.K. Soper, and J.A. Greathouse. 1999. Surface geochemistry of the clay minerals. *Proceedings National Academy of Science USA* 96: 3358–3364.

Sturini, M., A. Speltini, F. Maraschi et al. 2016. Removal of fluoroquinolone contaminants from environmental waters on sepiolite and its photo-induced regeneration. *Chemosphere* 150: 686–693.

Sugiura, M. and K. Fukumoto. 1992. Effect of relative humidity on the adsorption of acetaldehyde by sepiolite and sepiolite-2-aminobenzoic acid complex. *Clay Science* 8: 195–209.

Tabak, A., E. Eren, B. Afsin, and B. Caglar. 2009. Determination of adsorptive properties of a Turkish sepiolite for removal of reactive blue 15 anionic dye from aqueous solutions. *Journal of Hazardous Materials* 161: 1087–1094.

Tartaglione, G., D. Tabuani, and G. Camino. 2008. Thermal and morphological characterization of organically modified sepiolite. *Microporous and Mesoporous Materials* 107: 161–168.

Theng, B.K.G. 1972. Adsorption of ammonium and some primary *n*-alkylammonium cations by soil allophane. *Nature* 238: 150–151.

Theng, B.K.G. (Ed.). 1980. *Soils with Variable Charge*. Lower Hutt, NZ: New Zealand Society of Soil Science.

Theng, B.K.G. 2012. *Formation and Properties of Clay-Polymer Complexes*, 2nd edition. Amsterdam, The Netherlands: Elsevier.

Theng, B.K.G. 2018. *Clay Mineral Catalysis of Organic Reactions*. Boca Raton, FL: CRC Press.

Theng, B.K.G., M. Russell, G.J. Churchman, and R.L. Parfitt. 1982. Surface properties of allophane, halloysite, and imogolite. *Clays and Clay Minerals* 30: 143–149.

Thomas, B., T. Coradin, G. Laurent et al. 2012. Biosurfactant-mediated one-step synthesis of hydrophobic functional imogolite nanotubes. *RSC Advances* 2: 426–435.

Tian, G., G. Han, F. Wang, and J. Liang. 2019. Sepiolite nanomaterials: Structure, properties, and functional applications. In *Nanomaterials from Clay Minerals*, (Eds.) A. Wang and W. Wang, pp. 135–201. Amsterdam, The Netherlands: Elsevier.

Tian, G., Y. Kang, B. Mu, and A. Wang. 2014. Attapulgite modified with silane coupling agent for phosphorus adsorption and deep bleaching of refined palm oil. *Adsorption Science & Technology* 32: 37–48.

Tsai, Y.-L., P.-H. Chang, Z.-Y. Gao et al. 2016. Amitriptyline removal using palygorskite clay. *Chemosphere* 155: 292–299.

Undabeytia, T., F. Madrid, J. Vázquez, and J.I. Pérez-Martínez. 2019. Grafted sepiolites for the removal of pharmaceuticals in water treatment. *Clays and Clay Minerals* 67: 173–182.

van Olphen, H. 1966. Maya blue: A clay-organic pigment? *Science* 154: 645–646.

Varela, C.F., M.C. Pazos, and M.D. Alba. 2021. Organophilization of acid and thermal treated sepiolite for its application in BTEX adsorption from aqueous solutions. *Journal of Water Process Engineering* 40: 101949.

Vico, L.I. and S.G. Acebal. 2006. Some aspects about the adsorption of quinoline on fibrous silicates and Patagonian saponite. *Applied Clay Science* 33: 142–148.

Wada, K. and T. Henmi. 1972. Characterization of micropores of imogolite by measuring retention of quaternary ammonium chlorides and water. *Clay Science* 4: 127–136.

Wang, W. and A. Wang. 2019b. Palygorskite nanomaterials: Structure, properties, and functional applications. In *Nanomaterials from Clay Minerals*, (Eds.) A. Wang and W. Wang, pp. 21–133. Amsterdam, The Netherlands: Elsevier.

Wang, Q., J. Zhang, and A. Wang. 2014. Freeze-drying: A versatile method to overcome re-aggregation and improve dispersion stability of palygorskite for sustained release of ofloxacin. *Applied Clay Science* 87: 7–13.

Wang, Q., B. Mu, S. Li, X. Wang, L. Zong, and A. Wang. 2021. Facile fabrication of a stable fluorescent yellow X-10GFF/palygorskite hybrid pigment via semidry grinding. *Clay Minerals* 56: 37–45.

Weng, J., Z. Gong, L. Liao, G. Lv, and J. Tan. 2018. Comparison of organo-sepiolite modified by different surfactants and their rheological behavior in oil-based drilling fluids. *Applied Clay Science* 159: 94–101.

Wu, J., Y. Wang, Z. Wu, Y. Gao, and X. Li. 2020. Adsorption properties and mechanism of sepiolite modified by anionic and cationic surfactants on oxytetracycline from aqueous solutions. *Science of the Total Environment* 708: 134409.

Xi, Y., M. Mallavarapu, and R. Naidu. 2010. Adsorption of the herbicide 2,4-D on organo-palygorskite. *Applied Clay Science* 49: 255–261.

Xu, L., Y. Liu, J. Wang, Y. Tang, and Z. Zhang. 2021. Selective adsorption of Pb^{2+} and Cu^{2+} on amino-modified attapulgite: Kinetic, thermal dynamic and DFT studies. *Journal of Hazardous Materials* 404: 124140.

Xue, A., S. Zhou, Y. Zhao, X. Lu, and P. Han. 2010. Adsorption of reactive dyes from aqueous solution by silylated palygorskite. *Applied Clay Science* 48: 638–640.

Xue, A., S. Zhou, Y. Zhao, X. Lu, and P. Han. 2011. Effective NH_2-grafting on attapulgite surfaces for adsorption of reactive dyes. *Journal of Hazardous Materials* 194: 7–14.

Yah, W.O., K. Yamamoto, N. Jiravanichanun, H. Otsuka, and A. Takahara. 2010. Imogolite reinforced nanocomposites: Multifaceted green materials. *Materials* 3: 1709–1745.

Yamagishi, A. 1987. Optical resolution and asymmetric synthesis by use of adsorption on clay minerals. *Journal of Coordination Chemistry* 16: 131–211.

Yamamoto, K., H. Otsuka, S.-I. Wada, D. Sohn, and A. Takahara. 2005. Preparation and properties of [poly(methyl methacrylate)/imogolite] hybrid via surface modification using phosphoric acid ester. *Polymer* 46: 12386–12392.

Yamamoto, K., H. Otsuka, S.-I. Wada, and A. Takahara. 2001. Surface modification of aluminosilicate nanofiber 'imogolite'. *Chemistry Letters* 30: 1162–1163.

Yang, F. and A. Wang. 2022. Recent researches on antimicrobial nanocomposite and hybrid materials based on sepiolite and palygorskite. *Applied Clay Science* 219: 106454.

Yang, L., Y. Deng, D. Gong, H. Luo, X. Zhou, and F. Jiang. 2021. Effects of low molecular weight organic acids on adsorption of quinclorac by sepiolite. *Environmental Science and Pollution Research* 28: 9582–9597.

Yang, M., Y. Gao, J. Zhang, and F. Zhou. 2023. Hydrophobic modification of sepiolite by fatty acids for efficient oil removal: Influence of the alkyl chain length. *Applied Clay Science* 235: 106867.

Yang, R., D. Li, A. Li, and H. Yang. 2018. Adsorption properties and mechanisms of palygorskite for removal of various ionic dyes from water. *Applied Clay Science* 151: 20–28.

Yildiz, A. and A. Gür. 2007. Adsorption of phenol and chlorophenols on pure and modified sepiolite. *Journal of the Serbian Chemical Society* 72: 467–474.

Yin, S., J. Zhai, Z. Zhang et al. 2022. Study on the effects of cationic dyes species on the structure and performance of sepiolite hybrid pigments. *Materials Research Express* 9: 035001.

Youcef, L.D., L.S. Belaroui, and A. López-Galindo. 2019. Adsorption of a cationic methylene blue dye on an Algerian palygorskite. *Applied Clay Science* 179: 105145.

Yuan, G., B.K.G. Theng, G.J. Churchman, and W.P. Gates. 2013. Clays and clay minerals for pollution control. In *Handbook of Clay Science*, 2nd edition. Developments in Clay Science, Vol. 5A, (Eds.) F. Bergaya and G. Lagaly, pp. 587–644. Amsterdam, The Netherlands: Elsevier.

Yuan, G., B.K.G. Theng, R.L. Parfitt, and H.J. Percival. 2000. Interactions of allophane with humic acid and cations. *European Journal of Soil Science* 51: 35–41.

Yuan, G. and S.-I. Wada. 2012. Allophane and imogolite nanoparticles in soil and their environmental applications. In *Nature's Nanostructures*, (Eds.) A.S. Barnard and H. Guo, pp. 485–508. Singapore: Pan Stanford Publishing.

Yuan, P., A. Thill, and F. Bergaya (Eds.). 2016. *Nanosized Tubular Clay Minerals: Halloysite and Imogolite*. Developments in Clay Science, Vol. 7. Amsterdam, The Netherlands: Elsevier.

Zanzottera, C., A. Vicente, E. Celasco, C. Fernandez, E. Garrone, and B. Bonelli. 2012. Physico-chemical properties of imogolite nanotubes functionalized on both external and internal surfaces. *Journal of Physical Chemistry C* 116: 7499–7506.

Zhang, J., D. Cai, G. Zhang et al. 2013. Adsorption of methylene blue from aqueous solution onto multiporous palygorskite modified by ion beam bombardment: Effect of contact time, temperature, pH and ionic strength. *Applied Clay Science* 83–84: 137–143.

Zhang, Y., L. Fan, H. Chen, J. Zhang, Y. Zhang, and A. Wang. 2015c. Learning from ancient Maya: Preparation of stable palygorskite/methylene blue@SiO_2 Maya Blue-like pigment. *Microporous and Mesoporous Materials* 211: 124–133.

Zhang, J., Z. Yan, J. Ouyang, H. Yang, and D. Chen. 2018. Highly dispersed sepiolite-based organic modified nanofibers for enhanced adsorption of Congo red. *Applied Clay Science* 157: 76–85.

Zhang, Y., W. Wang, J. Zhang, P. Liu, and A. Wang. 2015b. A comparative study about adsorption of natural palygorskite for methylene blue. *Chemical Engineering Journal* 262: 390–398.

Zhang, Y., J. Zhang, and A. Wang. 2015d. Facile preparation of stable palygorskite/methyl violet@SiO_2 "Maya violet" pigment. *Journal of Colloid and Interface Science* 457: 254–263.

Zhang, Z., W. Gui, J. Wei et al. 2021. Functionalized attapulgite for the adsorption of methylene blue: Synthesis, characterization, and adsorption mechanism. *ACS Omega* 6: 19586–19595.

Zhang, Z., W. Wang, Y. Kang, L. Zong, and A. Wang. 2016. Tailoring the properties of palygorskite by various organic acids *via* a one-pot hydrothermal process: A comparative study for removal of toxic dyes. *Applied Clay Science* 120: 28–39.

Zhang, Z., W. Wang, and A. Wang. 2015a. Highly effective removal of methylene blue using functionalized attapulgite via hydrothermal process. *Journal of Environmental Sciences* 33: 106–115.

Zhao, Z., D. Fu, and Q. Ma. 2014. Adsorption characteristics of bisphenol A from aqueous solution onto HDTMAB-modified palygorskite. *Separation Science and Technology* 49: 81–89.

Zhao, S., G. Huang, S. Mu, C. An, and X. Chen. 2017. Immobilization of phenanthrene onto Gemini surfactant modified sepiolite at solid/aqueous interface: Equilibrium, thermodynamic and kinetic studies. *Science of the Total Environment* 598: 619–627.

Zhao, S., G. Huang, J. Wei, C. An, and P. Zhang. 2016. Phenanthrene sorption on palygorskite modified with Gemini surfactants: Insights from modeling studies and effects of aqueous solution chemistry. *Water Air Soil Pollution* 227: 1–14.

Zhu, J., P. Zhang, Y. Wang et al. 2018. Effect of acid activation of palygorskite on their toluene adsorption behaviors. *Applied Clay Science* 159: 60–67.

Zhuang, G., J. Gao, H. Chen, and Z. Zhang. 2018a. A new one-step method for physical purification and organic modification of sepiolite. *Applied Clay Science* 153: 1–8.

Zhuang, G., Z. Zhang, and M. Jaber. 2019. Organoclays used as colloidal and rheological additives in oil-based drilling fluids: An overview. *Applied Clay Science* 177: 63–81.

Zhuang, G., Z. Zhang, H. Yang, and J. Tan. 2018b. Structures and rheological properties of organo-sepiolite in oil-based drilling fluids. *Applied Clay Science* 154: 43–51.

Index

Note: Page numbers in *italics* indicate figures/schemes, and page numbers in **bold** indicate tables in the text.

B

C

D

H

I

L

M

N

O

Q

R

S

T

U

V